I0109882

WING PARTS

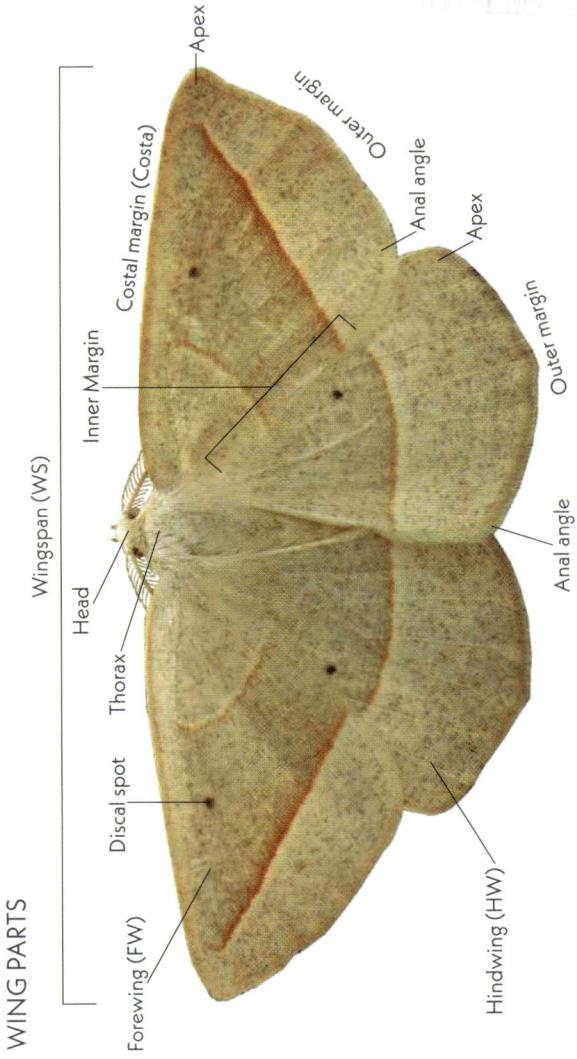

- Apex
- Costal margin (Costa)
- Outer margin
- Anal angle
- Apex
- Outer margin
- Anal angle
- Inner Margin
- Wingspan (WS)
- Head
- Thorax
- Discal spot
- Forewing (FW)
- Hindwing (HW)

Moths
of Western
North America

Moths
of Western
North America

Seabrooke Leckie

Princeton University Press

Princeton and Oxford

To my wonderful children, Coralie and Rowan.
May you never lose that beautiful spark of curiosity.
—SL

Copyright © 2026 by Seabrooke Leckie

Princeton University Press is committed to the protection of copyright and the intellectual
property our authors entrust to us. Copyright promotes the progress and integrity of
knowledge created by humans. By engaging with an authorized copy of this work, you are
supporting creators and the global exchange of ideas. As this work is protected by copyright, any
reproduction or distribution of it in any form for any purpose requires permission; permission
requests should be sent to permissions@press.princeton.edu. Ingestion of any PUP IP for any AI
purposes is strictly prohibited.

Published by Princeton University Press
41 William Street, Princeton, New Jersey 08540
99 Banbury Road, Oxford OX2 6JX
press.princeton.edu
GPSR Authorized Representative: Easy Access System Europe - Mustamäe tee 50, 10621
Tallinn, Estonia, gpsr.requests@easproject.com

All Rights Reserved
    ISBN (pbk.) 9780691232881
    ISBN (e-book) 9780691261454

British Library Cataloging-in-Publication Data is available

Editorial: Robert Kirk and Megan Mendonça
Production Editorial: Mark Bellis
Cover Design: Ben Higgins
Production: Steve Sears
Publicity: Matthew Taylor and Caitlyn Robson-Iszatt
Copyeditor: Charles J. Hagner

Designed by D & N Publishing, Wiltshire, UK

Cover Credit: Paul G. Johnson

Illustrated species: cover, Eight-barred Lygropia; title page, Mexican Tiger Moth;
p. iv, The Infant; p. v, Slanted Emerald; p. 13, White-lined Sphinx

This book has been composed in Garamond Premier Pro (main text) and Neue Kabel (headings)

Printed in China

10 9 8 7 6 5 4 3 2 1

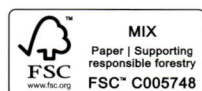

MIX
Paper | Supporting
responsible forestry
FSC® C005748
www.fsc.org

CONTENTS

Author's Note   vi
Acknowledgments   vi

Introduction   1

  How to Use This Book   1

    How to See Moths   1
    The Basics   1
    In "the Wild"   4
    Caterpillars and Cocoons   5

  How to Identify Moths   5

    Appearance   5
    Flight Periods   6
    Range Maps   8
    Habitat and Host Plants   9
    Abundance   10
    Moth Taxonomy   10
    Hodges Numbers and P-Numbers   11
    Common Names   11
    Micros versus Macros   12
    Moths and Conservation   12

Species Accounts   13

Glossary   658

Additional Resources   660

Photography Credits   662

Index   666

List of Taxonomic Groups   682

AUTHOR'S NOTE

A field guide is a monumental undertaking. This is my third, the first two being to the moths of northeastern and southeastern North America, co-authored with my friend David Beadle. From my beginnings as a novice moth-er at the start of our first book in 2008 to the completion of this one 17 years later, I have learned a lot. I've developed new computer skills, improved my project management and organization, and figured out how to do it all around raising a young family.

Most valuably, I've grown significantly in moth familiarity and knowledge. I've become proficient at navigating moth resources, and compiling and distilling that information into the field guide format. I've put my university degree to good use in digging into the scientific literature on moth taxonomy. And I've developed a knack for teasing out field marks that separate similar-looking species, and recognizing when a specimen is out of place.

That said, I am still a far cry from an expert; my formal education focused on birds and all my knowledge of moths comes from self-study. I have done my best to try to ensure the information presented in this guide is accurate and up to date, but it is inevitable, with a project of this scale, that mistakes will creep in. Many actual experts in the field provided assistance when I ran into tricky identification challenges or taxonomy queries, and reviewed the first draft of the manuscript, and I am deeply grateful for their support. Errors that remain in this work are my own. As they are identified, I will add them to a list of errata on my website, seabrookeleckie.com. If you would like to submit an error you have found for addition to this list and correction in future editions, please use the submission form also on my website.

ACKNOWLEDGMENTS

I personally have invested more than 3000 hours into this book, but I couldn't have done it without the help and contributions of a lot of other people. I don't have the space here to thank every single person by name, but I am incredibly indebted to everyone who offered their professional expertise, photographic work, and personal support to help make this book the best it can be.

My biggest thank you goes to all the photographers who provided images for consideration or use for the identification plates. Being myself a resident of the East, it is no exaggeration to say this guide would not exist without their generous contributions. A list of all the photographers whose photos were used is included separately.

A huge thank you to Steve Nanz, who helped me sort out several challenging species groups, and walked me through the use of the BOLD Systems website for reviewing DNA barcoding data. Chris Schmidt and Michael Sabourin also offered their expertise on some species. Greg Pohl provided feedback on the first draft of the manuscript.

Todd Gilligan and the Wedge Entomological Research Foundation provided me with an advanced copy of the 2023 *Annotated Taxonomic Checklist of the Lepidoptera of North America, North of Mexico* by Steve Nanz and Greg Pohl, so I could make sure the higher taxonomy and scientific names were up to date. Stephen Granderson offered suggestions for common names for some of the hundreds of Western species that lacked one.

Thank you to Lisa White for the start of this book, and Robert Kirk for its completion at Princeton University Press. The entire Princeton team did a wonderful job helping to turn my 4300 files into a beautiful and tidy guide—thank you especially to Mark Bellis, Chuck Hagner, and David Price-Goodfellow.

A special thank you to Russ Galen, who has been my advocate since the very first book, and without whom this series wouldn't have found either its first or second home.

And finally, to my husband Dan and our two kids: Thank you for the light and joy you add to my life daily, the smiles and laughter, every moth and dragonfly and snake you bring my way, every hike and paddle and camping trip full of wonder. You are my world.

INTRODUCTION

For years, moths have been plagued by bad press. The stereotype is of a drab brown creature that, at best, flutters aimlessly at your lights and, at worst, chews holes in your clothes. While there are species that do both things, they represent a tiny minority of the incredible diversity found in the world of moths.

Nearly 13,000 species of moths are currently recognized in North America. There are large moths and small ones, plain moths and bright ones, nocturnal moths and day-fliers, and moths inhabiting virtually every habitat niche you can think of. My goal with this guide is to introduce you to some of the remarkable species that may inhabit your backyard and neighborhood. Once you've seen a spectacular Elegant Sheepmoth or one of the incredible wasp-mimic clearwing borers, you won't be able to look at moths the same way again.

Thirteen thousand species is a far greater number than even the most ambitious field guide might be able to cover in one volume. In order to offer the most useful field guide to moths possible, I limited the scope to just western North America. Even this smaller region still contains thousands of species. Many of these are small, however, or very rare, or very localized in occurrence. I selected some 1,900 of the most common or most eye-catching moths in this area to include in this guide.

The majority of the species you are likely to encounter on any given night are present within these pages, and for those species that aren't, additional resources are provided at the end of the book. For readers residing in the eastern part of this book's range, within the Great Plains or boreal regions, I highly recommend also obtaining a copy of either of my other field guides, covering the moths of northeastern and southeastern North America. Many species that are common in the East range into eastern parts of our area, but I chose to omit some of these in order to include a fuller coverage of western residents and specialties.

The purpose of this guide is not to provide an exhaustive life history for each species, but to be an introduction to them. I selected photographs that depict representative individuals for each species and included basic information that will help in reaching an identification. If you wish to know more about any of these species, additional information can be found in the resources at the end of the guide.

While I made every effort to ensure that the content of this guide is as accurate as possible, there may still be errors present. Moths have not had the benefit of decades of hobbyists collecting data the same way some other taxonomic groups, such as birds and butterflies, have, so there are still many gaps in our knowledge. I hope that by providing a comprehensive, user-friendly guide, I may be able to fill in some of these gaps through introducing new enthusiasts to this amazing group of insects. I would be interested in receiving feedback on errors and omissions so that future editions of this guide might be corrected and improved.

HOW TO USE THIS BOOK

HOW TO SEE MOTHS

Moths are everywhere. You don't need to stray far from home to see them; even the tiniest urban lot has moths present, provided there are plants around for the caterpillars to feed on. Observing moths can be as easy as turning on the porch light on a warm summer evening and stepping out once in a while to see what might have come in.

Just as one's bird-watching experience is improved with a pair of binoculars, so, too, can a bit of equipment greatly increase the enjoyment you take in looking for moths. You don't need to spend very much, either—a simple outlay of $15 can dramatically increase the number of moths you might see. Of course, as with any hobby, if you discover a passion for it, it's definitely possible to spend quite a bit more.

THE BASICS

Lightbulbs While a simple incandescent light might draw in a few moths, the most effective bulbs project some light in the ultraviolet (UV) spectrum. A 2024 study determined that moths maintain flight stability by a reflex that orients their body relative to the sky, using light-detecting organs on their dorsal surface; when they near an artificial light, its brightness disorients their stability reflex, often trapping them in its orbit. However, it remains unclear whether this is also the mechanism behind long-range attraction, why UV wavelengths create a greater response, and why some moth species are rarely attracted to light. Whatever the reason, artificial lights will draw in the greatest number of species.

A blacklight is an inexpensive option for attracting moths. Blacklights are often sold at home renovation centers or other stores that provide a wide array of bulbs as party accessories, and they are available as compact fluorescent lights (CFL), fluorescent lights, and LED bulbs and fixtures. All can be effective for mothing, so which you choose depends primarily on how you intend to support the light (e.g., on a tripod, suspended, etc.). Where you plan to set up can also be a factor, as many fixtures must be plugged in to an outlet or generator. There are LED blacklight bars with USB plugs that can operate from a large power-bank battery, making portability a breeze. See figures 1 and 2 for different examples of blacklights and supports.

Similar in nature to blacklights are grow bulbs designed for plants or aquariums, and bug-zappers. (Make sure you disable the zapper if you purchase one of these!) Regular household CFL bulbs can be surprisingly effective, too, because of the mercury content, and they are easy to find. With these, a higher-wattage bulb will likely produce better results.

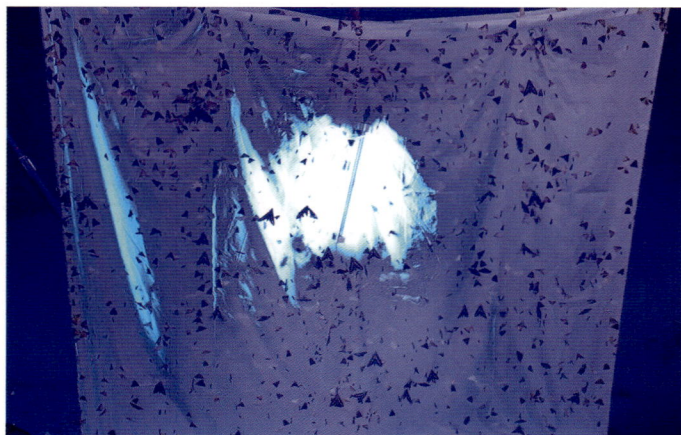

FIGURE 1 A full sheet near Lake Tahoe, CA. A USB-powered blacklight bar is suspended in front of the sheet by a thin cord secured to a clothes-peg. The sheet is hanging from a rope tied between two very tall tripods. CREDIT: PAUL G. JOHNSON

A moth-er checks her sheet. The sheet is hung from a thin rope strung between trees, and a blacklight is supported on a tripod in front. This setup works particularly well for CFL and mercury vapor bulbs using utility-light fixtures with clamps. CREDIT: ALLANAH VOKES AND SARAH BANNISTER

A more expensive, but definitely much more effective, option is a mercury vapor bulb. These powerful, high-wattage bulbs broadcast a very bright light in a broad spectrum of wavelengths, attracting more individuals and drawing moths in from farther away. They are the sorts of bulbs found in typical outdoor security lights and usually need to be ballasted in an appropriate fixture (although some are sold as self-ballasted). Be careful using these—they get very hot, so they can present a fire risk if knocked over or if your sheet falls across them.

White cotton sheet A lightbulb may be set up in front of a wall or other smooth surface that reflects the light and also provides a place for moths to settle. Pale surfaces work best, as they make it easier to detect the moths. A cotton sheet has the additional advantage of reflecting UV rays (synthetic fibers, such as polyester, typically do not), creating a much broader surface area for attracting moths. This is especially true when using a blacklight, which is much more effective when used in combination with a white sheet. See figures 1 and 2 for examples.

Light trap An essential piece of equipment for serious moth enthusiasts, a light trap functions to hold the moths drawn to the light until the moth-er can return to check them. This has the great benefit of allowing you to get some sleep overnight! Light traps can be purchased online but can also be made easily and inexpensively from household components. At its most basic, a trap consists of a container in which to hold the moths, egg cartons or some other textured structure for the moths to hide in while inside, a lid to keep them from escaping, a funnel in the lid to let moths in,

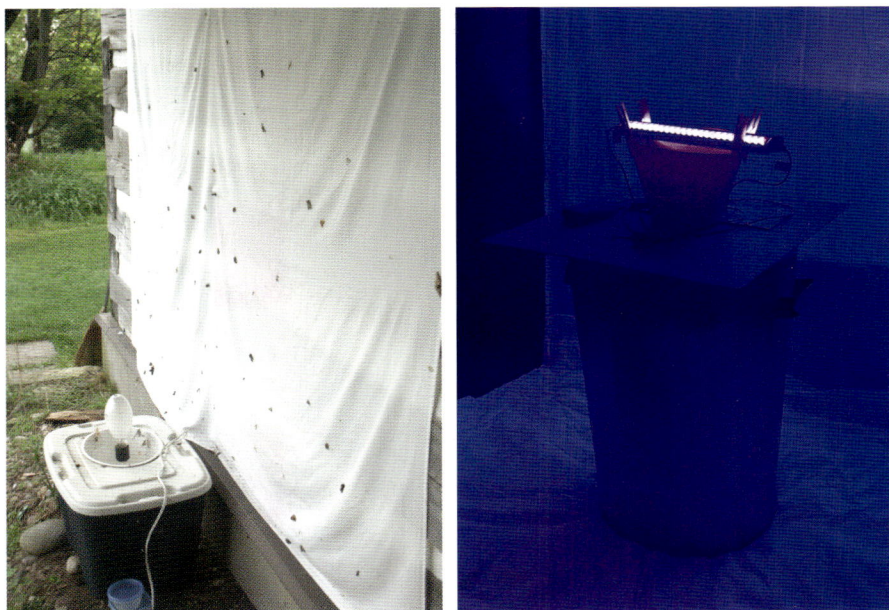

FIGURE 2 LEFT: The author's well-loved trap placed in front of a white sheet pinned to the side of the house, beneath an overhang. The trap consists of a ten-inch automotive funnel set into the lid of a plastic storage bin; the funnel has clear acrylic fins glued inside, which support the light fixture. The 250-watt mercury vapor bulb needs to be plugged into household power and protected from direct rain.

RIGHT: This inexpensive design was built from materials found around the house and consists of an old garbage bin, a flat sheet of cardboard with a hole cut in the center, a wide-mouthed cone of bristol board inserted into the cardboard, and two affordable USB-powered LED blacklights set on top of the cone and plugged into a portable power bank.

and a light source to sit in or above the funnel. Figure 2 shows two examples of different styles of trap, and an image search on the internet for "moth trap" will return many other styles and variations. The exact pieces you use to accomplish these goals are determined by what's available to you and how much you want to spend—a commercial trap can cost hundreds of dollars, but a DIY version might cost as little as $10. Take care not to run your trap on rainy nights, or otherwise provide some cover for it to keep the moths from getting wet. And be sure to check your trap first thing in the morning, or the birds will thank you for the breakfast buffet!

FIGURE 3 Sugar bait can be made from a variety of possible ingredients, depending on what you have in your kitchen. The goal of your recipe is to create a fermented, sweet liquid. Bait that will be painted onto trees or other surfaces should be thicker, while that used to soak ropes or cloth should be runnier. Shown here are some commonly used ingredients and supplies. Experiment to see what works best for you!

Sugar bait Some species of moths are not very attracted to artificial light but are nectar-feeders and will frequently come to sugar bait. This is usually a sugary, fermented liquid or paste. The combination of ingredients you use is flexible and can be determined by what you have on hand in the kitchen. A particularly effective recipe is to blend one soft banana, a scoop of brown sugar, a dollop of molasses, and a glug or two of beer or wine (flat or cheap is fine), but a simple mixture of wine and sugar can also work. You can experiment with what you have available to see what works best in your location. Allow the mixture to sit at room temperature to ferment for a day for improved results. Paint this sticky concoction onto tree trunks or logs with a brush. It may stain wood, so avoid using it on your deck or other structures. Another option is to soak a thick natural-fiber rope (at least 1.25 cm in diameter, so the moths can land on it) in the mixture and then string it between two supports. You will need to check the bait regularly, just as you would do with your cotton sheet, in order to see what it has attracted. This tool is particularly effective on cool nights.

Jars or containers Photographs taken during daylight hours allow for more evenly lit photos without any of the glare from a camera flash reflecting off the scales of a moth's wings. Moths can be cooled in the refrigerator (not the freezer!) for one night if necessary; the low temperature puts them into a state of torpor, which keeps them calm and lowers their metabolism, resulting in more cooperative subjects. Pill bottles such as those provided by a pharmacist are ideal containers, as they are small and don't take up much room in the fridge. Plus, they are clear, so that you can see what you have inside. You can order them from websites or ask at your local pharmacy.

Ring light or flash diffuser If you don't want to or can't hold moths to photograph in daylight, it's also possible to get evenly lit photographs at your moth sheet by using a source of wide illumination. Any light source that provides light from multiple directions at once will reduce both scale glare and harsh shadows. Ring lights, such as those used in home video studios, are a good option, and there are many clip-on styles that securely but gently affix to your smartphone. Long light wands or bar-style flashlights are also easy to hold by hand beside a smartphone or camera. If you have an SLR or other camera with a flash mount, a flash diffuser or other flash meant for portrait shots may be a more expensive but convenient option.

IN "THE WILD"

Not all moths are nocturnal. Some nectar-feeders can be seen supping at flowers during daytime, as well. A few species are almost exclusively diurnal and are rarely encountered at night. You can encourage day-fliers to visit your gardens by planting nectar-rich flowers, which are often sold at garden centers as appealing to butterflies or hummingbirds. Phlox, sage, and beebalm are particularly popular with sphinx moths; smaller moths with shorter tongues might benefit from shorter flowers, such as beggarticks or coneflower. Look for "butterfly-friendly" on the plant's label.

Don't forget to keep your eyes peeled while out hiking. The moths that visit your garden plants can also be found on wildflowers. Many species will rest in the grass or the leaf litter on the forest floor, so watch for small pale shapes that rise up ahead of you. All of those moths that come to your sheet at night have to find somewhere to spend the day; examine tree trunks, bark crevices, rock ledges, and other protected places. Don't overlook your house and garage—moths can turn up in the strangest places!

CATERPILLARS AND COCOONS

While this guide helps only with the identification of adults, moths actually spend a larger percentage of their lifetime in the larval stage. Some caterpillars can be very cryptic, blending in with the plants on which they feed and often going unnoticed. Others have behaviors or patterns that make them instantly recognizable, such as the springtime tents made by the Western Tent Caterpillar or the wandering habit of the fuzzy black-and-brown Woolly Bears, which are caterpillars of the Isabella Tiger Moth. This guide does not contain information on caterpillars or cocoons, but other publications are very helpful. If you are interested in learning to identify the caterpillars of moths, check out the resources at the end of the book.

HOW TO IDENTIFY MOTHS

With such a vast array of possibilities, it's easy to feel a bit overwhelmed when first learning to identify moths. Many of them appear to look the same, and even learning to tell one group from another can seem like a daunting challenge. It is only through constant practice and regular exposure that you will start to be able to discern differences, and practice takes time. Don't give up! It takes most moth enthusiasts a few years before they start feeling comfortable with their local species.

When you're beginning, it is easiest to select a subset of the moths at your lights and set about learning them first. Which subset you decide to start with is entirely up to you, but I recommend concentrating on learning the common species first, the ones that you see many of on a given night. Once you feel comfortable with the common species, it will be easier to spot the less common ones. Another easy approach is to begin with the flashy or brightly colored species, as these are easily remembered and tend to stand out from the rest. In subsequent seasons you can build on the skills you developed the year before, and each year you'll be able to identify a few more species without having to refer to your guidebooks. I recommend ignoring most of the smaller moths to begin with, and focusing just on the larger ones, which are usually easier to identify. When you feel comfortable with the larger moths, you can return to the smaller ones without feeling overwhelmed.

Another good way to familiarize yourself with the various species is simply to spend some time flipping through your field guide. If you've looked at an illustration in your book often, when the species finally turns up at your light, you'll already know what it is, or at least have a fair idea of where to find it in the guide.

APPEARANCE

When trying to identify a particular individual, first make sure that the lepidopteran you're looking at is in fact a moth. Some diurnal moths resemble butterflies in color and habit, and some drab butterflies might bring to mind a moth. The easiest way to tell them apart is by the antennae: if they are clubbed or swollen at the tip, it's a butterfly; if they are thread-like or feathery, it's a moth.

There are many questions you should ask yourself even before you begin looking at the patterns of lines and spots on the moth's wings. How large is it? How does it pose while at rest? What shape is it? What color is it? Answering these questions can help you narrow down which group a moth is in. A medium-sized moth that rests with its wings spread out to its sides is most likely in the family Geometridae, while one that sits with its wings folded tent-like over its back is probably in the Noctuidae. A noctuid that is long and thin is most likely in a different group than one that is more triangular. As with anything, there are exceptions, but it is a helpful place to start. The back endpapers show many different moth silhouettes for easy reference.

Size is a very important key in identifying moths. Not only can it help you determine which group of moths an individual belongs to, but it can also often help in separating two very similar species. All of the species within a single plate are sized relative to each other, so larger species will be shown larger. A life-sized gray silhouette of one of the moths is included on each plate to give you a sense of actual size. The scale might be different from one plate to the next, even within the same family, so make sure to refer to the silhouette on each page.

Once you have narrowed down the possibilities to a few families, you can look more closely at the patterns. In this guide, I often refer to the various spots and lines (the latter are often called "bands" when they are wider) that can be seen on the wing. Each has its own unique name, and knowing the terminology for the structure and patterns of the wing will be very helpful when reading the species accounts. Figure 4 shows the different markings and parts of the wing that I refer to when describing species; it also shows different antenna types. This diagram is repeated on the front endpapers for convenience.

To conserve space, abbreviations are frequently used when referring to commonly referenced parts or structures. Forewing and hindwing are shortened to FW and HW, respectively. Likewise, the antemedial line, postmedial line, and subterminal line are the AM, PM, and ST lines. These abbreviations are also used when referring to the areas of the wing with the same name. Measurements are given as either wingspan (WS) or total length (TL), with the former from wingtip to wingtip and the latter from head to outer margin.

Some additional moth-specific terminology is used throughout the accounts. These words and phrases have been defined in the glossary at the back of this book.

Some species of moths, such as Cinnabar Moth, are so distinctive that it's impossible to mistake them for any other species, while others, such as the underwings, are separable only by relatively subtle field marks. The markings that help distinguish a given species from similar ones are shown on the plates using black indicator lines. Clarification is also provided in the text. It is helpful to refer to both when trying to identify a moth.

Just as in humans, even within the same species of moth there is often quite a bit of individual variation. Some individuals might be lighter in color than others, or darker, or a different color altogether; or some might be well marked, while others are very faint. In some species, there are multiple color patterns; in others, females look very different from males. And finally, some species of moths may live for a few weeks or more, and over time some scales may be rubbed off the wings, affecting the appearance of their pattern. Some very worn individuals may not be identifiable at all.

It should also be noted that some species are so similar to one or more others that they simply cannot be identified visually with any confidence and require dissection or DNA analysis to determine species. This is particularly common among the micromoths (especially the Gracillariidae, Gelechiidae, and Coleophoridae), but many macromoth species also have near-identical look-alikes; this is usually noted in the species account. Additionally, this book treats only a small portion of the moth species that exist in western North America, and while most of what you encounter will be represented in these pages, it is always useful to consider the possibility that the moth you are looking at may be a less common species not included here. The additional resources at the end of this book may help you identify rarer species.

FLIGHT PERIODS

The species composition of the moths that come to your light will change over the course of a season. The majority of the moths you see in May will be different from those you see in September. Knowing when the adults of a certain species are on the wing can help narrow down an identification. If it's August and you are deciding between a species that flies in spring and one that flies in summer, it's clear which of the two your moth is.

The length of time adults might be present varies among species. Some may be encountered nearly all year, while others might fly only during a three- or four-week window. Several species raise two

FIGURE 4 Antenna types (top), forewing markings (middle), and wing parts (bottom) referred to in the species accounts.

ANTENNAE TYPES

Bipectinate

Pectinate

Filiform

FOREWING MARKINGS

Head

Thorax

Basal line

Basal dash

Antemedial (AM) line

Claviform spot

Orbicular spot

Median line

Subreniform spot

Reniform spot

Postmedial (PM) line

Subterminal (ST) line

Subapical patch

Apical dash

Adterminal line

Terminal line

Fringe

Anal dash

Basal area

Antemedial (AM) area

Postmedial (PM) area

Median area

Subterminal (ST) area

Terminal area

Total Length (TL)

WING PARTS

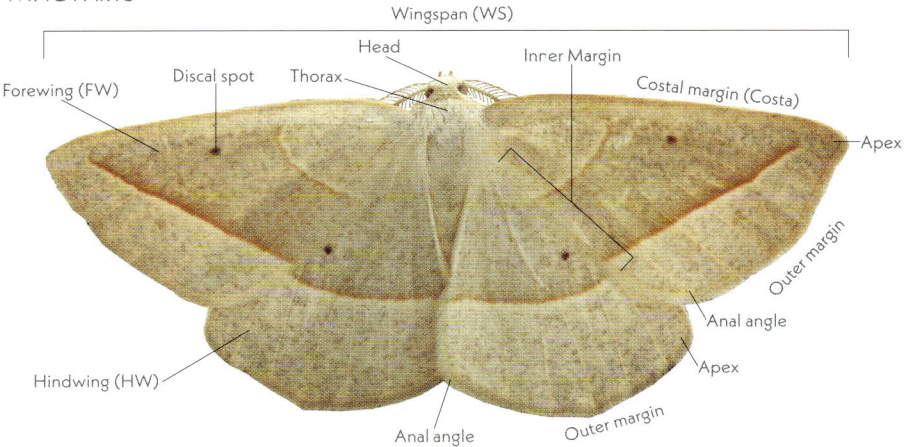

Wingspan (WS)

Head

Discal spot

Thorax

Inner Margin

Costal margin (Costa)

Forewing (FW)

Apex

Outer margin

Anal angle

Apex

Hindwing (HW)

Anal angle

Outer margin

FIGURE 5 Flight periods are illustrated using four colored icons representing the seasons. A fully colored icon indicates that the adult moth may be encountered during that season, while a gray icon means the moth usually is not encountered. A half-colored icon indicates the moth may be present in either the first or later half of that season, as appropriate. The seasonal phenology of a given area will determine how the icons correspond to calendar months.

(or more) broods, so you might encounter adults in two different nonoverlapping periods; others emerge from their cocoons in the fall and then overwinter as adults, waking from hibernation to fly again in the spring.

Flight periods are illustrated with a colored graphic beside the species account, as shown in figure 5. Each season is represented by an icon and colored according to whether adults may be found during that season. An icon that is fully colored means adults of the species can be found across the entire season, while a partly colored icon indicates the species can usually be found only in the first half or later half of the season. An icon that is entirely gray means adults are not typically encountered in that season.

Because the coverage area of this guide spans many degrees of latitude, you will need to use the phenology of your area to determine the months in which to expect a species. The timing of the arrival of spring will be different in Arizona than in Alberta, for example, and therefore so, too, will the timing of the emergence of spring species. Summer may be longer in California than in British Columbia, and some summer-flying species may be present for longer in the South.

As well, the southernmost parts of this guide's coverage area may have mild-enough winter weather that some species can be encountered year-round, or have a fall flight period that extends well into the winter months, or a spring one that starts early in the calendar year. Where a wide-ranging species has flight-period patterns that are notably different in the north and the south of its range, a flight period graphic is given for each.

RANGE MAPS

Knowing where a moth occurs is every bit as important as knowing when it occurs. Most previous field guides to moths have provided only a written description of the range for each species. Such descriptions are vague at best, as much is left to the interpretation of the user. In this guide, range maps are provided for as many species as possible. The maps are easy to read and interpret and can quickly tell you whether to expect a species in your area.

The relative paucity of data on moth species compared with other, more familiar groups, such as birds and trees, presents a challenge when preparing range maps. While the range of many common species is well known, for most moths it is difficult to draw a map with any level of precision.

To get around this difficulty, I relied heavily on the use of ecoregions in mapping ranges. Unlike vertebrates, moths are strongly tied to the food plants that their caterpillars eat—if the host plants aren't present, the moths won't be either. The host plants, in turn, are often restricted to certain environments, whether by temperature, soil type, drainage, and so on. Ecoregions are large areas that share similar environmental conditions and plant communities.

The ecoregion map used to create the range maps in this guide is shown in figure 6. This map was adapted from the North American Environmental Atlas, a project completed by the Commission for Environmental Cooperation in joint partnership with the governments of the United States, Canada, and Mexico.

While some plant species may be found in one part of an ecoregion but not another, in general this system allowed a range to be extrapolated from known data points. For instance, if a moth was recorded in central Washington and Oregon, there is a good chance it also occurs in northern California, and it is depicted as such, even if there is no data from the latter state. I did not try to replicate the intricate borders to the montane ecoregions, instead opting for an easy-to-read

FIGURE 6 The map of ecoregions used to create the range maps in this guide. Species ranges were inferred by matching known data points to ecoregions and extrapolating to those ecoregions' boundaries. The map is a simplified version adapted from the North American Environmental Atlas by the Commission for Environmental Cooperation and the governments of the United States, Canada, and Mexico. For a detailed explanation of the ecoregions, including region labels, visit www.cec.org.

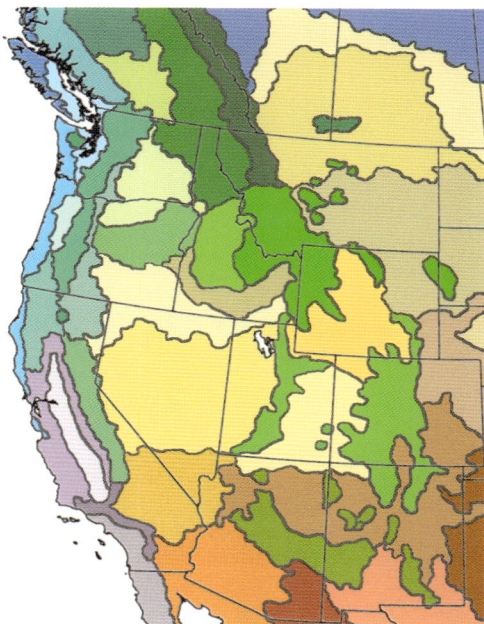

smoothed outline, and I included smaller montane ecoregion "islands" as range lobes, rather than disjunct patches. If your location falls within one of these transition areas, use the host-plant information to help you determine species presence.

This mapping system is, of course, not without its own problems, and undoubtedly there will be instances when the maps are inaccurate and either over- or underrepresent the actual distribution. Citizen science websites contributed greatly to the mapping data, but the potential for misidentifications on these websites may lead to incorrect range maps in some infrequent instances. Additionally, moths are winged creatures; many are capable of flying long distances, and they can also become caught up in traveling weather systems. A range map is never a hard-and-fast rule, as bird-watchers can attest—species may show up out of range from time to time.

Maps are usually depicted in one color but occasionally may show two. In these instances, the darker green shade indicates the typical range, where the moth can generally be encountered every year, while the lighter green shade represents the migratory or vagrant range, where the species is not necessarily present every year.

A note on introduced species North America's moth fauna includes many species that are not native to our region but were introduced deliberately or accidentally over the last two centuries. Where a species is known or believed to have originated on another continent, it is indicated with the code [IN] beside the abundance descriptor. Many introduced species are well established, effectively naturalized, and considered a "normal" part of the moth community. For the purposes of this guide, range maps for introduced species are treated the same as those for native ones.

HABITAT AND HOST PLANTS

A moth species is rarely distributed evenly across its range. With few exceptions, most species can be found only in or near habitats that contain their larval food plants. Even though a species may be shown to occur throughout an entire state, there will be locations within that state where the moth is abundant and other spots where it cannot be found at all.

Habitat is another important tool in moth identification. Knowing that one species is found only in bogs and marshes while another frequents upland forests can make identification easier. Likewise, if a species feeds on quaking aspen, but there are no quaking aspen trees anywhere near you, you are unlikely to have the species turn up. (But see the caveat above regarding winged organisms!) For space reasons, I did not include habitat in the species accounts, but the larval food plant is given for all species when known, and this information can be used to extrapolate in which habitats the species will likely occur. While the host-plant information given is the best currently known, in some rare instances the stated host species may be inaccurate as a result of assumptions made by early specimen collectors.

Some moth species are very specific in their requirements. Their larvae may feed exclusively on only one or two species of plants, and the distribution of such moths is often nearly identical to that of their host plants. Other species are generalists, whose caterpillars are not fussy about what they eat. These moths may be encountered across a broad range of habitats and ecoregions. Many (though not all) of the moths that are found throughout our coverage area are generalists, whereas those restricted to small ranges are often picky eaters.

ABUNDANCE

As you develop an interest in moths, one of the first things you are likely to notice is that some species are more abundant than others. This is true across all taxa—some species are just more common. An abundance label is provided for each species to give you a sense of the likelihood of occurrence. These labels are not a reflection of true abundance in the landscape but incorporate the probability of a species being encountered, as represented by records across multiple citizen science websites. Some flashy species may be recorded more often, and some common but secretive species may go overlooked. As this guide is intended to help casual and beginner observers determine the identification of an encountered specimen, I felt this was a more useful measure to use.

In addition to broad landscape-scale abundances, the relative abundance of the different species attracted to a light may change from one location to another; what is plentiful in one locale may not necessarily be plentiful in another. This is partly determined by the food plants the larvae favor, with generalists being on average more abundant than specialists, and with specialists being more abundant when their host plant is more abundant. Abundance is also affected by flight periods. A species will be more abundant in the middle of its flight period than at either end. Dozens may be seen in a night at the peak of the flight period, whereas a few weeks later you might find only one in an entire evening.

A large part of species identification in any taxon is knowing what to expect. By combining information on appearance, flight period, range, habitat, and abundance, it is usually possible to reach a probable or definitive identification.

MOTH TAXONOMY

There are two basic approaches to organizing a field guide: by grouping species of similar appearance, or by organizing species taxonomically. Although a beginner often finds it easier to use a book in which similar species are presented side by side, this approach is frustrating once you have enough experience to be able to recognize taxonomic groups. If you know the moth you are looking at is a carpet because of its size and shape but don't know which one, you want to be able to flip to the section on carpets and compare them all. Books organized by similar appearance will have you flipping back and forth and back again to compare the different carpets in order to determine which yours is. Becoming familiar with the different taxonomic groups may take some time, but in the long run it will greatly aid your ability to identify different species.

The use of the word *taxonomy* above refers to the organization of species in a field guide, but the word also can refer to an entire field of study that determines how different species are related to each other. As with all scientific disciplines, our understanding of species relatedness is always changing as new research reveals more about the relationships among species and species are moved from one group to another or are lumped or split to form new species. The organization of species accounts in this guide follows the order in the *Annotated Taxonomic Checklist of the Lepidoptera of North America, North of Mexico,* edited by Gregory Pohl and Stephen Nanz and published by the Wedge Entomological Research Foundation (2023), which includes all of the most recent taxonomic updates as of this writing. Major changes don't happen often, however, so the taxonomic organization used here should provide a firm base for learning.

HODGES NUMBERS AND P-NUMBERS

In addition to scientific and common names, each species of moth in North America is represented by two unique identification numbers. These are often useful to know, as a number usually remains with a species permanently once assigned, even through taxonomic revisions, renamings, and reorderings. The first series of numbers used originated from a 1983 checklist by lepidopterist Ronald Hodges in which he assigned a number to every known moth and butterfly species north of Mexico, and as such they are called Hodges numbers (or, less commonly, MONA numbers, for the *Moths of North America* reference series where the checklist was published). Many taxonomic changes have been made and new species have been described since 1983, however, and these numbers no longer accurately represent the full taxonomy of lepidoptera.

In the late 2000s, a new numbering system was created by Bob Patterson for the Moth Photographers Group (MPG) website. This system assigns a two-digit prefix to each superfamily, and to each species a four-digit suffix, unique within the superfamily. This allows superfamilies to be numbered independently and quickly helps identify to which group a species belongs. First adopted by Don Lafontaine and Christian Schmidt in their 2010 revision of the Noctuoidea superfamily, this new numbering system was applied to all groups in the 2016 annotated checklist by Gregory Pohl, Bob Patterson, and Jonathan Pelham. The 2023 version of this checklist updated by Pohl and Nanz, which I use for this guide, contains some revisions to the 2016 numbers; these are indicated with the appendage of the letter *a* to the two-digit superfamily prefix. In the first edition of my northeastern field guide to moths, we referred to this numbering system as "MPG numbers." In the southeastern field guide, we used "P3 numbers." However, the accepted terminology is now "P-numbers," a shortening of "phylogenetic species numbers."

In this guide, I present both the P-numbers and Hodges numbers (with the latter in parentheses) after a species' common and scientific names. Some websites and older publications still organize their lists by Hodges number, so these remain important to know, but most have switched or are in the process of switching to the new P-number system. I recommend adopting the P-numbers as your primary labeling tool, as these provide proper taxonomic order and will make searching online references much easier. You may choose to include the hyphen or not. Many websites and indeed the official checklist by Pohl and Nanz (2023) do not use a hyphen; however, because of the addition of the *a* in revised groups, computer filing systems may require the hyphen to keep numbers in proper order, and many users may find breaking the six digits into two groups makes the number easier to read.

COMMON NAMES

Common names serve an important function in field guides, inviting beginner enthusiasts to learn more about the organisms they're identifying. Common names often share information about the species, such as useful identification features, larval host plants, region or habitat preferences, or the organism's scientific name or taxonomic group. A standardized common name will typically remain the same through taxonomic revision, frequently providing consistency when the scientific name might change with genus reassignments or species mergers. And, perhaps most importantly, common names are easy to pronounce for beginners and those who have difficulty with foreign languages (such as Latin), making the names easier to learn and remember.

A few taxonomic groups, such as birds and butterflies, have publications providing standardized common names as decided by a committee of scientists and enthusiasts with expertise in the field. No such publication yet exists for moths. While some species, such as Polyphemus Moth, are so familiar that they are known everywhere by a single name, many species have no commonly accepted name. Some species may bear two or even three or more common names. The larval form of a species may have a different name than the adult.

Because of the potential for confusion, I tried to be mindful when selecting the common names to use in this guide. If a common name has previously been published or is in use on authoritative spaces on the web (such as iNaturalist, MPG, BugGuide, Butterflies and Moths of North America, various government agencies, etc.), I used it. Priority was usually given to the names used in my northeastern and southeastern guides to moths, and those on the North American websites BugGuide and MPG, in instances where there may be more than one common name in use. If there is a European name but no North American name, the European name is used. And if a species appears to have no common name present on the internet, then I assigned one.

While you may find the common names easiest to remember for casual use, many websites and publications use only the scientific name, so labeling your photos and data with both will make it easier if you ever need to find records or look information up.

MICROS VERSUS MACROS

If you happen to attend a moth night with experienced moth enthusiasts, you might hear them refer to "macromoths" and "micromoths" or simply "macros" and "micros." For the most part, micromoths are relatively small moths and macromoths relatively large, but each group has a few exceptions that are closer in size to those of the other group.

The definition of the two groups has more to do with taxonomy than it does with size. Taxonomy usually orders species according to their evolutionary age, with the oldest groups presented first and the youngest—those that diverged most recently—presented last. Strictly speaking, micromoths are those moths that appear on the first half of taxonomic lists; in this book, that means from Golden Oakminer (P-number 07-0002) through Pale-edged Grass-Veneer (P-number 80a-1583). The moths from Lettered Habrosyne (P-number 85-0003) through to the end of the list are macromoths.

MOTHS AND CONSERVATION

Although we don't often see moths, they are an important part of the environment. Adult moths are a valuable food source for bats, many species of which feed almost exclusively while in flight, and caterpillars are a vital part of the diet that adult songbirds feed their nestlings. Some species of moths have periodic "outbreak" years, and certain types of birds are so tied to these outbreaks for their breeding success that populations are seminomadic. As moth outbreaks have been suppressed for the benefit of the forestry industry, so, too, have we seen their dependent bird populations decline. There is evidence that insect populations as a whole have been declining for several decades, especially those of flying insects, including moths. Most species of birds that feed exclusively on the wing, such as swallows and nighthawks, are also in sharp decline. Although we typically think of bees as the pollinators of our fruits and vegetables, many species of moths are also valuable pollinators, and their decline could have repercussions for us, too.

A large part of the problem is that we simply don't know very much about our moths, or insects in general. Unlike for birds, which have millions of enthusiasts across the continent and for which there are several national and international monitoring schemes, there are few programs to track insect populations. Most of those that exist in North America are either local or limited to a few species. Great Britain has developed the National Moth Recording Scheme (NMRS), a citizen science project that invites anyone anywhere in Great Britain to submit observations to a national database used to help track moth populations and learn more about their distribution. There are many excellent guides to the moths of Great Britain, and mothing is a popular activity as a result. Since the NMRS's inception in 2007, more than forty million records have been collected for it; many of these records have been organized into maps and published as an atlas.

Currently in North America, we lack even a solid grasp of the distribution of many moth species. While state and county lists exist for birds and butterflies and even dragonflies, there are fewer such inventories for moths, and many are incomplete. The citizen science website iNaturalist is beginning to fill in some of these gaps, but we still have a long way to go to have a full understanding of North America's moth species. Because so little is known about our moths and their distribution, every time you put your lights out at night you have the potential to contribute something valuable—that dart could be new for the county, or your July date for that sallow could be the earliest on record.

The more we know, the better our understanding and the better able we are to design and implement conservation efforts. My hope is that this guide will not only open up the world of moths to budding naturalists and enthusiasts but also help ensure the future of these beautiful insects for generations to come.

SPECIES ACCOUNTS

METALLIC LEAF-MINERS SUPERFAMILY Eriocranioidea (07)
FAMILY Eriocraniidae

Very small metallic moths with broad rounded wings. Adults are mostly diurnal or crepuscular and are generally encountered near or on their host species in the early spring, when the foliage is beginning to leaf out. Rarely encountered at lights.

GOLDEN OAKMINER
Rare **RANGE-WIDE**
Dyseriocrania auricyanea
07-0002 (0004)
TL 5–7 mm Tiny moths with iridescent gold FW. Purplish spots form three loose transverse lines. **HOSTS** Oaks. **NOTE** Diurnal, typically found on or near host plants.

PURPLISH BIRCH-MINER
Uncommon **RANGE-WIDE**
Eriocrania semipurpurella
07-0003 (0005)
TL 5–7 mm Tiny moths with iridescent purple FW speckled with gold spots. Often has a large, pale yellow patch near anal angle of inner margin. **HOSTS** Birch and ocean spray. **NOTE** Diurnal, typically found on or near host plants.

GHOST MOTHS SUPERFAMILY Hepialoidea (11)
FAMILY Hepialidae

Medium to large moths with broad forewings that they fold tent-like against the body when at rest. Adults are typically crepuscular, and most species are rarely encountered at lights. Males congregate at dusk in mating swarms that can resemble ghostly clouds.

MATHEW'S GHOST MOTH
Uncommon **RANGE-WIDE**
Gazoryctra mathewi
11-0007 (0028)
TL 17–22 mm Resembles Smooth Ghost Moth, but median and ST areas bear scattered, often indistinct white spots. White Z stripe is often fragmented. **HOSTS** Unknown, but possibly roots of grasses.

SMOOTH GHOST MOTH
Rare **RANGE-WIDE**
Gazoryctra novigannus
11-0008 (0029)
TL 17–22 mm FW is warm brown, with a bold, white, Z-shaped stripe running from basal area to apex. Diffuse whitish marks are typically present along lower costa. **HOSTS** Unknown, but possibly roots of grasses.

ORANGE-LINED GHOST MOTH
Rare **NORTH**
Phymatopus hectoides
11-0015 (0036)
SOUTH
TL 15–18 mm Speckled gray FW has darker bands in basal, median, and ST areas speckled with orange. AM, PM, and ST lines are thinly edged in orange. Thorax is typically quite hairy. **HOSTS** Baccharis, horkelia, lupine, sneezeweed, woolly sunflower, figwort, and ferns.

GOLDEN OAKMINER

PURPLISH BIRCH-MINER

actual size

MATHEW'S GHOST MOTH

SMOOTH GHOST MOTH

actual size

ORANGE-LINED GHOST MOTH

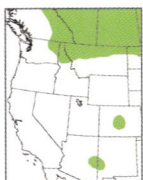

FOUR-SPOTTED GHOST MOTH

Uncommon

Sthenopis purpurascens

11-0017 (0019)

TL 40–55 mm Brown FW has paler, lavender-gray AM and PM bands that angle to meet at inner margin. Brown AM, median, and ST areas contain small, sometimes indistinct, white spots, with the one or two AM spots the largest. **HOSTS** Roots of poplar, alder, and willow.

PYGMY LEAFMINING MOTHS

SUPERFAMILY Nepticuloidea (16a)
FAMILY Nepticulidae

This group includes the smallest moth species, some as little as 2 mm long. Caterpillars are leaf-miners, creating serpentine tracks or blotches in their host plant's leaves. Adults will come to lights at night.

ORANGE-HEADED LEAFMINERS

Common

RANGE-WIDE

Stigmella sp.

16a-00xx (0062–0111)

TL 2–3 mm These very tiny moths are best known for, and identified by, the long serpentine leaf mines made by the larvae. In most cases, adults can be hard to separate to species without close inspection or dissection. Most have dark gray FW with one or two pale bands, and an orange tuft on the head. Some species are more tan or lack pale bands. Genus *Monopis* also has dark FW and orange head but is much larger, and white markings are patches, not bands. **HOSTS** Trees and herbaceous plants.

YUCCA and SHIELD-BEARING MOTHS

SUPERFAMILY Adeloidea (21a)
FAMILIES Prodoxidae and Heliozelidae

Small to very small moths with usually plain-colored wings. Caterpillars usually feed in the ovules or developing seeds of host plants, though *Coptodisca* are leaf-miners. Adults are diurnal and often encountered on or near their host plants.

POLITE GREYA

Rare

RANGE-WIDE

Greya politella

21a-0004 (0194)

TL 6–10 mm Silvery-gray FW is similar in shape to Obscure Greya and may be slightly peppery but otherwise unmarked. **HOSTS** Woodland stars and alumroot. **NOTE** Diurnal, typically found on or near host plants.

OBSCURE GREYA

Uncommon

RANGE-WIDE

Greya obscura

21a-0009 (0189.3)

TL 6–9 mm Light gray FW is quite narrow at base and flares out at outer margin. Paler inner margin is bisected by a darker median band. **HOSTS** Woodland stars. **NOTE** Diurnal, typically found on or near host plants.

actual size

FOUR-SPOTTED GHOST MOTH

actual size

ORANGE-HEADED LEAFMINERS

actual size

POLITE GREYA

OBSCURE GREYA

17

CALIFORNIA GREYA

Greya solenobiella

Rare

21a-0012 (0192)

RANGE-WIDE

TL 5–8 mm Peppery gray FW flares at anal angle. Broad white bands cross FW at AM, PM (broken), and ST areas. HOSTS California hedge parsley. NOTE Diurnal, typically found on or near host plants.

RETICULATED GREYA

Greya reticulatus

Rare

21a-0014 (0193)

RANGE-WIDE

TL 6–9 mm Sexually dimorphic. Male has peppery gray FW, with denser speckling forming a line at PM line, and a white patch near anal angle of inner margin. Female has golden-brown FW with bold white AM band and three long white patches at outsides of PM and ST areas. HOSTS Mountain sweet cicely. NOTE Diurnal, typically found on or near host plants.

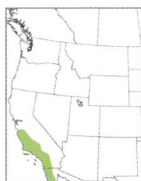

CHAPARRAL YUCCA MOTH

Tegeticula maculata

Rare

21a-0048 (0197)

RANGE-WIDE

TL 7–11 mm White FW has a single bold black spot (rarely three or four spots) in central wing. Terminal line is a row of smaller black dots that often form a dense black patch where outer margin comes to rounded point. A melanic morph (not shown) occurs that is entirely dark (or rarely light) gray. HOSTS Chaparral yucca.

YUCCA MOTH

Tegeticula yuccasella

Rare

21a-0057 (0198)

RANGE-WIDE

TL 12–13 mm Pure white FW has a rounded apex and five small (often indistinct or essentially absent) black spots evenly distributed across middle of wing. White antennae have black tips. HOSTS Yucca.

SILVER LEAFMINERS

Coptodisca sp.

Rare

21a-009x (0244–0260)

RANGE-WIDE

TL 2–3 mm Silver-gray FW is orange beyond median area, with large, black-edged, silver spots at outer ends of PM area. Black terminal band is interrupted by a few silver scales. Some species have black in inner half of ST area. All species are very similar, and identification is best made in combination with plant association. HOSTS Species are highly host-specific. *C. powellella* uses coast live oak; *C. arbutiella*, madrone; *C. saliciella*, willow; *C. cercocarpella*, mountain mahogany; and *C. quercicolella*, Gambel oak.

CALIFORNIA GREYA

male

RETICULATED GREYA

female

actual size

CHAPARRAL YUCCA MOTH

YUCCA MOTH

SILVER
LEAFMINERS

FAIRY (LONGHORN) MOTHS   SUPERFAMILY Adeloidea (21a)
FAMILY Adelidae

Small moths with very long, thread-like antennae. Most of our common species have black or metallic forewings, sometimes with bold white patterning. Genus *Cauchas* resemble *Adela* but have much shorter antennae. Adults are largely diurnal and often encountered at flowers but may rarely come to lights at night.

DESERT FAIRY MOTH   Rare   RANGE-WIDE
Adela punctiferella   21a-0104 (0219)

TL 4–6 mm Resembles plain-winged Gilia Fairy Moth, but antennae are shorter, about 2.5 times FW length. Some individuals may show a tiny white speck in central wing. **HOSTS** Gilia. **NOTE** Diurnal, typically found on or near host plants.

GILIA FAIRY MOTH   Uncommon   RANGE-WIDE
Adela singulella   21a-0105 (0220)

TL 4–5 mm Bronzy FW is crossed by a single white PM line. In some populations, the white line is absent or reduced. White antennae are extremely long, three to four times FW length. **HOSTS** Gilia. **NOTE** Diurnal, typically found on or near host plants.

OCEAN SPRAY FAIRY MOTH   Very Common   NORTH
Adela septentrionella   21a-0107 (0221)

TL 5–6 mm Sexually dimorphic. Black FW has white AM and   SOUTH
PM lines and a row of white dots at terminal line. White antennae are two to three times FW length. Females have orange head; males are dark. **HOSTS** Ocean spray. **NOTE** Diurnal, typically found on or near host plants.

THREE-STRIPED FAIRY MOTH   Very Common   RANGE-WIDE
Adela trigrapha   21a-0110 (0225)

TL 5–7 mm Sexually dimorphic. Black FW has three bold white lines; in some populations, these may be reduced or broken. Males have large eyes and antennae three times FW length; females have smaller eyes, antennae 1.5 times FW length, and orange head. **HOSTS** *Leptosiphon*. **NOTE** Diurnal, typically found on or near host plants.

ORANGE-HEADED FAIRY MOTH   Uncommon   RANGE-WIDE
Adela eldorada   21a-0111 (0226)

TL 6–8 mm Similar to Three-striped Fairy Moth with much overlap in FW pattern, but both sexes have orange heads. Where they co-occur, best separated by habitat and elevation: Orange-headed is found in shrubby areas above 1,500 ft; Three-striped is found in open meadows below 2,000 ft. **HOSTS** Unknown. **NOTE** Diurnal.

FLAMING FAIRY MOTH   Common   NORTH
Adela flammeusella   21a-0112 (0224)

TL 6–7 mm Sexually dimorphic. Coppery-bronze FW has two   SOUTH
large white dots along costa and one at inner margin, occasionally with two smaller dots in basal area. Females usually have reduced markings. In some populations, the spots may be very reduced or absent. Fringe is brown. Male antennae are more than three times FW length; female's are 1.5 times FW length. **HOSTS** Owl's-clover. **NOTE** Diurnal, typically found on or near host plants.

DESERT FAIRY MOTH

GILIA FAIRY MOTH

female

female

female

male

actual size

OCEAN SPRAY
FAIRY MOTH

male

male

male

THREE-STRIPED FAIRY MOTH

male

male

ORANGE-HEADED
FAIRY MOTH

FLAMING FAIRY MOTH

female

21

CALIFORNIA FAIRY MOTH
Adela thorpella

Rare 21a-0113 (0223)

RANGE-WIDE

TL 6–7 mm Resembles Flaming Fairy Moth, but fringe is pale and spots are reduced, sometimes indistinct. Male antennae are 2.5 times FW length; female's are 1.5 times FW length. **HOSTS** Creamcups; also fairypoppy. **NOTE** Diurnal, typically found on or near host plants.

DARK FAIRY MOTH
Cauchas simpliciella

Rare 21a-0124 (0216)

RANGE-WIDE

TL 4–6 mm Dark bronzy FW is unmarked. Head is orange. Antennae are shorter relative to other fairy moths. **HOSTS** Wallflower; also rockcress and *Thelypodium*. **NOTE** Diurnal, typically found on or near host plants.

BAGWORM MOTHS   SUPERFAMILY Tineoidea (30)
FAMILY Psychidae

Small to medium moths with broad wings and feathery antennae. Caterpillars construct cases of twigs or bits of leaves from their environment, which they carry with them as they feed. Females of many species are wingless and attract males using pheromones. Adult males are not usually encountered at lights.

CREOSOTE BUSH BAGWORM
Thyridopteryx meadii

Uncommon 30-0015.51 (0455)

RANGE-WIDE

TL 11–15 mm Densely hairy body of male is soft brown, with tapered abdomen ending in orange-brown claspers. Wings are clear, with brown veins and a sprinkling of dark scales along inner margin and costa. Antennae are bipectinate. Females are wingless and remain in their case for their entire life cycle. **HOSTS** Creosote bush; also mesquite. **NOTE** The eastern Evergreen Bagworm (*T. ephemeraeformis*) is similar but blackish and only ranges west as far as TX and NM.

DANCING, FUNGUS, TUBEWORM, and CLOTHES MOTHS
SUPERFAMILY Tineoidea (30)
FAMILIES Dryadaulidae and Tineidae

Small to medium moths, most of which have long narrow wings and small heads, though grass-tubeworms have broader wings and a typically thick, hairy thorax. Dancing moths are so-called for the adults' characteristic brief crab-like movements upon alighting. Larvae of many tineids are detritus-feeders, and some build silken cases. This group includes our notorious clothes moths. Adults of most species will come to lights in small numbers.

HAWAIIAN DANCING MOTH
Dryadaula terpsichorella

[IN] Uncommon 30-0033 (0307.1)

RANGE-WIDE

TL 4–5 mm Tan FW has white patch containing two tan blotches in inner median area. Whitish ST area has dark brown spots along ST line and solid black terminal line edged with tan shading. **HOSTS** Dead leaves. **NOTE** Adults briefly "dance" in rapid circular motions upon alighting.

DARK FAIRY MOTH

actual size

CALIFORNIA FAIRY MOTH

actual size

CREOSOTE BUSH BAGWORM

HAWAIIAN DANCING MOTH

actual size

ONE-BLOTCHED AMYDRIA

Amydria curvistrigella **Rare** 30-0044 (0332)

TL 10–11 mm Resembles Brown-blotched Amydria, but FW is less mottled and has only a single prominent blotch in PM area; often an indistinct blotch is present in AM area. **HOSTS** Unknown; likely a detritivore.

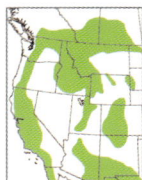

BROWN-BLOTCHED AMYDRIA

Amydria effrenatella **Uncommon** 30-0046 (0334)

RANGE-WIDE

TL 10–11 mm Mottled, light brown FW has dark brown blotches in outer basal, AM, and PM areas. Terminal line is a line of dark spots that curves around apex and along costa. Horn-like labial palps are densely hairy at base. **HOSTS** Unknown; likely a detritivore.

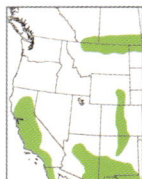

BANDED AMYDRIA

Amydria obliquella **Rare** 30-0048 (0336)

RANGE-WIDE

TL 6–9 mm Long, mottled, tan FW is banded with light and dark brown. Dark brown PM band widens around solid, dark brown spot. Light brown ST band is V-shaped. **HOSTS** Has been reared from root crowns of Menzies' goldenbush.

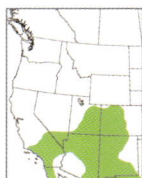

BOLD-DASHED GRASS-TUBEWORM

Acrolophus cockerelli **Uncommon** 30-0066.4 (0345)

RANGE-WIDE

TL 11–14 mm Resembles Short-dashed Grass-Tubeworm, but black median dash and PM spot are thicker, and spots along costa are indistinct or absent. Broad stripes of slightly lighter brown often pass behind black spots in inner and central FW. Fringe is usually strongly checkered. **HOSTS** Grass roots.

GRAY GRASS-TUBEWORM

Acrolophus griseus **Rare** 30-0075.1 (0355)

RANGE-WIDE

TL 11–14 mm Mottled gray FW has streaky blackish patches peppered with brown at central AM, inner median, central PM, and outer ST areas; sometimes patches are reduced to narrow dashes. Some individuals can be quite pale gray. **HOSTS** Grass roots.

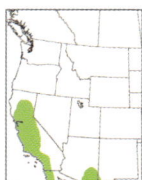

SHORT-DASHED GRASS-TUBEWORM

Acrolophus kearfotti **Uncommon** 30-0075.5 (0357)

RANGE-WIDE

TL 10–12 mm Light reddish- to grayish-brown FW has a short, thin (sometimes thick), black dash near inner margin of median area and usually a faint black spot in middle of PM area. Costa is densely marked with black spots. FW ground is evenly colored and often mottled with clumps of slightly longer, paler scales; some individuals have lighter brown streaks behind black spots. Fringe is only weakly checkered. **HOSTS** Grass roots.

DIVIDED GRASS-TUBEWORM

Acrolophus laticapitana **Uncommon** 30-0079 (0359)

RANGE-WIDE

TL 7–10 mm Tan FW has curving white PM line with patches of blackish shading at inner margin and midpoint. Angled white AM line lacks any black shading. ST line is a curving row of blackish dots narrowly edged with white. **HOSTS** Grass roots.

ONE-BLOTCHED AMYDRIA

BROWN-BLOTCHED AMYDRIA

BANDED AMYDRIA

BOLD-DASHED GRASS-TUBEWORM

GRAY GRASS-TUBEWORM

actual size

SHORT-DASHED GRASS-TUBEWORM

DIVIDED GRASS-TUBEWORM

RUSTY GRASS-TUBEWORM

Rare

Acrolophus pyramellus   30-0087.8 (0377)

TL 9–12 mm   Reddish-tan FW has dark brown blotch in outer PM and inner median areas. Dark spots of terminal line continue around apex and along costa. Row of white spots sometimes visible along ST line. Resembles One-blotched Amydria, but FW is wider and has inner median area blotch. **HOSTS** Grass roots.

VARIABLE GRASS-TUBEWORM

Uncommon

Acrolophus variabilis   30-0095 (0385)

TL 11–17 mm   Light gray to tan FW has black-edged brownish band that zigzags from costal AM area to anal angle, where it meets brownish terminal band; points of zags are usually blackish. Fringe and costa are checkered dusky. Resembles Gray and lighter Clemens' Grass-Tubeworms but has terminal band, and central markings are more defined and filled with brownish color. Some individuals may be quite pale with relatively indistinct markings. **HOSTS** Grass roots.

POPLAR FUNGUS MOTH

Rare

Nemapogon defectella   30-0110 (0264)

TL 6–9 mm   Narrow white FW is lightly speckled with black scales. Thick black median band is V-shaped and does not reach inner margin. Elongate black patches in basal and ST areas angle from costa medially. **HOSTS** Fungi on willow and poplar.

EUROPEAN GRAIN MOTH

[IN] Rare

Nemapogon granella   30-0112 (0266)

TL 7–9 mm   Resembles Poplar Fungus Moth, but median band is broken into two long spots, and basal and ST patches are reduced or absent. **HOSTS** Polypore fungi; in the home, also stored foods such as dried fruit, and cork. **NOTE** A native species, *N. molybdanellus*, is similar but has a tan FW and head and feeds on polypore fungi on hardwoods in coastal and montane CA.

PLASTER BAGWORM

Uncommon

Phereoeca praecox   30-0140.82 (0390.1)

TL 6–7 mm   Resembles Household Casebearer, but blackish patches are smaller and better defined. **HOSTS** Detritus. **NOTE** Larvae build oval or slightly hourglass-shaped, detritus-covered cases often found on building exteriors.

HOUSEHOLD CASEBEARER

Rare

Phereoeca uterella   30-0141 (0390)

TL 4–7 mm   Light brown FW is heavily peppered with dark scales, with bold blackish patches in basal and PM areas and two blackish spots in median area, the latter sometimes blurred together. Fringe is pale brown. Head is tawny. **HOSTS** Natural fibers, such as spiderwebs, silk, and wool. **NOTE** Larvae build oval sand-covered cases often found on building exteriors.

RUSTY GRASS-TUBEWORM

VARIABLE GRASS-TUBEWORM

actual size

POPLAR FUNGUS MOTH

EUROPEAN GRAIN MOTH

PLASTER BAGWORM

HOUSEHOLD CASEBEARER

YELLOW-HEADED CLOTHES MOTH

Rare   RANGE-WIDE

Tinea niveocapitella   30-0154 (0402)

TL 8–12 mm   Medium to dark gray FW has small blackish spots in inner median and central PM areas edged broadly on one side with tan. Terminal line is checkered. Head is tan. **HOSTS** Natural sources of keratin, such as bird feathers or fur in predator scat and owl pellets; in the house, wool.

WESTERN CLOTHES MOTH

Uncommon   RANGE-WIDE

Tinea occidentella   30-0155 (0403)

TL 8–12 mm   Medium to dark bluish-gray FW is mottled white in median and PM areas, with a small black spot in central PM area and usually a second, broadly white-edged spot in inner median area. Head is white. **HOSTS** Natural sources of keratin, such as bird feathers or fur in predator scat and owl pellets; in the house, wool.

CASE-BEARING CLOTHES MOTH

[IN] Uncommon   RANGE-WIDE

Tinea pellionella   30-0157 (0405)

TL 5–8 mm   Tan to brown, slightly peppery FW has two dusky spots (sometimes indistinct) in central median area and a larger spot in central PM area. Head is orange. **HOSTS** Feathers, wool, leather, fur, and other animal products.

EUROPEAN HOUSE MOTH

[IN] Rare   RANGE-WIDE

Niditinea fuscella   30-0164 (0411)

TL 7–8 mm   Speckly, light brown FW has a diffuse dusky band from dusky AM band down central wing (both bands sometimes indistinct), between two blackish median spots and through blackish PM area spot. **HOSTS** Commonly found in bird nests, including domestic coops, where it feeds on feathers and droppings. Also, less often, from household dry goods and other dry plant and animal remains.

BIRD NEST MOTH

Common   RANGE-WIDE

Monopis crocicapitella   30-0168 (0415)

TL 5–7 mm   Blackish FW is usually speckled with tan along costa and has a thick tan stripe along inner margin. Head and dorsal thoracic stripe are pale tan. **HOSTS** Bird nests, owl pellets, and dung in animal burrows. Also recorded from stored bulbs and potatoes.

WEBBING CLOTHES MOTH

[IN] Uncommon   RANGE-WIDE

Tineola bisselliella   30-0181 (0426)

TL 5–7 mm   Light tan FW is usually unmarked. Concolorous fringe is long. Head is slightly darker tan. **HOSTS** Natural sources of keratin, such as bird feathers or fur in predator scat and owl pellets; in the house, wool.

WHITE-HORNED RECYCLER MOTH

[IN] Uncommon   RANGE-WIDE

Opogona omoscopa   30-0195 (0433)

TL 7–11 mm   Dark gray FW has small, speckly, tan patches at base of thorax, inner margin of PM area, and costal ST area. Fringe is dark tan. Antennae usually pale. **HOSTS** Decaying plant material.

YELLOW-HEADED CLOTHES MOTH

actual size

WESTERN CLOTHES MOTH

CASE-BEARING CLOTHES MOTH

EUROPEAN HOUSE MOTH

BIRD NEST MOTH

WEBBING CLOTHES MOTH

WHITE-HORNED RECYCLER
MOTH

YELLOW-V RECYCLER MOTH
Common   RANGE-WIDE

Oinophila v-flava   30-0198 (0434)

TL 4–6 mm Dark gray FW has bold, tan, V-shaped median band and tan patches at inner margin and costa of ST area. Fringe is tan. Head is orange. **HOSTS** Decaying plant material.

GILT-EDGED MOTH
Rare   RANGE-WIDE

Daviscardia coloradella   30-0202 (0312)

TL 13–16 mm Narrow, dark brown FW has brown-speckled, light tan stripe along inner margin that curves around anal angle into terminal band and has points at median and PM lines. **HOSTS** Fungi on conifers and hardwoods.

CACTUS RECYCLER MOTH
Uncommon   RANGE-WIDE

Dyotopasta yumaella   30-0214 (0322)

TL 8–16 mm Light gray FW has dark gray AM, median, and PM bands that widen toward inner margin and contain small scale tufts; bands are sometimes edged with black. Basal and ST areas have brownish shading that is sometimes also adjacent to median band. **HOSTS** Dead cactus stems.

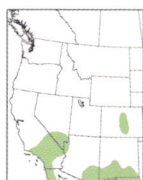

PALE XYLESTHIA
Uncommon   RANGE-WIDE

Xylesthia albicans   30-0222 (0316)

TL 6–8 mm Light brown FW has bands of raised scales that are double whitish lines filled with hoary grayish brown in between, as well as a patch of dark brown scales in central median and PM areas. **HOSTS** Unknown, but congener *X. pruniramiella* feeds on plum cankers.

RIBBED COCOON-MAKER and LEAF BLOTCH MINER MOTHS

SUPERFAMILY Gracillarioidea (33a)
FAMILIES Bucculatricidae and Gracillariidae
SUPERFAMILY Yponomeutoidea (36a)
FAMILY Bedelliidae

Small to very small moths with very long, narrow wings and long legs. Larvae are leaf-miners, with some species becoming leaf-rollers in later instars. Adults of *Caloptilia* characteristically rest with their thorax elevated on stilt-like legs, while those of *Cameraria* and *Phyllonorycter* usually elevate their abdomen. Morning-glory Leafminer (36a-0002) is not closely related to the others of this group but is similar in body shape and larval habit and so is presented alongside them. Adult moths of most species come to lights in small numbers. These groups are very large and diverse, with many very similar species that may require dissection or DNA barcoding to ID definitively; only a few representative species are presented here.

OAK RIBBED CASEMAKER
Uncommon   RANGE-WIDE

Bucculatrix albertiella   33a-0070 (0550)

TL 3–5 mm Tiny. Tan FW has paler bands at AM, median, and PM lines and dark brown patches at inner margin of median area and midpoint of outer margin. **HOSTS** California live oak.

YELLOW-V RECYCLER MOTH

actual size

GILT-EDGED MOTH

CACTUS RECYCLER MOTH

PALE XYLESTHIA

OAK RIBBED CASEMAKER

actual size

31

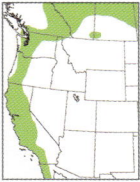

WILLOW LEAFBLOTCH MINER

Micrurapteryx salicifoliella

Rare   RANGE-WIDE

33a-0107 (0647)

TL 6 mm Dark brown FW has a jagged-edged white stripe along inner margin. Costa has narrow, acutely angled, white dashes, with a long thin dash crossing central wing. White dashes at outer edges of brown ST area frame black spot. Fringe has tuft of long scales at midpoint. Dorsal head and thorax are white. **HOSTS** Willow.

AZALEA LEAFMINER

Caloptilia azaleella

[IN] Rare   RANGE-WIDE

33a-0158 (0592)

TL 5–7 mm Dark brown FW has an elongated and poorly defined yellow patch along costa that widens in median area. Costa is usually speckled with blackish scales. Front of head is white. **HOSTS** Azalea.

POISON OAK CALOPTILIA

Caloptilia diversilobiella

Rare   RANGE-WIDE

33a-0169 (0602)

TL 7–8 mm Warm brown FW has pale tan costa below median line, with dark blackish spots along edge. Blackish spots in central wing are less visible. Face is pale tan. **HOSTS** Poison oak.

SUGAR BUSH CALOPTILIA

Caloptilia ovatiella

Rare   RANGE-WIDE

33a-0187 (0619)

TL 6–9 mm Darkish brown FW has paler costal half and is lightly speckled throughout with dark brown. Front of face has a small whitish patch. **HOSTS** Sugar bush.

RETICULATED CALOPTILIA

Caloptilia reticulata

Rare   RANGE-WIDE

33a-0196 (0628)

TL 7–9 mm Warm orangey-brown FW has bold, pale yellow spots along inner margin and mottled, pale yellow spots along costal half and outer margin. Top of head and thorax are pale yellow. **HOSTS** Oak.

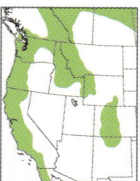

POPLAR CALOPTILIA

Caloptilia stigmatella

Rare   RANGE-WIDE

33a-0207 (0639)

TL 5–7 mm Warm brown FW has a pale yellow triangle at midpoint of costa; tip of triangle is sometimes hooked. Face is brown. **HOSTS** Poplar and willow.

WHITE-STRIPED LEAFBLOTCH MINERS

Cameraria sp.

Common   RANGE-WIDE

33a-0220-0273 (0735–0841)

TL 4–6 mm Golden-brown FW has black-edged white AM, median, PM, and ST lines; in most species these are V-shaped, but in some they are straight lines or angled bars at FW margins. Central terminal area is usually shaded black. A few species are pale, with broad peppery edging to white bands. **HOSTS** Leafminers on woody shrubs and trees. **NOTE** Most frequently encountered in larval form by observation of leaf damage.

WILLOW
LEAFBLOTCH
MINER

AZALEA LEAFMINER

SUGAR BUSH
CALOPTILIA

POISON OAK CALOPTILIA

actual size

RETICULATED
CALOPTILIA

POPLAR CALOPTILIA

WHITE-STRIPED LEAFBLOTCH MINERS

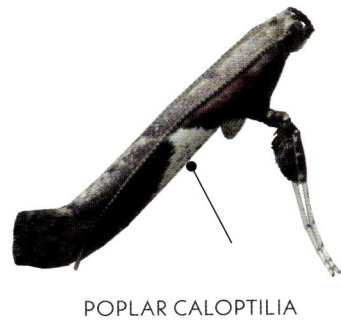

33

MANZANITA LEAFBLOTCH MINER

Rare

RANGE-WIDE

Phyllonorycter manzanita
33a-0335 (0768)

TL 3–4 mm Tiny. Golden-brown FW has bold, black-edged, white chevrons along costa and inner margin that curve at tip toward outer margin. White markings at basal half of inner margin join to isolate an elongate golden-brown patch. **HOSTS** Manzanita.

CITRUS LEAFMINER

[IN] Common

RANGE-WIDE

Phyllocnistis citrella
33a-0371 (0854.1)

TL 2–3 mm Tiny. Golden-tan FW has dusky stripe down central wing from base to median area, bold black spot at point of outer margin, and dark PM line indistinctly bordered with white. Costa has dark median and ST dashes. **HOSTS** Citrus.

ASPEN SERPENTINE LEAFMINER

Very Common

RANGE-WIDE

Phyllocnistis populiella
33a-0380 (0852)

TL 2–3 mm Tiny. Shining white FW has tan wash in outer third and bold black spot at point of outer margin. Costa and inner margin have dark gray dashes at PM, ST, and terminal lines. **HOSTS** Aspen and poplar.

MADRONE SKIN MINER

Very Common

RANGE-WIDE

Marmara arbutiella
33a-0385 (0703)

TL 2–3.5 mm Tiny. Peppery gray to brown FW has straight white AM and median bands and white triangles on costa at PM and ST lines and inner margin at PM line. **HOSTS** Arbutus; possibly also toyon.

MORNING-GLORY LEAFMINER

Uncommon

RANGE-WIDE

Bedellia somnulentella
36a-0002 (0466)

TL 4–6 mm Very long, narrow FW is peppery gray to tan, with paler strip along inner margin. Some individuals have dark spots along inner margin at AM, median, and PM lines. **HOSTS** Morning glory, sweet potato, and bindweed.

SUN MOTHS
SUPERFAMILY Yponomeutoidea (36a)
FAMILY Heliodinidae

Tiny moths with bright orange and silver wings bearing metallic spots. Many species rest with their spiny hindlegs elevated. Adults are diurnal and often encountered at flowers.

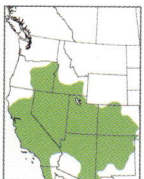

JEWEL-STUDDED SUN MOTH

Uncommon

RANGE-WIDE

Lithariapteryx abroniaeella
36a-0004 (2510)

TL 3–5 mm Striated silver FW has three large, raised, black-outlined, metallic gray circles at median band and costal AM line. A black-edged orange V shape frames triangular white apical patch. **HOSTS** Sand verbena and four o'clock.

MANZANITA LEAFBLOTCH MINER

CITRUS LEAFMINER

ASPEN SERPENTINE
LEAFMINER

actual size

MADRONE SKIN MINER

MORNING-GLORY LEAFMINER

actual size

JEWEL-STUDDED SUN MOTH

35

ORANGE-BANDED SUN MOTH

Lithariapteryx jubarella   36a-0006 (2511)   **Rare**

TL 3–5 mm Resembles Jewel-studded Sun Moth but has a straight orange PM band adjacent to orange V shape. HOSTS Sand verbena and four o'clock.

SPOTTED SUNSPUR

Neoheliodines vernius   36a-0015 (2502.7)   **Rare**

RANGE-WIDE

TL 4–5 mm Orange FW has three raised, metallic silver spots along inner margin and four metallic spots along costa. Fringe and thorax are metallic silver. Similar to *Aetole* species but rests with hindlegs down, and silver terminal band is much narrower. HOSTS Wishbone-bush.

ELEGANT SUNSPUR

Embola powelli   36a-0021 (2500.1)   **Rare**

RANGE-WIDE

TL 4–6 mm Orange FW has black-edged silver median band that connects to silver stripe along inner margin to anal angle. Wing margin has black-edged silver spot near anal angle and three along costa. Blackish shading edges costa in AM/basal area. Fringe and thorax are silver. Often rests with hindlegs raised. Resembles Banded Sunspur but has spots along costa and stripe along inner margin. HOSTS Four o'clock.

BEAUTIFUL SUNSPUR

Aetole bella   36a-0024 (2497)   **Uncommon**

RANGE-WIDE

TL 4–6 mm Orange FW has raised silvery spots edged with black at median and PM lines. Broad silver terminal band continues along inner margin to PM area, where it pinches slightly toward inner margin. Spiny gray hindlegs are held upward at rest. HOSTS Common purslane; native host unknown.

BRILLIANT SUNSPUR

Aetole tripunctella   36a-0030 (2505)   **Rare**

RANGE-WIDE

TL 4–6 mm Similar to Beautiful Sunspur, but terminal band is narrower along outer margin, continues along inner margin to median area, and is not pinched before terminus, and silvery spots are more basal, along AM line. HOSTS Heartleaf and four o'clock.

BANDED SUNSPUR

Aetole unipunctella   36a-0031 (2506)   **Rare**

RANGE-WIDE

TL 4–6 mm Orange FW has black-edged silver belt at median line and is entirely silver below black-edged PM line. Fringe and thorax are silver. Rests with spiny hindlegs raised. HOSTS Scarlet spiderling.

SPLENDID SUNSPUR

Aetole extraneella   36a-0032 (2499)   **Rare**

NORTH

SOUTH

TL 4–6 mm Resembles Brilliant Sunspur, but silver terminal band has more black shading along edging and tapers along inner margin as it passes behind large, raised, silver spots at PM and AM lines. Costa has four black-edged silver spots. HOSTS Clarkia and willowherb.

ORANGE-BANDED SUN MOTH

SPOTTED SUNSPUR

ELEGANT SUNSPUR

actual size

BEAUTIFUL SUNSPUR

BRILLIANT SUNSPUR

BANDED SUNSPUR

SPLENDID SUNSPUR

ERMINE and YPSOLOPHA MOTHS

SUPERFAMILY Yponomeutoidea (36a)
FAMILIES Attevidae, Yponomeutidae, Ypsolophidae, and Plutellidae

Small to very small moths with long narrow wings that are frequently broader at outer margin, often with a pronounced hooked or falcate anal angle. Ailanthus Webworm is a distinctive, primarily eastern species that enters our area in the South. The *Ypsolopha* and plutellid moths rest with their antennae held forward. Ailanthus Webworm may be encountered both at flowers during the day and at lights at night, but most of the remaining species are primarily nocturnal.

AILANTHUS WEBWORM
Atteva aurea

Uncommon
36a-0034 (2401)

RANGE-WIDE

TL 10–16 mm Orange FW has four bands of pale yellow spots thickly outlined with black. HOSTS *Ailanthus*; also other deciduous trees and shrubs.

PINE NEEDLE SHEATHMINER
Zelleria haimbachi

Uncommon
36a-0129 (2427)

RANGE-WIDE

TL 7–8 mm Tan FW has a soft-edged white stripe through central wing. Head and dorsal surface of thorax are white. HOSTS Pine.

WHITE HUCKLEBERRY MOTH
Eucalantica polita

Uncommon
36a-0137 (2350)

RANGE-WIDE

TL 6–8 mm Flared white FW has a small black discal dot and tiny black dot at inner margin of PM line. HOSTS Huckleberry.

SPECKLED HONEYSUCKLE MOTH
Euceratia castella

Uncommon
36a-0139 (2351)

NORTH

SOUTH

TL 8–11 mm Speckled white FW has small black spots at inner median and PM areas. Fringe sometimes tipped brown when fresh. Typically rests with banded antennae held forward. HOSTS Honeysuckle and snowberry.

BROWN HONEYSUCKLE MOTH
Euceratia securella

Rare
36a-0141 (2352)

NORTH

SOUTH

TL 9–11 mm Speckled tan FW is mottled brown (sometimes with blackish patches) in median and ST areas. Elongate, dark brown spots along inner median and PM lines are sometimes indistinct. Fringe is slightly checkered. Typically rests with antennae held forward. HOSTS Honeysuckle and snowberry.

AILANTHUS WEBWORM

PINE NEEDLE SHEATHMINER

WHITE HUCKLEBERRY MOTH

SPECKLED
HONEYSUCKLE MOTH

actual size

BROWN
HONEYSUCKLE MOTH

CANARY YPSOLOPHA

Ypsolopha canariella

Uncommon
36a-0146 (2371)

RANGE-WIDE

TL 10 mm Dimorphic. Brownish-yellow FW has brown V-shaped shading adjacent to slanting pale PM line, with a blackish dot at apex of V shape. Stripe-backed morph has a yellowish-brown FW, with wide, pale yellow stripe along inner margin that usually projects as tooth at PM line. Anal angle is falcate. Typically rests with antennae held forward. **HOSTS** Honeysuckle, snowberry, and willow. **NOTE** *Y. dentiferella* has recently been synonymized with Canary Ypsolopha.

FAWN YPSOLOPHA

Ypsolopha cervella

Uncommon
36a-0147 (2372)

NORTH

SOUTH

TL 7–10 mm Orange-brown FW has black speckles and black spots along inner margin at AM and median areas. Anal angle sometimes slightly falcate. Typically rests with antennae held forward. **HOSTS** Coast live oak and other oaks.

PENNANT YPSOLOPHA

Ypsolopha cockerella

Rare
36a-0148 (2373)

RANGE-WIDE

TL 10–11 mm White FW has golden-brown stripes along costa except at apex, and along inner and outer margins. Scales of fringe are longer at anal angle. Typically rests with antennae held forward. **HOSTS** Unknown.

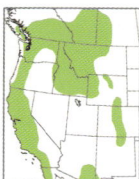

SCYTHED YPSOLOPHA

Ypsolopha falciferella

Uncommon
36a-0153 (2380)

RANGE-WIDE

TL 11–14 mm Light brown FW has indistinct brown AM, PM, and ST bands. Veins are slightly traced with brown. Black speckles across wing are raised tufts near inner margin. Apex is strongly falcate, and anal angle bulges to create concave curve along outer margin. Typically rests with antennae held forward. **HOSTS** Bitter cherry, chokecherry, and other *Prunus* species.

STREAKED YPSOLOPHA

Ypsolopha barberella

Rare
36a-0144 (2370)

RANGE-WIDE

TL 11–12 mm Hoary brown FW has a thick blackish streak down central wing, with a thin sliver-like tooth at median area; inner side is edged with hoary gray shading. FW is narrow and lacks falcate tips. Typically rests with antennae held forward. **HOSTS** Willow.

OCHRE YPSOLOPHA

Ypsolopha ochrella

Rare
36a-0159 (2391)

RANGE-WIDE

TL 7–9 mm Orange-brown FW has black dots at inner median and central PM lines and a dark streak at central terminal area that extends into fringe. Some individuals may also show a darker or dusky streak down length of central FW. White antennae are typically held forward at rest. **HOSTS** Barberry.

DAME'S ROCKET MOTH

Plutella porrectella

[IN] Rare
36a-0179 (2363)

RANGE-WIDE

TL 7–8 mm Tan FW is streaked with brown and has thin blackish basal dash along inner margin and short blackish dash in central median area. Terminal line is blackish and often curves onto inner margin. Forward-facing antennae are dusky near tips. **HOSTS** Dame's rocket.

FAWN YPSOLOPHA

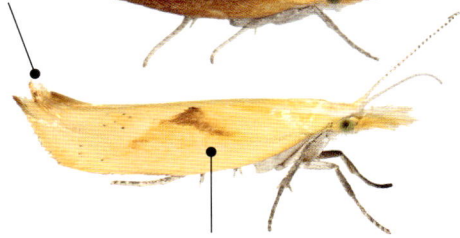

CANARY YPSOLOPHA

PENNANT YPSOLOPHA

SCYTHED YPSOLOPHA

actual size

STREAKED YPSOLOPHA

OCHRE YPSOLOPHA

DAME'S ROCKET MOTH

DIAMONDBACK MOTH
Plutella xylostella

Very Common
36a-0180 (2366)

NORTH

SOUTH

TL 7–8 mm Sexually dimorphic. Speckled, dark brown FW of male has a jagged, pale tan stripe along inner margin, edged with dusky shading. Paler female shows less contrast. **HOSTS** Brassicaceae.

INTERRUPTED SCHOLAR
Rhigognostis interrupta

Rare
36a-0182 (2360)

RANGE-WIDE

TL 8–9 mm Pale gray, speckled FW has thick, curving, black crescents at inner basal/AM and median areas. Inner and outer margins and costa have tiny black dots and two blackish patches on costa. Typically rests with antennae held forward. **HOSTS** Unknown.

SEDGE MOTHS
SUPERFAMILY Yponomeutoidea (36a)
FAMILY Glyphipterigidae SUBFAMILY Glyphipteriginae

Small moths with a broad, notched, outer margin to the forewing. Most species are brown to bronze, with bold patterns of white lines and metallic spots. Adults can be encountered both at flowers during daytime and at lights at night.

TWO-BANDED GLYPHIPTERIX
Glyphipterix bifasciata

Rare
36a-0194 (2339)

RANGE-WIDE

TL 5–9 mm Bronzy FW has white AM and PM bands and triangular white dashes with metallic silver tips along lower costa. Anal angle is black with four metallic silver spots. Narrow metallic PM line is sometimes indistinct. **HOSTS** Unknown. **NOTE** Primarily diurnal.

BRISTLE-LEGGED MOTHS
SUPERFAMILY Schreckensteinioidea (43a)
FAMILY Schreckensteiniidae

A unique superfamily with just three North American species. Small grayish moths with long pointed wings held in a V shape and long spiny hindlegs often raised above their wings when at rest. Adults can be encountered both at flowers during daytime and at lights at night.

BLACKBERRY SKELETONIZER
Schreckensteinia festaliella

Rare
43a-0003 (2509)

RANGE-WIDE

TL 4–7 mm Strongly pointed tan FW has two narrow brown central streaks and a broad brown PM band. Typically rests with its wings held slightly flared and its spiky hindlegs raised. **HOSTS** California blackberry, thimbleberry, salmonberry, and other *Rubus* species.

DIAMONDBACK MOTH

actual size

INTERRUPTED SCHOLAR

actual size

TWO-BANDED GLYPHIPTERIX

actual size

BLACKBERRY SKELETONIZER

METALMARK MOTHS

SUPERFAMILY Choreutoidea (47a)
FAMILY Choreutidae

Very small moths with broad squared wings typically bearing areas of metallic scales. Some species flick their wings during display on host-plant leaves. Adults can be encountered both at flowers during daytime and at lights at night.

JEWELED TEBENNA
Tebenna gemmalis

Rare · RANGE-WIDE

47a-0039 (2644)

TL 6–9 mm FW is yellow orange in basal half, with slightly metallic central streak. Lower half is hoary gray, with three large, black-edged, metallic silver spots set in a whitish patch and a few small metallic spots scattered around it. **HOSTS** Arrowleaf balsamroot.

EVERLASTING TEBENNA
Tebenna gnaphaliella

Rare · RANGE-WIDE

47a-0044 (2647)

TL 4–5 mm Resembles Jeweled Tebenna, but yellow-orange basal half has two metallic central streaks, and trio of black-edged metallic spots are arranged in a triangle, with central spot much larger than the others. Hoary gray lower half of FW lacks whitish patch and often appears to have a distinct gray median band. **HOSTS** Herbaceous plants, including cudweed, pussytoes, and strawflower.

APPLE LEAF SKELETONIZER
Choreutis pariana

[IN] Rare · RANGE-WIDE

47a-0046 (2650)

TL 7–9 mm Brown FW has a zigzag black AM line and thin, wavy, brown PM line set against broad grayish band. ST line is usually dusky or blackish, and terminal line is reddish. **HOSTS** Crab apple.

DIANA'S CHOREUTIS
Choreutis diana

Common (uncommon in South) · RANGE-WIDE

47a-0047 (2651)

TL 6–8 mm Grayish-tan FW has a wavy white AM line edged with blackish and narrow black patch at inner end of indistinct whitish PM area. Costal PM area has divided white patch. Terminal area is usually dusky. **HOSTS** Primarily alder, birch, willow, balsam poplar, and cherry.

BANYAN LEAF SKELETONIZER
Choreutis sexfasciella

[IN] [Uncommon] · RANGE-WIDE

47a-0048.1 (2652.01)

TL 6–8 mm Reddish-brown FW has brown, broad, straight AM and narrow wavy PM lines strongly edged with metallic gray. Central median area has irregular tan patch edged broadly in black below with small patches of metallic gray scales. **HOSTS** Fig and banyan.

JEWELED TEBENNA

EVERLASTING TEBENNA

actual size

APPLE LEAF SKELETONIZER

DIANA'S CHOREUTIS

BANYAN LEAF SKELETONIZER

TORTRIX LEAFROLLERS

SUPERFAMILY Tortricoidea (51a)
FAMILY Tortricidae SUBFAMILY Tortricinae,
TRIBES Torticini and Cnephasiini

Small flat moths with broad squared wings, sometimes with a curving costa. Many species have a dark triangular patch along the central costa, while other species are remarkably polymorphic. Caterpillars are typically leaf-rollers of woody plants. Adults are nocturnal and readily come to lights.

MAPLE LEAFTIER MOTH
Acleris forsskaleana
[IN] Uncommon
51a-0001 (3501)
RANGE-WIDE

TL 7–9 mm Pale yellow FW has fine reddish to orange net-like pattern. Median and terminal lines are usually darker, sometimes brown or dusky, with a pair of small dark dots near inner margin. Some individuals have a dusky patch in place of dots, while others have a large dusky patch across entire inner median area. HOSTS Maple and sycamore.

RED-EDGED ACLERIS
Acleris albicomana
Uncommon
51a-0002 (3502)
RANGE-WIDE

TL 8–9 mm Tan FW is heavily mottled with orange. Reddish-brown basal, AM, PM, and terminal bands are often metallic. Some individuals have black spot at inner median area. HOSTS Rose; also red columbine, red oak, and blueberry.

SNOWY-SHOULDERED ACLERIS
Acleris nivisellana
Uncommon
51a-0010 (3510)
RANGE-WIDE

TL 7–8 mm FW has white basal half heavily mottled with gray, with raised tuft of brown scales at inner AM area. Lower half is mottled brown and gray, with warm brown at inner PM area. Costa has large dusky triangle from AM to ST area. Head is dark brown. HOSTS Pin cherry.

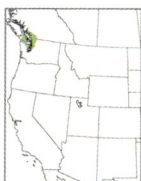

RHOMBOID TORTRIX
Acleris rhombana
[IN] Uncommon
51a-0011 (3511)
RANGE-WIDE

TL 9–12 mm Light orange-brown FW has fine net-like brown lines. Broad, dark brown, angled median bar usually has small white dot at inner end. Thin brown AM line is V-shaped, sometimes indistinct. In some individuals, median bar curves from costa to costal ST area, extending into large brown patch at inner median area. HOSTS Fruit trees and shrubs, including hawthorn, crabapple, cherry, and pear.

GARDEN ROSE TORTRIX
Acleris variegana
[IN] Uncommon
51a-0030 (3530)
NORTH
SOUTH

TL 8–10 mm Polymorphic. Most common form has whitish basal half, with tufted scales at inner AM area, and mottled, dark brown to dark gray lower half. Tufts of scales along median line and outer ST area are usually slightly metallic. Many individuals have a dark brown patch at inner AM area with raised tufts of slightly metallic scales, and grayish mottling at inner median area. Uncommonly, some individuals have the lower half partly or entirely white, with just a dark triangle along costa from median to ST areas. Other morphs not typically seen in western North America. HOSTS Rosaceae, including apple, pear, cherry, hawthorn, and firethorn.

MAPLE LEAFTIER MOTH

RED-EDGED ACLERIS

SNOWY-SHOULDERED
ACLERIS

RHOMBOID TORTRIX

actual size

GARDEN ROSE TORTRIX

WESTERN BLACK-HEADED BUDWORM

Common (uncommon in South)

Acleris gloveranus   51a-0048 (3547)

NORTH
SOUTH

TL 9–11 mm Polymorphic. Most common form has mottled, pale gray FW, with thick gray to black AM line, pale gray AM band below, and blackish crescents along costa from median to ST areas. Some individuals are hoary gray to brown, with indistinct AM band and costal crescent. Uncommonly, some individuals have mottled gray FW, with broad tan stripe down center from base to apex. Rare individuals are gray, with thin black striations, dusky AM band, and tan basal patch. **HOSTS** Fir, Douglas-fir, and hemlock. **NOTE** Can be an abundant forest pest in some regions.

MARBLED BUTTON

Uncommon

Acleris maccana   51a-0050 (3549)

INTERIOR
COASTAL

TL 10–13 mm Variable. Gray, tan, brown, or reddish FW has thin angled median line that may be tan or dusky and terminates in central wing. Outer basal area and triangle along costa from AM to ST areas are shaded darker. Thin PM line and small dark spot at inner basal area are often indistinct. **HOSTS** Generalist on woody shrubs and trees, including huckleberry, blueberry, rhododendron, apple, birch, willow, and poplar.

GREAT ACLERIS

Uncommon

Acleris maximana   51a-0057 (3557)

RANGE-WIDE

TL 12–15 mm Variable. Hoary gray FW has thin lines of slightly raised whitish scales along AM and median lines. Sometimes has dusky or blackish bands along AM and median lines, occasionally with mottling in ST/terminal area. Sometimes has just blackish spots at inner AM, median, and PM areas; rarely, this form may have entire AM and median areas also shaded darker gray. Some individuals have dusky spot at inner and costal AM line and fragmented dusky triangle at costa from median to PM areas. Large long wings for an acleris give the moth a particularly small-headed appearance. **HOSTS** Poplar and willow.

SILVER SHADE

Uncommon

Eana argentana   51a-0068 (3568)

RANGE-WIDE

TL 12–16 mm Long, satiny, white FW is unmarked. As wings become worn, they may develop a hoary gray appearance. **HOSTS** Roots of grass, willow, larch, and spruce. **NOTE** A complex of at least three cryptic species requiring more study.

GRAY-MARKED TORTRICID

Uncommon

Decodes basiplagana   51a-0072 (3573)

RANGE-WIDE

TL 8–10 mm Light gray FW is lightly striated with black. Pale, black-edged median band widens at inner margin and has short black dash in middle on inner margin. AM and PM bands are slightly darker gray. **HOSTS** Oaks.

ASHY TORTRICID

Rare

Decodes fragariana   51a-0073 (3574)

NORTH
SOUTH

TL 7–9 mm Resembles Gray-marked Tortricid, but markings are less contrasting and lack distinct black striations or edging, and inner margin of pale median band lacks short black dash. **HOSTS** Oaks.

WESTERN BLACK-HEADED BUDWORM

MARBLED BUTTON

GREAT ACLERIS

actual size

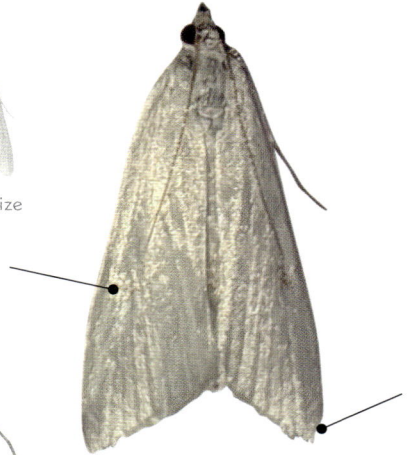

SILVER SHADE

GRAY-MARKED
TORTRICID

ASHY
TORTRICID

COCHYLID MOTHS
SUPERFAMILY Tortricoidea (51a)
FAMILY **Tortricidae** SUBFAMILY Tortricinae TRIBE Cochylini

Small to very small moths with long, slightly flared wings held folded against the sides of the body and bushy labial palps that are typically held pointed downward at rest. This group was once considered its own family (Cochylidae). Caterpillars are typically borers. Adults are nocturnal and readily come to lights.

SULPHUR KNAPWEED MOTH
Agapeta zoegana

[IN] Common
51a-0089.1 (3762)

RANGE-WIDE

TL 8–13 mm Bright yellow FW has lumpy brown ST line in the shape of an inverted V and lumpy brown spot at inner median area. Terminal line and fringe are brown. Thorax has a brown tuft at base and brown stripes laterally. **HOSTS** Knapweeds, especially spotted knapweed. **NOTE** Intentionally released in 1984 for biological control of spotted knapweed.

CONTRASTING HENRICUS
Henricus edwardsiana

Rare
51a-0122.2 (3797)

RANGE-WIDE

TL 6–7 mm White FW has silvery-gray basal area mottled with dark brown that extends along inner margin to PM area. ST/terminal area has mottled grayish shading. **HOSTS** Associated with fly galls in fleabane.

CONE COCHYLID
Henricus fuscodorsana

Rare
51a-0123.1 (3798)

RANGE-WIDE

TL 8–10 mm White FW has warm brown basal area that extends along inner margin, often mottled with silver, to PM area, where brown PM band reaches to central FW. Apex and costal PM area are mottled with brown. **HOSTS** Cones of Douglas-fir, alpine fir, western larch, and coast redwood.

BROWN-SHOULDERED HENRICUS
Henricus umbrabasana

Uncommon
51a-0126.1 (3801)

NORTH

SOUTH

TL 6–10 mm Tan FW is lightly striated with gray or brownish. Dusky to brownish patch at inner margin of median area is mottled with metallic scales and has a small black dot adjacent. Upper half of basal area and thorax are brown. **HOSTS** Oaks.

SMEATHMANN'S AETHES
Aethes smeathmanniana

Uncommon
51a-0155.1 (3760.3)

RANGE-WIDE

TL 6–8 mm Pale tan FW has warm brown angled AM and split PM bands (sometimes edged with black) that terminate at central wing, and patches along costa at AM and PM lines. Basal, median, and ST areas are shaded tan along center. **HOSTS** Yarrow, chamomiles, knapweeds, and lettuces.

actual size

SULPHUR KNAPWEED MOTH

CONTRASTING HENRICUS

CONE COCHYLID

BROWN-SHOULDERED HENRICUS

SMEATHMANN'S AETHES

BLACK-TIPPED RUDENIA

Rudenia leguminana

Uncommon
51a-0161.1 (3839)

TL 6–9 mm Light gray to whitish FW has gray basal area with sparse black striations and dark gray ST/terminal area and inner median area. ST area has a small, black-outlined, dark brown patch at midpoint. Inner median and PM lines along gray median patch are black, usually with short raised scales. Gray thorax has brown scale tufts. **HOSTS** Legumes, including acacia, locust, and mesquite.

CHRYSANTHEMUM FLOWER BORER

Lorita scarificata

Uncommon
51a-0169.1 (3803)

TL 4–6 mm Light tan FW has a mottled pattern of dark tan and light blackish patches. Inner AM and (often) PM, costal median, and central ST areas have blackish patches. Central basal and median areas and most of ST area have dark tan patches. In some individuals, some of the patches may be fainter or indistinct. **HOSTS** Dodder; also cultivated species, including bell pepper and chrysanthemum.

HAPPY TORTRIX

Platphalonidia felix

Uncommon
51a-0206.1 (3831)

TL 7–10 mm Light tan FW has brown, angled median and curving ST bands; median is darker brown at inner margin, and ST band has a blackish spot at midpoint. Terminal line is checkered. **HOSTS** Groundsel, including threadleaf groundsel and Blochman's ragwort.

BROAD-PATCH COCHYLID

Thyraylia nana

Rare
51a-0221.1 (3778)

TL 6–7 mm FW is mottled dark gray from base to median area, with darker striations, and dusky brown to blackish patch at inner median band and paler patch at inner basal area. Lower wing is ivory, with a warm brown wash and dusky gray patch at costa of ST area. **HOSTS** Unknown.

FERRUGINOUS EULIA

Eulia ministrana

Uncommon
51a-0227 (3565)

TL 9–11 mm Yellowish-tan FW has large rusty bands at basal, median, and terminal areas. Center of FW is often washed with rusty brown. Central PM area has a small whitish discal dash. Thorax is tufted with brown scales. **HOSTS** Deciduous trees and shrubs, including birch, rose, and willow.

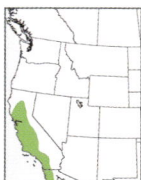

TRIANGULAR ANOPINA

Anopina triangulana

Common
51a-0232 (3583)

TL 6–9 mm Tan to grayish FW is usually peppered with rusty scales and has triangular dusky patch in outer half of median area. Pale AM and PM bands become double at costa. Base of FW is narrowly dusky. In some individuals, entire lower half of wing is washed gray. **HOSTS** Willow.

BLACK-TIPPED RUDENIA

CHRYSANTHEMUM
FLOWER BORER

HAPPY TORTRIX

BROAD-PATCH COCHYLID

FERRUGINOUS EULIA

actual size

TRIANGULAR
ANOPINA

ARCHIPS LEAFROLLERS
SUPERFAMILY Tortricoidea (51a)
FAMILY **Tortricidae** SUBFAMILY **Tortricinae** TRIBE Archipini

Small moths with broad squared wings typically bearing a pattern of transverse brown lines and bands or a reticulated pattern. Caterpillars are mostly leaf-rollers of woody plants; some can be serious crop pests of orchards or the forestry industry. Adults are nocturnal and readily come to lights.

BARRED FRUIT-TREE TORTRIX
Pandemis cerasana

[IN] Rare

51a-0246 (3592)

RANGE-WIDE

TL 8–13 mm Light brown FW has a darker basal area bounded by curving AM line. Darker median band slants from costa of median area to inner margin of ST area. Costal ST area has a small darker semicircle. HOSTS Generalist on many deciduous trees; also several shrubs, including blueberry and European honeysuckle, and a few flowering plants, including avens and *Impatiens*.

PANDEMIS LEAFROLLER
Pandemis pyrusana

Rare

51a-0251 (3596)

RANGE-WIDE

TL 8–14 mm Resembles Barred Fruit-tree Tortrix, but AM line is wavy, with a distinct bulge near inner margin. HOSTS Willow; also cultivated fruit trees.

PASQUEFLOWER TORTRIX
Argyrotaenia coloradana

Rare

51a-0267 (3608)

RANGE-WIDE

TL 9–12 mm Tan FW has broad reddish-brown basal and median bands and dark brown semicircle at costa of PM area. Basal half of FW is heavily suffused with rosy purple to warm brown. HOSTS Pasqueflower; possibly also other species.

CLOUDY-BANDED TORTRIX
Argyrotaenia provana

Rare

51a-0268 (3609)

RANGE-WIDE

TL 9–12 mm Light gray FW has gray basal and median bands and gray semicircle at costal PM area (sometimes also a small patch at central ST area), edged with blackish lines. Pale areas often show fine gray striations. HOSTS Fir and Douglas-fir.

ORANGE TORTRIX
Argyrotaenia franciscana

Common

51a-0270 (3612)

NORTH

SOUTH

TL 6–9 mm Highly variable. Pale tan, light brown, or reddish-brown FW usually has faint, short, brown striations. Broad median band is shaded slightly darker than ground color (sometimes faint) and slants from costal AM to inner margin of PM area; many individuals have a dusky to blackish bar at inner end of median band that flares along inner margin. Lower costa has a semicircular patch of a darker shade of ground color (sometimes faint). Outer margin is straight, and apex does not flare. HOSTS Broad generalist.

actual size

BARRED FRUIT-TREE
TORTRIX

PANDEMIS
LEAFROLLER

PASQUEFLOWER TORTRIX

CLOUDY-BANDED
TORTRIX

ORANGE TORTRIX

CHAMISE TORTRIX

Rare   RANGE-WIDE

Argyrotaenia niscana   51a-0273 (3610)

TL 6–10 mm   Warm brown FW has narrow, pale tan AM and PM bands, edged in central wing with blackish color. PM band connects to tan fringe at anal angle, and costal half of PM band is often whitish. Subapical area has two short, pale tan bars. HOSTS Chamise.

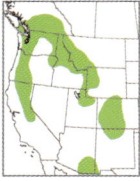

VARIABLE CONIFER TORTRIX

Uncommon   RANGE-WIDE

Argyrotaenia dorsalana   51a-0277 (3618)

TL 8–13 mm   Variable. Pale tan to yellowish FW has a brown bar (sometimes widened into a patch) at inner margin of median area and two smaller brown patches on costa at AM and PM areas. Some individuals have reduced or absent patches on costa, while others have larger AM and median patches that almost join to form a band. Occasionally, individuals have brown reticulations below small median patch and in terminal area. HOSTS Fir, Douglas-fir, larch, spruce, juniper, and whitecedar; also oak.

OBLIQUE-BANDED LEAFROLLER

Common   NORTH

Choristoneura rosaceana   51a-0300 (3635)

SOUTH

TL 8–14 mm   Light brown FW has short brown striations and a thin, brown, slanting median line broadly bordered by brown shading. Thin brown PM line curves basally to costa, framing a dark brown costal patch. Basal and ST/terminal areas are sometimes faintly shaded brown. Outer margin is slightly wavy, and rounded apex flares outward. HOSTS Woody plants, including apple, blueberry, oak, and pine.

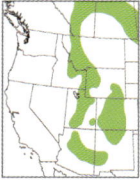

LARGE ASPEN TORTRIX

Uncommon   RANGE-WIDE

Choristoneura conflictana   51a-0302 (3637)

TL 13–18 mm   Grayish brown FW is very broad, with a wide whitish AM band bordered by darker gray. Dark gray striations are primarily visible in ST and terminal areas. HOSTS Primarily quaking aspen; also other poplars and willow. NOTE Can be an abundant forest pest in some regions.

SPRUCE BUDWORM

Common   RANGE-WIDE

Choristoneura fumiferana   51a-0303 (3638)

TL 11–15 mm   Variable reddish-brown to gray, and virtually identical to Western Spruce Budworm; best identified by range and host species. Populations tend to average more gray individuals than rusty. HOSTS Balsam fir and white spruce; also hemlock, tamarack, and other spruce and fir.

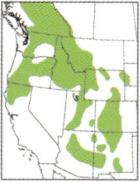

WESTERN SPRUCE BUDWORM

Common   RANGE-WIDE

Choristoneura occidentalis   51a-0305.1 (3640)

TL 12–18 mm   Reddish-brown to gray FW has a lighter AM band and terminal area, usually lighter patch at central PM area, and pale square on costa at median area. Central median area has a vertical black bar, sometimes reduced to patchy black scales. Wing is usually covered with brown to black reticulations, sometimes reduced to short striations. Best separated from Spruce Budworm by range and host species. HOSTS Douglas-fir; also hemlock, spruce, fir, and larch. NOTE Part of a complex of multiple, very similar western species that can be difficult to separate; one of the most common, Modoc Budworm (*C. retiniana*, not shown), is found on fir in CA, OR, ID, and UT and is similar but tan, with stronger pale patches in PM area.

CHAMISE TORTRIX

VARIABLE
CONIFER TORTRIX

actual size

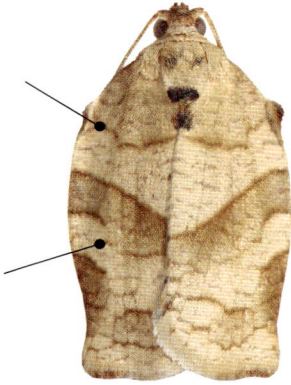

OBLIQUE-BANDED
LEAFROLLER

SPRUCE BUDWORM

LARGE ASPEN TORTRIX

WESTERN SPRUCE BUDWORM

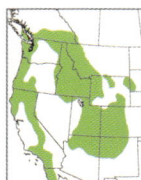

SUGAR PINE TORTRIX
Choristoneura lambertiana

Uncommon
51a-0309 (3644)

TL 11–15 mm Resembles Western Spruce Budworm, but rusty FW lacks any black scales in median area or along reticulations, and reticulations are weaker, particularly along pale bands. Southern populations are darker reddish orange. Northern populations may have a small black dash in median area. **HOSTS** Sugar and ponderosa pine in North; lodgepole and Jeffrey pine in South. **NOTE** Similar Jack Pine Budworm (*C. pinus*, not shown), found on jack pine in AB, SK, MB, and eastward, has strong rusty reticulation.

JUNIPER BUDWORM
Choristoneura houstonana

Uncommon
51a-0313 (3647)

TL 10–12 mm Orange-brown FW has a reticulated pattern of dark brown lines, shaded slightly darker in between at basal, median, and ST areas. Paler areas are often slightly raised, creating a bumpy appearance. **HOSTS** Juniper.

ROSE TORTRIX
Archips rosana

[IN] Uncommon
51a-0320 (3650)

TL 7–11 mm Lightly striated, light to golden-brown FW has angled brown median band that broadens at inner margin and rectangular brown patch along lower costa that tails out along ST line. Inner basal area is shaded brown. Brown areas are sometimes edged with thin, dark brown lines. **HOSTS** Broad generalist.

FRUIT-TREE LEAFROLLER
Archips argyrospila

Common
51a-0323 (3648)

TL 8–13 mm Reddish-brown FW is mottled reddish gray along indistinct AM and PM bands and has squarish ivory patches on costa of bands. Median band and costa adjacent to ivory patches is usually darker reddish brown. Some individuals appear more tan, with brown markings. **HOSTS** Fruit-bearing trees and plants, including apple, blueberry, peach, and pear.

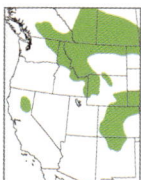

UGLY-NEST CATERPILLAR MOTH
Archips cerasivorana

Uncommon
51a-0334 (3661)

TL 11–13 mm Purplish pink FW is crossed with orange to orange-brown striations. Costa at AM and PM bands, and inner median area, have squarish, dark reddish patches; sometimes also has dark reddish marks at inner basal area. **HOSTS** Wild cherries; also cultivated cherry, chokecherry, apple, hawthorn, and roses. **NOTE** Larvae form silken webs in their host plant resembling those of tent caterpillars but present later in the season.

CARNATION TORTRIX
Cacoecimorpha pronubana

[IN] Uncommon
51a-0349 (3678)

TL 8–12 mm Orange-brown FW has thick, angled, brown median band that is connected along inner margin to brown ST/terminal area. Inner basal area has a small brown patch. Brown areas are sometimes indistinctly edged with thin brown lines. FW often has short brown striations and sometimes has a rosy wash over lower half. **HOSTS** Generalist on a wide variety of herbaceous plants.

SUGAR PINE TORTRIX

JUNIPER BUDWORM

ROSE TORTRIX

FRUIT-TREE LEAFROLLER

actual size

UGLY-NEST CATERPILLAR MOTH

CARNATION TORTRIX

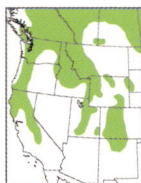

WHITE TRIANGLE TORTRIX

Common (uncommon in South)

Clepsis persicana   51a-0357 (3682)

RANGE-WIDE

TL 10–11 mm   FW has yellow-orange basal half, reddish-brown lower half, and bold white triangle at costa of median area. Fringe is pale. In some individuals, white costal patch is more rectangular, and inner end seems to diffuse into central wing. **HOSTS** Various trees, including alder, apple, birch, maple, and spruce.

PRIVET TORTRIX

[IN] Common

Clepsis consimilana   51a-0359 (3683)

RANGE-WIDE

TL 8–10 mm   Sexually dimorphic. Yellowish- to reddish-brown FW has an indistinct net-like pattern. Female FW is otherwise unmarked. Male has indistinct darker reddish-brown median band that angles down toward inner margin and an indistinct small reddish-brown patch at costal ST are. Costal edge of wing is convex at midpoint. **HOSTS** Privet, lilac, and English yew.

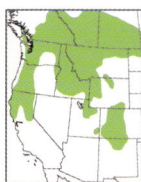

CLEMENS' CLEPSIS

Uncommon

Clepsis clemensiana   51a-0360 (3684)

RANGE-WIDE

TL 10–11 mm   Tan FW has darker tan along veins (sometimes indistinct). Outer margin relatively square, with pointed apex. **HOSTS** Grasses; also aster and goldenrod.

GARDEN TORTRIX

Very Common (rare in North)

Clepsis peritana   51a-0364 (3688)

NORTH
SOUTH

TL 6–8 mm   Lightly striated, light brown FW has dark brown patches at costal median and ST areas and slanting brown median band that ends at dark brownish shading at inner margin. **HOSTS** Herbaceous plants, including strawberry.

GREENISH APPLE MOTH

Uncommon

Clepsis virescana   51a-0366 (3689)

RANGE-WIDE

TL 6–9 mm   Resembles Garden Tortrix, but median band is usually lighter and lacks dark shading at inner margin (sometimes also at costa). **HOSTS** Fresh and decaying leaves of *Prunus* and *Rosa*.

LIGHT BROWN APPLE MOTH

[IN] Common

Epiphyas postvittana   51a-0368 (3693.1)

RANGE-WIDE

TL 7–12 mm   Variable. Brown FW has an indistinct, darker brown, net-like pattern, a thin, brown, slanting median line, and a thin brown crescent along lower costa. Broad band of reddish-brown shading along inner and outer margins is sometimes indistinct. Some individuals are entirely dark reddish brown below median line. Uncommonly, sometimes all markings except net-like pattern are nearly absent. **HOSTS** Generalist on many herbaceous plants and trees. **NOTE** Considered a serious pest of apples and other fruit.

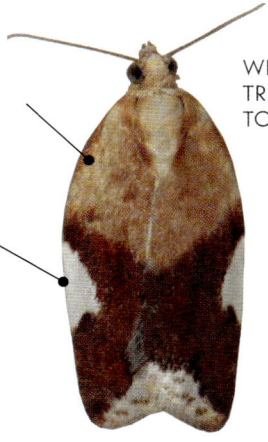

WHITE
TRIANGLE
TORTRIX

male

PRIVET TORTRIX

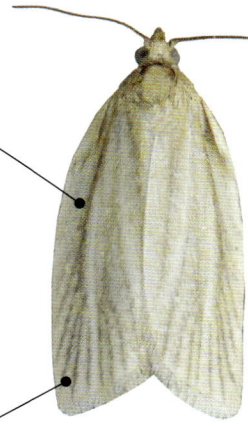

female

actual size

CLEMENS' CLEPSIS

GARDEN TORTRIX

GREENISH APPLE MOTH

LIGHT BROWN APPLE MOTH

RED-BARRED TORTRIX

Ditula angustiorana

[IN] Uncommon
51a-0371 (3692)

RANGE-WIDE

TL 6–8 mm Reddish-brown (sometimes olive) FW has paler basal half, with slanting reddish-brown patch at inner basal area. Hoary, white-edged band slants from white patch at costa of median area to anal angle. Hoary, white-edged ST band slants from costa to midpoint of outer margin but, when wings are folded at rest, appears as a straight band across both wings. Some individuals have a dusky patch at outer half of basal area. **HOSTS** Broad generalists at northern latitudes; recorded only from ornamental yew in CA.

SPARGANOTHID LEAFROLLERS

SUPERFAMILY Tortricoidea (51a)

FAMILY Tortricidae SUBFAMILY Tortricinae TRIBE Sparganothini

Small moths with broad, slightly pointed wings and relatively long labial palps. Many species are yellow or orange, with rusty bands or reticulations. Males of most species are often more strongly marked. Caterpillars are leaf-rollers of woody plants, and most are host generalists. Adults are nocturnal and readily come to lights.

WESTERN AVOCADO LEAFROLLER MOTH

Amorbia cuneanum

Uncommon
51a-0373 (3749)

NORTH

SOUTH

TL 10–15 mm Broad, rounded, tan FW has fine, light brown striations, thin, dark brown edging along costa and outer margin, and a hooked, dark brown triangle at costal median area. Apex is often slightly falcate. Some individuals have diffuse, dark brown patches in inner median area. Rare individuals may be dark brown instead of tan. **HOSTS** Generalists, but particularly on manzanita, madrone, toyon, and *Prunus*.

SPARGANOTHIS FRUITWORM

Sparganothis sulfureana

Uncommon
51a-0390 (3695)

RANGE-WIDE

TL 7–12 mm Yellow FW has a loose network of fine orange lines (occasionally absent) and straight brownish-orange AM and PM lines that join at inner margin to form a distinct X shape when wings are folded at rest. Terminal line has brownish-orange band. Some individuals have dark reddish or brown patches at midpoint and inner margin of AM and PM lines, and sometimes also costa. **HOSTS** Various trees and plants, including apple, clover, corn, cranberry, pine, and willow.

ONE-LINED SPARGANOTHIS

Sparganothis unifasciana

Rare
51a-0396 (3711)

RANGE-WIDE

TL 10–12 mm Yellow FW has a slanting reddish-brown bar from costal AM to inner margin of PM areas and reddish-brown patches at costal PM and central ST areas. **HOSTS** Various trees and herbaceous plants, including ash and pine.

ANCIENT SPARGANOTHIS

Sparganothis senecionana

Uncommon
51a-0406 (3714)

NORTH

SOUTH

TL 8–13 mm Variable. Most commonly, yellowish-brown (rarely whitish or brown) FW has a slanting brown AM bar and curving PM line bounding brown ST/terminal area. Rare individuals have a pale apical bar bisecting ST/terminal shading or only a shaded triangle at costa of ST area. **HOSTS** Broad generalist.

RED-BARRED TORTRIX

actual size

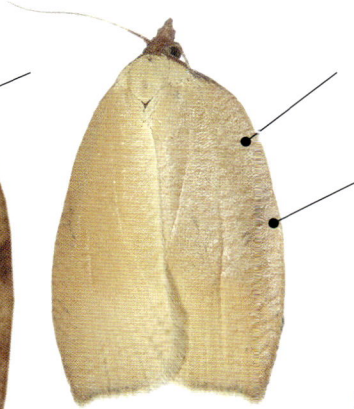

WESTERN AVOCADO
LEAFROLLER MOTH

ONE-LINED
SPARGANOTHIS

ANCIENT
SPARGANOTHIS

SPARGANOTHIS
FRUITWORM

actual size

RETICULATED FRUITWORM
Cenopis reticulatana

Uncommon (rare in South)
51a-0419 (3720)

RANGE-WIDE

TL 9–12 mm Resembles Sparganothis Fruitworm, but network of orange lines is dense, sometimes making FW appear more orange than yellow. AM and PM lines have brown patches at costa and inner margin but not at midpoint. Terminal line is a thin brown line. HOSTS Various trees, shrubs, and herbaceous plants, including alder, apple, aster, blueberry, maple, and oak.

TOASTED PLATYNOTA
Platynota labiosana

Uncommon
51a-0440 (3738)

RANGE-WIDE

TL 6–10 mm Brown FW has darker brown patches at costal median and ST areas. Angled median band is indistinctly shaded darker. Major transverse FW lines are marked by raised scales, with scale tufts (sometimes colored dark brown) at inner AM and median and central PM lines. HOSTS Generalist, including prickly poppy, okra, boneset, and penstemon.

OMNIVOROUS LEAFROLLER
Platynota stultana

Common
51a-0449 (3736)

RANGE-WIDE

TL 4–8 mm FW is mottled warm brown in basal half and tan in lower half, with large dusky patch at central median area and mottled, warm brown patch at costal ST area. Major transverse FW lines are marked by silvery raised scales, with scale tufts (sometimes colored dusky) at inner AM and central median lines. Rare individuals appear in grayscale, with dark gray basal and medium gray lower half, and pale scale tuft at inner median line. HOSTS Broad generalist.

OLETHREUTINE MOTHS
SUPERFAMILY Tortricoidea (51a)
FAMILY Tortricidae SUBFAMILY Olethreutinae

A large and varied group of small moths with long squared wings typically held near or folded against the body. Most have a slight crease in the central ST/terminal area of the forewing. Adults are primarily nocturnal and readily come to lights, though some species can be encountered at flowers or in foliage during daytime.

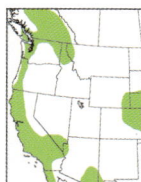

VERBENA BUD MOTH
Endothenia hebesana

Uncommon
51a-0466 (2738)

RANGE-WIDE

TL 6–9 mm FW has a complexly mottled appearance. Dusky AM, PM, and ST bands are irregularly shaped; PM band has two prominent projections on lower edge. Outer median and lower basal area have bluish-gray patches. Terminal band and thoracic tuft are rusty. HOSTS Seeds of herbaceous plants, including iris, speedwell, and vervain.

RUSH BACTRA
Bactra furfurana

Rare
51a-0477 (2706)

NORTH

SOUTH

TL 6–8 mm Light brown FW has a warm brown, black-speckled, hook-shaped median band with a blackish tip, a curving, warm brown AM band, and elongate, warm brown apical patch speckled black. Curving, metallic silver lines pattern the areas in between. Costa has paired white to tan dashes interspersed with dark brown spots. HOSTS Rush and bulrush.

TOASTED PLATYNOTA

actual size

RETICULATED FRUITWORM

OMNIVOROUS
LEAFROLLER

VERBENA BUD MOTH

actual size

RUSH BACTRA

JAVELIN MOTH

Bactra verutana

Uncommon
51a-0478 (2707)
RANGE-WIDE

TL 6–10 mm Light brown FW has dusky patches at central AM and PM areas (sometimes indistinct). Costa and outer margin are checkered. Some individuals are shaded slightly darker along inner half; others have sparse striations in central FW. **HOSTS** Flatsedge.

SUMAC LEAFTIER

Episimus argutana

Rare
51a-0485 (2701)
RANGE-WIDE

TL 7–8 mm Mottled brown FW has black-edged brown patches (sometimes indistinct) at inner AM and outer PM areas; the latter has a narrow, black-edged, brown line crossing to inner margin. Lower half of wing has dark bluish-gray shading between mottling, particularly dense in inner ST area. Inner terminal area has three or four black spots set inside a brown patch. **HOSTS** Spurge, poison ivy, sumac, and witch hazel.

GREEN ASPEN LEAFROLLER

Apotomis removana

Uncommon
51a-0523 (2768)
RANGE-WIDE

TL 10–13 mm Mottled gray FW has irregular, black-edged, dark gray AM, PM, and ST bands; PM band splits near midpoint to become two narrower bands in inner half. Some individuals have dusky fill in costal half of PM band, and sometimes at costa of AM band. Rare individuals have a dark, diffuse streak from base to PM area along inner margin. **HOSTS** Quaking aspen.

POPLAR LEAFROLLER

Pseudosciaphila duplex

Common (uncommon in South)
51a-0538 (2769)
RANGE-WIDE

TL 9–13 mm Mottled, light gray FW has a white hourglass-shaped AM band, with a tooth near midpoint projecting into median area. White PM band runs diagonally to anal angle, becoming diffuse in inner half. Lower costa has white dashes, and brown apex usually contains a black spot. Thoracic tuft is brown or gray. Rare dark morph is charcoal, with blackish basal and median bands and brown terminal line and apex; some individuals have a white wash across ST/terminal area. **HOSTS** Poplar; also alder, birch, maple, and willow.

CELYPHA MOTH

Celypha cespitana

Common (uncommon in South)
51a-0629 (2859)
RANGE-WIDE

TL 7–9 mm Mottled brown FW has white AM band with gray striations. Often-indistinct PM band crosses to anal angle, with a basal-pointing bulge near midpoint. Costa has paired white dashes along lower half. **HOSTS** Clover, cottonwood, and strawberry.

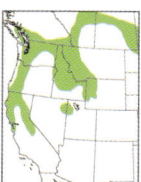

OFF-WHITE HEDYA

Hedya ochroleucana

Uncommon
51a-0635 (2861)
RANGE-WIDE

TL 8–12 mm FW is ivory below PM line and marbled brown, black, and white above, with large black-speckled patches of dark bluish gray at inner basal, outer median, and central PM areas. **HOSTS** Apples, mountain-ashes, and roses.

JAVELIN MOTH

SUMAC LEAFTIER

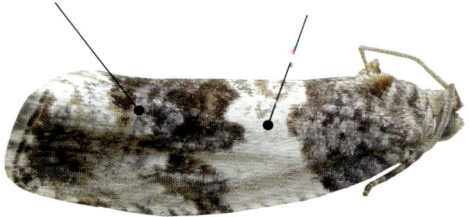

POPLAR LEAFROLLER

actual size

GREEN ASPEN LEAFROLLER

CELYPHA MOTH

OFF-WHITE HEDYA

GREEN BUDWORM MOTH

Hedya nubiferana

[IN] Uncommon
51a-0636 (2862)

RANGE-WIDE

TL 9–11 mm Resembles Off-white Hedya, but ivory section has gray patch at anal angle and brownish shading across terminal area, and blue-grayish color extends into entire basal and median bands. **HOSTS** Hawthorn and blackthorn.

SIMILAR ANCYLIS COMPLEX

Ancylis columbiana/simuloides

Uncommon
51a-0650/1 (3362/3)

NORTH

SOUTH

TL 8–11 mm Ivory FW has a large, warm to dark brown patch in inner basal, AM, and median areas that creates a prominent oval dorsally when wings are folded at rest. Thick median bar slants down to connect to triangular patch in central PM area. Inner PM, ST, and terminal area is speckled gray. Costa has squat brown triangles below median bar and brown spot at falcate apex. Similar (*A. simuloides*) and Columbia (*A. columbiana*) Ancylis can be separated only by DNA or genitalia. **HOSTS** Ceanothus.

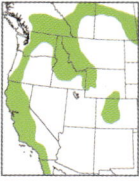

SERVICEBERRY LEAFFOLDER

Ancylis mediofasciana

Rare
51a-0673 (3384)

NORTH

SOUTH

TL 8–12 mm Mottled brownish-gray FW has a wide white costa, with a thick gray median bar. Central AM area has a warm brown patch mottled with black. Falcate apex has a black spot surrounded by warm brown dashes. Costa edge has small blackish dots. Head is brown. **HOSTS** Not fully known; has been reared from chokeberry, serviceberry, and *Prunus*.

SNOWY TORTRICID

Hystrichophora vestaliana

Rare
51a-0689 (3399)

RANGE-WIDE

TL 10–13 mm White FW has a crisp black terminal line, sometimes edged with warm brown shading, and short black dashes along lower half of costa. Inner PM area has a small black dot. Apex is pointed. **HOSTS** Unknown.

KNOTTY RETINIA

Retinia sabiniana

Uncommon
51a-0725 (2896)

RANGE-WIDE

TL 10–14 mm Golden-orange FW has white median and ST bands, with patches of raised scales in central wing along usual positions of cross-lines. Thorax has a white band at posterior. **HOSTS** Pines.

FIR TWIG RETINIA

Retinia picicolana

Rare
51a-0731 (2901)

RANGE-WIDE

TL 13–18 mm Mottled gray FW has curving black crescent in ST area connecting to brown shading along outer margin and black spot on inner margin of PM area. Head is orange. **HOSTS** Firs.

GREEN BUDWORM MOTH

SIMILAR ANCYLIS COMPLEX

SERVICEBERRY LEAFFOLDER

SNOWY TORTRICID

actual size

KNOTTY RETINIA

FIR TWIG RETINIA

DOUGLAS-FIR CONE MOTH

Barbara colfaxiana

Rare

51a-0733 (2903)

TL 9–10 mm Gray FW has warm brown along outer margin and around apex to costal PM area and is sometimes streaked through central wing. Irregular, darker gray, transverse bands are sometimes thinly outlined with black. HOSTS Cones of Douglas-fir and fir.

EYE-SPOTTED BUD MOTH

Spilonota ocellana

[IN] Uncommon

51a-0735 (2906)

TL 6–9 mm FW has mottled brown and gray basal/AM area, and whitish median/PM area mottled with gray and brown in costal half. Inner margin of PM area has a smallish gray and black triangle. Inner ST area has a gray stripe. Outer ST area has a patch of thick black dashes. HOSTS Generalist, primarily of deciduous trees.

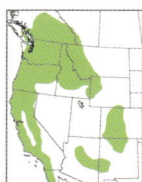

WESTERN PINE SHOOT BORER

Eucopina sonomana

Rare

51a-0737.3 (3065)

TL 9–12 mm Orange FW has reddish shading along costa and central wing and silvery AM, PM, and terminal bands mottled with orange. Basal area has scattered silvery scales. Fringe, head, and thorax are ivory. HOSTS Ponderosa, lodgepole, and Bishop pine and Engelmann spruce.

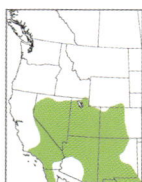

PINYON CONEBORER

Eucopina bobana

Rare

51a-0737.4 (3067)

TL 9–14 mm Golden-orange FW has rusty-orange basal, AM, PM, and ST bands edged with white lines; rusty bands typically contain scattered blackish scales, particularly in ST band. PM band often appears as connected but slightly offset squares. ST band has a broad squarish section at midpoint. Costa has two small rounded semicircles between PM and ST bands, the most basal sometimes thinly connected to PM band. Thorax is golden orange. HOSTS Pinyon pine.

PONDEROSA CONEBORER

Eucopina ponderosa

Rare

51a-0737.45 (3068)

TL 9–12 mm Resembles Pinyon Coneborer, but rusty-orange bands have very little black shading and the edging along bands is silvery gray, sometimes peppery. Thorax is golden orange, shaded rusty at anterior. HOSTS Ponderosa and Jeffrey pine.

LODGEPOLE CONEBORER

Eucopina rescissoriana

Rare

51a-0737.6 (3071)

TL 9–12 mm Resembles Ponderosa Coneborer, but rusty basal and AM bands are close together, with the narrow band between them often also rusty, and the silvery lines tend to be more densely peppered. The dark bands are frequently deeper reddish orange. Thorax is reddish, with gray posterior. HOSTS Lodgepole and western white pine.

DOUGLAS-FIR CONE MOTH

EYE-SPOTTED BUD MOTH

WESTERN PINE
SHOOT BORER

actual size

PINYON CONEBORER

PONDEROSA CONEBORER

LODGEPOLE
CONEBORER

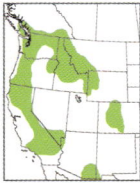

FIR CONEBORER
Eucopina siskiyouana   **Rare**   RANGE-WIDE

51a-0737.8 (3075)

TL 9–12 mm Light gray to pale brown FW has a reticulated pattern of black lines, with short black dashes along the central veins connecting the zigzag transverse lines. Some black lines in lower half of wing, particularly along costa and outer margin, widen and are filled with brown. Some individuals have a pinkish wash over entire wing. **HOSTS** Cones of white fir and possibly other firs.

ECCENTRIC EUCOSMA
Eucosma offectalis   **Rare**   RANGE-WIDE

51a-0823 (3005)

TL 12–16 mm Highly variable. Bicolored morph has golden-tan to warm or reddish-brown FW, with peppery gray along inner margin, sometimes extending well into central wing, often edged with brown shading in inner AM area. Inner ST area has a pale tan line (often raised and rounded) and long, thin, blackish dashes along veins; sometimes has a tan patch below raised scales or second tan line below black dashes. Streaked morph is pale tan to brown, with diffuse dusky patches in inner AM and PM areas of variable extent. Veins are marked with dusky gray. Fringe is checkered gray and white, and costa has many dark brown dashes. Rounded tan patch at inner ST area may be indistinct against ground color. Dark morph has a golden- to reddish-brown FW, with dark brown along inner margin and an irregular, hoary, gray patch at inner margin of median area. Inner ST area shows pale lines and black dashes as in others. Costa and apex have irregular brown semicircles outlined in gray or whitish. Fringe is speckled gray. **HOSTS** Mugwort, sagebrush, and groundsel.

PALE-STRIPED EUCOSMA
Eucosma pallidarcis   **Rare**   NORTH

51a-0845 (2976)   SOUTH

TL 6–8 mm Basal half of FW is broadly striped tan and white longitudinally. Lower half of costa has dark brown dashes that become tan as they extend into central wing. Inner ST area has tan mottling with small black dots. Fringe is speckled brown. **HOSTS** Big and California sagebrush.

SADDLE-BACKED EUCOSMA
Eucosma apacheana   **Rare**   RANGE-WIDE

51a-0871 (2946)

TL 5–9 mm Gray FW has rounded, rectangular, dark brown patch in inner AM area, indistinctly outlined with light gray, and an angled PM band from costal median area to inner ST area, gray toward costa and brown at inner margin. Central ST area has a round brown spot. Lower FW from PM area to outer margin washed brown. Costa has dusky semicircles from PM band to apex. Inner terminal area usually has a few short, thin, black streaks or dots. **HOSTS** Cudweeds.

PIED PELOCHRISTA
Pelochrista agassizii   **Rare**   RANGE-WIDE

51a-0885.1 (3026)

TL 9–10 mm Tan FW has satiny white patches edged with black. Basal patch connects to white patch laterally on shoulder of thorax, and white stripe along inner margin connects to white dorsal stripe on thorax. Median patch turns at inner margin to curve sinuously toward anal angle. Exact pattern of white patches may be variable, connected in some individuals and unconnected in others. **HOSTS** Unknown.

FIR CONEBORER

actual size

ECCENTRIC EUCOSMA

PALE-STRIPED EUCOSMA

SADDLE-BACKED EUCOSMA

PIED PELOCHRISTA

SKEWBALD PELOCHRISTA

Pelochrista bolanderana   **Rare**   RANGE-WIDE

51a-0886.1 (3029)

TL 9–10 mm Resembles Pied Pelochrista, but basal patch touches inner margin, inner margin lacks white stripe, and thorax has a transverse white band that rarely connects to FW markings. **HOSTS** Hoary tansyaster.

THIN-LINED PELOCHRISTA

Pelochrista agricolana   **Common**   RANGE-WIDE

51a-0897.1 (3037)

TL 8–10 mm White FW has thin tan lines running longitudinally in basal half and angled to margins in lower half. Inner half of ST/ terminal area has tiny black dots along tan lines. Terminal band is speckled tan. **HOSTS** Unknown.

MORRISON'S PELOCHRISTA

Pelochrista morrisoni   **Uncommon**   RANGE-WIDE

51a-0898.1 (3035)

TL 8–10 mm Brown to tan FW has a bold white stripe down center from base to median area. Lower half of FW has peppery shading in between curving white marks along costa and in inner PM/ST area. Thorax has broad white stripes bordering a tan dorsal stripe. **HOSTS** Unknown.

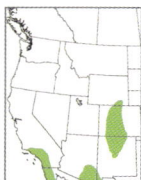

GALENA PELOCHRISTA

Pelochrista galenapunctana   **Rare**   RANGE-WIDE

51a-0904.1 (3045)

TL 9–12 mm FW is marbled ochrous yellow, sprinkled between with small silvery-gray dots. Costa has short white dashes; white costal dashes at median, PM, and ST lines are much longer than the others and curve toward outer margin, becoming silvery at tips. Inner ST/terminal area has a pale yellow patch bordered by two raised, pale yellow bars, with three black dashes perpendicularly between. Fringe is peppered with dark brown. **HOSTS** Unknown.

CURLY PELOCHRISTA

Pelochrista comatulana   **Rare**   RANGE-WIDE

51a-0905.1 (3041)

TL 8–10 mm White FW is heavily mottled with a network of orange-brown lines; white spaces are variably filled with peppery, dark brown speckles, sometimes extending across entire wing and other times only across lower half. Markings run roughly transverse from base to PM area and angle toward costa from PM line to apex. Inner ST/terminal area has shiny lines connected by two very thin blackish dashes and peppery brown shading behind. Terminal line is speckled blackish. **HOSTS** Unknown. **NOTE** Individuals with reduced or blotchy peppery speckling have commonly been mislabeled as Galena Pelochrista in online resources.

AVALON PELOCHRISTA

Pelochrista avalona   **Rare**   RANGE-WIDE

51a-0923.1 (3020)

TL 12–16 mm Golden-brown FW has darker brown shading along costa to central outer margin. White stripes run from base to anal angle, and from outer PM area to apex, with two shorter dashes in the central FW. Costa and inner margin are white. **HOSTS** California sagebrush.

SKEWBALD PELOCHRISTA

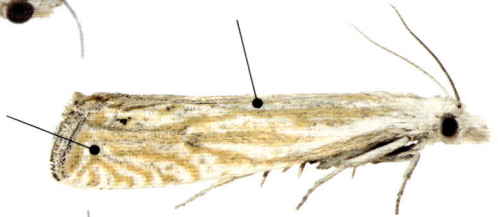

THIN-LINED PELOCHRISTA

MORRISON'S
PELOCHRISTA

actual size

GALENA
PELOCHRISTA

CURLY
PELOCHRISTA

AVALON PELOCHRISTA

MARBLED PELOCHRISTA

Pelochrista curlewensis   **Rare**   RANGE-WIDE

51a-0933.1 (3061.1)

TL 8–13 mm   Golden-brown FW has multiple irregularly shaped white patches: a band in basal area, a large patch in inner median area, a small patch in inner ST area, a crescent at costa of PM area, and a small angled subapical spot. **HOSTS** Unknown.

SNAKEWEED BORER

Pelochrista ridingsana   **Uncommon**   RANGE-WIDE

51a-0938.1 (3014)

TL 8–14 mm   Golden-brown FW has silvery-white markings outlined in dark brown: a stripe from base to central median area, inner margin and basal costa, from costal median area to central PM area, a semicircle at costal ST area, a crescent in inner PM/ST area, and a small mark at apex. **HOSTS** Hairy false goldenaster, California aster, hairy gumweed, snakeweeds, and goldenbushes.

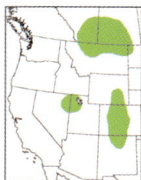

FERNALD'S PELOCHRISTA

Pelochrista fernaldana   **Rare**   RANGE-WIDE

51a-0940.1 (3015)

TL 8–14 mm   Pink (sometimes orangey) FW has bold, black-edged, satiny, white central streak from base to median area, irregular white patch in inner ST/terminal area, and triangular white patch at costal ST area. Inner margin has narrow white stripe that connects to parallel white thoracic stripes. **HOSTS** Unknown.

BLACK-CIRCLE PELOCHRISTA

Pelochrista eburata   **Uncommon**   RANGE-WIDE

51a-0951.1 (3085)

TL 12–14 mm   White FW has a small black-outlined circle filled with mottled brown in outer ST area, and a larger patch in inner PM area, with blackish spot at lower edge. Black AM line is fragmented. Basal and median areas are mottled with gray, sometimes heavily. Inner ST area is washed rosy brown. Head is brown. **HOSTS** Coyote brush.

TWO-BARRED PELOCHRISTA

Pelochrista canana   **Rare**   RANGE-WIDE

51a-0955.1 (3141)

TL 8–9 mm   Speckly brownish-gray to whitish FW has triangular brown patch at inner margin of AM area that has crisp lower edge and soft basal edge, as well as roundish brown patch at inner margin of PM area that is often sprinkled with black scales. Inner ST area has two peach to gray bars. Brown apical spot has blackish scales. Costa has many short, dark brown dashes. **HOSTS** Showy goldeneye.

REVERSED PELOCHRISTA

Pelochrista reversana   **Rare**   RANGE-WIDE

51a-0964.1 (3157)

TL 9–12 mm   Ivory FW has large brown patches at central basal, AM, PM, and ST areas and a triangular patch at inner margin between PM and ST area patches. Fainter patches at inner margin of AM area and costa of median area can sometimes be indistinct. FW usually has light brown striations and dark brown dashes along costa. **HOSTS** Unknown.

MARBLED PELOCHRISTA

SNAKEWEED BORER

actual size

FERNALD'S PELOCHRISTA

BLACK-CIRCLE
PELOCHRISTA

TWO-BARRED
PELOCHRISTA

REVERSED PELOCHRISTA

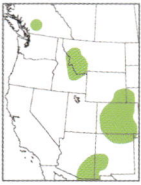

VARIEGATED PELOCHRISTA
Pelochrista ragonoti

Rare

51a-0969.1 (3030)

RANGE-WIDE

TL 11–12 mm Golden-brown FW has irregularly shaped white patches: two angled white bands in basal/AM and AM/median areas, a white crescent at costal PM area, a short angled subapical dash, a small triangle at center of outer margin, and a roundish patch at inner PM area (sometimes with a golden-brown center). HOSTS Unknown.

BROWN-CHEVRON PELOCHRISTA
Pelochrista corosana

Rare

51a-0972.1 (3162)

RANGE-WIDE

TL 9–11 mm Peppery ivory FW has elongate, rounded, dusky brown patch at inner AM area that angles toward central median area, creating a broken V shape with dark brown bar from costal median area to inner margin of PM area; lower bar is bordered outwardly with an ivory line. Outer ST area is shaded dusky. HOSTS Unknown.

SPANGLED PELOCHRISTA
Pelochrista scintillana

Uncommon

51a-0973.1 (3151)

RANGE-WIDE

TL 7–15 mm Peppery brownish-gray FW has large golden-tan patch in lower half crossed by narrow, silvery median, PM, and ST lines, and three rows of black spots in inner ST area. Central ST area has many thin, brown lines parallel to costa. In some individuals, the golden-tan patch is reduced or indistinct. HOSTS Sunflower heads.

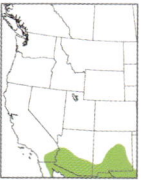

WONDERFUL PELOCHRISTA
Pelochrista mirosignata

Rare

51a-0984.1 (3138)

RANGE-WIDE

TL 7–11 mm Peppery grayish-brown FW has white-edged, dark brown patches at inner margin of median and ST areas. Apex has a small brown spot. HOSTS Unknown.

MORNING PELOCHRISTA
Pelochrista matutina

Uncommon

51a-0989.1 (3091)

RANGE-WIDE

TL 7–8 mm White FW has dark brown to blackish bands from costal median to inner PM area, and on inner half of AM area. Apical patch is dark brown to blackish, with paired white dashes along costa. Basal area and thorax have many small dark spots. Inner median and ST areas are often washed brownish. HOSTS Unknown.

SILVER-LINED PELOCHRISTA
Pelochrista argenteana

Rare

51a-1015.1 (3149)

RANGE-WIDE

TL 9–10 mm Golden-brown FW has white stripe from base to central ST area, often containing a thin, dark brown line along costal edge. A white crescent (sometimes fragmented) curves from costal PM area to apex, sometimes with white spot at costa within curve. Thin, dark brown, parallel lines mark central ST area. Small white dashes flank central stripe in median/PM area. Costa and inner margin are narrowly edged with white. Thin terminal line is dark brown. HOSTS Unknown.

VARIEGATED PELOCHRISTA

actual size

BROWN-CHEVRON PELOCHRISTA

SPANGLED PELOCHRISTA

WONDERFUL
PELOCHRISTA

MORNING PELOCHRISTA

SILVER-LINED PELOCHRISTA

RAGWEED EPIBLEMA
Epiblema strenuana

Uncommon

51a-1065 (3172)

TL 6–9 mm Hoary brown FW has a light tan terminal area, sometimes with dark brown marks, and large, triangular, metallic brown (sometimes washed with tan) patch at inner margin of median area. Dark brown dashes along lower costa and apex are backed by golden-brown shading. HOSTS Stems of annual ragweed.

BROKEN-SPOTTED SONIA
Sonia vovana

Rare

51a-1121.1 (3220)

TL 8–13 mm Light gray FW has irregularly edged, dark brown AM band and roundish, dark brown patch at inner PM area that is cut through by a diagonal gray stripe, as though the patch were cracked in two. Costa has short, dark brown dashes and small, dark brown spot at apex. HOSTS Snakeweeds and goldenbushes.

SUNFLOWER BUD MOTH
Suleima helianthana

Uncommon

51a-1121.5 (3212)

TL 5–7 mm Striated, pale gray FW has bold blackish patch at inner median area edged by indistinct, thin, white median line and smaller dusky patch at inner ST area. Inner terminal area has a whitish patch. Costa is marked with alternating fine pale and dusky dashes, with bold pair of white dashes at apex. HOSTS Stems and buds of sunflower.

SUNFLOWER STEM MOTH
Suleima baracana

Uncommon

51a-1122.1 (3216)

TL 8–11 mm Mottled brown FW has bold white median and terminal areas edged narrowly with black. Inner basal area and posterior of thorax are mottled whitish. Costa has short white dashes, with bolder pair at apex. HOSTS Stems and buds of sunflower.

MAPLE TWIG BORER
Proteoteras aesculana

Uncommon

51a-1133 (3230)

TL 7–10 mm Light gray FW is washed with green and striated with dusky lines. Dusky crescent curves from costal median area to apex; middle section has a thin black line along inner edge. Dusky AM band is sometimes indistinct. FW has raised scale tufts at inner AM and PM areas, as well as posterior thorax. HOSTS Terminal shoots of maple.

SPRUCE BUD MOTH
Zeiraphera canadensis

Uncommon

51a-1143 (3240)

TL 6–8 mm Resembles Larch Needleworm but is often warmer brown, with less silvery edging to patches. Silver-edged brown ST band tapers toward anal angle. HOSTS White spruce; also other spruce, Douglas-fir, fir, and western hemlock.

RAGWEED EPIBLEMA

actual size

BROKEN-SPOTTED SONIA

SUNFLOWER BUD MOTH

SUNFLOWER STEM MOTH

MAPLE TWIG BORER

SPRUCE BUD MOTH

LARCH NEEDLEWORM
Zeiraphera improbana

Rare · RANGE-WIDE · 51a-1144 (3241)

TL 6–10 mm Light brownish-gray FW is unevenly washed with rusty shading and has dark brown triangle in outer basal/AM area, dark brown band from costal median area to inner margin of PM area, and small, round, dark brown patch in outer ST area. Costa has many small, dark brown spots. **HOSTS** Larch.

EARLY ASPEN LEAFROLLER
Pseudexentera oregonana

Common · RANGE-WIDE · 51a-1159 (3248)

TL 9–11 mm Variable. Iron-gray FW has fractured, thin, dusky cross-lines that are commonly weak or absent except along costa. Lines are often edged with warm brown, especially in inner half. Some individuals have a dusky patch at inner margin of AM, median, and/or ST areas. **HOSTS** Aspen and willow. **NOTE** Often one of the earliest micromoths encountered in spring.

LIVE-OAK LEAFROLLER
Pseudexentera habrosana

Rare · RANGE-WIDE · 51a-1163 (3256)

TL 8–11 mm Peppery gray FW has bronzy patch at inner half of AM and PM areas and bronzy shading in terminal and costal ST areas. **HOSTS** Coast and interior live oak.

COTTON TIPWORM MOTH
Crocidosema plebejana

Common · RANGE-WIDE · 51a-1186.2 (3274)

TL 6–9 mm Light brown FW has a dark to dusky brown patch on inner margin at basal/AM area, often overlaid with gray or brown striations, a brown to dusky triangle at inner margin of PM area, and light gray in inner ST area containing medium gray and black spots. Central PM area has a subtle, curving, brown patch. Costa has short brown dashes, with paired white dashes at ST area. Males have a dark brown to dusky wash along costal half of FW. **HOSTS** Mallow and hibiscus, and other members of Malvaceae, including cotton, okra, and hollyhock.

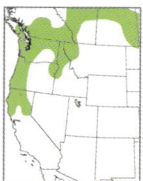

RED-STRIPED NEEDLEWORM
Epinotia radicana

Uncommon · RANGE-WIDE · 51a-1191 (3269)

TL 6–8 mm Light gray FW has reddish-brown AM band and brown band from inner PM to costal median area. Outer ST area has a triangular brown patch. Some individuals are washed with light brown in pale areas. **HOSTS** Spruce, Douglas-fir, and fir; also, occasionally, pine, hemlock, and juniper. **NOTE** Can be a common forest pest in some regions.

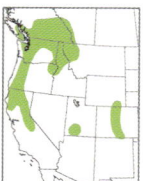

CURRANT EPINOTIA
Epinotia castaneana

Rare · RANGE-WIDE · 51a-1200 (3288)

TL 6–9 mm Light gray to pale brown FW has warm brown basal and ST/terminal areas. Warm brown median band is faded except at costa. ST area has small pale subapical spot on costa. **HOSTS** Currants and gooseberries.

LARCH NEEDLEWORM

EARLY ASPEN LEAFROLLER

LIVE-OAK LEAFROLLER

COTTON TIPWORM MOTH

actual size

RED-STRIPED NEEDLEWORM

CURRANT EPINOTIA

OCEAN SPRAY EPINOTIA

Epinotia johnsonana
51a-1202 (3289)

Uncommon

RANGE-WIDE

TL 6–9 mm Lightly speckled, light gray to ivory FW is orange brown in inner half of basal/AM area, bordered by a thick black median line. Costa is broadly washed brown. **HOSTS** Ocean spray.

PACIFIC EPINOTIA

Epinotia subviridis
51a-1210 (3298)

Uncommon

NORTH

SOUTH

TL 8–11 mm Light gray to greenish-gray FW has black V-shaped AM line and black median and PM lines that form an hourglass shape, and incomplete black basal and ST lines. Inner AM and PM lines and outer median and ST lines are shaded basally with gray. FW is often entirely overlaid with light brown striations. **HOSTS** Monterey cypress, false cypress, whitecedar, and juniper.

MANZANITA EPINOTIA

Epinotia subplicana
51a-1212 (3300)

Rare

RANGE-WIDE

TL 8–11 mm Variable. Most common form has gray to light brown FW, with short, dark striations, often washed brownish overall. Sometimes has an indistinct, angled, darker PM band, as well as an indistinct darker shade in inner basal area, often with scattered black dots. Pale form is light gray to ivory, with scattered small black dots in inner basal area. Brown-patched form is pale gray, with speckly, dark brown patches in inner basal/AM and PM areas. **HOSTS** Manzanita.

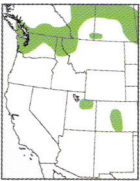

POPLAR BRANCHLET BORER

Epinotia nisella
51a-1218 (3306)

Uncommon

RANGE-WIDE

TL 8–10 mm Variable. Light to medium gray FW is lightly to heavily mottled with dusky gray, sometimes presenting as fractured lines or bands. V-shaped AM line is usually defined by dusky shading in basal area. Inner margin frequently has a warm brown patch in median area, often extending to basal area. Inner ST area has three or four thin, short, parallel, black lines, sometimes hidden within a small dusky patch. Some individuals are yellowish brown in outer median and ST areas. **HOSTS** Balsam poplar, quaking aspen, and willow; also birch.

ALDER EPINOTIA

Epinotia albangulana
51a-1220 (3308)

Uncommon

RANGE-WIDE

TL 6–9 mm FW is mottled with brown, black, and metallic gray. Costal AM/median area is ivory, with faint gray striations. Inner AM line is black, edged outwardly with ivory and basally with brown speckled with black scales. PM area is predominantly brown spotted with black. Inner terminal area is ivory, with black dots between veins. Apex is warm brown crossed diagonally by ivory-edged, metallic gray lines. **HOSTS** Alder.

OCEAN SPRAY EPINOTIA

PACIFIC EPINOTIA

actual size

MANZANITA EPINOTIA

POPLAR BRANCHLET
BORER

POPLAR BRANCHLET
BORER

POPLAR BRANCHLET
BORER

ALDER EPINOTIA

MADRONE EPINOTIA

Epinotia nigralbana

Uncommon

51a-1231 (3319)

RANGE-WIDE

TL 6–8 mm White FW has brown patch in inner AM area and brown band from costal median area to inner margin of PM area; both are crossed by black striations. Thorax and inner basal area have blackish striations. ST/terminal area is hoary gray, with a small, rectangular, black patch toward apex. Apex is warm brown crossed diagonally by white and gray lines. Costa has paired white dashes in lower half. **HOSTS** Manzanita and madrone.

VARIABLE EPINOTIA

Epinotia emarginana

Uncommon

51a-1235 (3323)

NORTH

SOUTH

TL 8–9 mm Highly variable. Two basic FW patterns exist with many color variations. Solid-patched form has a large colored patch along inner margin of FW that creates an oval when viewed dorsally when at rest. FW may be gray or brown, pale or dark, and patch may be brown, tan, whitish, or blackish. FW and patch are usually of contrasting colors. Some individuals of this morph show dark bars along costal half of FW. Checker-patched form has alternating dark and light patches along inner margin, usually dark brown with either light brown, tan, or whitish in basal, median, and ST areas. Many individuals of this form also have dark brown bars along costal half of FW. **HOSTS** Oaks, manzanita, and madrone.

SUMMER HOLLY EPINOTIA

Epinotia arctostaphylana

Rare

51a-1240 (3328)

RANGE-WIDE

TL 8–11 mm Variable. In most common form, striated reddish, tan, or brown FW is often indistinctly marked. Some individuals have a broad dusky basal dash and may have a darker wash across costal half of FW. Lighter individuals usually show a large, slightly angled, brown patch in central PM area and brownish shading in basal area, sometimes indistinct; when viewed dorsally when at rest, pale areas along inner margin form a diamond shape in median area. **HOSTS** Summer holly, greenleaf manzanita, and madrone.

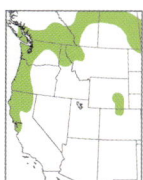

DELTA EPINOTIA

Epinotia lomonana

Rare

51a-1250 (3342)

RANGE-WIDE

TL 9–10 mm FW is mottled dark gray, brown, and black in basal/AM area and in a broad triangle in outer PM area; inner point of triangle is often browner, while outer section is gray. Inner median area and ST/terminal areas are white. Anterior thorax has a mottled white band. Head is brown. **HOSTS** Vine maple, coffeeberry, quaking aspen, hollyleaf cherry, and pin cherry.

SERPENTINE EPINOTIA

Epinotia kasloana

Rare

51a-1258 (3350)

NORTH

SOUTH

TL 8–11 mm Similar to Sigmoid Epinotia, but curving band is more distinctly marked with black, and basal end turns to central median area instead of costa. Band separates paler inner FW from darker costal half; in some individuals inner and outer FW are of subtle contrast, while in others the difference is marked: either pale and dark gray, or tan and dark brown. Head and thorax are usually concolorous with FW, rarely contrastingly tan. **HOSTS** Blueblossom and jojoba.

MADRONE EPINOTIA

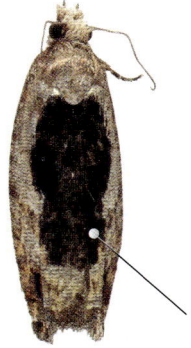

VARIABLE EPINOTIA

actual size

SUMMER HOLLY EPINOTIA

DELTA EPINOTIA

SERPENTINE EPINOTIA

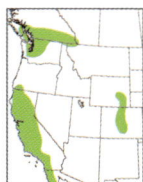

SIGMOID EPINOTIA
Epinotia signiferana

Uncommon
51a-1259 (3350.1)

TL 7–9 mm Speckled, light gray FW has a dark brown S-shaped band from costal median area to outer ST area; center of band is shaded blackish. Basal area is shaded dusky. Costa has small dusky dashes. Head and thorax are brown. **HOSTS** Blueblossom.

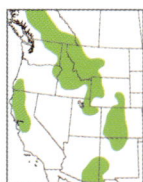

SADDLED TORTRICID
Dichrorampha simulana

Uncommon
51a-1291.7 (3404)

TL 6–8 mm Grayish-brown FW has bold white to pale yellow bar at inner median area and short white dashes along lower half of costa. Bronzy metallic PM, ST, and terminal lines are marked with black dots. **HOSTS** Unknown; possibly lupine.

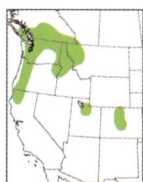

TANSY ROOT MOTH
Dichrorampha vancouverana

Rare
51a-1291.8 (3408)

TL 6–8 mm Rusty FW has a broad yellowish to orange bar at inner median area, sometimes indistinct. Costa has small white dashes along lower half. Metallic silvery PM and ST lines are often most visible in costal half. Terminal area has black dots toward anal angle. **HOSTS** Yarrow, tansy, and ox-eye daisy.

ONE-BARRED GRAPHOLITA
Ephippiphora lunatana

Rare
51a-1316 (3437)

TL 6–8 mm Bronzy FW has bold white crescent at inner median line and metallic ST/terminal area marked with white dashes at costa and patch of white and black scales near anal angle. **HOSTS** Sweet pea and vetchling.

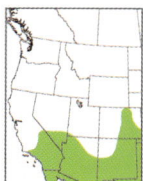

TWELVE-LINED OFATULENA
Ofatulena duodecemstriata

Uncommon
51a-1327 (3444)

TL 6–9 mm Brownish-gray FW has multiple thin, parallel, V-shaped, white lines filling AM and median areas. Metallic PM line is straight, edging pale tan to grayish ST/terminal area filled with thin black dashes. Costa has many paired white dashes along length. **HOSTS** Seedpods of mesquite and screw bean.

LUMINOUS OFATULENA
Ofatulena luminosa

Rare
51a-1328 (3445)

TL 6–9 mm Resembles Twelve-lined Ofatulena, but lines in AM/median area and dashes along costa are poorly defined. ST/terminal area is a stronger brownish tan. **HOSTS** Probably palo verde.

ASPEN CYDIA
Cydia populana

Uncommon
51a-1351 (3463)

TL 7–8 mm White FW has black patches from basal costa to central AM area, in central median area, and from inner margin of PM area to apex, forming an uneven dusky crescent. Black striations are prominent along costa and inner margin. **HOSTS** Aspen and cottonwood.

SIGMOID EPINOTIA

SADDLED TORTRICID

TANSY ROOT MOTH

actual size

ONE-BARRED GRAPHOLITA

TWELVE-LINED OFATULENA

LUMINOUS OFATULENA

ASPEN CYDIA

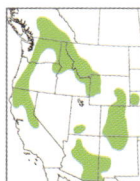

PONDEROSA PINE SEEDWORM MOTH
Uncommon

RANGE-WIDE

Cydia piperana   51a-1375 (3489)

TL 9–11 mm Speckly silvery-gray FW has metallic silver median, PM, and adterminal bands (the latter two slightly raised) and silvery fringe. Basal area is unspeckled. **HOSTS** Ponderosa pine seeds.

CODLING MOTH
[IN] Common

NORTH

Cydia pomonella   51a-1380 (3492)

SOUTH

TL 8–12 mm Striated gray FW has large brown patch at inner ST/ terminal area, bounded by black PM line that is double near inner margin. **HOSTS** Apple, pear, and plum.

FILBERTWORM MOTH
Very Common

NORTH

Cydia latiferreana   51a-1383 (3494)

SOUTH

TL 7–11 mm Variable. Tan to reddish FW has metallic silver median band and PM and ST lines. Costa has short pale dashes. ST area often speckled with small black dashes. **HOSTS** Beech, filbert, hazelnut, and oak.

PSYCHEDELIC LEAFROLLERS
SUPERFAMILY Tortricoidea (51a)

FAMILY Tortricidae SUBFAMILY Chlidanotinae

Very small metallic moths with long narrow wings striped with bold oranges and yellows. The three species in this subfamily are associated with conifer forests, mostly pine. Adults are largely diurnal but will also come to lights at night.

PSYCHEDELIC YOUNG MOTH
Rare

RANGE-WIDE

Thaumatographa youngiella   51a-1396 (3752)

TL 6–7 mm Dark brown to blackish FW has three curving lines (silver, orange, and silver) at AM line and single silvery-white PM and ST lines. ST/terminal area is orange, with small black dots in inner ST area. **HOSTS** Unknown; possibly pine.

PSYCHEDELIC REGAL MOTH
Rare

RANGE-WIDE

Thaumatographa regalis   51a-1397 (3753)

TL 8–11 mm Orange FW has large white median/PM patch with thin black streaks, and bold black bar with silver spots that curves around anal angle. Thick, metallic silver lines mark basal half and apical half of ST/terminal area. **HOSTS** Pine; possibly other conifers.

PONDEROSA PINE
SEEDWORM MOTH

actual size

CODLING MOTH

FILBERTWORM MOTH

PSYCHEDELIC YOUNG MOTH

actual size

PSYCHEDELIC REGAL MOTH

CARPENTERWORM MOTHS
SUPERFAMILY Cossoidea (53a)
FAMILY Cossidae

Medium to very large moths with broad wings that are held folded against their chunky, hairy body at rest; their size and shape make them somewhat resemble prominent moths. Some species have very hairy forelegs that are stretched out in front of the body at rest. Most species are strongly sexually dimorphic, with females notably larger than males. Caterpillars bore in trunks and branches of trees, sometimes weakening the tree when abundant. Adults are nocturnal and will come to lights in small numbers.

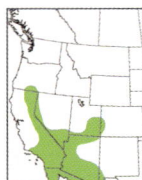

DESERT CARPENTERWORM
Uncommon   RANGE-WIDE
Hypopta palmata   53a-0007 (2656)
TL 15–22 mm Hoary gray FW has black veins that flare at outer margin, where they have whitish edges. Veins at inner median and central PM areas are bordered by patches of white and blackish shading. **HOSTS** Unknown.

SMUDGED CARPENTERWORM
Rare   RANGE-WIDE
Givira mucidus   53a-0008 (2660)
TL 18–20 mm Peppery gray FW has narrow blackish striations and dusky patch at costal PM area. Costa and fringe are checkered. **HOSTS** Unknown.

THEODORE CARPENTERWORM
Rare   RANGE-WIDE
Givira theodori   53a-0010 (2662)
TL 12–18 mm White FW has a large, dark brown patch mottled with black in outer half of PM area, and mottled gray shading in ST/terminal area. White thorax is grayish brown in posterior half. **HOSTS** Unknown.

MARGARET'S CARPENTERWORM
Uncommon   RANGE-WIDE
Givira marga   53a-0017 (2669)
TL 14–20 mm Hoary gray FW has small, white, comma-shaped dash in dusky-shaded outer PM/ST area and narrow black striations. Basal area is often indistinctly washed warm brown; AZ/NM populations have a stronger brown wash and a narrow white edge to inner margin of basal area. **HOSTS** Unknown.

PINE CARPENTERWORM
Rare   RANGE-WIDE
Givira lotta   53a-0018 (2670)
TL 13–15 mm Resembles Margaret's Carpenterworm, but white dot on FW is smaller or indistinct and grayish shading is reduced or absent. ST/terminal area is a slightly paler gray. Inner margin of basal area has a thin white edge. **HOSTS** Ponderosa pine.

DESERT CARPENTERWORM

SMUDGED CARPENTERWORM

THEODORE
CARPENTERWORM

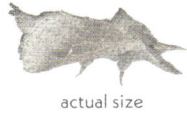

actual size

MARGARET'S CARPENTERWORM

PINE CARPENTERWORM

93

BLACK-VEINED CARPENTERWORM
Uncommon

Inguromorpha itzalana   53a-0024 (2657)

TL 19–22 mm Black-peppered, pale gray FW has veins traced in black. Typically rests with abdomen held well above wings. **HOSTS** Unknown.

POPLAR CARPENTERWORM
Rare
RANGE-WIDE

Acossus centerensis   53a-0025 (2675)

TL 22–27 mm Dark gray FW is crossed by thick black striations. ST/terminal area is lighter gray. Gray thorax has narrow white transverse lines at both anterior and posterior ends. **HOSTS** Poplar.

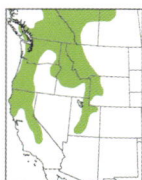

ASPEN CARPENTERWORM
Uncommon
RANGE-WIDE

Acossus populi   53a-0026 (2676)

TL 30–45 mm Light gray FW has slightly scalloped, fractured, black striations and a thickened black line at median area. PM area often has light dusky shading. Light gray thorax has indistinct, white, transverse anterior and posterior lines. **HOSTS** Quaking aspen and other poplars.

ROBIN'S CARPENTERWORM
Common
NORTH

SOUTH

Prionoxystus robiniae   53a-0029 (2693)

TL 27–45 mm Light gray FW has network of scalloped lines that connect into a pattern of rounded spots that resemble foamy bubbles. Inner median area, and sometimes basal and/or PM areas, is shaded dusky. Pale gray thorax has two thin black lines laterally. Black HW has large orange patch at inner half of ST/terminal area. **HOSTS** Ash, chestnut, locust, oak, poplar, and willow.

HENRY'S CARPENTERWORM
Rare
RANGE-WIDE

Comadia henrici   53a-0035 (2679)

TL 14–22 mm White-speckled tan FW has white stripe along costal margin, with tan spots at costa and shorter stripes in central and inner wing. Lower part of FW has white along veins and white interveinal patch in central ST area. Females are often noticeably larger than males. **HOSTS** Unknown.

WHITE-STRIPED CARPENTERWORM
Rare
RANGE-WIDE

Comadia albistriga   53a-0046 (2690)

TL 13–16 mm Hoary brownish-gray FW has white stripe along costal margin, with dark spots along costa and shorter white stripe in basal half of central wing. Central PM area has an elongate, angled, white wedge. Thin white ST line is faulted. Thick, brown-edged, white striations across FW are often indistinct. **HOSTS** Unknown.

BLACK-VEINED
CARPENTERWORM

POPLAR CARPENTERWORM

actual size

ASPEN CARPENTERWORM

ROBIN'S CARPENTERWORM

HENRY'S CARPENTERWORM

WHITE-STRIPED CARPENTERWORM

CLEARWING MOTHS SUPERFAMILY Sesioidea (55a)
FAMILY Sesiidae

These distinctive small to large moths are wasp mimics; some are quite convincing, though identifiable as moths by their antennae and lack of a pinched waist. The caterpillars are borers of stems, trunks, or roots of their host plants, and a few, such as the Squash Vine Borers familiar to gardeners, can be crop pests. Adults are diurnal; while some may occasionally be encountered at their host plants or while nectaring at flowers, commercial pheromone lures are available for many species and are the most reliable way to see these moths.

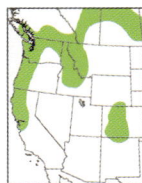

RASPBERRY CROWN BORER
Pennisetia marginatum

Uncommon
55a-0057 (2513)

RANGE-WIDE

TL 13–16 mm Resembles Fireweed Clearwing, but anal segment of abdomen is yellow, basal section has two yellow rings at anterior of segments, and thorax has a transverse yellow line and short yellow dashes at collar. Middle unbanded abdominal segment often has a yellow patch. Wings have a brown margin and are sometimes tinted brown. **HOSTS** Blackberry, raspberry, and boysenberry. **NOTE** Diurnal.

WESTERN POPLAR CLEARWING
Paranthrene robiniae

Uncommon
55a-0061 (2526)

NORTH

SOUTH

TL 16–23 mm Thick yellow abdomen has black basal segments, with a single thin ring in middle. Thorax has curving yellow line at posterior and yellow-brown patches laterally. Collar has two yellow lines. FW is orange brown, with dark veins and a dark brown border; HW is clear, with dark veins and border. Rocky Mountain form has rusty-brown thorax and abdomen, with just two thin rings. **HOSTS** Willow and poplar. **NOTE** Diurnal.

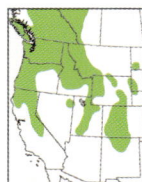

FIREWEED CLEARWING
Albuna pyramidalis

Common
55a-0069 (2533)

RANGE-WIDE

TL 11–18 mm Thick black abdomen has yellow bands along posterior of segments, one ring basally and three at middle, with black segment between. Anal segment is black, sometimes with sparse yellow scales. Thorax has yellow transverse line (sometimes indistinct) at posterior. Clear wings are bordered reddish brown. **HOSTS** Fireweed, evening primrose, and other members of Onagraceae. **NOTE** Diurnal.

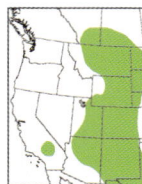

NEBRASKA CLEARWING
Euhagena nebraskae

Rare
55a-0072 (2535)

RANGE-WIDE

TL 9–15 mm Striking reddish-orange wings have black margin and veins and thick black PM bar. Head, thorax, and basal half of abdomen are gray; thorax is sometimes hairy. Lower half of abdomen is black, with thin white rings (sometimes yellowish, rarely broad). Some individuals have pearly-white or semitranslucent wings. **HOSTS** Evening primrose. **NOTE** Diurnal.

PACIFIC HORNET MOTH
Sesia pacificum

Common
55a-0075.01 (2543.01)

RANGE-WIDE

TL 16–23 mm Thick black abdomen has broad yellow bands at every segment excepting the second and fourth, where the yellow is narrow or nearly absent. Abdomen tip is warm brown. Black thorax has paired yellow patches at posterior connected to thin lateral stripes, which connect to yellow lines at anterior. Wings have dark brown margin and veins. **HOSTS** Poplar and willow. **NOTE** Diurnal. Previously considered part of American Hornet Moth (*S. tibiale*).

male

RASPBERRY CROWN
BORER

female

actual size

WESTERN POPLAR
CLEARWING

Rocky
Mountain
form

FIREWEED
CLEARWING

NEBRASKA CLEARWING

PACIFIC HORNET
MOTH

97

GRAY SQUASH VINE BORER
Eichlinia snowii

Uncommon
55a-0083 (2538)

TL 11–16 mm Dusty gray FW has pale yellowish margin. Abdomen has dusty gray dorsal stripe and bright orange-red sides. Thorax is dusty gray. Legs have thickly scaled orange-red basal segments. Eyes are orange. **HOSTS** Gourds. **NOTE** Diurnal.

GLORIOUS SQUASH VINE BORER
Eichlinia gloriosa

Uncommon
55a-0085 (2540)

TL 20–34 mm Thick abdomen has black bands separating brightly patterned segments; the most basal is orange, followed by alternating orange and gray, and pale yellow. Gray FW has black terminal line and orange fringe. Gray thorax has a narrow orange margin. Densely scaled hindlegs are red orange at base, and eyes are orange. **HOSTS** Gourds and wild cucumber. **NOTE** Diurnal.

CURRANT CLEARWING
Synanthedon tipuliformis

[IN] Rare
55a-0094 (2553)

TL 8–12 mm Resembles Strawberry Crown Moth, but anal tuft is entirely black, collar and face are black, and abdominal rings are thinner. FW often has yellowish-brown scales in margin border and terminal band. **HOSTS** Cultivated currant, gooseberry, and raspberry. **NOTE** Diurnal.

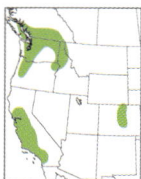

WESTERN WILLOW CLEARWING
Synanthedon albicornis

Rare
55a-0112 (2570)

TL 14–18 mm Blue-black abdomen is unmarked or occasionally has one or two thin yellow rings, with broad anal tuft. Black thorax is unmarked or has two thin yellow lines laterally. Black antennae are pale yellow to whitish in distal third. Clear wings have black margin and veins, and black PM bar. **HOSTS** Willow. **NOTE** Diurnal.

STRAWBERRY CROWN MOTH
Synanthedon bibionipennis

Common
55a-0118 (2576)

TL 13–18 mm Black abdomen has three or four evenly spaced, thin, yellow rings and broad black anal tuft with yellow outer stripes (sometimes also central stripe). Black thorax has two thin yellow stripes laterally and thin yellow collar. Blackish head has yellow face. Wings have dark brown margin and veins and a thick brown PM band on FW. **HOSTS** Strawberry, blackberry, rose, and cinquefoil. **NOTE** Diurnal.

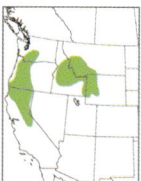

KNOTWEED ROOT BORER
Synanthedon chrysidipennis

Rare
55a-0120 (2578)

TL 13–16 mm Abdomen has alternating thick bands of black and yellow, and yellow anal tuft. Black thorax has thin yellow dorsal and lateral lines. Transparent wings have black veins thickly edged with orange. **HOSTS** Knotweed and smartweed. **NOTE** Diurnal.

GRAY SQUASH
VINE BORER

GLORIOUS
SQUASH VINE
BORER

CURRANT
CLEARWING

mating pair
(not to scale)

actual size

WESTERN WILLOW
CLEARWING

STRAWBERRY CROWN MOTH

KNOTWEED ROOT BORER

BUCKWHEAT BORER
Synanthedon polygoni

Uncommon

55a-0122 (2581)

TL 10–15 mm Blue-black abdomen is orange red laterally, with broad orange-red bands at middle segment and near tip and orange-red anal tuft. Blue-black thorax has thick orange-red lateral stripes that connect to orange-red stripe along inner margin of blue-black FW. HW is orange red, with black terminal band. Rare individuals have entire lower half of abdomen orange red or have white patches on central FW. **HOSTS** Buckwheat and knotweed. **NOTE** Diurnal.

SYCAMORE BORER
Synanthedon resplendens

Uncommon

55a-0123 (2582)

TL 11–15 mm Abdomen has thick yellow bands separated by thin black rings, with one or two thick black bands in basal half, and yellow anal tuft. Black thorax has thick yellow lateral stripes. Transparent wings have dark brown veins thickly edged with yellow scales. **HOSTS** Western sycamore; also coast live oak and avocado. **NOTE** Diurnal.

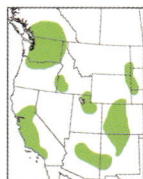

PEACHTREE BORER
Synanthedon exitiosa

Uncommon

55a-0124 (2583)

TL 15–20 mm Sexually dimorphic and variable. Female is entirely blue black, with a broad orange band at middle of abdomen and black FW. Male typically has thin yellow rings on all segments of black abdomen and an arrowhead-shaped anal tuft edged laterally with yellow; thin yellow lateral stripes on black thorax; and yellow-tinted transparent wings with dark brown margins and veins. Some males are entirely blue black with no yellow markings or with a reduced number; these latter are separable from similar species by the dart-shaped anal tuft. **HOSTS** Fruit-bearing trees, including cherry, peach, and plum. **NOTE** Diurnal.

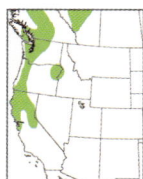

DOUGLAS-FIR PITCH MOTH
Synanthedon novaroensis

Rare

55a-0125 (2584)

TL 15–18 mm Narrow abdomen has three or four evenly spaced, thin, orange rings and broad black anal tuft edged with orange; underside of abdomen is entirely orange. Black thorax has thick orange lateral stripes and short orange dorsal spot at posterior. Transparent wings have black outer margin and veins. Legs are orange. Some males are entirely blue-black with very faint or nearly absent orange markings; these are separable from similar species by the rounded anal tuft with a pale spot at its center. **HOSTS** Douglas-fir, spruce, and pine. **NOTE** Diurnal.

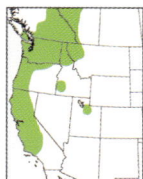

SEQUOIA PITCH MOTH
Synanthedon sequoiae

Rare

55a-0127 (2586)

TL 15–21 mm Resembles Sycamore Borer, but yellow rings on black abdomen are narrower and anal tuft usually has black center and yellow band along fringe (sometimes entirely yellow). Yellow lateral lines on black thorax are narrow or indistinct. **HOSTS** Pine, spruce, and Douglas-fir. **NOTE** Diurnal. Despite the name, the species has no proven association with sequoia.

actual size

BUCKWHEAT BORER

SYCAMORE BORER

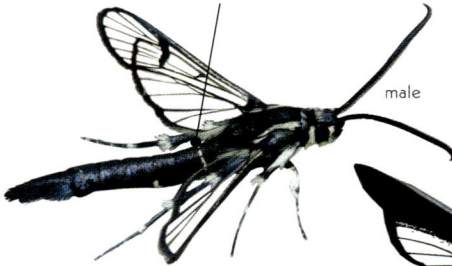

male

PEACHTREE BORER

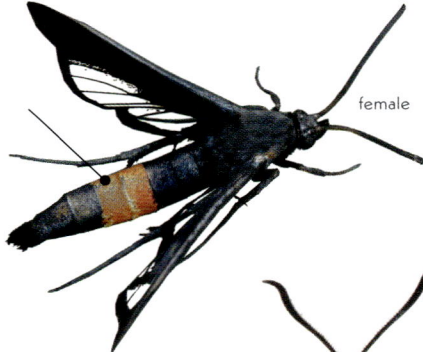

female

black form

DOUGLAS-FIR
PITCH MOTH

SEQUOIA
PITCH MOTH

CANYON CLEARWING

Rare

Carmenta engelhardti   55a-0141 (2598)

TL 12–14 mm Resembles Currant Clearwing but is nonsympatric and has small yellow spot beneath PM bar on FW. May resemble male Peachtree Borer, but anal tuft is fan-shaped and PM bar has yellow spot. **HOSTS** Unknown; likely brickellbush. **NOTE** Diurnal; often encountered at flowers or on foliage of host during daytime.

WILD GERANIUM BORER

Rare

Carmenta giliae   55a-0143 (2599)

TL 13–16 mm Variable. Resembles Currant and Engelhardt Clearwings, but transparent FW has reddish-brown scales along margins and PM bar has a reddish-brown spot. Anal tuft has yellow central stripe. Some individuals may have yellow bands on most abdominal segments. **HOSTS** Wild geranium. **NOTE** Diurnal.

CORONOPUS BORER

Rare

Carmenta mimuli   55a-0149 (2602)

TL 12–17 mm Variable. Resembles Wild Geranium Borer, but anal tuft is yellow with black central stripe, and thorax has yellow right-angled lines at base of wings instead of thin stripes. Transparent FW often has a more blackish margin and reddish PM bar. Antennae are sometimes tinted reddish. Number of bands on abdomen is variable. **HOSTS** Unknown; likely greenleaf five-eyes. **NOTE** Diurnal.

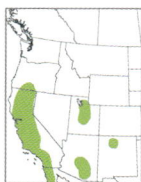

BEARDTONGUE BORER

Rare

Penstemonia clarkei   55a-0167 (2618)

TL 9–14 mm Abdomen has a mix of broad and thin yellow and black bands. Anal tuft has two yellow stripes (sometimes folded together to look like a single broad central stripe). FW margin and PM bar are brown, with yellow scales in central wing. Thorax has thin yellow lateral stripes. **HOSTS** Beardtongue and penstemon. **NOTE** Diurnal.

LEAF SKELETONIZER MOTHS

SUPERFAMILY Zygaenoidea (57a)
FAMILY Zygaenidae

Small to medium moths in combinations of black, orange, and red, with long, bipectinate antennae. The *Harrisina* hold their long rounded wings at a 45° angle from their body, while the others typically keep them folded. Caterpillars are typically found in groups and generally skeletonize the leaves of their host plant. Both larvae and adults produce hydrogen cyanide to make them unpalatable to predators and bear aposematic coloration as warning. Adults can be encountered both at flowers during daytime and at lights at night.

GRAPELEAF SKELETONIZER

Uncommon

Harrisina americana   57a-0006 (4624)

TL 8–12 mm Resembles red-collared form of Western Grapeleaf Skeletonizer, but FW is matte black and collar is orange. Primarily an eastern species. Similar to Yellow-collared Scape Moth, which has thin yellow edge to costa and typically holds wings folded. **HOSTS** Grapes. **NOTE** Often encountered at flowers during daytime.

CANYON CLEARWING

WILD GERANIUM
BORER

CORONOPUS BORER

BEARDTONGUE BORER

actual size

GRAPELEAF SKELETONIZER

actual size

WESTERN GRAPELEAF SKELETONIZER

Common

Harrisina metallica   57a-0008 (4623)

TL 11–13 mm Elongate, rounded, blue-black FW is unmarked; wings are held away from body. Blue-black abdomen widens slightly toward tufted tip. Antennae are pectinate in female, bipectinate in male. Some individuals in AZ/NM have orange-red collar. **HOSTS** Grapes. **NOTE** Often encountered at flowers during daytime.

BLACK-CLOAKED SKELETONIZER

Rare

Neoilliberis fusca   57a-0010 (4635)

TL 13–14 mm Rounded black FW is unmarked. Head, thorax, and abdomen are yellow orange. **HOSTS** Unknown.

CANYON GRAPELEAF SKELETONIZER

Rare

Neoalbertia constans   57a-0014 (4626)

TL 14–17 mm Rounded yellow-orange to red-orange FW has wide black terminal band. Head and thorax are orange, and abdomen is black. **HOSTS** Canyon wild grape.

FLANNEL MOTHS   SUPERFAMILY Zygaenoidea (57a)
FAMILY Megalopygidae

Chunky, medium-sized moths with a densely hairy thorax and abdomen and broad rounded wings that have a velvety texture. Caterpillars are typically hairy, with urticating spines, and often feed in groups. Adults are nocturnal and readily come to lights.

BROWN-PATCHED FLANNEL MOTH

Rare

Megalopyge lapena   57a-0039 (4645)

TL 17–23 mm Satiny ivory FW has a rippled texture. Inner AM area has a wavy brown patch, and two small brown dots mark outer PM and ST areas. Apex often translucent. Densely hairy thorax has slightly curly appearance. **HOSTS** Unknown.

PINK-SPOTTED FLANNEL MOTH

Rare

Trosia obsolescens   57a-0042 (4640)

TL 13–19 mm Ivory FW has a peach tinge and row of small black dots along median line. Hairy white thorax has pinkish-red spots. Abdomen and underside of wings are pinkish orange. White legs have pinkish red along basal segments. **HOSTS** Mexican blue oak and Emory oak.

MESQUITE STINGER MOTH

Uncommon

Norape tener   57a-0043 (4648)

TL 12–17 mm Brown FW has broad white central stripe from base to apex, with brown bulge at inner margin of median area. Hairy thorax has broad brown dorsal stripe. Antennae are yellow. **HOSTS** Mesquite, acacia, and willow.

WESTERN GRAPELEAF
SKELETONIZER

CANYON GRAPELEAF
SKELETONIZER

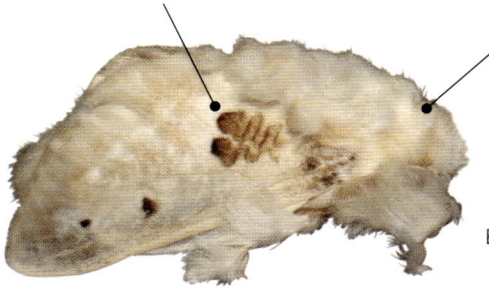

actual size

BLACK-CLOAKED
SKELETONIZER

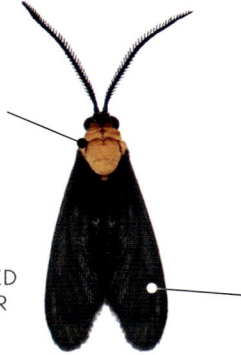

BROWN-PATCHED
FLANNEL MOTH

actual size

PINK-SPOTTED FLANNEL MOTH

MESQUITE STINGER MOTH

SLUG MOTHS
SUPERFAMILY Zygaenoidea (57a)
FAMILIES Dalceridae and Limacodidae

Small moths with very broad wings that are folded against their body when at rest. Some species will curl the tip of the abdomen above the wings, resembling a thorn. The caterpillars are often brightly colored and either spiny or covered in hairs; the spines of many species are stinging. The group's common name comes from the slug-like movement of the caterpillars. Adults are nocturnal and readily come to lights.

JELLY SLUG MOTH
Dalcerides ingenita   57a-0047 (4702)   **Rare**   RANGE-WIDE

TL 10–16 mm Broad, rounded, yellow-orange FW is unmarked. Edge of inner margin is hairy. Head appears very small. Typically rests with forelegs extended out front. HOSTS Manzanita and oak.

EARLY BUTTON SLUG MOTH
Tortricidia testacea   57a-0064 (4652)   **Uncommon**   RANGE-WIDE

TL 8–12 mm Short, rounded, tan FW has diffuse band of brown across median area to apex, with veins traced with darker brown in lower half of wing. Typically rests with tip of abdomen raised above folded wings. HOSTS Deciduous trees, including beech, birch, black cherry, chestnut, and oak.

WESTERN SKIFF MOTH
Prolimacodes trigona   57a-0066 (4670)   **Uncommon**   RANGE-WIDE

TL 13–16 mm Brown FW has a broad white patch in outer basal area that curves and tapers toward inner PM area. Terminal band is slightly lighter brown, curving to become broad lighter band along inner side of white patch. HOSTS Manzanita and oak.

PLAIN SLUG MOTH
Isa schaefferana   57a-0079 (4680)   **Uncommon**   RANGE-WIDE

TL 12–14 mm Unmarked tan FW has a slightly rippled appearance. Sometimes has sparse, dark brown scales peppered across FW and dark brown scales in fringe. HOSTS Hackberry and soapberry.

WHITE-LINED OAK-SLUG MOTH
Euclea obliqua   57a-0087 (4690)   **Rare**   RANGE-WIDE

TL 12–14 mm Brown FW has wavy white inner AM line edged below with dark brown and thin white outer ST line. Often has an indistinct diffuse dark median band between the white lines. HOSTS New Mexico locust and narrowleaf willow.

GREEN OAK-SLUG MOTH
Euclea incisa   57a-0088 (4696)   **Rare**   RANGE-WIDE

TL 10–15 mm Mint-green FW has broad brown terminal band that curves around anal angle to terminate with a squared end at PM line. Inner margin and costa are narrowly brown. Fluffy thorax is brown. HOSTS Unknown, but presumably deciduous trees.

JELLY SLUG MOTH

EARLY BUTTON SLUG MOTH

WESTERN SKIFF
MOTH

PLAIN SLUG MOTH

actual size

WHITE-LINED OAK-SLUG MOTH

GREEN OAK-SLUG MOTH

SMALLER PARASA

Parasa chloris

Uncommon
57a-0093 (4698)

TL 10–14 mm Mint-green FW has broad brown terminal band that does not curve onto inner margin. Outer half of basal area is brown. Fluffy thorax is mint green. **HOSTS** Deciduous trees, including apple, dogwood, elm, and oak. **NOTE** Southwestern population is disjunct from main population in East.

CONCEALER and SCAVENGER MOTHS

SUPERFAMILY Gelechioidea (59a)
FAMILIES Autostichidae and Oecophoridae

Small to very small moths with long rounded wings and, usually, thin palps that curve around the front of the head. Caterpillars feed on dead leaves, detritus, and fungi. Adults of most species are nocturnal and will come to lights, but a few species are primarily or entirely diurnal.

FOUR-SPOTTED YELLOWNECK

Oegoconia novimundi

Very Common
59a-0002 (1134)

RANGE-WIDE

TL 7–9 mm Blackish FW has broad tan median band, fragmented tan ST band, and small tan dots in basal area. Thorax has a narrow tan band across base. **HOSTS** Detritus in leaf litter.

SIGNATE SYMMOCA

Symmoca signatella

[IN] Uncommon
59a-0006 (1133)

RANGE-WIDE

TL 6–9 mm Peppery, light gray FW has angled blackish AM line that does not touch wing margins, and dark blotch at inner margin of PM line. AM and PM lines are bordered by tan shading. Median area has small black dot, and ST area has diffuse dusky shading. **HOSTS** Scavengers of decaying plant matter.

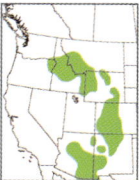

HOOKED TAYGETE

Taygete decemmaculella

Uncommon
59a-0011 (1844)

RANGE-WIDE

TL 6–8 mm Pale tan FW has slightly hooked, dark brown triangles at costa of AM and PM areas. Basal patch, thorax, and head form a dark brown collar. **HOSTS** Unknown.

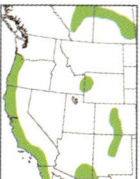

FOUR-SPOTTED GLYPHIDOCERA

Glyphidocera septentrionella

Rare
59a-0024 (1142)

RANGE-WIDE

TL 8–9 mm Peppery tan FW has a dusky basal spot, two AM spots, and a dusky PM bar. Long pointed palps curve backward over head. **HOSTS** Unknown.

CHALKY INGA

Inga cretacea

Uncommon
59a-0030 (1035)

RANGE-WIDE

TL 8–9 mm Pale tan, lightly speckled FW has three dusky dots at AM and PM lines and a thin ST line of often-indistinct tiny dots. **HOSTS** Unknown.

actual size

SMALLER PARASA

FOUR-SPOTTED YELLOWNECK

SIGNATE SYMMOCA

HOOKED TAYGETE

FOUR-SPOTTED
GLYPHIDOCERA

actual size

CHALKY INGA

109

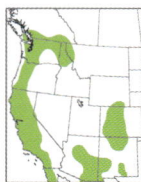

YELLOW-SPOTTED CONCEALER MOTH

Uncommon

RANGE-WIDE

Decantha stonda　　59a-0040 (1045)

TL 4–7 mm Dark brown FW has large patches of white-edged yellowish tan in inner basal/AM, outer median, inner PM, and outer ST areas. Fringe and thorax are yellowish tan; head is dark brown. **HOSTS** Varied; has been reared from *Hypoxylon* oak canker, *Polyporus* on lodgepole pine, and pine bark.

LESSER TAWNY CRESCENT

[IN] Common

RANGE-WIDE

Batia lunaris　　59a-0044 (1049)

TL 4–6 mm Warm brown FW has long, dark brown triangle at inner margin of PM line, edged thinly with white basally, and thin white basal dash. Costa, ST/terminal area, and fringe are slightly darker. Head is dark brown, with thin white stripes laterally. **HOSTS** Decaying wood under bark of dead shrub limbs.

FOUR-SPOTTED CONCEALER MOTH

Uncommon

RANGE-WIDE

Brymblia quadrimaculella　　59a-0051 (1055)

TL 7–8 mm Blackish FW has pale yellow, triangular patches at AM and PM areas of inner margin and median and PM areas of costa. Head is yellowish orange. **HOSTS** Fungi in cracks of bark.

SULPHUR TUBIC

[IN] Common

NORTH

Esperia sulphurella　　59a-0053 (1057)

SOUTH

TL 7–8 mm Hoary dusky to blackish FW has black-bordered tan streaks from base to median area and small, black-bordered, tan triangles at inner margin and costa of PM line. Thorax has tan streaks laterally. Head and palps are orange. **HOSTS** Dead and fallen oak trunks. **NOTE** Diurnal.

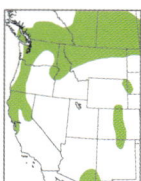

SKUNK MOTH

Uncommon

RANGE-WIDE

Polix coloradella　　59a-0054 (1058)

TL 8–11 mm Black FW has pale yellow streak along inner margin that connects to curving, pale yellow triangle at inner PM line and to paired, pale yellow lateral stripes on thorax. **HOSTS** Dead and decaying wood and bark.

BROWN HOUSE MOTH

Very Common

RANGE-WIDE

Hofmannophila pseudospretella　　59a-0060 (1064)

TL 7–14 mm Mottled brown FW has four black spots: one basal, two AM, and one median. Dusky spots of curving adterminal line often blend into ground color. **HOSTS** Generalist on harvested and dead plant and animal products, including dried fruit, stored grains, cork, wools, furs, and leathers. **NOTE** Cosmopolitan; native range unknown. Cannot survive in desert environments.

WHITE-SHOULDERED HOUSE MOTH

Common

NORTH

Endrosis sarcitrella　　59a-0061 (1067)

SOUTH

TL 5–9 mm FW is mottled brown, with indistinct blackish spots. Thorax and head are bold white. **HOSTS** Dead and decaying plant matter; sometimes also stored grains and seeds. **NOTE** Found on most continents; native range unknown.

YELLOW-SPOTTED CONCEALER MOTH

LESSER TAWNY CRESCENT

FOUR-SPOTTED
CONCEALER MOTH

SULPHUR TUBIC

actual size

SKUNK MOTH

BROWN HOUSE MOTH

WHITE-SHOULDERED
HOUSE MOTH

GOLD-BASE TUBIC

Oecophora bractella

[IN] Uncommon

59a-0063 (1069.1)

RANGE-WIDE

TL 7–9 mm FW is bright yellow in basal half and shiny black in outer half, with a yellow circle at costa. Costa of basal area, median band, central PM line, and terminal line curving onto inner margin are all metallic gray. Apical fringe is tipped white. **HOSTS** Fungi on decaying wood.

EUCALYPTUS CONCEALER MOTH

Tachystola hemisema

[IN] Uncommon

59a-0063.6 (1069.5)

RANGE-WIDE

TL 6–8 mm Tan FW has black marks at inner margin of PM line and along terminal band. Many individuals also have a blackish spot at inner median area. **HOSTS** *Eucalyptus*.

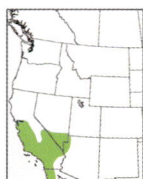

WHITE-EDGED CONCEALER MOTH

Pleurota albastrigulella

Rare

59a-0064 (1074)

RANGE-WIDE

TL 7–11 mm Peppery, light gray FW has broad white costa with dark edge from base to PM area. Central FW has three small black dots at AM and PM areas, and terminal line is a row of black dots. Bushy palps are very long and held forward. **HOSTS** Possible association with *Adenostoma*. **NOTE** Often encountered diurnally.

FLAT MOTHS and ALLIES

SUPERFAMILY Gelechioidea (59a)
FAMILY Depressariidae

Small moths with thin palps that curve around the front of the head. The *Agonopterix* and *Carcina* are relatively flat, with rectangular, rounded wings, while the *Ethmia* hold their wings snug against their body. Caterpillars of most species are leaf-tiers, though a few are borers. Adults are nocturnal and will come to lights.

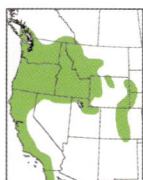

POISON HEMLOCK AGONOPTERIX

Agonopterix alstroemeriana

[IN] Common

59a-0087 (874.1)

RANGE-WIDE

TL 10–12 mm Pale brownish gray FW has dusky patch at outer median area, with adjacent elongate rusty spot and tiny black dot above. Terminal line is black dashes. **HOSTS** Poison hemlock.

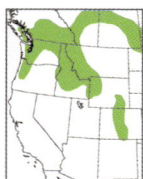

CANADIAN AGONOPTERIX

Agonopterix canadensis

Uncommon

59a-0091 (878)

RANGE-WIDE

TL 10–12 mm Mottled brown FW has dusky blotch in central median area, with two small black dots basally and one black dot below. Some individuals have two tiny white dots beside dusky patch. Thorax and inner basal area are pale and edged by crisply defined dark shading. **HOSTS** Ragwort.

GORSE TIP MOTH

Agonopterix nervosa

[IN] Uncommon

59a-0108 (895)

RANGE-WIDE

TL 7–11 mm Speckled tan FW has dusky patch in median area, with two white dots set in a reddish wash on inner side. Terminal line is diffuse dusky shading. **HOSTS** Gorse, French and Scotch brooms, and golden chain tree.

GOLD-BASE TUBIC

actual size

EUCALYPTUS
CONCEALER MOTH

WHITE-EDGED CONCEALER MOTH

POISON HEMLOCK
AGONOPTERIX

CANADIAN
AGONOPTERIX

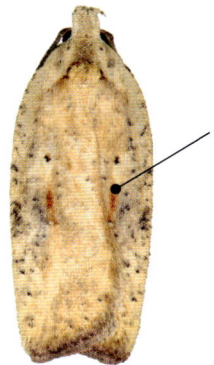

actual size

GORSE TIP
MOTH

MOURNING ETHMIA
Ethmia semilugens

Uncommon   59a-0183 (976)   RANGE-WIDE

TL 9–13 mm Black FW has an irregular white stripe along inner margin and white patch at apex. Curved terminal line is a row of black dots. Inner AM area and white thorax and head have bold black spots. **HOSTS** Unknown.

STONE ETHMIA
Ethmia arctostaphylella

Uncommon   59a-0186 (979)   RANGE-WIDE

TL 9–14 mm Silvery-gray FW has white stripe along inner margin, with thin black dashes along edge at AM, median, and PM lines. Terminal line is a row of tiny black dots. White thorax has a row of black spots. **HOSTS** Yerba santa. **NOTE** Rests along leaf midrib, mimicking bird droppings. Spring broods darker than summer.

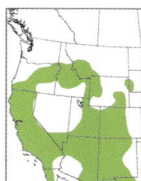

MOUNTAIN-MAHOGANY ETHMIA
Ethmia discostrigella

Very Common   59a-0187 (980)   RANGE-WIDE

TL 10–15 mm Gray FW has mottled white along inner margin. Thin black dashes mark basal dash, inner AM line, inner median area, and central FW from AM to PM areas. Terminal line is short black dashes. Grayish thorax has a row of black spots. **HOSTS** Mountain mahogany.

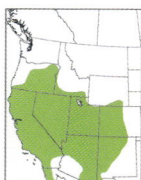

SHADED ETHMIA
Ethmia semitenebrella

Uncommon   59a-0188 (981)   RANGE-WIDE

TL 10–15 mm Inner half of gray FW is white with an irregular edge. Thick black dashes mark inner AM line and inner median area; thinner black dashes mark basal dash and central FW. Terminal line is short black dashes. White thorax has a row of black spots. **HOSTS** Mountain mahogany. **NOTE** Originally considered part of *E. discostrigella*.

GRAY ETHMIA
Ethmia monticola

Uncommon   59a-0194 (987)   RANGE-WIDE

TL 11–15 mm Silvery-gray FW has long black dashes in ST area, long black basal dash, central FW dash with small black spot at each end, and round black spots at inner AM, median, and ST areas. Terminal line is bold black spots. Gray thorax has a row of black spots. Some individuals have a dusky wash across outer half of FW; other individuals may lack all streaks and show just round spots in central FW. **HOSTS** Possibly *Lithospermum* and others in the borage family.

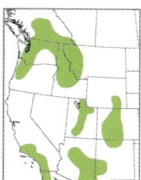

MARBLED ETHMIA
Ethmia marmorea

Uncommon   59a-0216 (1004)   RANGE-WIDE

TL 9–13 mm Resembles Mourning Ethmia but has additional white patches along costa at AM and PM areas, the latter containing a black spot. **HOSTS** Unknown.

OAK SKELETONIZER
Carcina quercana

[IN] Common   59a-0223.5 (1069)   RANGE-WIDE

TL 8–11 mm Reddish-brown FW has yellow square at costal median area and two pale spots at inner AM area. Fringe is yellow, edged by dark reddish-brown terminal line. **HOSTS** Deciduous trees, including oak, beech, and apple.

MOURNING ETHMIA

STONE ETHMIA

MOUNTAIN-MAHOGANY
ETHMIA

SHADED ETHMIA

GRAY ETHMIA

actual size

MARBLED ETHMIA

OAK SKELETONIZER

COSMET MOTHS
SUPERFAMILY Gelechioidea (59a)
FAMILY Cosmopterigidae

Small to very small moths with extremely long, narrow wings held either flat or tube-like around the body. Caterpillars are largely miners or feed on flower buds or seeds; the fluffy cattail heads seen in winter are the result of the silken webs of seed-eating Shy Cosmet caterpillars. Adults are nocturnal and will come to lights.

SWEETCLOVER ROOT BORER
Walshia miscecolorella   Uncommon   59a-0321 (1615)

NORTH
SOUTH

TL 6–10 mm FW has straight, angled AM line and tan median area that blends into warm brown, or mottled tan and gray, lower half. Basal half is gray. FW is covered in round metallic patches of raised scales that give moth a lumpy appearance. **HOSTS** Lupine, sweetclover, and other legumes; also thistle.

COFFEEBERRY MIDRIB GALL MOTH
Sorhagenia nimbosus   Common   59a-0339 (1633)

NORTH
SOUTH

TL 4–6 mm Gray FW has round tuft of blackish scales at inner margin of AM area, two tufts edged narrowly in white along median line, and small dusky tuft at inner PM line. Small dusky dots of terminal line are also slightly raised. **HOSTS** Coffeeberry and cascara.

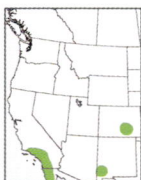

WHITE-LINED ERALEA
Eralea albalineella   Rare   59a-0385 (1501)

RANGE-WIDE

TL 4–5 mm Black FW has bold white lines along central wing and white V-shaped ST line. Terminal area and fringe are tan at anal angle and apex. Long, curving, black palps are white along front side. Antennae are white. **HOSTS** Unknown.

SILVER-SPOTTED ETEOBALEA
Eteobalea iridella   Rare   59a-0394 (1510)

RANGE-WIDE

TL 4–7 mm Tan FW has triangular white patches along costa at AM, median, and PM areas and conspicuous silvery bumps over inner half of wing. Long tan fringe is white at midpoint. **HOSTS** Vinegar weed.

FLORIDA PINK SCAVENGER MOTH
Anatrachyntis badia   Uncommon   59a-0399 (1513)

RANGE-WIDE

TL 4–6 mm Brown FW has tan-edged black dashes at AM and median lines and angled black-and-tan dashes at outer PM and inner ST areas. Black-tipped brown fringe has a distinctive fish-tail shape. Eyes are bright red. Antennae, palps, and legs are banded. **HOSTS** Seedpods of coffee senna and fruit of grapefruit, lime, and peach; also pinecones.

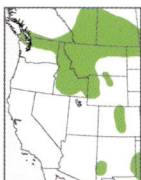

SHY COSMET
Limnaecia phragmitella   Uncommon   59a-0401 (1515)

RANGE-WIDE

TL 7–11 mm Light brown FW has a darker central streak crossing behind two white-ringed blackish spots in median and PM areas and an indistinct third spot inside from median spot. Apex is pointed. **HOSTS** Flowers and developing seeds of cattail.

SWEETCLOVER
ROOT BORER

COFFEEBERRY MIDRIB GALL MOTH

WHITE-LINED ERALEA

SILVER-SPOTTED ETEOBALEA

FLORIDA PINK
SCAVENGER MOTH

actual size

SHY COSMET

TWIRLER MOTHS
SUPERFAMILY Gelechioidea (59a)
FAMILY Gelechiidae

Small to very small, relatively flat moths with long narrow wings and curving labial palps that are fluffy at the base. The feeding habits of caterpillars are varied; they may be leaf-tiers, leaf-miners, or detritus feeders. Adults are nocturnal and will come to lights, though a few may also be encountered during daytime, often resting on foliage. This is a large group with a substantial number of undescribed species, and many look similar and are difficult to ID to species without dissection or DNA barcoding.

CALIFORNIA TWIRLER
Leucogoniella californica

Uncommon
59a-0455 (1848)

RANGE-WIDE

TL 4–6 mm Brownish-gray FW has dark brown PM/ST area with white V-shaped PM line. Inner basal and AM areas and costal median area have a dusky spot. Curving, dark brown terminal band frames a whitish patch at outer margin. Anal angle has a black spot near point of white V that is often hidden within ST shading. **HOSTS** Unknown.

ELEGANT BATTARISTIS
Battaristis concinnusella

Uncommon
59a-0466 (2225)

RANGE-WIDE

TL 6–7 mm Gray FW has thin, whitish, V-shaped ST line and black-edged white dash at costal PM line. Whitish ST/terminal area has warm brown along costa and contains small black dot near center of outer margin. Central FW contains a few indistinct blackish dots. Eyes are red. **HOSTS** Goldenrod and aster.

SOYBEAN WEBWORM MOTH
Mesophleps adustipennis

[IN] Uncommon
59a-0500 (2272)

RANGE-WIDE

TL 5–9 mm Narrow tan FW has dark brown edge to costa and three brown spots ringed indistinctly in white, placed in a spaced-out line. **HOSTS** White leadtree and mesquite.

SANDY HELCYSTOGRAMMA
Helcystogramma badia

Rare
59a-0503 (2263)

RANGE-WIDE

TL 7–8 mm Tan FW has pointed outer margin, with dashed black terminal line and three black spots arranged in a wide triangle in central wing. Veins often indistinctly traced with brown in ST/terminal area. Tan ST line is indistinct. **HOSTS** Unknown.

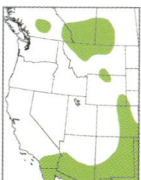

SUNSET DICHOMERIS
Dichomeris simpliciella

Rare
59a-0568 (2303)

RANGE-WIDE

TL 8–10 mm Dusky FW has a broad tan band along costa. Along inner edge of tan, narrow blackish stripes have small tan dots. Terminal area is dusky. Head is tan, and thorax has tan stripes laterally. **HOSTS** Unknown.

WHITE CROSS MOTH
Epilechia catalinella

Rare
59a-0589 (2259)

RANGE-WIDE

TL 9–11 mm Black FW has bold white stripe along inner margin connected to straight white PM line. AM and ST lines are bold, white, angled lines at costa. Head and thorax are white dorsally. **HOSTS** Unknown.

CALIFORNIA TWIRLER

ELEGANT BATTARISTIS

SOYBEAN
WEBWORM MOTH

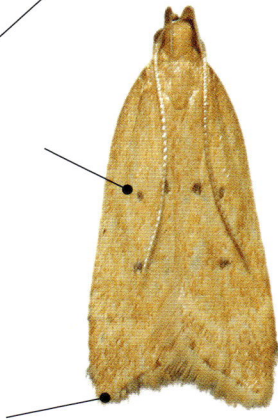

SANDY
HELCYSTOGRAMMA

SUNSET
DICHOMERIS

actual size

WHITE CROSS MOTH

COTTON STEM MOTH
Platyedra subcinerea

[IN] Uncommon
59a-0591 (2262)

NORTH

SOUTH

TL 8–9 mm Black-speckled tan FW has diffuse dusky stripe down center, passing behind three tan spots (sometimes with black dots at center) at AM, median, and PM areas. **HOSTS** Hollyhock, cheeseweed, and other mallows.

HOOK-WINGED CALLIPRORA
Calliprora sexstrigella

Rare
59a-0594.5 (2212)

RANGE-WIDE

TL 3–5 mm Gray to grayish tan FW is blackish in lower half, with white spots at costal median band, inner margin, and costa of PM line, and paired spots on costa at ST line and apex. Orange-brown dashes run just inside costa from median to PM line and from PM line to apex. Terminal line is tricolored with black, tan, and brown. Apex has a pronounced hook. **HOSTS** Probably mesquite.

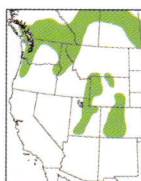

BURDOCK SEEDHEAD MOTH
Metzneria lappella

[IN] Rare
59a-0600 (1685)

RANGE-WIDE

TL 7–9 mm Pale golden-brown FW has highly jagged AM and PM lines that create a streaked appearance and small black dots at median and PM areas. Large palps curve over the head. **HOSTS** Burdock seeds.

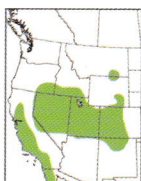

BLACK-DASHED ISOPHRICTIS
Isophrictis magnella

Rare
59a-0608 (1694)

RANGE-WIDE

TL 7–8 mm Tan FW has brown ST/terminal area, with long, white, V-shaped PM line and short white dashes along costa and anal angle. Central FW has elongate black dots in median and PM areas. White fringe has three parallel, thin, dark brown lines. **HOSTS** Sawtooth goldenbush.

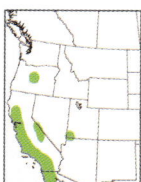

SILVER-BANDED ARISTOTELIA
Aristotelia argentifera

Rare
59a-0643 (1730)

RANGE-WIDE

TL 5–8 mm Tan FW has broad white AM, median, and PM bands that are metallic silver along inner half and a thin, metallic silver basal line. Central FW and costa are dark brown. Palps and legs are striped. **HOSTS** Goldenbush and coyote brush.

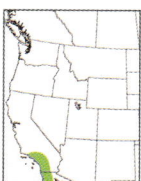

GRAY-BANDED ARISTOTELIA
Aristotelia calens

Uncommon
59a-0645 (1732)

RANGE-WIDE

TL 5–6 mm Resembles larger Sloped Aristotelia, but FW is lighter brown and hoary gray basal, AM, median, and PM bands are narrower, especially in inner half. Dark outer AM band is more rectangular. **HOSTS** Unknown.

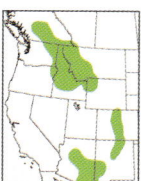

SLOPED ARISTOTELIA
Aristotelia devexella

Uncommon
59a-0648 (1734)

NORTH

SOUTH

TL 7–9 mm Patchwork FW has broad gray AM, median, and PM bands. Black shading of costal half of AM area forms a hooked shape. Basal half of basal area, inner median area, and central PM area are all pinkish orange. **HOSTS** Unknown. **NOTE** DNA barcoding suggests this may represent multiple cryptic species; further study is needed.

HOOK-WINGED
CALLIPRORA

COTTON STEM MOTH

BURDOCK SEEDHEAD MOTH

BLACK-DASHED ISOPHRICTIS

SILVER-BANDED ARISTOTELIA

actual size

GRAY-BANDED ARISTOTELIA

SLOPED ARISTOTELIA

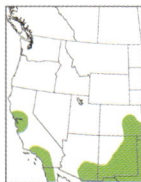

ELEGANT ARISTOTELIA

Aristotelia elegantella

Uncommon

59a-0650 (1736)

TL 6–7 mm FW has bands of rusty orange separated by crisp bands of black-edged white that have slightly raised, metallic silver scales toward inner margin. Central FW has five bold black spots set against a strip of metallic silver scales. **HOSTS** Willowherb; possibly also fireweed.

SIX-LINED ARISTOTELIA

Aristotelia hexacopa

Uncommon

59a-0653 (1739)

TL 5–6 mm Dark bronzy FW has narrow, white, V-shaped AM and median lines that are split at midpoint, and white triangles at inner margin and costa of ST area. Top of head and thorax is white, with two black patches at center. **HOSTS** Unknown.

PINK-WASHED ARISTOTELIA

Aristotelia roseosuffusella

Uncommon

59a-0670 (1761)

TL 6–8 mm Brown FW has broad grayish basal, AM, and median bands, split grayish PM band, and grayish patch at apex. Lighter brown along inner margin is usually washed pink, sometimes quite brightly. Dark brown of AM area is bluntly hooked. **HOSTS** Legumes, such as clover, bushclover, prairie clover, and ticktrefoil.

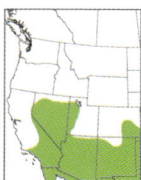

BLUSHING DIAMONDBACK

Ornativalva erubescens

[IN] Uncommon

59a-0681.5 (1928.1)

TL 6–7 mm FW has a hoary gray costa, dark brown center, and broad tawny inner margin that curves around outer margin and across basal area. Border of tawny band is zigzagged and narrowly edged with white. White costal ST line bends before touching tawny band. Central thorax and head are tawny. **HOSTS** Unknown; possibly tamarisk.

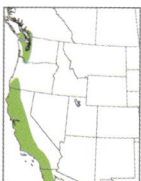

CALIFORNIA TELPHUSA

Telphusa sedulitella

Rare

59a-0691.9 (1859)

TL 6–8 mm Variable. Blackish basal area has straight AM border; AM area can be anywhere from broadly whitish to mostly brown with a thin pale edge along AM line; when present, pale shading usually curves along inner margin to PM area. Inner ST line is very white, and curving, sometimes bordered by white shading outwardly. Inner margin has small scale tufts at AM, median, PM, and ST lines. In paler-marked individuals, head has a broad central stripe of matching color. **HOSTS** Live oaks.

CONIFER COLEOTECHNITES COMPLEX

Coleotechnites spp.

Uncommon

59a-0708-59a0751 (1789–1832)

TL 4–6 mm Light grayish-brown FW has white costal dashes at AM and median lines and white V-shaped PM line, each bordered basally by broadly diffuse, blackish shading. Inner AM and PM areas have a small black dot of raised scales. **HOSTS** Conifers. **NOTE** A group of look-alike species that are difficult to separate visually; *Coleotechnites* species are host-specific on their particular conifer species.

ELEGANT ARISTOTELIA

SIX-LINED ARISTOTELIA

PINK-WASHED
ARISTOTELIA

actual size

BLUSHING DIAMONDBACK

CALIFORNIA TELPHUSA

CONIFER COLEOTECHNITES
COMPLEX

WHITE-LINED PSEUDOCHELARIA
Pseudochelaria manzanitae **Rare** 59a-0802 (1861)

TL 8–10 mm Gray FW has a thick, black, angled AM line and straight white ST line. Central median and ST/terminal areas are dark brown. Inner median area sometimes has a light tan dot. **HOSTS** Manzanita.

RANGE-WIDE

SADDLED PSEUDOCHELARIA
Pseudochelaria scabrella **Uncommon** 59a-0804 (1863)

TL 8–10 mm Thorax and basal area together have a roughly circular, dark brown patch edged narrowly in white. Thin, straight, white ST line borders dark brown shading in costal half of median/PM area. Inner margin has small scale tufts at AM, median, and PM lines. **HOSTS** Island manzanita, summer holly, and mission manzanita.

NORTH

SOUTH

SIX-SPOTTED GROUNDLING
Prolita sexpunctella **Rare** 59a-0809 (1898)

TL 8–9 mm Peppery grayish FW has thick, dark brown, straight AM and bent PM bands and angled basal and ST bands. Heavily marked individuals may appear to be dark brown, with three peppery grayish bands. **HOSTS** Unknown in North America; in Europe, heather, heath, bilberry, and other blueberry relatives.

RANGE-WIDE

VARIABLE PROLITA
Prolita variabilis **Rare** 59a-0810 (1902)

TL 7–10 mm Brown FW has thick, white-edged, black dashes in basal and median areas. Some individuals also have black shading or thin black dashes in terminal area. **HOSTS** Possibly narrowleaf goldenbush.

RANGE-WIDE

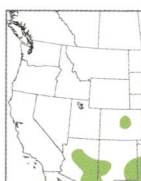

RUSTY EUDACTYLOTA
Eudactylota iobapta **Uncommon** 59a-0837 (1913)

TL 5–7 mm Warm brown FW has straight whitish PM band edged basally with dark brown shading. Terminal area usually slightly mottled with dusky shading. **HOSTS** Unknown.

RANGE-WIDE

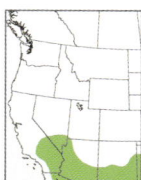

MESQUITE WEBWORM MOTH
Friseria cockerelli **Uncommon** 59a-0840 (1916)

TL 6–8 mm Brown to light tan FW has dark brown patches along costa and small blackish dashes at central AM and PM lines. Terminal area is shaded dusky, and terminal line is black dots, sometimes indistinct. **HOSTS** Honey mesquite.

RANGE-WIDE

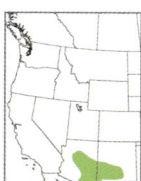

WHITE-DUSTED FRISERIA
Friseria caieta **Rare** 59a-0841 (1915)

TL 6–9 mm Pale gray FW has bright orange, black-tipped, raised scales along dusky AM and PM lines and costal median area. Central median and ST/terminal areas usually have dusky patches. **HOSTS** Unknown.

RANGE-WIDE

WHITE-LINED
PSEUDOCHELARIA

SADDLED
PSEUDOCHELARIA

SIX-SPOTTED GROUNDLING

VARIABLE PROLITA

RUSTY
EUDACTYLOTA

MESQUITE WEBWORM
MOTH

actual size

WHITE-DUSTED FRISERIA

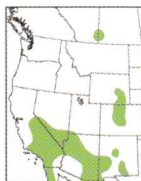

RIBBONED RIFSERIA

Rare RANGE-WIDE

Rifseria fuscotaeniaella 59a-0844 (1918)

TL 5–8 mm White FW has warm brown basal band that crosses white thorax. Costa has narrow brown spots at median and ST areas. Some individuals have brown mottling in ST/terminal area. **HOSTS** Pearly everlasting and cudweed.

WHITE-LINED GELECHIA

Rare RANGE-WIDE

Gelechia desiliens 59a-0856 (1938)

TL 8–10 mm Brown FW has narrow white ST line and small, white-edged, black dots at central AM and PM areas; third dot at inner AM area is often indistinct. **HOSTS** Western sycamore.

COASTAL CHIONODES

Rare NORTH

Chionodes lophosella 59a-0906 (2089)

SOUTH

TL 4–6 mm Dark gray FW has square white patch at costal ST area and usually smaller one at inner margin. Three spots of mottled brown, black, and tan mark outer AM and inner AM and PM areas. Edge of thorax has a crescent of brown. **HOSTS** Yellow and dune bush lupines.

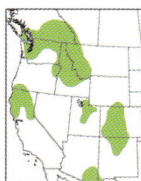

WHITE-PATCHED CHIONODES

Rare RANGE-WIDE

Chionodes abella 59a-0920 (2055)

TL 6–8 mm Dimorphic. Most common morph has mottled blackish FW, with slanting white AM band, and white ST/terminal area. Some individuals have a partial, narrow, white median line. Both morphs have a tan spot at posterior of thorax. Rare dark morph (not shown) has dusky FW, with angled white AM band, straight white PM band, and several white spots in median area. Head is pale. **HOSTS** White fir and Douglas-fir. **NOTE** A few other *Chionodes* (also not shown) resemble dark morph but can usually be separated by combination of head, thorax, and AM band.

BANDED CHIONODES

Rare RANGE-WIDE

Chionodes pinguicula 59a-0938 (2109)

TL 6–8 mm Resembles Large-toothed Chionodes, but AM tooth does not extend past central wing and tan ST band is solid. In some individuals, gray ST area may be washed brownish. **HOSTS** Unknown.

SMALL-TOOTHED CHIONODES

Rare RANGE-WIDE

Chionodes dentella 59a-0944 (2071)

TL 6–8 mm Black FW has a whitish stripe along inner margin, with a tooth at AM and PM areas, the latter sometimes becoming a thin PM line. ST area has white triangles at costa and inner margin. Head and thorax are white dorsally. **HOSTS** Unknown.

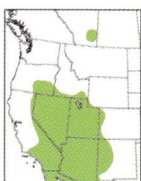

LARGE-TOOTHED CHIONODES

Rare NORTH

Chionodes fructuaria 59a-0948 (2078)

SOUTH

TL 6–8 mm Resembles Small-toothed Chionodes, but teeth of band at inner margin are broader, often reaching costa. White marks at ST line often join to form a line. **HOSTS** Unknown.

RIBBONED RIFSERIA

COASTAL CHIONODES

WHITE-LINED GELECHIA

WHITE-PATCHED
CHIONODES

actual size

BANDED
CHIONODES

SMALL-TOOTHED
CHIONODES

LARGE-TOOTHED
CHIONODES

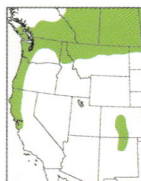

BLACK-SMUDGED CHIONODES

Chionodes mediofuscella

Common

59a-0971 (2093)

RANGE-WIDE

TL 5–8 mm Brown FW has an angled black AM line bordering large dusky patch in outer median and PM areas. Wavy brown ST line cuts across dusky ST/terminal area. Central median area has small black spots that are often obscured by dark patch. **HOSTS** Giant ragweed.

YELLOW-HEADED CHIONODES

Chionodes phalacra

Rare

59a-1001 (2108)

RANGE-WIDE

TL 4–6 mm Blackish FW has tan spots at inner margin and costa of ST area. Thorax and head are tan dorsally. **HOSTS** Mallows.

TWO-BANDED CHIONODES

Chionodes lugubrella

Rare

59a-1009 (2090)

RANGE-WIDE

TL 7–9 mm Black FW has a straight white PM band and angled white AM band that does not reach inner margin. **HOSTS** Unknown.

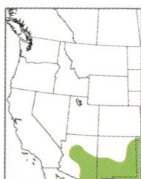

BENT-LINE FILATIMA

Filatima albilorella

Uncommon

59a-1079 (2129)

RANGE-WIDE

TL 7–10 mm Black FW has white, angled AM, hook-shaped median, and wavy PM lines. Head and thorax are white dorsally. **HOSTS** Unknown.

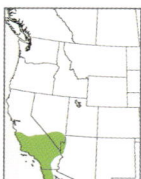

STRIPED AROGA

Aroga morenella

Uncommon

59a-1146 (2192)

RANGE-WIDE

TL 7–9 mm Black FW has white stripes along inner margin and costa and hoary gray band around curved outer margin. Head and thorax are white dorsally. **HOSTS** Possibly California buckwheat.

TAN-BACKED AROGA

Aroga paraplutella

Uncommon

59a-1147 (2193)

NORTH

SOUTH

TL 5–7 mm Dark gray FW has a tan stripe along inner margin from base to PM area and two indistinct black spots in central PM area. Fringe is hoary gray. Some individuals have a narrow pale ST band. Head and thorax are tan dorsally. **HOSTS** Seacliff buckwheat.

STARING AROGA

Aroga paulella

Uncommon

59a-1148 (2194)

RANGE-WIDE

TL 9–12 mm Black FW has white edging along inner and outer margins, a white patch at costa of median area, and straight white AM and ST bands. Head and thorax are white dorsally. **HOSTS** Desert trumpet.

actual size

TWO-BANDED
CHIONODES

BLACK-SMUDGED
CHIONODES

YELLOW-HEADED
CHIONODES

BENT-LINE FILATIMA

STRIPED AROGA

TAN-BACKED AROGA

STARING AROGA

SKUNK-BACKED AROGA
Aroga unifasciella

Uncommon
59a-1153 (2200)

TL 7–9 mm Black FW has white stripe along inner margin from base to straight white ST band. Head and thorax are white dorsally. HOSTS Unknown.

PALO VERDE WEBWORM
Faculta inaequalis

Uncommon
59a-1159 (2206)

RANGE-WIDE

TL 5–7 mm Mottled brown to brownish-gray FW has bold, black, straight AM band that does not touch inner margin and black spot at costa of PM line. Some individuals also have a smaller dot in inner PM line. HOSTS Palo verde.

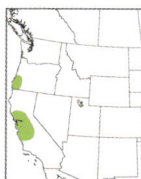

CLOVER TWIRLER
Mirificarma eburnella

[IN] Rare
59a-1163 (2208)

RANGE-WIDE

TL 7–8 mm Orange-brown FW has paired white streaks in basal area, a crescent of short white streaks in median area, and white inner-margin dash and costal patch at ST area. HOSTS Clover.

COYOTE BRUSH STEM GALL MOTH
Gnorimoschema baccharisella

Very Common
59a-1180 (1972)

RANGE-WIDE

TL 9–10 mm Variable. Gray to brown FW has warm brown and tan basal patch. Central AM and PM and inner AM areas have dark brown spots. Veins often marked with pale tan. Thorax and head are tan to gray. HOSTS Galls on coyote brush.

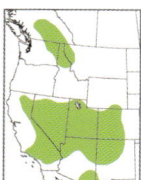

RABBITBRUSH STEM GALL MOTH
Gnorimoschema octomaculella

Uncommon
59a-1231 (1995)

RANGE-WIDE

TL 6–11 mm Hoary gray FW has brownish-orange basal, AM, and PM patches. HOSTS Sagebrush, rabbitbrush, and goldenhead.

CASEBEARER MOTHS
SUPERFAMILY Gelechioidea (59a); FAMILY Coleophoridae

Small moths with long narrow wings and forward-pointing antennae. Many species are strikingly metallic. Caterpillars make distinctive cases of leaf material and frass. Some 150 species have been described in North America, and dozens more are yet to be described, but most are poorly known. Many species are extremely similar and require dissection or DNA barcoding for definitive ID. Adults are nocturnal and will come to lights.

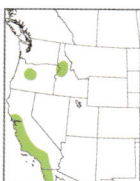

WHITE-EDGED COLEOPHORA
Coleophora accordella

Rare
59a-1510.1 (1289)

RANGE-WIDE

TL 6–8 mm Tawny FW has crisp white stripe along costa. Inner basal area shaded slightly paler. HOSTS Legumes, bird's-foot trefoil, and sweetvetch.

actual size

SKUNK-BACKED AROGA

PALO VERDE
WEBWORM

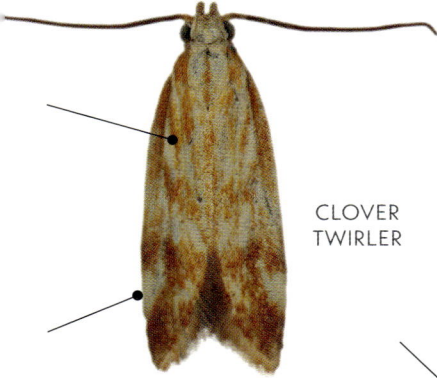

CLOVER
TWIRLER

COYOTE BRUSH
STEM GALL MOTH

RABBITBRUSH STEM
GALL MOTH

actual size

WHITE-EDGED COLEOPHORA

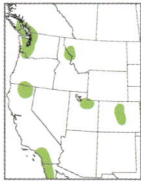

STREAKED COLEOPHORA
Coleophora cratipennella   **Rare**   59a-1553.1 (1365)   RANGE-WIDE

TL 6–8 mm Pale tan FW has white veins. Antennae are banded at base. **HOSTS** Seeds of rushes. **NOTE** The most common of a few look-alike species.

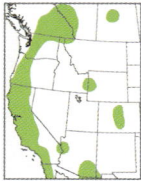

PEPPERED COLEOPHORA
Coleophora glaucella   **Rare**   59a-1569.1 (1305)   NORTH

SOUTH

TL 5–7 mm Pale whitish FW is lightly peppered with dark brown scales and pale brownish streaks along veins. **HOSTS** Manzanitas, summer holly, and arbutus.

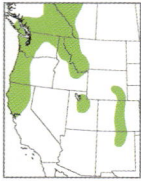

METALLIC COLEOPHORA
Coleophora mayrella   **[IN] Rare**   59a-1599.1 (1387)   RANGE-WIDE

TL 6–8 mm Metallic bronzy-gold FW is reddish in ST/terminal area. Bases of white-tipped antennae are thickened with long black scales. **HOSTS** Seeds of clover.

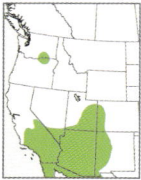

RUSSIAN THISTLE STEM MINER
Coleophora parthenica   **[IN] Rare**   59a-1611.1 (1398.8)   RANGE-WIDE

TL 7–9 mm Whitish-tan FW has slightly darker tan along veins. Antennae bases are thickened with long tan scales. **HOSTS** Russian thistle. **NOTE** Intentionally introduced from the Mediterranean as biological control for Russian thistle.

LARGE CLOVER CASEBEARER
Coleophora trifolii   **[IN] Uncommon**   59a-1646.1 (1388)   RANGE-WIDE

TL 8 mm Resembles Metallic Coleophora, but antennae lack thickened basal section. Outer FW often less reddish. **HOSTS** Sweetclover.

SCYTHRID FLOWER MOTHS and ALLIES
SUPERFAMILY Gelechioidea (59a)

FAMILIES Scythrididae, Blastobasidae, and Momphidae

An assorted group of small to very small moths with long narrow wings and curving palps. The forewings of blastobasids have a distinctive curved appearance at rest. Momphids have raised scale tufts along the inner margin. Adults of all three families are nocturnal and will come to lights; scythrids and blastobasids are also commonly encountered at flowers in daytime.

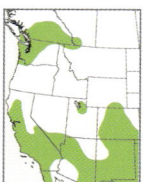

BANDED SCYTHRIS
Scythris trivinctella   **Common**   59a-1713 (1678)   RANGE-WIDE

TL 6–7 mm Brown FW has straight white basal and median bands and Y-shaped PM band. **HOSTS** Smooth pigweed.

STREAKED COLEOPHORA

PEPPERED COLEOPHORA

METALLIC COLEOPHORA

actual size

RUSSIAN THISTLE
STEM MINER

LARGE CLOVER CASEBEARER

actual size

BANDED SCYTHRIS

OSO FLACO FLIGHTLESS MOTH

Uncommon

RANGE-WIDE

Areniscythris brachypteris   59a-1731 (1680)

TL 4–6 mm Both sexes flightless. Wings are shorter than the rounded, pale brown abdomen. FW has a whitish inner/costal margin bordered by brownish shading. Central wing has checkered spots of white and dark brown. Eyes are yellow. **HOSTS** Generalist on herbaceous plants of sand dunes, including woolly sunflower, lupine, ragweed, and coyote mint. **NOTE** Diurnal. Adults run on the sand surface and use their long hindlegs to leap into the wind to travel farther distances; may climb onto vegetation to escape hot sand. Similar species of *Areniscythris* are also found in western inland dune systems.

CONTRASTING APHID MOTH

Uncommon

RANGE-WIDE

Asaphocrita aphidiella   59a-1733 (1171)

TL 9–11 mm Dimorphic. Most commonly, hoary, pale gray FW has blackish V-shaped AM and ST lines, with two connected blackish dots forming a short bar at central PM line; upper basal area and costal median area shaded dusky. Some individuals have an iron-gray FW sprinkled with metallic bronzy scales and a bronzy thorax. **HOSTS** Aphid galls on hickory. **NOTE** Likely represents a species complex, based on DNA barcoding data.

ACORN MOTH

Common

RANGE-WIDE

Blastobasis glandulella   59a-1766 (1162)

TL 8–13 mm Gray FW has a curving black AM line edged with diffuse white shading basally and broad dusky shading medially. Two black dots in central PM area are usually well defined; another small dot below AM line is sometimes noticeable. **HOSTS** Larvae feed inside acorns, hickory nuts, and chestnuts.

WHITE-BASED MOMPHA

Uncommon

RANGE-WIDE

Mompha albocapitella   59a-1813.1 (1448)

TL 5–7 mm Peppery brown FW has an angled white basal patch adjacent to white thorax and head. Inner AM and PM areas have white patches with tufts of raised scales. Often has small white subapical patch. **HOSTS** Evening primrose. **NOTE** This species was previously known as *M. murtfeldtella*, actually a junior synonym.

RED-STREAKED MOMPHA

Rare

RANGE-WIDE

Mompha eloisella   59a-1833 (1443)

TL 6–8 mm White FW has bold black spots in inner AM and PM and outer median areas, reddish V-shaped PM and ST lines, and tufts of brown scales at inner margin of ST area. Dark fringe has a long tuft at center. Thorax has several smaller black spots. **HOSTS** Stems of evening primrose.

ONE-LINED MOMPHA

Uncommon

RANGE-WIDE

Mompha unifasciella   59a-1856 (1458)

TL 5–7 mm Dark brown to dusky FW has narrow, straight, whitish AM and PM bands; PM band is broken in some individuals. Thorax and inner basal area may be either dark or pale brownish. **HOSTS** Fireweed. **NOTE** DNA barcode data suggests this species label likely represents three distinct species.

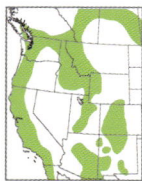

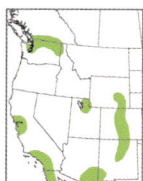

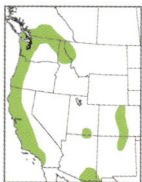

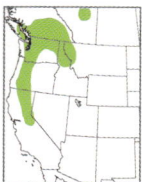

OSO FLACO
FLIGHTLESS MOTH

CONTRASTING
APHID MOTH

ACORN MOTH

actual size

WHITE-BASED MOMPHA

RED-STREAKED MOMPHA

ONE-LINED MOMPHA

MANY-PLUMED MOTHS
SUPERFAMILY Alucitoidea (61a)
FAMILY Alucitidae

Small moths with distinctive wings divided into many feather-like plumes. Three species are recognized as resident in North America, with two more occurring as vagrants, but by far the most common is Montana Six-plume Moth. These are sometimes encountered on walls and woodpiles during daytime but are generally nocturnal and usually seen at lights at night.

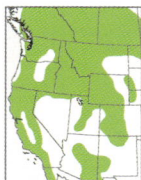

MONTANA SIX-PLUME MOTH
Alucita montana

Very Common
61a-0001 (2313)

NORTH

SOUTH

WS 12–14 mm Unique wings are composed of multiple feather-like plumes. FW has dark brown median and terminal bands edged thinly with white and separated by tan bands. **HOSTS** Honeysuckle and snowberry.

PLUME MOTHS
SUPERFAMILY Pterophoroidea (63a)
FAMILY Pterophoridae

Small to large moths with a thin body, long spindly legs, and long narrow wings that are held perpendicular to the body when at rest. Nearly all are shades of tan or brown with subtle markings, though a few species are more boldly patterned; the pattern on the abdomen is often just as useful for identification as that on the forewings. Caterpillars are often host specialists, with the species named after the host plant. Adults of most species are nocturnal and will come to lights, but many species can also be found resting in vegetation during daytime.

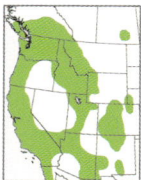

ARTICHOKE PLUME MOTH
Platyptilia carduidactylus

Common
63a-0005 (6109)

RANGE-WIDE

WS 18–27 mm Brown to brownish-gray FW has a large, dark brown triangle on costa at PM area and a smaller one at AM area. ST/terminal area is warmer brown. Light brown or gray abdomen has two inverted, dark brown chevrons in basal half and is often washed warmer brown on lower half. **HOSTS** Artichoke and thistle.

YARROW PLUME MOTH
Gillmeria pallidactyla

Uncommon
63a-0012 (6107)

RANGE-WIDE

WS 24–27 mm Pale to golden-tan FW has brown shading at AM and PM areas and brown shading (usually faint) in ST/terminal area, with pale washes in costal median and ST areas. PM area often has a small brown spot. Pale ST line is often indistinct. Pale to golden-tan abdomen has a thin brown dorsal line. **HOSTS** Yarrow, possibly tansy.

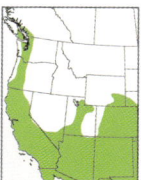

SAGE PLUME MOTH
Anstenoptilia marmarodactyla

Very Common
63a-0014 (6117)

NORTH

SOUTH

WS 15–20 mm Lightly striated, grayish-brown FW has dark brown triangle on costa at PM area and dark brown ST area, with a warm tan patch in between. ST line is pale. Terminal area is grayish brown. Base of thorax has a dark brown U shape bordered posteriorly with ivory. Grayish-brown abdomen has two brown diamond or triangle shapes on middle segments. **HOSTS** Mints, including rosemary, sage, pitcher sage, *Monardella*, and bluecurls.

actual size

MONTANA SIX-PLUME MOTH

actual size

ARTICHOKE
PLUME MOTH

YARROW
PLUME MOTH

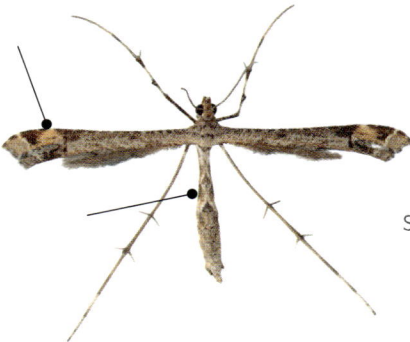

SAGE PLUME
MOTH

LANTANA PLUME MOTH

Lantanophaga pusillidactylus

Uncommon

63a-0015 (6119)

RANGE-WIDE

WS 11–14 mm Resembles larger Snapdragon Plume Moth, but FW is darker without a warm brown patch between dark brown PM and ST patches, and pale ST line is thin. Abdomen has a dark brown band on middle segment. HOSTS Lantana.

SNAPDRAGON PLUME MOTH

Stenoptilodes antirrhina

Rare

63a-0017 (6123)

RANGE-WIDE

WS 15–19 mm Resembles Sage Plume Moth, but has golden-brown patch between triangle and ST area, extending as streaks to outer margin. Warm brown abdomen has a light tan base and squarish, dark brown marks dorsally on lower half. HOSTS Snapdragon.

GERANIUM PLUME MOTH

Amblyptilia pica

Very Common

63a-0051 (6118)

RANGE-WIDE

WS 15–22 mm Mottled, warm brown FW has chestnut-brown triangle on costa at PM area and chestnut-brown ST area, separated by an ivory to tan patch. ST line is pale. Thorax has dark brown triangle bordered by ivory V shape posteriorly. Middle segments of brown abdomen have curving pale marks at posterior end. HW projects as dark brown triangles from below inner margin of FW at AM and PM areas when at rest. An uncommon white morph has blackish triangle and ST/terminal area with white ST line, and blackish mottling along inner margin; thorax has black marks at collar and posterior, and abdomen has black patches laterally on middle segments connected by gray to blackish rings at anterior of segments. HOSTS Generalist; annual geranium (*Pelargonium*), snapdragon, paintbrush, penstemon, figwort, and others.

HOURGLASS PLUME MOTH

Michaelophorus indentatus

Uncommon

63a-0053 (6103)

RANGE-WIDE

WS 12–16 mm Dark brownish-gray FW has whitish wash at costa of PM area, pale ST line, and dusky apex. Warm brown abdomen is whitish at base and has thin, white, scalloped lines dorsally on middle segments. HOSTS Unknown.

GUMWEED PLUME MOTH

Dejongia californicus

Rare

63a-0071 (6100)

RANGE-WIDE

WS 12–16 mm Golden-brown FW has pale bands at PM and terminal areas, occasionally also AM area, sometimes only evident at costa. Fringe along inner margin of outer wing appears checkered. Posterior of thorax has a white line that crosses even with inner margin of FWs. Brown abdomen is paler at base, with a darker band at middle segment and fine whitish striations in segments in between. HOSTS Gumweed and goldenbush.

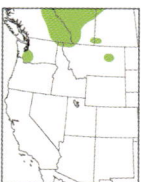

PLAIN PLUME MOTH

Hellinsia homodactylus

Common

63a-0104 (6203)

RANGE-WIDE

WS 23–25 mm Similar in shape and size to Morning-glory Plume Moth but entirely white, without distinct markings. Sometimes has indistinct grayish smudges in AM and PM areas. Abdomen sometimes tinged yellowish or pale greenish. HOSTS Goldenrod.

LANTANA PLUME MOTH

SNAPDRAGON PLUME MOTH

actual size

GERANIUM
PLUME MOTH

HOURGLASS PLUME MOTH

GUMWEED PLUME MOTH

PLAIN PLUME MOTH

COYOTE BRUSH BORER PLUME MOTH

Hellinsia grandis

Uncommon 63a-0111 (6211)

RANGE-WIDE

WS 26–35 mm Similar in shape and size to Morning-glory Plume Moth but entirely tan, without distinct markings. Abdomen has a thin brown dorsal stripe. **HOSTS** Coyote brush.

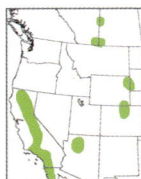

WORMWOOD PLUME MOTH

Oidaematophorus grisescens

Rare 63a-0142 (6171)

RANGE-WIDE

WS 20–29 mm Peppery gray FW has a paler costa and warm brown shading in inner ST/terminal area (the latter sometimes hidden when FW is rolled at rest). A blackish dot at base of cleft is bordered outwardly with white. Thorax is dusky at posterior, bordered by a white U shape. Gray abdomen has a dusky V-shaped band at lower middle segment, with small brown dots laterally. **HOSTS** Wormwood and sagebrush.

MORNING-GLORY PLUME MOTH

Emmelina monodactyla

Very Common 63a-0150 (6234)

NORTH

SOUTH

WS 25–28 mm Variable. FW and body color may be tan or grayish, with variably extensive markings. FW is long and of even width when curled at rest, with a hooked tip, and usually a black dot about one-third out from thorax and brown smudge at costa of PM area. Thorax usually has a pale triangle dorsally, tapering into very thin, black-dotted dorsal line on thorax. **HOSTS** Convolvulaceae, including morning glory and bindweed.

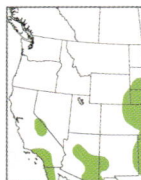

RAGWEED PLUME MOTH

Adaina ambrosiae

Uncommon 63a-0157 (6160)

RANGE-WIDE

WS 15–16 mm Peppery grayish-brown FW has black dashes on costa at PM and ST areas, blackish patch in median area at base of cleft, and indistinct dusky spot in AM area. Abdomen has a row of dusky chevrons down center. Thoracic collar is dark brown. **HOSTS** *Ambrosia* species, aster, and ragweed.

FRUITWORM MOTHS and ALLIES

SUPERFAMILY Carposinoidea (65a)

FAMILIES Copromorphidae and Carposinidae

Small to very small moths with longish, slightly pointed wings that are held flat at rest and create the appearance of a notched triangle. *Bondia* species have rows of raised scales on the forewing. Caterpillars are usually miners, and some species can be fruit pests. Adults are nocturnal and come to lights.

MADRONE FRUITWORM

Lotisma trigonana

Uncommon 65a-0001 (2312)

NORTH

SOUTH

TL 8–12 mm Mottled gray FW has angled black AM line with paler area above containing two small black spots (inner spot sometimes indistinct against mottled gray inner margin). Outer median area is shaded dusky. Central PM area has small black dot. **HOSTS** Madrone, manzanita, huckleberry, and salal.

actual size

COYOTE BRUSH
BORER PLUME
MOTH

WORMWOOD
PLUME MOTH

MORNING-GLORY
PLUME MOTH

RAGWEED PLUME MOTH

actual size

MADRONE FRUITWORM

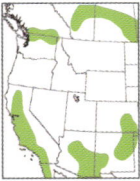

PRUNE LIMB BORER

Bondia comonana   **Rare**   65a-0010 (2318)

TL 7–9 mm   Gray FW has dark basal and ST/terminal areas and pale gray AM area; median area is mottled, with white-ringed brown spots. Reniform spot is fractured, white, kidney-shaped outline shaded gray inside. **HOSTS** Deciduous tree branch galls.

NORTH

SOUTH

WINDOW-WINGED MOTHS

SUPERFAMILY Thyridoidea (70)
FAMILY Thyrididae

Small chunky moths with broad dark wings bearing semitranslucent white spots; wings are usually held spread and raised at an angle when at rest. Caterpillars are typically leaf-rollers. Adults of *Thyris* are diurnal and commonly encountered at flowers or while mudpuddling in groups at wet dirt patches along forest edges or tracks, while those of *Dysodia* are nocturnal and come to lights.

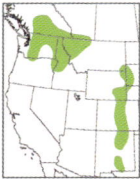

SPOTTED THYRIS

Thyris maculata   **Rare**   70-0003 (6076)

RANGE-WIDE

WS 12–15 mm   Calico wings are mottled black and orange, with single white spot on FW and two-lobed white spot on HW. Fringe is broadly checkered black and white. Rests with wings held in a dihedral over body. **HOSTS** Clematis and *Houstonia* species. **NOTE** Diurnal; often encountered while mudpuddling or at flowers.

NETTED DYSODIA

Dysodia granulata   **Rare**   70-0006 (6079)

RANGE-WIDE

WS 20–26 mm   Orange to brown wings have a net-like pattern of thin lines and translucent, whitish, round spot on FW and double-lobed spot on HW. Median area is slightly darker. Thorax is densely hairy. Typically rests with FW raised in a dihedral. **HOSTS** Probably Chihuahuan brickellbush.

ASSORTED PYRALIDS

SUPERFAMILY Pyraloidea (80a)
FAMILY Pyralidae EXCLUDING SUBFAMILY Phycitinae

Moths in the family Pyralidae are a varied assemblage, but most are fairly flat, deltoid-shaped moths with smallish heads and relatively large eyes. A number of species curl the abdomen above their wings at rest. Members of the subfamily Chrysauginae habitually rest with the thorax elevated on stilt-like legs. Caterpillars in this family feed on a wide variety of hosts, including dead plants, detritus, and beehives; several species are pests of stored food products. Adults are typically nocturnal and come to lights. Household pests can potentially be encountered indoors year-round.

ROSY CAPHYS

Caphys arizonensis   **Uncommon**   80a-0001 (5537)

RANGE-WIDE

TL 12–15 mm   Hoary rosy to purplish FW has pale yellow, straightish, narrow AM and PM lines. Median area and fringe often shaded darker. Some individuals are tan with a pinkish wash in ST/terminal area. Typically rests with thorax elevated on long forelegs; forelegs have pink scale tufts in basal half. **HOSTS** Unknown.

actual size

PRUNE LIMB BORER

actual size

SPOTTED THYRIS

NETTED DYSODIA

actual size

ROSY CAPHYS

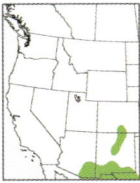

OCHRE PARACHMA
Parachma ochracealis

Uncommon

80a-0002 (5538)

RANGE-WIDE

TL 8–12 mm Resembles larger Rosy Caphys, but FW is tan to orangish, broader across basal area, and with thin, blackish-purple terminal line. Purple legs are more densely scaled, especially hindlegs. **HOSTS** Unknown.

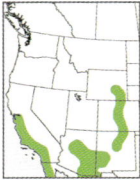

ORANGE-LINED ACALLIS
Acallis griphalis

Rare

80a-0004 (5541)

RANGE-WIDE

TL 12–15 mm Purplish-brown FW is often hoary with tan scales. Curving tan AM and PM lines are indistinctly edged medially with blackish or dark purple; PM line straightens to meet costa perpendicularly. **HOSTS** Unknown.

GREEN ANEMOSELLA
Anemosella viridalis

Uncommon

80a-0011 (5549)

RANGE-WIDE

TL 9–11 mm Light green FW has thick, white, straightish AM and curving PM lines framing dark green median area. Lines sometimes curve to meet along inner margin. **HOSTS** Unknown.

BOXWOOD LEAFTIER
Galasa nigrinodis

Uncommon

80a-0017 (5552)

RANGE-WIDE

TL 9–11 mm Reddish-purple FW has orange basal area and whitish shading in costal median area. Thin AM and PM lines sometimes present as fractured or indistinct white lines. Costal edge of FW is slightly concave at median area. Typically rests with body held at an angle, elevated on four long tufted legs. **HOSTS** Boxwood and olive.

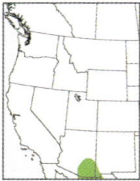

RUSTY LEAFTIER
Galasa nigripunctalis

Rare

80a-0018 (5553)

RANGE-WIDE

TL 9–11 mm Resembles Boxwood Leaftier, but FW color is uniformly rusty brown. Ranges do not overlap. **HOSTS** Unknown.

DESERT WILLOW DESTROYER
Satole ligniperdalis

Uncommon

80a-0027 (5562)

RANGE-WIDE

TL 8–10 mm Brownish to gray FW has thin, white, wavy AM and PM lines edged in black. Median area usually lightly washed rusty, sometimes strongly orange. Dusky median line usually fragmented, sometimes indistinct, and does not reach inner margin. Female has straight costa; male has concave costa and broad basal area with rounded scale tuft. **HOSTS** Desert willow.

POSTURING ARTA
Arta statalis

Rare

80a-0031 (5566)

RANGE-WIDE

TL 8–10 mm Resembles Gallant Arta, but FW is reddish brown and terminal line is a row of black dashes. Legs have long scales on middle segment. **HOSTS** Unknown.

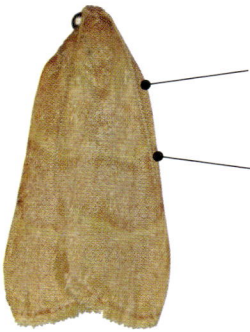

OCHRE PARACHMA

actual size

ORANGE-LINED
ACALLIS

GREEN
ANEMOSELLA

BOXWOOD
LEAFTIER

RUSTY LEAFTIER

DESERT WILLOW DESTROYER

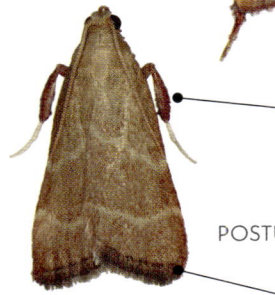

POSTURING ARTA

GALLANT ARTA

Common

Arta epicoenalis 80a-0032 (5567)

NORTH
SOUTH

TL 7–11 mm Resembles Ochre Parachma, but tan AM and PM lines are slightly wavy and wider apart at inner margin than at costa. Terminal line is usually absent. Legs have reduced or short scales on middle segment. **HOSTS** Unknown.

GREATER WAX MOTH

[IN] Common

Galleria mellonella 80a-0041 (5622)

NORTH
SOUTH

TL 13–18 mm Gray FW has brown along inner margin from central base to anal angle, and brown thorax. ST line is a curving row of black interveinal dashes. FW is falcate at anal angle. **HOSTS** Beeswax, dried fruit, pollen, and dead insects.

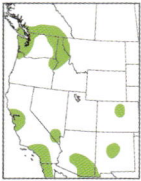

LESSER WAX MOTH

[IN] Common

Achroia grisella 80a-0042 (5623)

RANGE-WIDE

TL 8–12 mm Plain mousy-gray FW has a rounded outer margin. Small head is yellow orange. **HOSTS** Old wax and debris of vacated beehives.

BEE MOTH

[IN] Uncommon

Aphomia sociella 80a-0049 (5629)

RANGE-WIDE

TL 14–19 mm Sexually dimorphic. Male has whitish to greenish FW peppered with black scales. Outer median area is shaded warm brown and bordered by jagged brown AM line. Jagged dusky PM line often indistinct. ST/terminal area is olive. Female has peppery grayish-green to greenish-gray FW with indistinct, jagged, dusky AM and PM lines and two black dots in outer median area. Median area sometimes washed slightly brownish. **HOSTS** Nest material and stored food of bumblebees and wasps; sometimes beeswax in beehives.

TRANSPOSED ALPHEIAS

Uncommon

Alpheias transferens 80a-0067 (5646)

RANGE-WIDE

TL 10–13 mm Peppery gray FW has narrow, zigzag, white AM line edged medially with black, and curving dusky PM line. Black reniform dash is edged with white. Costal PM area is washed whitish. AM area is sometimes diffusely shaded brownish along AM line. **HOSTS** Rabbitbrush.

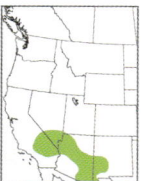

FOX-FACED DECATURIA

Uncommon

Decaturia pectinalis 80a-0071 (5650)

RANGE-WIDE

TL 6–7 mm Tan FW has a pointed black oval in outer-central median area. Angled white lines run from inner median area to costal AM and PM areas, forming a white V around black spot. Terminal area is hoary whitish gray, bordered by black-edged white terminal line. **HOSTS** Unknown.

YELLOW-BASED CACOZELIA

Uncommon

Cacozelia basiochrealis 80a-0620 (5580)

RANGE-WIDE

TL 9–12 mm FW has a curving double AM line, golden-brown basal area, and light gray median area. Dusky PM line is indistinct except for a whitish patch at costa. ST area is reddish, often with a grayish wash at center. Terminal area is gray. **HOSTS** Unknown.

GREATER
WAX MOTH

GALLANT ARTA

LESSER WAX
MOTH

male

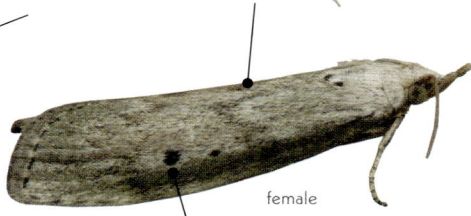

female

BEE MOTH

actual size

TRANSPOSED ALPHEIAS

FOX-FACED DECATURIA

YELLOW-BASED CACOZELIA

147

HOARY PYRALID

Common

Toripalpus trabalis

80a-0627 (5585)

TL 12–16 mm FW has hoary gray basal half and whitish median area, with broad tan shading along toothed PM line. Black double median line is gently wavy with whitish fill. ST/terminal area is hoary brownish to grayish, often with diffuse white wash at center. Thorax is tan. **HOSTS** Redflower, seacliff, and naked buckwheat.

COMMON MEAL MOTH

[IN] Very Common

Pyralis farinalis

80a-0667 (5510)

TL 14–16 mm Reddish-brown FW has tan median area shaded reddish brown in inner half. White, curving AM and sinuous PM lines widen at costa. Sometimes has a dusky discal spot. **HOSTS** Stored grain products.

ELECTED TABBY

Uncommon

Aglossa electalis

80a-0671 (5513)

TL 11–13 mm Resembles Jacketed Tabby, but AM and PM lines are broader and paler tan and ST/terminal area is darker brown. **HOSTS** Unknown.

JACKETED TABBY

Uncommon

Aglossa cacamica

80a-0672 (5514)

TL 10–13 mm Dark brown FW has tan, zigzag AM and curving PM lines that widen toward costa. Median area has a tan wash at center and multiple tan dots along costa. ST/terminal area is brownish tan, often with brown shading along veins. **HOSTS** Possibly plant detritus in rodent burrows.

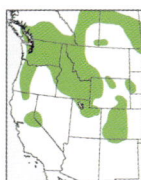

LARGE TABBY

[IN] Common

Aglossa pinguinalis

80a-0673 (5516)

TL 15–20 mm Tan FW is heavily peppered with dark brown scales. Thick, wavy, dark brown AM and PM lines are broadly edged outwardly with pale tan. Squarish, dark brown discal spot is edged with pale tan patch below. **HOSTS** Animal dung and hay refuse in barns.

STORED GRAIN MOTH

[IN] Uncommon

Aglossa caprealis

80a-0674 (5517)

TL 12–14 mm Dark brown FW has wavy tan AM and ST lines and a blurry tan patch (sometimes defined as a tan circle with dark brown center) in outer median area. **HOSTS** Stored foods, grain chaff, fungi, and dead animals.

WIDE-BANDED TABBY

Rare

Aglossa acallalis

80a-0676 (5519)

TL 8–10 mm Hoary, light brown FW has thick, wavy, dark brown AM and PM lines broadly edged with pale tan. Hoary median area is slightly darker brown. Costa lacks tan dots. **HOSTS** Unknown.

HOARY PYRALID

COMMON MEAL MOTH

ELECTED TABBY

JACKETED TABBY

actual size

LARGE TABBY

STORED GRAIN MOTH

WIDE-BANDED TABBY

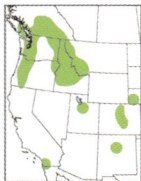

CLOVER HAYWORM
Hypsopygia costalis

[IN] Uncommon
80a-0681 (5524)

RANGE-WIDE

TL 9–10 mm Reddish-purple FW has yellow triangles along costa at the ends of usually indistinct yellow AM and PM lines. HW has wavy yellow AM and PM lines. Yellow of fringe bleeds into terminal area. Head is yellowish. **HOSTS** Mostly dried plant material and stored hay.

BROWN HAYWORM
Hypsopygia phoezalis

Uncommon
80a-0684 (5527)

RANGE-WIDE

TL 8–13 mm Brown FW has black AM and PM lines edged with yellowish tan that widens at costa. Median area is slightly darker, with black discal spot and a light sprinkling of scales in inner half. **HOSTS** Cypress and citrus.

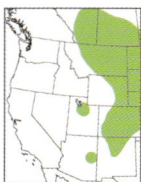

YELLOW-FRINGED HYPSOPYGIA
Hypsopygia olinalis

Uncommon
80a-0689 (5533)

RANGE-WIDE

TL 10–12 mm Resembles Clover Hayworm, but yellow costal triangles are narrower and terminal area and base of yellow fringe are purple. Head is purplish. **HOSTS** Oaks. **NOTE** Similar Spruce Needleworm (*H. thymetusalis*, not shown) has a pink fringe and blackish edging along inside of AM and PM lines.

PHYCITINE MOTHS
SUPERFAMILY Pyraloidea (80a)
FAMILY Pyralidae SUBFAMILY Phycitinae

Phycitines are distinctive moths, with long wings that are held slightly curved over the abdomen and proportionally small heads with relatively large eyes. The extremely common household pest Indian Meal Moth is a member of this group, along with a few other species that feed on stored grain products. *Laetilia* are among the few predaceous lepidopteran species, with caterpillars feeding on scale insects. Adult phycitines are primarily nocturnal and will come to lights, but a few species, such as Dusky Raisin Moth and American Sunflower Moth, can sometimes be found at flowers in daytime. This group contains many very similar species that may require dissection or DNA barcoding for definitive ID.

TRICOLORED ACROBASIS
Acrobasis tricolorella

Common
80a-0076 (5655)

RANGE-WIDE

TL 9–12 mm Gray FW has narrow black AM and PM lines edged outwardly thinly with white and broadly with brick reddish brown, which in the inner AM band is edged in white and in the costal PM band becomes black. Reniform spot is a short black dash; PM area costal to it is shaded whitish. Separable from Brown-banded Catastia by white line basal to AM band and lack of black dashes in ST area. **HOSTS** Fruit-bearing trees, including apple, cherry, and plum.

SMALL OAK ACROBASIS
Acrobasis comptella

Uncommon
80a-0077 (5656)

RANGE-WIDE

TL 9–12 mm Peppery gray FW has narrow black AM and PM lines. AM is bordered basally with a broad brown band, then a thickish straight or wavy black line, often with white edging; black line basal to brown patch reaches inner margin. Reniform spot is two black dots. Median area costal to dots is shaded white. Inner margin is sometimes washed brownish. Thorax is brownish. **HOSTS** Oaks.

actual size

CLOVER HAYWORM

BROWN HAYWORM

YELLOW-FRINGED
HYPSOPYGIA

TRICOLORED ACROBASIS

actual size

SMALL OAK
ACROBASIS

LARGE OAK ACROBASIS
Uncommon RANGE-WIDE

Acrobasis caliginella 80a-0112 (5695)

TL 12–16 mm Resembles smaller Small Oak Acrobasis, but black line on basal side of brown AM patch does not reach inner margin or becomes thin and faint as it approaches inner margin. **HOSTS** Oaks.

MINUTE MYELOPSIS
Uncommon RANGE-WIDE

Myelopsis minutulella 80a-0126 (5719)

TL 9–11 mm Peppery, light gray FW has thick black AM band and narrow black PM line edged outwardly with pale gray. Two black reniform dots are often indistinct among black speckles. Base of inner margin has small tan to brownish patches. **HOSTS** Unknown.

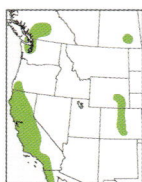

GRAY MYELOPSIS
Uncommon NORTH

Myelopsis subtetricella 80a-0128 (5718)

SOUTH

TL 11–12 mm Resembles Minute Myelopsis, but FW is brownish gray and AM and PM lines are more soft-edged; PM line is sometimes indistinct. Base of inner margin is gray, without brown patches. **HOSTS** Acorns.

TWO-STRIPED APOMYELOIS
Rare RANGE-WIDE

Apomyelois bistriatella 80a-0134 (5721)

TL 9–12 mm Resembles Minute Myelopsis, but FW is darker gray and white edging along PM line more pronounced. Base of inner margin is gray. **HOSTS** Fungi on deadwood.

NAVEL ORANGEWORM MOTH
Uncommon RANGE-WIDE

Amyelois transitella 80a-0137 (5724)

TL 12–24 mm Light gray FW has a warm brown wash along inner margin and onto thorax, and is pale gray along costa. Curving black AM line has a wide section just before costa. White PM line has dusky blackish edge on both sides. Reniform spot is two black dots. **HOSTS** Citrus, walnuts, pistachios, and almonds; also numerous others, including date palm, apple, and acacia.

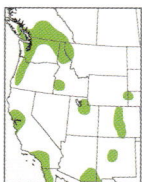

AMERICAN PLUM BORER
Uncommon RANGE-WIDE

Euzophera semifuneralis 80a-0138 (5995)

TL 9–13 mm Reddish-brown FW has a dark gray band in PM area, bordered by wavy white median and PM lines. Reniform spot is a white dash. Peppery costa and terminal area are washed pale gray. **HOSTS** Deciduous trees, including apple, cherry, peach, pear, sweetgum, and walnut.

BROAD-BANDED EULOGIA
Uncommon RANGE-WIDE

Eulogia ochrifrontella 80a-0145 (5999)

TL 7–10 mm Resembles larger American Plum Borer, but golden-brown to warm brown FW lacks a pale costa and ST/terminal area is narrower. Median line is evenly curving. **HOSTS** Oak, apple, pecan, and possibly serviceberry.

LARGE OAK ACROBASIS

MINUTE
MYELOPSIS

GRAY MYELOPSIS

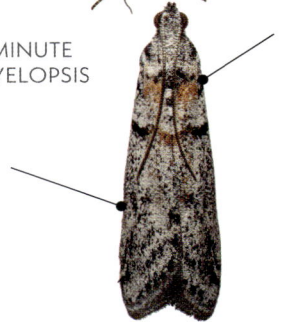

TWO-STRIPED
APOMYELOIS

actual size

NAVEL ORANGEWORM MOTH

AMERICAN PLUM
BORER

BROAD-BANDED
EULOGIA

DUSKY RAISIN MOTH
Ephestiodes gilvescentella

Very Common
80a-0146 (6000)

TL 5–9 mm Peppery FW has a blackish median area bordered by straightish white AM and PM lines and crossed by broad brown veins. Basal and ST/terminal areas are peppery grayish with brown veins. Reniform spot is two black dots. Similar, larger Beehive Honey Moth has wavy AM and PM lines. **HOSTS** Possibly a detritus scavenger in galls and mines of other moth larvae.

RED-STREAKED SNOUT MOTH
Ephestiodes erythrella

Rare
80a-0150 (6002)

TL 6–9 mm Resembles Dusky Raisin Moth, but basal and terminal areas are concolorous with median area, and veins and thorax are a deep reddish brown. **HOSTS** Unknown.

BEEHIVE HONEY MOTH
Vitula serratilineella

Uncommon
80a-0164 (6007.1)

TL 9–12 mm Peppery, light gray FW is washed brown along inner margin. Wavy blackish AM and PM lines are indistinctly edged outwardly with pale gray. Black reniform dash curves around a white dot. Similar smaller Dusky Raisin Moth has smooth AM and PM lines. **HOSTS** Pollen and honey in bee and wasp nests; also dried fruit.

GRAY VITULA
Vitula insula

Rare
80a-0167 (6009.2)

TL 8–10 mm Resembles Beehive Honey Moth, but AM line is straight and inner margin lacks brown shading. Median area is sometimes shaded dusky. **HOSTS** Unknown.

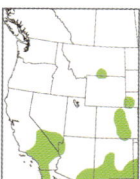

BLACK-DOTTED SOSIPATRA
Sosipatra rileyella

Uncommon
80a-0171 (6015)

TL 6–7 mm Lightly speckled, ivory to whitish FW has two bold black dots at AM line and several smaller black dots at PM line, as well as a small black dot at reniform spot. Veins are indistinctly traced with very pale tan. **HOSTS** Debris left behind by yucca moth larvae in yucca seed capsules.

INDIAN MEAL MOTH
Plodia interpunctella

Very Common
80a-0181 (6019)

TL 6–9 mm FW is tan in basal half and rusty brown with metallic gray lines in lower half. Appears clearly bicolored to the unaided eye. **HOSTS** Stored food products, including flour, oatmeal, and seeds. **NOTE** Originally from South America, now found worldwide. A common household pest.

DUSKY RAISIN MOTH

RED-STREAKED SNOUT MOTH

BEEHIVE HONEY
MOTH

actual size

GRAY VITULA

BLACK-DOTTED SOSIPATRA

INDIAN MEAL MOTH

MEDITERRANEAN FLOUR MOTH
Uncommon

Ephestia kuehniella   80a-0183 (6020)

NORTH
SOUTH

TL 8–14 mm Mottled gray to light gray FW has broad dusky blackish AM band edged with wavy, pale gray line basally, and narrow, sometimes indistinct, wavy, pale gray PM line edged on both sides with dusky shading. Reniform spot is two dusky blackish dots. HOSTS Stored food products, including seeds of barley, corn, oats, and rice.

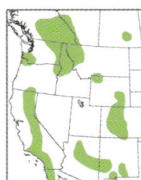

WHITE-EDGED PIMA
Rare

Pima fosterella   80a-0202 (5748)

NORTH
SOUTH

TL 15–17 mm Tan FW has a bold white stripe along costa, edged with peppery blackish shading. Inner margin is peppered with black scales. Inner AM and central PM areas each have a small black dot. HOSTS Unknown.

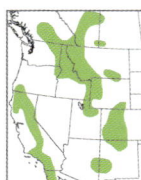

RED-PATCHED AMBESA
Rare

Ambesa laetella   80a-0214 (5759)

RANGE-WIDE

TL 13–16 mm Whitish FW has a squarish brick-red patch on costa between black orbicular and reniform dots and brick-red ST area. PM line and inner AM line are crisp black, edged outwardly with white. Inner margin is washed with tan, and reddish at median area and AM line. HOSTS Roses.

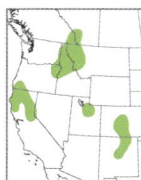

BROWN-BANDED CATASTIA
Rare

Catastia actualis   80a-0221 (5764)

RANGE-WIDE

TL 13–15 mm Peppery, light gray FW has narrow, zigzag, black AM and PM lines edged outwardly with white, then warm brown bands; warm brown in ST area has black dashes at veins. Black reniform dots are often edged with white. Central veins are often washed with streaks of warm brown. HOSTS Unknown.

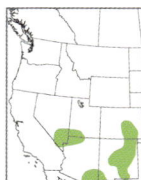

HOLLOW-SPOTTED OLYBRIA
Rare

Olybria aliculella   80a-0225 (5768)

RANGE-WIDE

TL 10–12 mm Peppery, light gray FW has narrow black AM and PM lines edged outwardly with white, then broad bands of rusty brown. Reniform spot is a black-outlined circle, often with brownish scales inside. Sometimes has a streak of brownish wash along inner margin. HOSTS Unknown.

CROSS-BANDED SNOUT MOTH
Rare

Quasisalebria admixta   80a-0254 (5779)

RANGE-WIDE

TL 8–10 mm Peppery gray FW has thin white AM line bordered basally with a broad black band that does not reach costa, and below with a squarish black patch at costa, so it looks like the white line crosses the black band. Costal PM area is washed white. Reniform spot is two black dots; costal-side dot is often indistinct. Indistinct white ST line bisects blackish subapical patch. HOSTS Unknown.

MEDITERRANEAN
FLOUR MOTH

actual size

WHITE-EDGED PIMA

RED-PATCHED AMBESA

BROWN-BANDED CATASTIA

HOLLOW-SPOTTED
OLYBRIA

CROSS-BANDED SNOUT MOTH

LESSER ASPEN WEBWORM

Meroptera pravella

Uncommon
80a-0270 (5787)

TL 10–12 mm Peppery, light gray FW has broad, diffuse, dusky AM band, usually containing a thin, angled, pale gray AM line at inner margin. Zigzag, pale gray ST line is broadly edged with dusky gray. A narrow smudge of dusky gray crosses from costal ST line to inner margin of median area. Abdomen has a whitish patch dorsally at base, often visible as a pale square between base of wings. **HOSTS** Quaking aspen; also willow, birch, and alder.

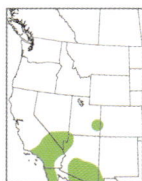

DOUBLE-BANDED SCIOTA

Sciota bifasciella

Rare
80a-0281 (5801)

TL 11–12 mm Peppery, light gray FW has straight, double, black AM band and wavy, double, dusky gray ST line. Reniform spot is two black dots, sometimes indistinct. Base of inner margin is tinged with brown. **HOSTS** Skunkbush sumac and laurel sumac.

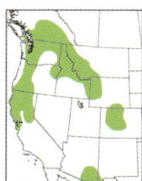

OVAL TELETHUSIA

Telethusia ovalis

Rare
80a-0301 (5812)

TL 12–14 mm Peppery, light gray FW has a black AM band interrupted by a brown patch in central wing and edged medially with narrow white and black lines. Reniform spot is two indistinct black dots. Narrow, wavy, dusky PM line is often indistinct. Inner half of wing sometimes has indistinct streaks of brown wash. **HOSTS** Pussytoes and woolly sunflower.

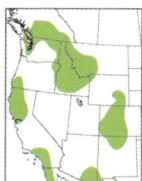

EVERGREEN CONEWORM

Dioryctria abietivorella

Uncommon
80a-0332 (5841)

TL 12–15 mm Dusky gray FW has bold, zigzag, black-edged, white AM and PM lines bordered outwardly with broad bands of dusky shading. Reniform spot is a small white crescent. **HOSTS** Fir, pine, and spruce.

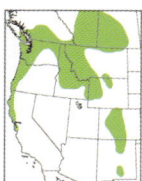

SPRUCE CONEWORM

Dioryctria reniculelloides

Common (uncommon in South)
80a-0334 (5843)

TL 12–15 mm Hoary gray FW has black scalloped AM and zigzag PM lines edged outwardly with white. Dusky band basal to AM has a brownish patch in inner half and is bordered basally by white. Median area is often darker, hoary blackish. Reniform spot is a small white crescent. **HOSTS** Spruce; also Douglas-fir, hemlock, fir, and lodgepole pine.

DOUGLAS-FIR CONEWORM

Dioryctria pseudotsugella

Rare
80a-0335 (5844)

TL 10–13 mm Very similar to Spruce Coneworm, but brown patch bordering AM line is more strongly colored and often reaches costa and lacks whitish band basally. Sometimes has brownish streaks in inner median area and ST/terminal area. **HOSTS** Douglas-fir; also spruce, hemlock, fir, and lodgepole pine.

DOUBLE-BANDED SCIOTA

OVAL TELETHUSIA

LESSER ASPEN
WEBWORM

EVERGREEN
CONEWORM

SPRUCE CONEWORM

actual size

DOUGLAS-FIR CONEWORM

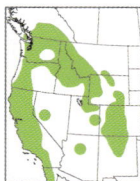

PONDEROSA PINE CONEWORM

Uncommon

Dioryctria auranticella   80a-0336 (5846)

TL 11–15 mm Brick-red FW is heavily suffused with orange. AM and PM lines are white. Long white dash of orbicular connects to white crescent of reniform spot. Terminal line is a row of white dots. Thorax is striped white and brick red. **HOSTS** Pines.

ZIMMERMAN PINE MOTH

Uncommon

RANGE-WIDE

Dioryctria zimmermani   80a-0353 (5852)

TL 13–18 mm Gray FW has white zigzag AM and PM lines edged on both sides with black, and a broad reddish-brown band basal to AM line, sometimes extending to base of FW. Reddish band and dusky median band have rows of raised scales. Inner median and ST areas are washed reddish brown. Reniform spot is white. **HOSTS** Pines.

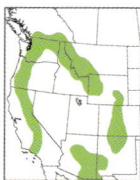

WESTERN PINE MOTH

Rare

RANGE-WIDE

Dioryctria cambiicola   80a-0355 (5854)

TL 14–16 mm Resembles Zimmerman Pine Moth, but FW is extensively rusty-brown. Black median band usually has a rusty-brown wash or patch at inner margin. **HOSTS** Pines.

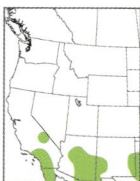

LESSER CORNSTALK BORER

Common

RANGE-WIDE

Elasmopalpus lignosella   80a-0409 (5896)

TL 7–11 mm Variable and sexually dimorphic. Male has tan FW, with gray along inner and outer margin and often costa, and small blackish dots at inner margin of median area and central PM area. Gray edging sometimes bleeds inward along veins. Some individuals are warm brown instead of tan. Females are largely gray with tan or brown as a narrow stripe or patch at center of FW, or they are entirely dusky with some red scaling at base of FW; black dots at median and PM areas may or may not be visible. **HOSTS** Grasses and legumes.

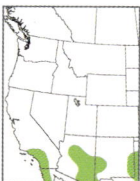

RUSTY-BANDED SNOUT MOTH

Rare

RANGE-WIDE

Macrorrhinia aureofasciella   80a-0427 (5912)

TL 7–11 mm Speckled gray FW has orange-brown AM band edged with white medially and a thick blackish patch along inner half basally. Reniform spot is two small black dots. Base of FW is washed orange brown. **HOSTS** Unknown.

WHITE-NOTCHED SNOUT MOTH

Uncommon

RANGE-WIDE

Promylea lunigerella   80a-0436 (5727)

TL 8–12 mm Peppery gray FW has straight black AM line bordered basally by broad brown band; inner end of band has a small white spot against AM line and is edged basally with black. Black PM line is edged outwardly with pale gray, bordering brown ST/terminal area. Curved reniform dash is thicker at inner end but often indistinct. Median area has a dusky patch against inner margin. **HOSTS** Unknown.

PONDEROSA PINE
CONEWORM

actual size

ZIMMERMAN
PINE MOTH

WESTERN
PINE MOTH

male

LESSER
CORNSTALK
BORER

female

female

RUSTY-BANDED
SNOUT MOTH

WHITE-NOTCHED
SNOUT MOTH

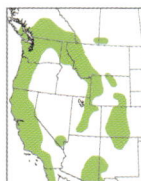

GRAY-BASED DASYPYGA
Dasypyga alternosquamella

Uncommon
80a-0439 (5730)

RANGE-WIDE

TL 9–15 mm Tan FW has gray basal and terminal areas. Bold straight AM band is tan, with blackish edges on both sides. Two dark rusty-brown streaks, the outer thick and inner thin, run from central median area to ST line. **HOSTS** Dwarf-mistletoe on pines and other conifers.

GRAY-SHOULDERED DASYPYGA
Dasypyga salmocolor

Rare
80a-0440 (5731)

RANGE-WIDE

TL 9–11 mm Resembles Gray-based Dasypyga, but gray basal and terminal areas are narrower and AM band lacks black edging. **HOSTS** Unknown.

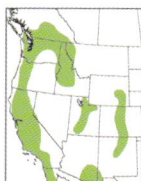

GOLD-BANDED ETIELLA
Etiella zinckenella

Uncommon
80a-0447 (5744)

NORTH

SOUTH

TL 9–13 mm Gray FW has bold white stripe along costa. Raised rusty AM band is bordered medially with golden tan. Long palps create a snouted appearance. **HOSTS** Immature seeds of legumes, including lupine, milkvetch, and locust.

PALE-PATCHED TACOMA
Tacoma feriella

Rare
80a-0452 (5901)

RANGE-WIDE

TL 8–11 mm Hoary gray FW has a white and tan patch at inner AM area. Wavy grayish AM and PM lines are usually indistinct. Many individuals have a tan stripe from base to anal angle, crossing AM patch. **HOSTS** Mistletoe.

CALICO ANCYLOSIS
Ancylosis morrisonella

Uncommon
80a-0462 (5916)

RANGE-WIDE

TL 6–11 mm Variable. FW has rusty-brown AM and ST bands and stripe of rusty brown from outer basal to central median area, sometimes as far as ST area. FW ground color is usually hoary gray, sometimes hoary black, often uniform in color but sometimes with median area darker or shaded blackish, particularly along AM and PM lines. Fringe may be rusty brown or gray. Thorax is usually rusty brown, sometimes with black at center. **HOSTS** Silver beachweed; possibly other *Ambrosia*.

SUGARBEET CROWN BORER
Ancylosis undulatella

Uncommon
80a-0464 (5918)

NORTH

SOUTH

TL 8–11 mm Hoary brownish-gray FW has wavy white AM and PM lines edged medially with black and outwardly with broad brown bands that are lightly streaked with black. Reniform spot is two black dots. **HOSTS** Recorded from sugar beet but thought perhaps to be a generalist.

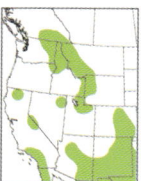

HONEYED HONORA
Honora mellinella

Uncommon
80a-0467 (5919)

NORTH

SOUTH

TL 8–12 mm Dusky gray FW has a tan patch in inner AM area edged basally by partial narrow white AM line and indistinct dusky band. Costa has whitish wash. Reniform spot is two black dots, often indistinct. Thorax is tan. **HOSTS** Spanish needle.

actual size

GRAY-BASED DASYPYGA

GRAY-SHOULDERED DASYPYGA

GOLD-BANDED ETIELLA

PALE-PATCHED TACOMA

CALICO
ANCYLOSIS

SUGARBEET
CROWN BORER

HONEYED HONORA

TRICOLORED HONORA

Honora montinatatella

Rare   80a-0471 (5923)

TL 13–15 mm Deep red FW has white broadly along costa and small tan patch at inner margin adjacent to narrow white AM line. PM line is a row of white dots. Reniform spot is a black dot at inner edge of white patch. **HOSTS** Unknown.

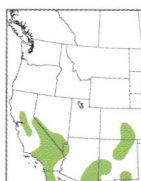

WESTERN SCALE-FEEDING SNOUT

Laetilia dilatifasciella

Uncommon   80a-0481 (5949.1)

CALIFORNIA

INLAND

TL 7–9 mm Peppery white FW has tan along inner margin, bisected by white AM band. Costa has a large black patch below AM band. Often-indistinct pale PM line is edged on both sides with diffuse blackish shading. Reniform spot is two black dots. **HOSTS** Larvae are predaceous on scale insects. **NOTE** Scale-feeding Snout (*L. coccidivora*, not shown) is similar but reddish brown instead of tan and does not occur west of TX.

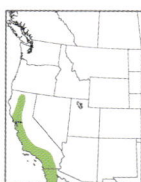

MONTEREY SCALE-FEEDING SNOUT

Laetilia zamacrella

Rare   80a-0482 (5950)

RANGE-WIDE

TL 8–14 mm Peppery gray FW has zigzag gray AM and PM lines bordered on both sides by diffuse, dusky gray shading. Reniform spot is two conjoined, dusky gray dots. Inner margin of median area has a diffuse dusky spot. Inner basal and median areas are usually sprinkled with tan scales. **HOSTS** Larvae are predaceous on scale insects on Monterey pine, and sometimes other conifers.

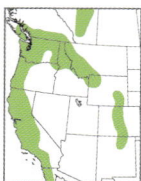

GOOSEBERRY FRUITWORM MOTH

Zophodia grossulariella

Uncommon   80a-0503 (5968)

RANGE-WIDE

TL 12–17 mm Hoary gray FW is lighter gray along costa. Straightish to slightly wavy white AM line is bordered medially by a broad black band. Slightly zigzag gray PM line is edged on both sides with diffuse black shading. Reniform spot is two black dots, sometimes fused as one. Most individuals have blackish streaks in basal and ST/terminal areas and occasionally through inner median area. Separable from smaller *Myelopsis* species by distinct PM line and streaked gray basal area. **HOSTS** Currants and gooseberries.

BLUE CACTUS BORER

Cactobrosis fernaldialis

Rare   80a-0510 (5989)

RANGE-WIDE

TL 15–21 mm Variable. Speckled to hoary gray FW has a light gray to white AM band edged with soft dusky shading (sometimes indistinct). ST/terminal area has thin black lines along veins and black dots along terminal line. Zigzag pale PM line is usually indistinct. Sometimes has a strong black streak from base to central median area. Rare individuals have basal area entirely dusky to blackish. **HOSTS** Fishhook barrel cactus and saguaro; likely also other barrel cactuses.

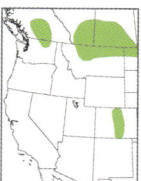

TOOTHED PRICKLYPEAR BORER

Melitara dentata

Rare   80a-0514 (5971)

RANGE-WIDE

TL 17–25 mm Lightly speckled, pale brownish FW is washed white along costa. Thin black AM line is deeply V-shaped, and PM line is strongly zigzagged, becoming double at costa. Reniform spot is a small blackish crescent. AM and PM lines and reniform spot are often indistinct. Veins are traced with soft brown. Base of thorax has a dusky brown band extending to FW base. **HOSTS** Plains, prairie, and brittle pricklypear.

actual size

TRICOLORED
HONORA

WESTERN
SCALE-FEEDING SNOUT

MONTEREY SCALE-FEEDING
SNOUT

GOOSEBERRY
FRUITWORM MOTH

BLUE CACTUS BORER

TOOTHED PRICKLYPEAR BORER

165

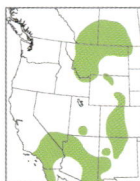

THIN-LINED PRICKLYPEAR BORER

Melitara subumbrella

Rare

80a-0519 (5973)

RANGE-WIDE

TL 17–25 mm Mousy-brown FW has a whitish wash along costa and thin blackish veins. Deeply scalloped, thin, black AM line connects at multiple spots to PM line, often creating outlines of shapes in median area; in some individuals, AM and PM lines are indistinct. Reniform spot is two dusky dots, sometimes joined into a smudgy dash. **HOSTS** Pricklypear.

REDDISH CHOLLA BORER

Alberada franclemonti

Rare

80a-0523 (5975.2)

RANGE-WIDE

TL 8–10 mm Tan to reddish-brown FW has a whitish wash along costa. Zigzag black AM and PM lines are double, with whitish fill. Inner median area has a dusky to blackish smudge. Reniform spot is a single black dot. ST area is usually shaded dusky. Posterior of thorax has a dusky blackish band with rounded patches of raised scales. **HOSTS** Cholla.

CAHELA MOTH

Cahela ponderosella

Rare

80a-0526 (5977)

RANGE-WIDE

TL 12–20 mm Hoary gray FW has thick black veins in central basal/AM and outer median/PM areas and thin black veins in ST/terminal area. Some individuals have faint dusky shading in inner half of FW. **HOSTS** Chollas, especially tree cholla.

AMERICAN SUNFLOWER MOTH

Homoeosoma electella

Very Common

80a-0548 (5935)

NORTH

SOUTH

TL 8–13 mm Peppery gray FW has two black dots at outer PM area and two in AM area; costal PM dot is slightly more basal than inner one, and costal AM dot is often indistinct. Basal half of costa is tan to pale gray. **HOSTS** Mostly aster; sometimes a pest of commercial sunflowers.

WHITE-EDGED PHYCITODES

Phycitodes mucidella

Uncommon

80a-0571 (5946)

NORTH

SOUTH

TL 7–11 mm Peppery white FW has thick tan lines along inner veins. Outer median and PM areas have bold black spots; costal PM dot is slightly more apical than inner one. Median spot is usually enlarged in males. **HOSTS** Groundsel, plume thistle, cudweed, and gumweed.

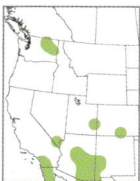

WESTERN COENOCHROA

Coenochroa californiella

Rare

80a-0575 (6039)

RANGE-WIDE

TL 10–11 mm Pale tan FW has speckled gray veins. Central PM area has a dusky spot. **HOSTS** Unknown.

THIN-LINED PRICKLYPEAR BORER

REDDISH CHOLLA BORER

AMERICAN
SUNFLOWER MOTH

CAHELA MOTH

male

female

HOARY PHYCITODES

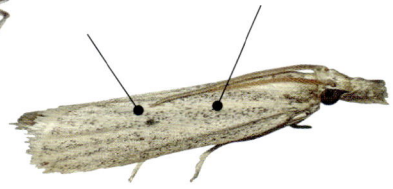

actual size

WESTERN COENOCHROA

PYRAUSTINE MOTHS
SUPERFAMILY Pyraloidea (80a)
FAMILY Crambidae SUBFAMILY Pyraustinae

Small moths that usually rest with their broad wings held flat over the abdomen in a triangular shape; a few genera, notably the *Anania*, typically rest with wings spread. The head is small with large eyes and long palps that give a snouted appearance. While most pyraustine species are shades of tan or brown, many *Pyrausta* are brightly colored with pink and yellow. Adults are nocturnal and come to lights, but some species are also commonly encountered in low vegetation during daytime.

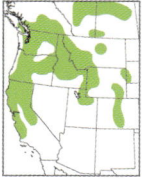

DUSKY SAUCROBOTYS
Saucrobotys fumoferalis   Uncommon   RANGE-WIDE
80a-0695 (4935)

TL 12–17 mm   Hoary brown FW has wavy brown AM line and deeply scalloped brown PM line edged with pale tan. Dusky reniform crescent has faint tan wash beneath. Dark brown terminal line has tan dots at veins. **HOSTS** Hickory; possibly spreading dogbane.

DOGBANE SAUCROBOTYS
Saucrobotys futilalis   Uncommon   RANGE-WIDE
80a-0696 (4936)

TL 13–14 mm   Tan, brown, or rusty FW is often peppery. Thin, brown, wavy AM and deeply scalloped PM are sometimes faint or indistinct; PM is often edged below with tan. Indistinct brown reniform spot is crescent-shaped; orbicular spot is usually absent. Terminal line usually has pale dots at veins. **HOSTS** Dogbane and milkweed.

TITIAN PEALE'S PYRALID
Perispasta caeculalis   Uncommon   RANGE-WIDE
80a-0715 (4951)

TL 7–10 mm   Sexually dimorphic. Dark brown FW has thin, blackish, curving AM and bulging PM lines. Costa is convex. Subapical outer margin is concave with white fringe, and apex is slightly pointed. Male has translucent whitish patch in outer median area; female does not. **HOSTS** Unknown.

SMALL MAGPIE
Anania hortulata   [IN] Common   RANGE-WIDE
80a-0716 (4952)

WS 24–28 mm   White FW has black costa and outer margin, spotted black PM and terminal bands, and large dense spots in basal area. Head and thorax are orange, with black spots dorsally. **HOSTS** Nettle; also bindweed and mint.

CROWNED ANANIA
Anania tertialis   Uncommon   RANGE-WIDE
80a-0717 (4953)

WS 22–26 mm   Tan FW is heavily hoary with dark brown and has large tan spots at central median area between indistinct orbicular and reniform spots, and at central PM area, where dark brown zigzag PM line bulges. PM line widens into dark brown patch at costa and is bordered below with zigzag tan line that broadens into tan patch at costa. HW has two tan patches separated by thin, dark brown, zigzag median line. **HOSTS** Deciduous trees and plants, including alder, elderberry, hickory, and viburnum.

DUSKY SAUCROBOTYS

actual size

DOGBANE SAUCROBOTYS

TITIAN PEALE'S PYRALID

SMALL MAGPIE

CROWNED ANANIA

BROWN ANANIA
Rare

Anania mysippusalis   80a-0723 (4957)

WS 25–30 mm Tan FW is shaded with peppery brown. Wavy brown PM line bulges widely in costal half, meeting costa at right angle and indistinctly edged below with tan. Straightish brown AM line and short brown reniform dash are usually indistinct. ST/terminal area is shaded slightly darker. Apex is pointed. **HOSTS** Possibly aster.

WHITE-SPOTTED SABLE
Uncommon

Anania funebris   80a-0724 (4958)

WS 18–22 mm Black FW and HW have large white spots in AM and PM areas; FW also has tiny white dot in outer AM area. Fringe is white when fresh. Black thorax has broad orange scale tufts laterally. Black abdomen has white rings. **HOSTS** Goldenrod.

GOLDEN ANANIA
Rare

Anania labeculalis   80a-0725 (4959)

WS 17–20 mm Tan FW has crisp, wavy, dark brown, straightish AM line and bulging PM line that makes a right-angle turn near midpoint, where it almost touches dark brown reniform dash. Diffuse ST and thin terminal lines are dark brown. Central FW often lightly peppered with dark brown scales. Apex is pointed. **HOSTS** Unknown.

PERSISTENT HAHNCAPPSIA
Uncommon

Hahncappsia pergilvalis   80a-0735 (4968)

WS 18–28 mm Pale yellowish-tan FW has wavy, warm brown AM and PM lines; PM line bulges in outer half, making a right-angle turn in central wing but not approaching warm brown reniform smudge. Terminal line is indistinct. Apex is pointed. **HOSTS** Unknown; possibly corn.

GARDEN WEBWORM
Very Common

Achyra rantalis   80a-0742 (4975)

TL 10–11 mm Variable. FW may be grayish brown, light brown, rusty brown, or tan, with well-defined or faded markings. Slightly darker median area is bounded by thick, dark brown AM line and zigzag, dark brown PM line narrowly edged outwardly with paler brown; AM line usually reaches only to midpoint of wing, sometimes with a disjunct, diffuse spot in outer half. Small, dark brown orbicular dot and rounded reniform spot have a lighter brown patch between. **HOSTS** Herbaceous plants and crops, including alfalfa, bean, corn, and strawberry.

WESTERN GARDEN WEBWORM
Uncommon

Achyra occidentalis   80a-0743 (4976)

TL 11–13 mm Very similar to Garden Webworm, but tan to brown or brownish-gray FW is usually sprinkled with reddish scales, particularly along veins, dark brown AM line reaches past midpoint of FW, and dark brown PM line has broader pale edging. Reniform spot is small or a short curving bar. Spring individuals are typically much darker than summer individuals. **HOSTS** Unknown; probably herbaceous plants and crops, possibly also grasses. **NOTE** Prone to wandering and periodically but regularly encountered outside its normal range.

BROWN ANANIA

WHITE-SPOTTED SABLE

GOLDEN ANANIA

actual size

PERSISTENT HAHNCAPPSIA

GARDEN
WEBWORM

WESTERN
GARDEN
WEBWORM

ARIZONA NEOHELVIBOTYS
Neohelvibotys arizonensis

Uncommon
80a-0745 (4978)

RANGE-WIDE

TL 10–13 mm Translucent yellow wings have brown, straight AM and curving PM lines; PM turns to meet inner margin perpendicularly. Brown reniform spot is a long crescent; orbicular dot is often indistinct. Costa and outer margin are solid golden brown. **HOSTS** Unknown.

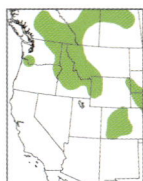

DIMORPHIC SITOCHROA
Sitochroa chortalis

Uncommon
80a-0754 (4987)

RANGE-WIDE

TL 13–15 mm Light tan to light brown FW has deeply scalloped, brown PM, ST, and terminal lines connected by brownish veins to create a lacy pattern in lower half of FW. Markings in upper FW usually indistinct, except for brown along costa and veins and a light peppering of black scales. Rare individuals are faintly or indistinctly marked. **HOSTS** Amaranth and pigweed; probably also other herbaceous plants. **NOTE** Commonly seen in meadows and pastures in daytime.

PINK-BORDERED XANTHOSTEGE
Xanthostege plana

Rare
80a-0757 (4990)

RANGE-WIDE

TL 7–10 mm Golden-yellow FW is unmarked except for pink terminal area and fringe. **HOSTS** Unknown.

GENISTA BROOM MOTH
Uresiphita reversalis

Very Common
80a-0759 (4992)

RANGE-WIDE

TL 15–18 mm Rusty to dark brown FW has small dusky orbicular and reniform spots, thin, zigzag, straightish AM line, and curving PM line that is a row of dusky dots; AM and PM lines are sometimes indistinct. Apex is pointed. HW and abdomen are yellow orange. **HOSTS** Pea family shrubs, including acacia, *Genista*, and Texas mountain laurel.

TAN-EDGED LOXOSTEGE
Loxostege albiceralis

Uncommon
80a-0761 (4993)

RANGE-WIDE

TL 14–19 mm Hoary gray FW has tan costal stripe that widens at median area, edged with black and shaded chestnut brown in PM area. Terminal area is tan. Inner basal area has small tuft of gray scales. Thorax is tan. **HOSTS** Christmas berry.

CHARMING LOXOSTEGE
Loxostege lepidalis

Uncommon
80a-0763 (4995)

RANGE-WIDE

TL 14–16 mm FW has hoary gray basal, median, and ST areas cut through by thin black AM and PM lines broadly edged with white, and vertical stripes of brown wash along costa, inner margin, and down central FW. Small, dark brown orbicular and reniform spots have a white patch between. Terminal area is tan. Veins in ST/terminal area are traced with brownish. Separable from similar Beet Webworm by white along AM and PM lines. **HOSTS** Greasewood.

ARIZONA
NEOHELVIBOTYS

DIMORPHIC
SITOCHROA

PINK-BORDERED
XANTHOSTEGE

GENISTA
BROOM MOTH

TAN-EDGED LOXOSTEGE

actual size

CHARMING LOXOSTEGE

WOLFBERRY LOXOSTEGE

Loxostege allectalis

Uncommon

80a-0768 (5000)

TL 10–13 mm Light brown to brown FW has hoary gray median area containing small brown to dark brown orbicular and reniform spots; spots have a tan patch between and dark brown shading below reniform spot. Thin brown PM line curves basally near inner margin, with a small medial-pointing bulge just inside from inner margin. Thicker brown AM line is usually indistinct. ST/terminal area shaded grayish. **HOSTS** Berlandier's wolfberry.

GRAY-BANDED LOXOSTEGE

Loxostege typhonalis

Rare

80a-0769 (5001)

TL 9–11 mm Resembles Wolfberry Loxostege, but FW is brown, orbicular and reniform spots are brown and often indistinct, with no pale spot between, AM and basal areas are brown and relatively unmarked, and ST/terminal area is brown. **HOSTS** Unknown.

NEARCTIC BEET WEBWORM

Loxostege munroealis

Uncommon

80a-0772 (5004)

TL 13 mm FW has hoary gray basal, inner median, and PM areas crossed by vertical brown stripes at costa and central FW. Brown orbicular and reniform spots have bold, squarish, tan patch between. Thin black PM line cuts through hoary gray PM area. ST area is washed brown, and terminal area is crisply tan. Some individuals bear this pattern of markings but in monotones of brown or brownish gray. Separable from Charming Loxostege by absence of white along AM/PM lines. **HOSTS** Herbaceous plants, including beet, flax, spinach, and wormwood.

ALFALFA WEBWORM

Loxostege cerealis

Very Common

80a-0785 (5017)

TL 10–13 mm Light brown to brownish-tan FW has small, dark brown orbicular and reniform spots with squarish tan patch between, and dark brown, strongly zigzag AM and PM lines that often connect in inner median area, sometimes creating brown-outlined circles. Costa, inner margin, and central vein are traced with brown. Basal area has strong blackish basal dash. Terminal area is tan. Rare individuals have indistinct AM and PM lines and median area concolorous with rest of FW; black basal dash is still evident. Separable from Nearctic Beet Webworm by zigzag AM/PM lines and basal dash. **HOSTS** Herbaceous plants and crops, including alfalfa. **NOTE** Often encountered during daytime.

FULVOUS-EDGED PYRAUSTA

Pyrausta nexalis

Uncommon

80a-0787 (5019)

TL 7–10 mm Hoary brown FW has an orange costa, thin, white, curving PM line, and branching pattern of thin white lines in central and inner median area. **HOSTS** Unknown.

WOLFBERRY
LOXOSTEGE

GRAY-BANDED
LOXOSTEGE

NEARCTIC BEET
WEBWORM

actual size

ALFALFA WEBWORM

FULVOUS-EDGED
PYRAUSTA

FROSTED PYRAUSTA

Uncommon

80a-0791 (5023)

RANGE-WIDE

Pyrausta napaealis

TL 9–11 mm Brown FW has very thin, black-edged, white PM line bordering a hoary ST/terminal area. Inner AM line and adjacent inner margin has short, hoary, white patch. Dusky orbicular and reniform dots are usually indistinct. Some individuals have a lighter brown outer and blackish inner half of median area. **HOSTS** Unknown.

LETHAL PYRAUSTA

Uncommon

80a-0795 (5027)

RANGE-WIDE

Pyrausta lethalis

TL 8–10 mm Rusty-brown FW has dark-edged tan PM line bordering hoary ST/terminal area. Sometimes has indistinct pale AM line. **HOSTS** Unknown.

VOLUPIAL PYRAUSTA

Very Common

80a-0797 (5029)

NORTH

SOUTH

Pyrausta volupialis

TL 9–11 mm Pink FW has bold, ivory, zigzag AM, smoothly curving PM lines, and small ivory orbicular spot. Fringe and thorax are tan. **HOSTS** Rosemary and other members of mint family.

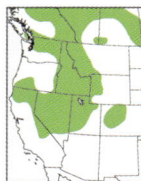

CHESTNUT PYRAUSTA

Uncommon

80a-0800 (5032)

RANGE-WIDE

Pyrausta nicalis

TL 10–12 mm Dark reddish-brown FW is lightly hoary with tan scales and sometimes shaded with dark brown in inner half of median and ST areas. Curving ivory PM line widens at costa and inner margin and is often fragmented along middle portion. Orbicular and reniform spots are indistinct dusky blotches. **HOSTS** Unknown; possibly mint.

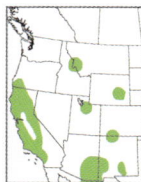

ELEGANT PYRAUSTA

Uncommon

80a-0801 (5033)

RANGE-WIDE

Pyrausta grotei

TL 10–12 mm Pink FW has thin ivory PM line that widens at costa and is usually fractured into spots in middle portion. Sometimes has short ivory AM line at inner margin. Thorax is orange brown. **HOSTS** Unknown.

RASPBERRY PYRAUSTA

Uncommon

80a-0802 (5034)

RANGE-WIDE

Pyrausta signatalis

TL 8–11 mm Resembles Elegant Pyrausta, but AM and PM lines are bold, and median area has a large pale discal spot. Fringe and thorax are orangish. Separable from Volupial Pyrausta by PM line. **HOSTS** Beebalm; possibly also other mints.

FROSTED PYRAUSTA

LETHAL PYRAUSTA

actual size

VOLUPIAL PYRAUSTA

CHESTNUT PYRAUSTA

ELEGANT PYRAUSTA

RASPBERRY PYRAUSTA

INORNATE PYRAUSTA
Pyrausta inornatalis

Very Common
80a-0805 (5037)

RANGE-WIDE

TL 9–11 mm Rich pink FW is unmarked except for ivory fringe. Thorax is pink and head is tan. **HOSTS** Sage. **NOTE** Diurnal; often seen at flowers.

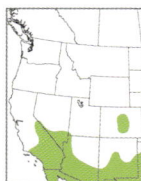

DESERT PYRAUSTA
Pyrausta pseudonythesalis

Uncommon
80a-0811 (5043)

RANGE-WIDE

TL 10–12 mm Orange FW has thin, wavy, dark pink AM and PM lines and reniform dash. Median area has light rosy wash, with an orange patch above reniform dash. ST area is orange, and terminal area is strongly reddish pink. **HOSTS** Unknown; possibly sage.

CALIFORNIA PYRAUSTA
Pyrausta californicalis

Very Common
80a-0820 (5052)

RANGE-WIDE

TL 8–10 mm Reddish-brown FW has thin, wavy, dark brown PM line set against yellowish-brown band. Indistinct dusky AM line is usually edged basally with subtle yellowish-brown shading. Orbicular and reniform spots are dusky, often indistinct. **HOSTS** Mints; also coyote mint.

HOARY PYRAUSTA
Pyrausta dapalis

Uncommon
80a-0822 (5054)

RANGE-WIDE

TL 8–10 mm Hoary brownish-gray FW has thick blackish AM and PM lines. PM line has sharp white triangle at costa and sometimes thin white edging along entire line. Basal and ST areas are brownish. Black orbicular spot is ringed with brown scales. HW is bright reddish pink with black central spot. **HOSTS** Sage. **NOTE** Diurnal.

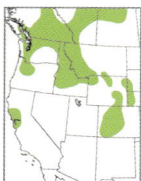

ORANGE MINT MOTH
Pyrausta orphisalis

Uncommon
80a-0826 (5058)

RANGE-WIDE

TL 8–10 mm Purple FW has orange basal area and is often hoary with blackish scales. Wavy orange PM line is often indistinct, curving around large orange patch in outer PM area. Orange orbicular spot is often indistinct. Blackish-purple HW has orange median band. **HOSTS** Mint.

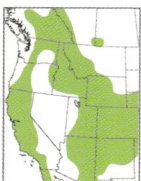

MOTTLED PYRAUSTA
Pyrausta subsequalis

Very Common
80a-0828 (5060)

NORTH

SOUTH

TL 9–15 mm FW appears orange heavily mottled with black. Broad orange AM and PM lines frame blackish median area containing dusky orbicular and reniform spots separated by orange patch. Basal area is mottled, ST area is blackish, and terminal area is orange. Orange HW has blackish AM and PM lines and fringe. **HOSTS** Reported on thistle. **NOTE** Primarily diurnal.

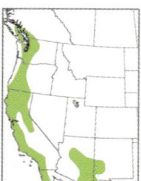

SHASTA PYRAUSTA
Pyrausta perrubralis

Uncommon
80a-0833 (5064)

NORTH

SOUTH

TL 10–14 mm Yellowish-orange FW has broad pink ST/terminal band and wavy, thin, pink AM line and orbicular dot. Large pink reniform spot is connected to terminal band. Basal half of costa is pink. Median area sometimes peppered with pink scales. **HOSTS** Coyote mint.

INORNATE PYRAUSTA

DESERT PYRAUSTA

CALIFORNIA PYRAUSTA

HOARY PYRAUSTA

ORANGE MINT MOTH

MOTTLED
PYRAUSTA

actual size

SHASTA PYRAUSTA

PINK-BANDED PYRAUSTA
Rare   RANGE-WIDE
Pyrausta scurralis   80a-0834 (5065)

TL 12–14 mm Yellow FW has broad pink ST band connected to short pink costal PM bar. Basal section of costa is pink. Pink orbicular dot is often indistinct. Separable from Shasta Pyrausta by AM and terminal areas. **HOSTS** Unknown.

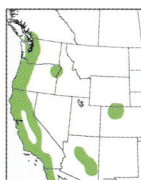

RUSTY PYRAUSTA
Uncommon   RANGE-WIDE
Pyrausta semirubralis   80a-0836 (5067)

TL 10–12 mm Pale yellowish-tan FW has reddish-brown reniform spots connected to reddish-brown ST/terminal area, and small reddish-brown orbicular dot. In some individuals, terminal area is yellowish tan, and brown of ST area may be reduced. **HOSTS** Unknown.

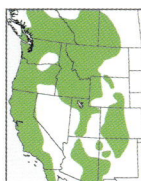

ONE-BANDED PYRAUSTA
Common   RANGE-WIDE
Pyrausta unifascialis   80a-0837 (5068)

TL 8–15 mm Tan FW is hoary with black, with hoary ivory PM and terminal lines and orbicular spot. Pale markings may be sometimes indistinct. **HOSTS** Generalists, including pussytoes, groundsmoke, and buckwheat. **NOTE** Females average 3 mm smaller.

COFFEE-LOVING PYRAUSTA
Uncommon   RANGE-WIDE
Pyrausta tyralis   80a-0838 (5069)

TL 7–8 mm Wine-purple FW has thick, wavy, yellow AM and PM lines; AM line does not reach costa. Median area has large yellow dot. Rare individuals have yellow markings strongly reduced or indistinct. Dark-morph individuals have dusky rose ground color and dark pink markings. **HOSTS** Wild coffee and probably other plants.

SOUTHERN PURPLE MINT MOTH
Very Common   CALIFORNIA
Pyrausta laticlavia   80a-0839 (5070)
INLAND

TL 8–12 mm Yellowish-orange FW has pink median area, with orange spot in outer half, and broad pink terminal band. Costa is pink. Rare individuals may entirely have a dusky wash. **HOSTS** Mint.

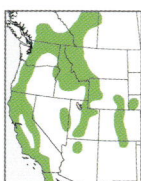

COYOTE-MINT PYRAUSTA
Uncommon   NORTH
Pyrausta fodinalis   80a-0843 (5074)
SOUTH

TL 11–16 mm Tan FW has broad, diffuse, brown ST band that curves to apex and thin brown PM line that makes a right turn near midpoint. Thin brown AM line, diffuse brown median band, and brown orbicular and reniform spots are sometimes indistinct. **HOSTS** Coyote mint and mountain coyote mint.

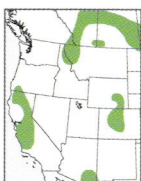

SOCIABLE PYRAUSTA
Rare   CALIFORNIA
Pyrausta socialis   80a-0844 (5075)
INLAND

TL 11–13 mm Resembles Coyote-mint Pyrausta, but FW is orangish and wing markings and ST band are reddish to maroon, the latter usually bleeding into terminal area. **HOSTS** Unknown; possibly mint.

actual size

PINK-BANDED PYRAUSTA

RUSTY PYRAUSTA

ONE-BANDED
PYRAUSTA

COFFEE-LOVING
PYRAUSTA

SOUTHERN
PURPLE MINT
MOTH

COYOTE-MINT
PYRAUSTA

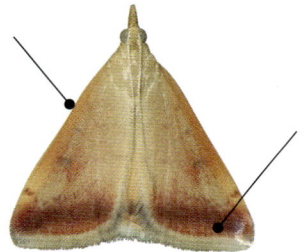

SOCIABLE PYRAUSTA

SPILOMELINE MOTHS SUPERFAMILY Pyraloidea (80a)
FAMILY Crambidae SUBFAMILY Spilomelinae

Previously considered a tribe within the subfamily Pyraustinae, these small, broad-winged, large-eyed, and long-palped moths are similar in appearance but typically rest with wings partly or entirely spread open. Many have particularly long legs that stick out well beyond the wing edges. The *Lineodes* and a few other species are unusually shaped for this group, with long narrow wings and abdomen held curled above the thorax. Adults regularly come to lights at night but may also be flushed from low vegetation during daytime.

LEAD-MARKED CHORISTOSTIGMA
Uncommon   RANGE-WIDE
Choristostigma plumbosignalis   80a-0851 (5128)

TL 10–12 mm Yellow FW has silvery-brown reniform spot connected to silvery-brown ST band, edged by brown lines. Brown ST line angles to apex. Thin brown AM and median lines are scalloped and nearly connecting to form a chained appearance in inner wing. Round orbicular spot is silvery brown with brown outline. Basal half of costa is brown. **HOSTS** Unknown.

ZEPHYR CHORISTOSTIGMA
Rare   RANGE-WIDE
Choristostigma zephyralis   80a-0852 (5129)

TL 11–13 mm Yellow FW has brown costa often touching silvery-brown orbicular and reniform spots. Diffuse silvery-brown ST band has large blotch in inner half; band does not usually touch reniform spot. Some individuals are entirely silvery brown below PM line. **HOSTS** Unknown; possibly coyote mint.

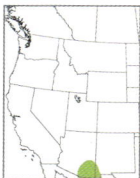

PINK-BORDERED CHORISTOSTIGMA
Uncommon   RANGE-WIDE
Choristostigma roseopennalis   80a-0853 (5130)

TL 10–12 mm Yellow FW has bright pink PM and terminal bands that connect at inner margin. Pink costa has small bulge at AM line. **HOSTS** Unknown.

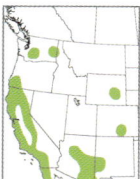

ELEGANT CHORISTOSTIGMA
Rare   RANGE-WIDE
Choristostigma elegantalis   80a-0855 (5132)

TL 10–12 mm Yellow FW has brown-outlined orbicular and reniform spots shaded lightly grayish. Reniform spot connects to brown PM line. ST/terminal area, and sometimes median area, are lightly peppered brown; ST area often has grayish shading. Thin, wavy, brown AM and median lines are curving. **HOSTS** Unknown.

ZEBRA CONCHYLODES
Rare   RANGE-WIDE
Conchylodes ovulalis   80a-0871 (5292)

WS 23–30 mm White wings have crisp black orbicular and reniform rings, curving black PM line, and black basal, AM, and ST lines. White abdomen has black stripes in basal half and large orange dorsal patch adjacent to black tip. **HOSTS** Aster.

EIGHT-BARRED LYGROPIA
Common   RANGE-WIDE
Conchylodes octonalis   80a-0873 (5251)

TL 8–10 mm White FW has large, black-edged, orange spots along costa at basal, AM, PM, and ST lines, usually connected to thin, sometimes indistinct, dark brown transverse lines. **HOSTS** *Heliotropium curassavicum*.

LEAD-MARKED
CHORISTOSTIGMA

ZEPHYR CHORISTOSTIGMA

PINK-BORDERED
CHORISTOSTIGMA

ELEGANT
CHORISTOSTIGMA

ZEBRA CONCHYLODES

actual size

EIGHT-BARRED LYGROPIA

CELERY LEAFTIER
Udea rubigalis

Very Common
80a-0877 (5079)

RANGE-WIDE

TL 7–10 mm Orange-brown to brownish-tan FW has brown-outlined, round orbicular and figure-eight reniform spots and thin brown AM and PM lines; PM line has a squarish, medial-pointing bulge near inner margin and bends slightly to meet costa. Some intermediate individuals may not be reliably separated from False Celery Leaftier visually. HOSTS Herbaceous plants and crops, including bean, beet, celery, and spinach.

FALSE CELERY LEAFTIER
Udea profundalis

Common
80a-0878 (5080)

NORTH

SOUTH

TL 9–13 mm Strongly resembles smaller Celery Leaftier, but reniform spot is not fully pinched closed at center and PM line is more strongly angled where it bends to meet costa. Some intermediate individuals may not be reliably identified to species visually. HOSTS Generalists on herbaceous plants.

WASHINGTON UDEA
Udea washingtonalis

Uncommon
80a-0879 (5081)

RANGE-WIDE

TL 9–14 mm Pale brown FW has brown orbicular and reniform spots outlined in dark brown; reniform spot connects to costa. Thin, dark brown AM and PM lines are sometimes indistinct. Inner median area and apex are washed brownish. Terminal line is a row of dark brown dots that continues up lower costa. HOSTS Unknown.

EIGHT-MARKED UDEA
Udea octosignalis

Rare
80a-0880 (5082)

RANGE-WIDE

TL 10–12 mm Light brown FW has rounded, brown-outlined orbicular and reniform spots and thin brown AM and PM lines; PM line makes a right-angle curve to inner margin and does not bend to meet costa. HOSTS Unknown.

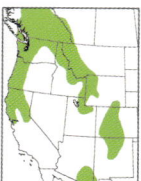

GRAY-EYED UDEA COMPLEX
Udea turmalis/abstrusa

Uncommon
80a-0896/8 (5098/5100)

RANGE-WIDE

TL 12–15 mm Peppered, light brown FW has dark gray to brown orbicular and reniform spots outlined in dark brown and with a short, dark brown line at center. Dark brown AM line is indistinct. PM line is dark brown triangles edged below and above with pale tan, with a large dark patch at costa. Terminal line is dark brown dots that continue up lower costa. *U. turmalis* and *U. abstrusa* have much overlap in characteristics, but *turmalis* typically has a pale wash in PM area while in *abstrusa* the PM and ST areas are usually concolorous, and *abstrusa* has a dark brown costa while *turmalis* usually does not. Similar Lucerne Moth has distinctly separate orbicular and claviform spots. HOSTS Thistle.

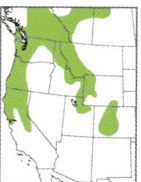

SHADED UDEA
Udea itysalis

Rare
80a-0897 (5099)

RANGE-WIDE

TL 11–13 mm Resembles Gray-eyed Udea, but PM line dots are rounded and darker ST/terminal area is interrupted by a pale streak adjacent to apex (sometimes indistinct). AM line is usually absent. HOSTS Bluebells, stickseed, and hound's-tongue. NOTE Phenotypes given here for Shaded and Gray-eyed Udea are based on characteristics of DNA-barcoded individuals that fall into separate barcode groups; however, the barcode groups are weakly defined, and there is overlap in genital morphology of these species, so further study of this genus may be required.

CELERY LEAFTIER

FALSE CELERY
LEAFTIER

WASHINGTON UDEA

EIGHT-MARKED UDEA

U. abstrusa

GRAY-EYED
UDEA COMPLEX

U. turmalis

actual size

SHADED
UDEA

EGGPLANT LINEODES

Lineodes integra

Common

80a-0904 (5107)

TL 9–10 mm Long, narrow, grayish-brown FW has broad brown band that curves S-shaped from inner margin of median area to subapical costa; lower half of band is narrowly edged in white. Costal ST area is brown, with a blackish patch adjacent to band. Rests propped up on long legs, with narrow abdomen usually held curled above wings, sometimes dramatically so. **HOSTS** Eggplant, ground cherry, pepper, and others in nightshade family.

INTERRUPTED LINEODES

Lineodes interrupta

Uncommon

80a-0905 (5108)

TL 9–10 mm Resembles Eggplant Lineodes, but curving band is broken into two sections: a straight median band and an angled crescent. Costal ST area often entirely dark brown. Separable from Erythrina Borer by width of lower band and black shading in outer ST area. **HOSTS** Unknown.

PIMA APILOCROCIS

Apilocrocis pimalis

Rare

80a-0912 (5113)

WS 20–30 mm Translucent tan wings have an iridescent purplish sheen. Median and ST/terminal areas have a purplish-gray wash. White orbicular and reniform patches are separated by dark brown square. Basal area is dark brown, and costa is pale tan. **HOSTS** Unknown.

WHITE-SPOTTED BROWN

Diastictis ventralis

Rare

80a-0917 (5255)

TL 12–15 mm Brown FW has groups of two or three white spots ringed in dark brown along median and PM lines; lines themselves usually absent. **HOSTS** Unknown.

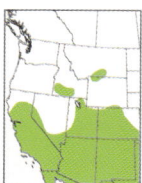

FRACTURED WESTERN SNOUT

Diastictis fracturalis

Very Common

80a-0918 (5256)

TL 11–16 mm Brownish-tan FW has bold, brown-edged white AM, median, and ST bands that do not touch costa or inner margin. Median band has tooth near inner margin, and ST band is fractured and offset into two sections. **HOSTS** Unknown. **NOTE** In similar *D. sperryorum* (not shown), the median band is broken into two spots.

SKY-POINTING MOTH

Agathodes monstralis

Uncommon

80a-0943 (5240)

TL 14–20 mm Olive FW has white-edged maroon median band that curves down central wing to touch outer margin. Inner ST/terminal area is rusty brown. Costa is edged with white. Abdomen is white at base, with a rusty band below. **HOSTS** Coral trees and ice-cream-bean.

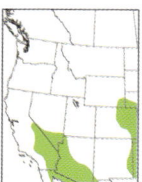

SESAME LEAFROLLER

Antigastra catalaunalis

Uncommon

80a-0944 (5181)

TL 11–13 mm Tan FW has orange veins, orange costa and outer margin, and thin orange transverse lines that are thicker at ends. **HOSTS** Yellow trumpet flower.

actual size

EGGPLANT
LINEODES

INTERRUPTED
LINEODES

PIMA APILOCROCIS

WHITE-SPOTTED BROWN

FRACTURED
WESTERN
SNOUT

SKY-POINTING MOTH

SESAME
LEAFROLLER

MELONWORM MOTH

Diaphania hyalinata

Rare

80a-0956 (5204)

RANGE-WIDE

WS 27–30 mm Satiny white FW has broad brown bands along outer margin and costa that connect across thorax. Tip of white abdomen has orange scale tuft. **HOSTS** Cucumber, melon, and squash.

SWAN PLANT FLOWER MOTH

Chabulina onychinalis

[IN] Common

80a-0966 (5199.1)

RANGE-WIDE

WS 14–16 mm Pale tan wings have thick-lined, dark brown, double basal, AM, median, and PM lines. Interior of double median line on FW has a thin brown dash near costa and brown spot near inner margin. Terminal area is a row of brown spots. Thorax is longitudinally striped with brown and tan, with a brown spot at posterior. **HOSTS** Balloonplant, jasmine, and oleander.

FOUR-SPOTTED PALPITA

Palpita quadristigmalis

Uncommon

80a-0978 (5218)

RANGE-WIDE

WS 30–34 mm Satiny, white, translucent FW has brown band along costa and four small blackish dots, three of which touch costal band. **HOSTS** Unknown

GRACILE PALPITA

Palpita atrisquamalis

Uncommon

80a-0981 (5220)

RANGE-WIDE

WS 22–26 mm Satiny, white, translucent FW has brown band along costa and dark brown speckled median and terminal bands. Dark brown spots at median, AM, and basal areas touch costal band. Abdomen has dark brown dorsal patches at middle. **HOSTS** Privet.

ERYTHRINA BORER

Terastia meticulosalis

Uncommon

80a-0996 (5239)

RANGE-WIDE

TL 17–24 mm Long, narrow, brown FW has broad brown median band, and slanting brownish patch in lower half of wing. Outer ST area is pale. Wavy HW costa usually projects beyond costa of FW. Separable from Interrupted Lineodes by broad lower band and lack of black shading in outer ST area. **HOSTS** Coral trees.

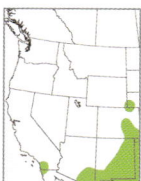

SOUTHERN BEET WEBWORM

Herpetogramma bipunctalis

Uncommon

80a-1019 (5272)

RANGE-WIDE

WS 22–24 mm Tan wings have a dark brown wash. Brown, wavy AM and irregular PM lines are broadly edged with tan; PM line has a large bulge near midpoint. Small orbicular and reniform dots are dark brown. Pointed abdomen has single pair of tiny brown dots on a middle segment. **HOSTS** Generalist on aquatic plants.

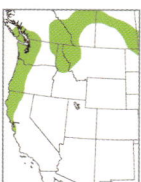

NORTHERN HERPETOGRAMMA

Herpetogramma aquilonalis

Very Common

80a-1025 (5277.1)

RANGE-WIDE

WS 23–27 mm Ivory wings have veins lightly traced with brown and scalloped brown PM and ST lines. Orbicular and reniform spots are thickly outlined with brown; pale center is often small dot. Terminal area is sometimes lightly washed brown. Paler individuals may resemble Southern Beet Webworm but are separable by hollow spots and lack of brown wash. **HOSTS** Generalist of herbaceous plants and low shrubs, including brambles, nettles, and violets.

MELONWORM
MOTH

SWAN PLANT
FLOWER MOTH

FOUR-SPOTTED PALPITA

GRACILE PALPITA

SOUTHERN BEET
WEBWORM

ERYTHRINA
BORER

actual size

NORTHERN
HERPETOGRAMMA

MAGICIAN MOTH

Hileithia magualis

Uncommon
80a-1030 (5187)

WS 15–17 mm Tan wings have a broad, curving, double PM line filled with white with brown edges, and dark brown AM line edged broadly with white basally. Orbicular and reniform spots are thickly outlined with dark brown. Midpoint of terminal area has brown patch. HOSTS Unknown.

SPOTTED BEET WEBWORM

Hymenia perspectalis

[IN] Uncommon
80a-1036 (5169)

TL 9–11 mm Resembles Hawaiian Beet Webworm, but median line on FW is fractured, PM line lacks white dots beside bold dash, and median band on HW is toothed. HOSTS Generalist on crops such as beets, chard, potatoes, amaranth, and others.

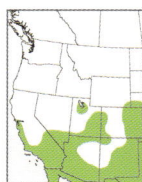

HAWAIIAN BEET WEBWORM

Spoladea recurvalis

[IN] Common
80a-1037 (5170)

TL 9–11 mm Dark brown FW has thin white AM line, thick white dash at outer PM line with three small white dots to inside, and bold white median line that visually connects to wide, smooth-edged, white median band on HW when wings held open at rest. Fringe is checkered. HOSTS Herbaceous plants, including beet, chard, and spinach.

SPIDERLING PSARA

Psara dryalis

Uncommon
80a-1045 (5269)

TL 13-15 mm Orange-brown wings have thin, brown, zigzag, curving AM and sinuous PM lines. Brown orbicular and reniform spots are usually hollow. Costa and terminal area of FW are shaded brownish, sometimes indistinctly. HOSTS Scarlet spiderling.

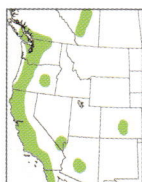

EUROPEAN PEPPER MOTH

Duponchelia fovealis

[IN] Uncommon
80a-1051 (5156.5)

TL 9–12 mm Dark brown FW has straight white AM and PM lines; PM line has rounded tooth near midpoint. Median area is hoary, with a large, dark brown spot toward costa. Male has a translucent patch basal to dark spot. HOSTS Generalist on herbaceous plants and crops.

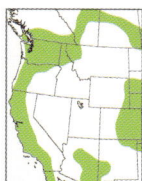

WHITE-HEADED GRAPE LEAFFOLDER

Desmia maculalis

Rare
80a-1065 (5160)

WS 20–25 mm Black FW has two roundish white spots and broadly checkered fringe. Black HW has broad white median band. Black abdomen has three spaced-out white marks on both dorsal and ventral sides. HOSTS Evening primrose, grape, and redbud.

HARLEQUIN WEBWORM

Diathrausta harlequinalis

Uncommon
80a-1083 (5175)

WS 13–18 mm Dark brown wings have thin orange PM line and squarish white spots in central wing. Basal area of FW has two orange bands. Fringe is checkered. HOSTS Unknown.

MAGICIAN MOTH

actual size

SPOTTED BEET
WEBWORM

HAWAIIAN BEET
WEBWORM

SPIDERLING PSARA

EUROPEAN PEPPER MOTH

WHITE-HEADED
GRAPE LEAFFOLDER

HARLEQUIN
WEBWORM

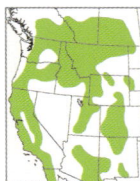

PALE-SPOTTED MECYNA

Mecyna mustelinalis

Uncommon
80a-1089 (5137)

NORTH

SOUTH

TL 13–17 mm Light brown FW has dark brown PM line of small chevrons, the outer- and innermost (sometimes all) edged below with pale yellowish. Round orbicular, claviform, and reniform spots are dusky brown ringed with dark brown, often with a small pale patch between. Wavy brown AM line is sometimes indistinct. **HOSTS** Unknown.

BROWN-SHADED MIMORISTA

Mimorista subcostalis

Common
80a-1092 (5139)

RANGE-WIDE

TL 16–20 mm Translucent yellow FW has faintly brown veins and brown terminal band and costa. Brown AM and PM lines are very slightly scalloped. Orbicular, claviform, and reniform spots are brown outlines, shaded brownish inside. **HOSTS** Unknown.

LUCERNE MOTH

Nomophila nearctica

Very Common
80a-1097 (5156)

NORTH

SOUTH

TL 14–18 mm Narrow brown FW has large, dark brown to dusky orbicular, claviform, and reniform spots outlined with black. Transverse lines are thin and extremely zigzagged, giving a streaky look to FW. **HOSTS** Herbaceous plants, including alfalfa, celery, clover, and smartweed. **NOTE** Frequently flushed from grass and low vegetation during daytime.

ASSORTED CRAMBIDS

SUPERFAMILY Pyraloidea (80a)

FAMILY Crambidae SUBFAMILIES Odontiinae and Glaphyriinae

Small moths with broad pointed forewings that are usually held slightly or strongly tented over the abdomen when at rest. Many Glaphyriinae also raise their abdomen and extend their forelegs in front of their head. As in other pyraloids, they have small heads and large eyes, but the palps are not always pronounced. Caterpillars of a few species can be crop pests, particularly of brassicas. Adults will come to lights; several species may also be encountered in low vegetation or at flowers during daytime.

BROWN BANTAM

Noctueliopsis brunnealis

Rare
80a-1109 (4830)

RANGE-WIDE

TL 9–12 mm Light brown FW has thin, dark brown, wavy AM and scalloped PM lines. ST area and lower half of basal area are darker brown, and central median area has a diffuse brown wash. Reniform spot is blackish. **HOSTS** Unknown.

ZIGZAG BANTAM

Noctueliopsis aridalis

Uncommon
80a-1113 (4834)

RANGE-WIDE

TL 6–7 mm Pale yellowish FW has brownish-red terminal band and basal area, the latter containing a large, pale yellowish patch. Costa and inner margin of PM area have large, diffuse, triangular, brownish-red patches. **HOSTS** Unknown. **NOTE** Often encountered at flowers.

actual size

PALE-SPOTTED MECYNA

BROWN-SHADED
MIMORISTA

LUCERNE MOTH

ZIGZAG
BANTAM

actual size

BROWN BANTAM

TAWNY BANTAM
Noctueliopsis bububattalis
Rare
80a-1116 (4837)
RANGE-WIDE

TL 7–9 mm Tan FW has white, straight median and zigzag PM lines thinly edged outwardly with dark brown, and white terminal line. PM and terminal areas are washed orange brown. Fringe is tan. HOSTS Unknown.

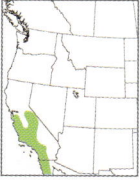

SUNSHINE BANTAM
Nannobotys commortalis
Rare
80a-1124 (4822)
RANGE-WIDE

TL 4–5 mm Tan FW has thin blackish median and ST lines broadly bordered inwardly with diffuse white. Reniform spot is paired black dots. Terminal line and fringe are white. HOSTS Unknown. NOTE Diurnal; often encountered at flowers, especially yellow ones.

PINK BANTAM
Mojavia achemonalis
Uncommon
80a-1131 (4839)
RANGE-WIDE

TL 6–7 mm Bright pink FW has large, pale yellow patch across median and ST areas, bordered by pink costal and terminal bands. Thorax is tan. HOSTS Unknown.

WHIP-MARKED SNOUT MOTH
Microtheoris vibicalis
Uncommon
80a-1132 (4795)
RANGE-WIDE

TL 5–7 mm Pale yellow FW has bold pink ST band and slanting pink band from inner AM to costal PM areas. Inner margin and fringe are pink. In some individuals, markings are brown, occasionally narrow or indistinct. HOSTS Unknown.

YELLOW-VEINED MOTH
Microtheoris ophionalis
Common
80a-1133 (4796)
NORTH

SOUTH

TL 6–7 mm Reddish-brown to brownish-tan FW has thick, pale yellow ST band, thin dusky PM line, and yellowish-traced veins in basal half. Blackish orbicular and reniform spots are sometimes indistinct. HOSTS Unknown.

SUNFLOWER BANTAM
Frechinia helianthiales
Uncommon
80a-1135 (4798)
RANGE-WIDE

TL 7–9 mm Pale FW has curving white ST line bordering dusky brownish ST area containing black-traced veins. Median area is brown, speckled with black scales. Diffuse white PM line forms a large white patch near costa. Thick white basal dash terminates in fork at AM area. HOSTS Sunflower.

TAWNY
BANTAM

SUNSHINE BANTAM

PINK BANTAM

WHIP-MARKED
SNOUT MOTH

SUNFLOWER
BANTAM

YELLOW-
VEINED MOTH

actual size

AMBER BANTAM
Frechinia laetalis

Rare
80a-1138 (4800)

TL 5–7 mm FW has curving tan terminal area edged by thin, black-edged, white ST line. Median and ST areas are mottled brown, with white patch at outer PM line extending as spikes along veins. Tan AM band is broadly bordered on both sides with white. Thorax is tan. **HOSTS** Flatspine bursage, burrobush, and western ragweed.

CLOAKED BANTAM
Gyros muirii

Uncommon
80a-1149 (4811)

TL 6–8 mm Dark brown to maroon FW has straight white median band and white ST line (sometimes indistinct) that does not reach inner margin. Central PM area has black spot. Some individuals have a sprinkling of white scales between median and ST lines. HW is orange. Fringe is black on all wings. Typically holds wings partly open. **HOSTS** Unknown. **NOTE** Diurnal; usually encountered at flowers or mudpuddling.

VENERABLE BANTAM
Anatralata versicolor

Rare
80a-1152 (4814)

TL 6–8 mm Hoary gray FW has brownish-red median and terminal bands; all areas may show more of black ground color when worn. Fringe is black basally, tipped with white when fresh. Thorax is hoary gray when fresh. **HOSTS** Narrowleaf mule-ears.

RUFOUS-BANDED CRAMBID
Mimoschinia rufofascialis

Very Common
80a-1156 (4826)

TL 7–12 mm White FW has rusty-brown, angled, straight AM and straight PM lines shaded broadly outwardly with diffuse rusty-brown bands; PM line has outward-pointing tooth near midpoint. Costa of median area has triangular rusty-brown patch. In some individuals, terminal area is also rusty brown, and costal patch bleeds into central median area, sometimes nearly entirely filling the space with brown. **HOSTS** Seeds of mallow.

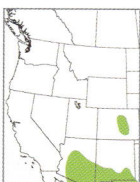

BELTED BANTAM
Jativa castanealis

Rare
80a-1158 (4827)

TL 8–10 mm Orange FW has narrow white median area edged thinly with dark brown and containing diffuse orange patch at costa. Orange thorax has white stripe laterally that creates a U shape with white basal area. **HOSTS** Unknown.

DESERT BANTAM
Pseudoschinia elautalis

Uncommon
80a-1159 (4828)

TL 11–14 mm Pale FW has wavy, dark brown AM and PM lines edged broadly in tan. Reniform spot is a small dark crescent. Median and terminal areas and thorax are mottled with tan. **HOSTS** Unknown. **NOTE** Often seen at flowers during daytime.

AMBER
BANTAM

CLOAKED BANTAM

VENERABLE BANTAM

RUFOUS-BANDED
CRAMBID

actual size

BELTED BANTAM

DESERT BANTAM

CABBAGE WEBWORM
Hellula rogatalis

Very Common
80a-1169 (4846)

TL 11–15 mm Unevenly tan FW has narrow, wavy, brown AM and PM lines edged with white, and a dusky kidney-shaped reniform spot outlined in white. Inner median area and occasionally outer ST area sometimes shaded dusky blackish. Terminal line a thin row of black dots. HOSTS Brassicaceae, including cabbage, kale, rape, and horseradish.

BRASSICA WEBWORM
Hellula aqualis

Rare
80a-1172 (4849)

TL 8–10 mm Resembles a pale Cabbage Webworm, but median area is usually washed white (especially basally), reniform spot is a narrow blackish bar or sometimes small dot, and only a few black dots of terminal line usually present. HOSTS Brassicaceae.

WHITE-TRIMMED BROWN PYRALID COMPLEX
Abegesta reluctalis/remellalis

Uncommon
80a-1188/9 (4866/7)

TL 8–10 mm Orange-brown FW has thick, brown-edged, white median line that tapers toward costa, then turns sharply basally to meet costal margin. Thin, brown-edged, white ST line curves basally near costa and meets costal margin in PM area. Terminal area is gray, with tiny black dots. Orange thorax has narrow white transverse stripes. Species *A. reluctalis* and *A. remellalis* cannot be separated by genitalia, and phenotypic differences are subtle and overlapping; they may represent forms of the same species, but further study is needed. At the phenotypic extremes, *reluctalis* is golden orange, and *remellalis* is brown, but intermediate golden-brown variations occur. Other suggested phenotypic markers, such as apical fringe or ST line, are inconclusive. HOSTS Unknown.

WESTERN STEGEA
Stegea salutalis

Uncommon
80a-1197 (4865)

TL 7–10 mm Light brown FW has thin, wavy, white AM and PM lines thinly edged inwardly with dark brown. Terminal line is black dashes. HOSTS Unknown.

THREE-SPOTTED KIDNEY MOTH
Nephrogramma separata

Uncommon
80a-1200 (4858)

TL 9–11 mm Peppery brown FW has thin white AM and PM lines thinly edged inwardly with dark brown. White subreniform spot is round; white reniform bar is usually broken into two spots. Median area is sometimes shaded darker brown. HOSTS Unknown.

CRESCENT-SPOTTED SCYBALISTODES
Scybalistodes periculosalis

Uncommon
80a-1201 (4853)

TL 9–11 mm Lightly peppery brown FW has thin, curving, dark brown AM and PM lines, with a small tooth near midpoint. Outer median area has strong white comma shape, bordered basally by an indistinct, dark brown spot. Apex is falcate. HOSTS Unknown.

CABBAGE WEBWORM

BRASSICA WEBWORM

WHITE-TRIMMED BROWN
PYRALID COMPLEX

actual size

WESTERN STEGEA

THREE-SPOTTED
KIDNEY MOTH

CRESCENT-SPOTTED SCYBALISTODES

FANCY LIPOCOSMA

Lipocosma albinibasalis

Rare

80a-1212 (4886)

RANGE-WIDE

TL 8–9 mm White FW has warm brown lower half, with thin, blackish, slightly scalloped PM line and dusky reniform spot outlined in black. Inner PM area sometimes has dusky shading. Terminal area is unevenly shaded white. Inner margin usually has brownish scale tufts at AM and PM areas. HOSTS Possibly lichens.

PEPPERY DICYMOLOMIA

Dicymolomia opuntialis

Uncommon

80a-1217 (4891)

RANGE-WIDE

TL 8–11 mm Resembles Dusky-patched Dicymolomia, but inner PM area has well-defined peppery gray patch, median line lacks white, and reniform bar and dusky shading are absent. White patch at anal angle of HW is completely filled with short black striations and has single row of black spots along margin. HOSTS Pricklypear.

DUSKY-PATCHED DICYMOLOMIA

Dicymolomia metalliferalis

Common

80a-1218 (4892)

NORTH

SOUTH

TL 8–11 mm Warm brown FW has peppery gray AM, median, and PM lines; inner half of median line has white center. PM area is washed peppery gray, with clean white reniform bar above dusky brown patch. Orange-brown basal area has peppery gray longitudinal stripes, and brown ST area has long dusky dashes along veins. Gray HW has white patch at anal angle, with band of long black striations and double row of black dots along margin. HOSTS Seed pods of bush lupine.

SOOTY-WINGED CHALCOELA

Chalcoela iphitalis

Common

80a-1221 (4895)

RANGE-WIDE

TL 10–12 mm Tan FW has large peppery gray patch in inner lower half, with curving line of small black dots along PM line. Tan stripe costal to gray patch has white PM and ST lines. Anal angle of peppery gray HW has double row of black spots along margin. HOSTS Larvae of paper wasps.

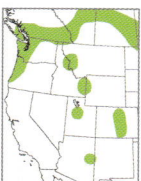

PURPLE-BACKED CABBAGEWORM

Evergestis pallidata

Rare

80a-1223 (4897)

RANGE-WIDE

TL 11–13 mm Pale yellow, lightly speckled FW has distinctive pattern of three triangles created by thin, brown, W-shaped AM line touching brown outline of hourglass-shaped reniform spot. Thin brown PM line is curving. Terminal area is shaded brown, with pale yellow spot at midpoint. HOSTS Brassicaceae, including bittercress, cabbage, and horseradish.

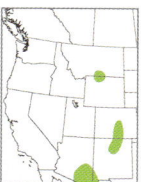

SINGED EVERGESTIS

Evergestis consimilis

Rare

80a-1225 (4899)

RANGE-WIDE

TL 15–16 mm Pale yellow FW has large, warm brown patch with dark brown center in central ST/terminal area, bleeding across PM line into central PM area. Thin, slightly scalloped AM and PM lines and outline of reniform spot are warm brown, sometimes indistinct. HOSTS Unknown.

FANCY LIPOCOSMA

PEPPERY DICYMOLOMIA

DUSKY-PATCHED
DICYMOLOMIA

actual size

SOOTY-WINGED CHALCOELA

PURPLE-BACKED
CABBAGEWORM

SINGED EVERGESTIS

SLEEPING EVERGESTIS
Evergestis lunulalis
Rare
80a-1228 (4902)

TL 10–12 mm Pale FW has light brown AM and PM lines and light brown bands in AM and apical areas. Angled inner PM line is black and thickened, crossing a large brown patch. **HOSTS** Unknown.

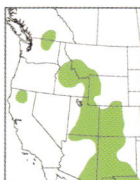

STREAKED EVERGESTIS
Evergestis simulatilis
Uncommon
80a-1230 (4904)

TL 13–15 mm Pale brown to grayish FW has brown shading in inner median and terminal areas and is dark brown streaked with black in ST area. Narrow brown median and PM lines are edged with white outwardly, widening into brown patches at costa; median line angles basally to meet inner margin and has a large tooth in costal half. Wavy ST line is white only. **HOSTS** Unknown.

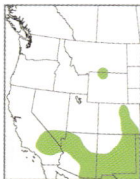

GRAY EVERGESTIS
Evergestis vinctalis
Uncommon
80a-1232 (4906)

TL 12–14 mm Resembles Streaked Evergestis, but ST area is brown and median and terminal areas are gray. Inner ST area usually has a small dusky patch. Median and PM lines lack brown patches at costa. Terminal line is edged with white. **HOSTS** Unknown.

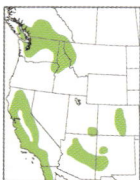

FROSTED EVERGESTIS
Evergestis funalis
Uncommon
80a-1235 (4909)

TL 13–15 mm Resembles Brown-patched Evergestis, but brown in ST area is reduced, and thin AM line is indistinct. PM line becomes indistinct toward costa. Basal area is sometimes diffusely shaded dusky brownish-gray. **HOSTS** Unknown.

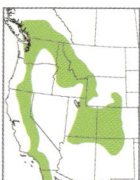

BROWN-PATCHED EVERGESTIS
Evergestis subterminalis
Uncommon
80a-1236 (4910)

TL 13–15 mm Hoary gray FW is mottled with black and has wide, warm brown ST band with black apical dashes and a black patch near inner margin. Jagged AM line has a blackish patch basally near costa. Thin black PM line is widened and broadly bordered by white at inner margin. **HOSTS** Unknown.

SLENDER EVERGESTIS
Evergestis obliqualis
Rare
80a-1238 (4912)

TL 14–19 mm Resembles Frosted Evergestis, but thin, black PM line is straight to costa, brown in ST area is narrower, and shading in ST area bleeds to outer margin in central FW. Median line is indistinct to absent. In some individuals, the brown of ST area does not cross ST line, and ST line has large black apical dashes and blackish patches near inner margin. **HOSTS** Purslane.

ASSORTED CRAMBIDS

SLEEPING EVERGESTIS

STREAKED EVERGESTIS

actual size

GRAY
EVERGESTIS

FROSTED
EVERGESTIS

BROWN-PATCHED EVERGESTIS

SLENDER
EVERGESTIS

DONACAULAS SUPERFAMILY Pyraloidea (80a)
FAMILY Crambidae SUBFAMILY Schoenobiinae

Brown with long narrow wings and long palps, the donacaulas can be mistaken for grass-veneers. In most species, males have a rounded forewing and females a sharply pointed apex; a few species also have sexually dimorphic forewing patterns. Most North American species can be difficult to separate visually, as patterns are often similar across species and variable within species. Additionally, DNA barcoding suggests there may be many species yet to be identified and described; a 2010 unpublished thesis outlines a few of these, possible representatives of which have been included here as "n. sp." (new species) with temporary numbers. Adults are nocturnal and come to lights in small numbers.

DELIGHTFUL DONACAULA   Rare   RANGE-WIDE
Donacaula melinellus   **Rare** 80a-1284 (5316)
TL 15–20 mm Lightly speckled, light to grayish-brown FW is shaded slightly darker in costal half. Central PM area has a solid black dot, and inner basal, AM, and PM areas have smaller, indistinct, blackish dots. An indistinct dusky apical dash crosses diagonally into central ST area. Female sometimes has an indistinct brownish stripe from base to ST area. **HOSTS** Unknown.

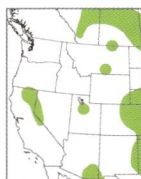

WANDERING DONACAULA   Rare   RANGE-WIDE
Donacaula dispersellus   80a-1285 (5316.1)
TL 14–18 mm Male resembles male Delightful Donacaula but lacks the indistinct dots in inner wing. Female resembles female Delightful and Dark-striped Donacaulas, but FW is golden brown with a dusky brown to grayish central stripe that is stronger than Delightful but weaker than Dark-striped. **HOSTS** Unknown.

WHITE-EDGED DONACAULA   Rare   RANGE-WIDE
Donacaula albicostellus   80a-1286 (5316.2)
TL 13–18 mm Light brown FW has an ivory stripe along costa, shaded lightly with brown toward apex and edged along inner side with darker brown shading. **HOSTS** Unknown.

BROWN-EDGED DONACAULA   Rare   RANGE-WIDE
Donacaula n. sp.   80a-128xA (n.sp.)
TL 14–17 mm Golden-tan FW has a broad, light brown stripe along costa, edged along inner side by a narrow line of white. **HOSTS** Unknown.

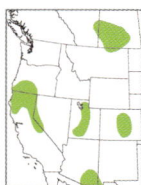

DARK-STRIPED DONACAULA   Rare   RANGE-WIDE
Donacaula n. sp.   80a-128xB (n.sp.)
TL 14–20 mm Golden-tan to light brown FW has a dark brown stripe through central wing from base to apex with an adjacent, small, blackish dot in PM area, a dark brown apical dash that crosses stripe diagonally in ST area, and usually two or three indistinct, dark brown dots in inner AM and median areas. **HOSTS** Unknown.

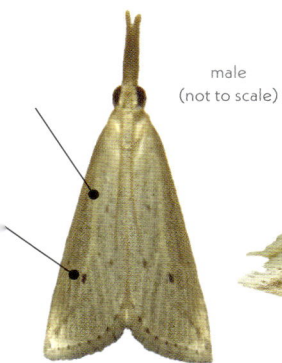

male
(not to scale)

female

DELIGHTFUL DONACAULA

female

WANDERING DONACAULA

male

WHITE-EDGED
DONACAULA

actual size

BROWN-EDGED
DONACAULA

DARK-STRIPED
DONACAULA

AQUATIC CRAMBIDS

SUPERFAMILY Pyraloidea (80a)
FAMILY Crambidae SUBFAMILY Acentropinae

Small moths with broad wings that are held either spread or in a triangular shape when at rest. Many species have bold, contrasting wing patterns. Caterpillars are aquatic, adapted to underwater life through presence of external gills or by trapping air in caddisfly-like cases. Adults will commonly come to lights at night, but many species can also be found among lily pads or water's-edge vegetation during daytime.

PONDSIDE CRAMBID
Elophila icciusalis

Uncommon
80a-1308 (4748)

RANGE-WIDE

TL 9–13 mm Yellow-orange FW has irregularly wavy, white cross-lines edged with dark brown. Round white patch in inner median area is outlined with a semicircle of dark brown. **HOSTS** Aquatic plants, including duckweed, pondweed, and eelgrass.

WATERLILY LEAFCUTTER
Elophila obliteralis

Uncommon
80a-1313 (4755)

RANGE-WIDE

TL 7–11 mm Dark brown FW has tan AM and PM bands, usually narrowly edged with white, and white reniform spot with dark brown center. Zigzag ST line is white. Inner median area sometimes has whitish patch. HW is dark brown, with orange patch near anal angle and indistinct whitish to tan dashes along inner margin. Separable from Burhead Leafcutter by HW and presence of white AM/PM lines. **HOSTS** Aquatic plants, including duckweed, pondweed, and water lily.

KEARFOTT'S PETROPHILA
Petrophila kearfottalis

Uncommon
80a-1337 (4773)

RANGE-WIDE

TL 8–12 mm Ivory FW has paired golden-brown median bands, peppery scales in PM area, and three golden-brown bars at apex separated by two long white darts. White HW has golden-brown median band, thin, dark brown, looping PM line, and a row of black spots along outer margin. **HOSTS** Algae on rocks in flowing streams.

JALISCO PETROPHILA
Petrophila jaliscalis

Very Common
80a-1341 (4775)

RANGE-WIDE

TL 8–12 mm Peppery, light gray FW has bronzy AM band bordered by white, and three long bronzy bars at apex separated by two long white darts. Peppery, light gray HW has a warm brown median band and white terminal band set with black spots with metallic blue centers. **HOSTS** Algae on rocks in slow-moving streams.

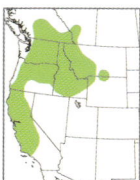

CONFUSING PETROPHILA
Petrophila confusalis

Very Common
80a-1346 (4780)

RANGE-WIDE

TL 6–12 mm Resembles Jalisco Petrophila, but markings are yellow brown. AM band is broader, bordered by white-filled, brown-lined, double median line. Basal of two white apical darts is thinner and edged with dark brown. Black spots on HW margin are backed by tan shading. **HOSTS** Algae on rocks in flowing streams.

SCHAEFFER'S PETROPHILA
Petrophila schaefferalis

Rare
80a-1350 (4784)

RANGE-WIDE

TL 11–15 mm Hoary gray FW has broad, dusky AM band and thin median line. Apex has thick blackish dart edged basally with thin gray line and second, much thinner black dart. Inner PM area has a dusky smudge, sometimes connected to outer margin. Basal area sometimes has black wash. **HOSTS** Unknown; aquatic.

WATERLILY
LEAFCUTTER

PONDSIDE CRAMBID

KEARFOTT'S
PETROPHILA

JALISCO
PETROPHILA

actual size

CONFUSING PETROPHILA

SCHAEFFER'S PETROPHILA

207

MOSS-EATING CRAMBIDS

SUPERFAMILY Pyraloidea (80a)
FAMILY Crambidae SUBFAMILY Scopariinae

Small gray to brownish moths with long narrow wings held flat over the body at rest. Most species have peppered forewings with black AM and PM lines and wing spots, and differences between species can be subtle. Caterpillars are unique among lepidoptera in feeding on mosses and clubmosses. Adults commonly come to lights at night.

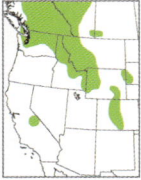

SMOKY GESNERIA
Gesneria centuriella

Uncommon
80a-1356 (4703)

RANGE-WIDE

TL 12–15 mm Hoary, dark gray to brownish-gray FW has diffuse whitish median patch and thin, wavy, black AM and PM lines edged outwardly with white shading. Round orbicular and figure-of-eight reniform spots are outlined in black; claviform spot is a black dot. **HOSTS** Unknown.

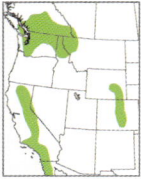

TRICOLORED COSIPARA
Cosipara tricoloralis

Rare
80a-1358 (4705)

RANGE-WIDE

TL 10–11 mm Resembles Smoky Gesneria, but AM line is straighter and reniform spot is washed with warm brown. **HOSTS** Unknown.

PALE SCOPARIA
Scoparia palloralis

Uncommon
80a-1364 (4711)

RANGE-WIDE

TL 6–10 mm Pale gray FW is lightly speckled. White AM and PM lines are often indistinct and sometimes narrowly edged with dusky gray. Reniform spot is marked by a dusky X shape. AM area has two weak dusky spots. **HOSTS** Unknown.

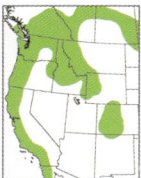

DOUBLE-STRIPED SCOPARIA
Scoparia biplagialis

Common
80a-1369 (4716)

RANGE-WIDE

TL 7–9 mm Speckly gray FW has widely spaced, pale gray, curving AM and V-shaped PM lines; AM line is shaded faintly brownish to grayish medially with two vertical black dashes. Faintly brownish to grayish reniform spot has a short black vertical dash at center. ST/terminal area is dusky with a white ST line. **HOSTS** Unknown.

MANY-SPOTTED SCOPARIA
Scoparia basalis

Uncommon
80a-1372 (4719)

RANGE-WIDE

TL 6–8 mm Resembles Double-striped Scoparia, but FW is brownish, black dashes in median area are shorter, PM line is more smoothly curving, and terminal area has a row of blackish spots. **HOSTS** Unknown.

SMOKY GESNERIA

TRICOLORED COSIPARA

PALE SCOPARIA

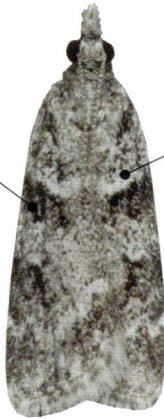

DOUBLE-STRIPED
SCOPARIA

MANY-SPOTTED
SCOPARIA

actual size

HOOK-LINED EUDONIA
Eudonia rectilinea

Uncommon
80a-1375 (4722)

NORTH

SOUTH

TL 9–11 mm Resembles Double-striped Scoparia, but FW is brownish gray, angled PM line is straightish and curves slightly basally to meet costa, and terminal area has a row of blackish spots. Dusky ST area often has indistinct blackish dashes. PM area is sometimes washed whitish. **HOSTS** Unknown.

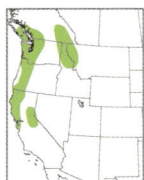

STRAIGHT-LINED EUDONIA
Eudonia commortalis

Uncommon
80a-1376 (4723)

NORTH

SOUTH

TL 8–11 mm Resembles Hook-lined Eudonia, but FW is grayish brown, PM line remains straight to meet costa, ST area is not significantly darker, and black dashes of ST area are more pronounced and often connect to those in terminal area. **HOSTS** Unknown.

FOUR-EYED EUDONIA
Eudonia torniplagalis

Uncommon
80a-1379 (4726)

RANGE-WIDE

TL 9–13 mm Hoary gray to brownish FW has white, strongly angled AM and curving PM lines with blackish to dark brown edging medially. Small orbicular and reniform spots have dark brown to blackish outline. PM area is often washed white. Dusky ST/terminal area has black dashes at veins. White ST line is often diffuse. **HOSTS** Unknown.

LONG-LINED EUDONIA
Eudonia spenceri

Uncommon
80a-1383 (4730)

NORTH

SOUTH

TL 9–10 mm Hoary gray FW has a thin black line vertically through central wing from AM to PM areas; lower end often appears to fork. Terminal area has a row of short black dashes, with longest at midpoint. Whitish V-shaped AM and PM lines are sometimes indistinct or nearly absent. Some individuals have a darker median area and dark triangular patch behind longest dashes in terminal area. **HOSTS** Unknown.

GOLD-LINED EUDONIA
Eudonia echo

Uncommon
80a-1388 (4735)

RANGE-WIDE

TL 8–10 mm Peppery gray FW has golden brown along veins. Wavy white AM and PM lines are edged with dusky shading; PM has a small bulge below reniform spot that touches (or nearly touches) wavy white ST line and curves to meet inner margin perpendicularly. Inner AM area has a medium-sized black spot. **HOSTS** Unknown.

HOOK-LINED EUDONIA

STRAIGHT-LINED
EUDONIA

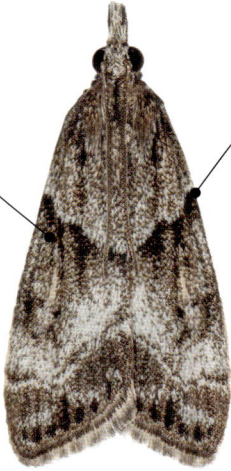

FOUR-EYED EUDONIA

LONG-LINED EUDONIA

actual size

GOLD-LINED
EUDONIA

GRASS-VENEERS

SUPERFAMILY Pyraloidea (80a)
FAMILY Crambidae SUBFAMILY Crambinae

Small narrow moths with long fuzzy palps that give them a snouted appearance. Most species are shades of tan or brown, many with white markings. The host plants of nearly all species in this subfamily are grasses or related species. Adults commonly come to lights at night but are also easily flushed from low vegetation in grassy meadows, urban lawns, and woodlands in daytime.

DIMORPHIC GRASS-VENEER
Hemiplatytes epia

Uncommon
80a-1393 (5507)

RANGE-WIDE

TL 7–11 mm Sexually dimorphic. Males have golden-tan FW with triangular white patch in central basal area and white median and lower PM areas that are sometimes broken into patches. PM area has peppery brown dashes between veins. Narrow ST area is peppery gray. Females have a similar pattern, but the white areas are very broad and the golden-tan areas pale, giving the appearance of a white moth with golden markings. **HOSTS** Grasses.

WHITE-VEINED GRASS-VENEER
Eufernaldia cadarellus

Rare
80a-1396 (5338)

RANGE-WIDE

TL 10–12 mm Golden-tan FW has broad white stripes along veins edged with peppery brown scales. Fringe is white with thin brown line. Often rests with wings rolled around abdomen. **HOSTS** Unknown.

SNOWY UROLA
Urola nivalis

Uncommon
80a-1456 (5464)

RANGE-WIDE

TL 10–12 mm Satiny white FW has golden-tan fringe with thin, scalloped, black terminal line and tiny black dot at midpoint of inner margin. Sides of head are golden tan. **HOSTS** Grasses.

BELTED GRASS-VENEER
Euchromius ocellea

[IN] Very Common
80a-1464 (5454)

NORTH

SOUTH

TL 10–13 mm Speckled tan to brownish-gray FW has two broad, straight to slightly curved, parallel, golden-tan median bands separated by a silvery stripe, and three angled golden-tan bars at apex. Inner terminal area has a series of bold black dots on a white background. A very thin brown line runs parallel to the edge of the speckled PM area; the distance between the thin line and black dots is no more than 1.5 times that between the line and speckled area. Some individuals with broader white may be best separated from California Grass-veneer by genitalia or DNA. **HOSTS** Grasses and cereal grains.

CALIFORNIA GRASS-VENEER
Euchromius californicalis

Uncommon
80a-1465 (5455)

NORTH

SOUTH

TL 12–14 mm Virtually identical to Belted Grass-veneer, but the distance between the black dots and thin brown line is twice or more than that between the thin line and speckled PM area. **HOSTS** Grasses. **NOTE** Generally much less common than the introduced Belted Grass-veneer, except in Canada, where Belted Grass-veneer is rarely found.

DIMORPHIC
GRASS-VENEER

WHITE-VEINED
GRASS-VENEER

SNOWY UROLA

actual size

BELTED GRASS-VENEER

CALIFORNIA GRASS-VENEER

HARLEQUIN MOUNTAIN MOTH
Diptychophora harlequinalis

Rare   80a-1460 (5458)

RANGE-WIDE

TL 5–7 mm Orange FW has thick black basal and AM lines and black double PM line that has long fingers pointing into inner and outer median area. Orange thorax has black collar. Some individuals have a partial, diffuse, black median band. **HOSTS** Unknown.

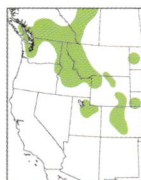

TWO-BANDED CATOPTRIA
Catoptria latiradiellus

Uncommon   80a-1469 (5408)

RANGE-WIDE

TL 13–15 mm Golden-brown FW has a broad white stripe down central wing, crossed by thick golden-brown median and PM lines. **HOSTS** Grasses.

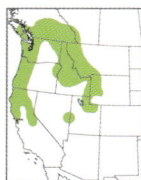

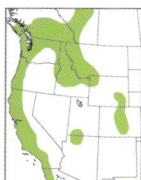

OREGON CATOPTRIA
Catoptria oregonicus

Uncommon   80a-1470 (5409)

RANGE-WIDE

TL 11–13 mm Brown FW has a broad white stripe from base to brown median line. Central and inner median line and brown PM area have black dashes at veins. Curving, thin, white PM line has black dots at veins. **HOSTS** Grasses.

WOOLLY GRASS-VENEER
Thaumatopsis pexellus

Uncommon   80a-1472 (5439)

RANGE-WIDE

TL 12–18 mm Tan FW has a thick white central vein bordered on inner side in upper half, and costal side on lower half, by narrow, dark brown patches. PM line is thick, dark brown scallops but is often indistinct. Some individuals have weak markings. **HOSTS** Grasses.

WESTERN LAWN MOTH
Tehama bonifatella

Rare   80a-1485 (5453)

NORTH

SOUTH

TL 11–15 mm Light tan FW has smudgy brown markings that make the moth appear worn even when fresh. Brown smudges are variable but are usually concentrated as a stripe in central wing, with a pale tan mark where central vein forks. Veins in ST/terminal area are pale tan. **HOSTS** Grasses.

PROFANE GRASS-VENEER
Fissicrambus profanellus

Rare   80a-1488 (5431)

RANGE-WIDE

TL 9–11 mm Resembles Intermediate Grass-veneer but has three or four smudgy gray spots in inner AM and PM areas. **HOSTS** Grasses. **NOTE** This is an eastern species that appears to have been recently introduced to southern CA and is becoming established there.

INTERMEDIATE GRASS-VENEER
Fissicrambus intermedius

Uncommon   80a-1489 (5432)

RANGE-WIDE

TL 9–11 mm Golden-brown FW has thick white stripe in central wing from base to median area, tapering and splitting into thin white veins to central outer margin. Lower half of wing is brownish gray on inner half, crossed by indistinct golden-brown median and PM lines. **HOSTS** Grasses. **NOTE** This is the only *Fissicrambus* species commonly found in the West. A few eastern species range as far west as CO and central TX. Four-spotted Grass-veneer (*F. quadrinotellus*, not shown) is also occasionally encountered in southern AZ.

HARLEQUIN MOUNTAIN
MOTH

actual size

TWO-BANDED CATOPTRIA

OREGON CATOPTRIA

WOOLLY
GRASS-VENEER

WESTERN LAWN MOTH

PROFANE
GRASS-VENEER

INTERMEDIATE
GRASS-VENEER

CHEVRON GRASS-VENEER
Microcrambus copelandi

Rare   80a-1495 (5418)   RANGE-WIDE

TL 7–9 mm Satiny white FW has large brown spots along inner margin at median and PM lines; lines are V-shaped and become golden brown in outer half. Thin terminal line is dark brown. White thorax has a large brown spot at posterior. **HOSTS** Grasses.

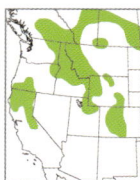

SOD WEBWORM
Pediasia trisecta

Common   80a-1522 (5413)   RANGE-WIDE

TL 14–18 mm Peppery grayish-brown FW is tan in outer basal/AM area, branching into tan veins in lower wing. Dark brown median and PM lines are strongly angled in inner half, curving deeply and turning basally to meet costa; in many individuals, the costal half of the line is indistinct, while in some, the entire line is indistinct but for dark spots in central median and PM areas. **HOSTS** Grasses.

SPECKLE-BACKED GRASS-VENEER
Pediasia dorsipunctellus

Uncommon   80a-1526 (5417)   RANGE-WIDE

TL 14–16 mm Golden-tan FW is peppered with dark brown scales in inner half and has pale tan veins. Dark tan PM line is sometimes visible across inner FW. **HOSTS** Grasses.

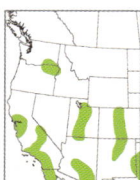

BLUEGRASS WEBWORM
Parapediasia teterrellus

Common   80a-1529 (5451)   RANGE-WIDE

TL 10–13 mm Dusky brown FW has warm brown, curving PM and V-shaped median lines; median is edged basally with blackish shading, sometimes broadly. Dusky terminal line has black dots at veins. Veins are traced with tan, usually faintly. **HOSTS** Grasses, including bluegrass and tall fescue.

STRAW GRASS-VENEER
Agriphila straminella

Rare   80a-1537 (5396)   RANGE-WIDE

TL 10–13 mm Golden-tan FW is slightly paler in inner half. Dark brown scales are arranged in narrow lines between veins. Terminal line is a row of black dots. Fringe is silvery. **HOSTS** Grasses.

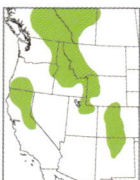

ORANGE-MARGINED GRASS-VENEER
Agriphila plumbifimbriellus

Uncommon   80a-1538 (5397)   RANGE-WIDE

TL 10–12 mm Resembles larger Vagabond Crambus, but FW terminal area is yellowish orange and bordered by thin brown ST line, always present but sometimes only visible at costa. Peppery brown scales are usually sparser than on Vagabond Crambus. **HOSTS** Grasses.

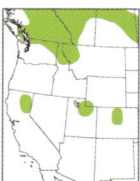

LESSER VAGABOND CRAMBUS
Agriphila ruricolellus

Uncommon   80a-1540 (5399)   RANGE-WIDE

TL 10–12 mm Resembles Orange-margined Grass-veneer, but terminal area is tan and FW has a diffuse brown median line, sometimes indistinct or only partially visible. **HOSTS** Grasses and common sheep sorrel.

CHEVRON GRASS-VENEER

SOD WEBWORM

SPECKLE-BACKED
GRASS-VENEER

BLUEGRASS WEBWORM

STRAW GRASS-VENEER

ORANGE-MARGINED
GRASS-VENEER

actual size

LESSER VAGABOND
CRAMBUS

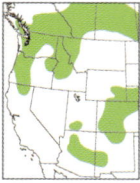

VAGABOND CRAMBUS

Common

RANGE-WIDE

Agriphila vulgivagellus　80a-1544 (5403)

TL 15–20 mm FW is heavily peppered with dark brown scales between light tan veins. Costa is sometimes golden tan, and terminal area may be peach orange. Terminal line is a row of blackish dots at veins. Fringe is dusky brown. Cross-lines are absent. **HOSTS** Grasses and cereal grains.

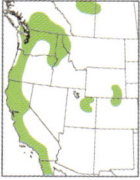

ATTENUATED GRASS-VENEER

Uncommon

NORTH

Agriphila attenuatus　80a-1545 (5404)

SOUTH

TL 13–17 mm Light tan FW has a pale streak from base to central median area. Entire wing is peppered with dark brown scales, densely along costal side of pale streak and very sparsely in center of pale streak; costa usually lacks dark scales. Dark brown PM and ST lines are usually visible only at costa. Terminal line is a row of black dots. Fringe is silvery. **HOSTS** Grasses.

TOPIARY GRASS-VENEER

Very Common

RANGE-WIDE

Chrysoteuchia topiarius　80a-1548 (5391)

TL 11–14 mm Variable. FW ranges from dark olive, to tan, to almost white. Golden-orange ST/terminal area is crossed by silvery V-shaped ST line thinly edged basally with dark brown. Dark brown shading between veins gives moth a streaky appearance. Thin black terminal line has black dots in inner half. **HOSTS** Grasses and herbaceous plants, including blueberry and cranberry.

INLAID GRASS-VENEER

Uncommon

RANGE-WIDE

Crambus pascuella　80a-1549 (5339)

TL 13–15 mm Golden-tan FW has a wide, brown-edged, white central band that comes to a point in PM area, with a rectangular white patch at inner side of point. PM area has gray veins edged with dark brown scales. Tan V-shaped ST line is double at costa. Terminal area has a white costal patch, with a tan triangle at center, and tan inner half with black dots at veins. **HOSTS** Grasses.

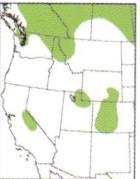

IMMACULATE GRASS-VENEER

Common

RANGE-WIDE

Crambus perlella　80a-1553 (5343)

TL 12–13 mm Satiny white to ivory FW is completely unmarked. Long fluffy palps have a tan stripe along each side. **HOSTS** Grasses.

WIDE-STRIPED GRASS-VENEER

Rare

RANGE-WIDE

Crambus unistriatellus　80a-1554 (5344)

TL 12–17 mm Golden-tan FW has a broad, brown-edged, white stripe from base to outer margin, flaring in terminal area. Inner terminal area has a row of small black dots. **HOSTS** Grasses.

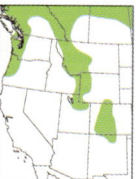

WHITMER'S GRASS-VENEER

Uncommon

RANGE-WIDE

Crambus whitmerellus　80a-1555 (5345)

TL 12–15 mm Resembles Inlaid Grass-veneer, but inner edge of white central stripe has a prominent tooth, costa has a narrow white stripe in basal half, and white apical patch has a thick, angled, tan apical dash. **HOSTS** Grasses.

VAGABOND CRAMBUS

ATTENUATED
GRASS-VENEER

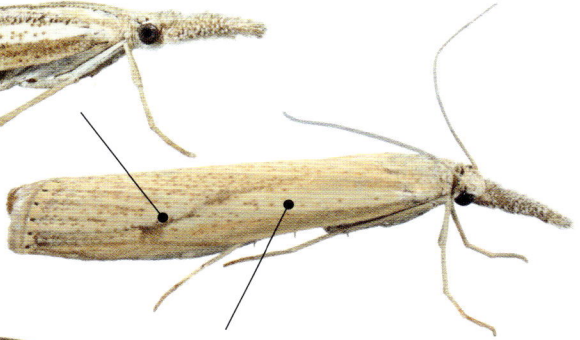

actual size

TOPIARY GRASS-VENEER

INLAID GRASS-VENEER

IMMACULATE GRASS-VENEER

WIDE-STRIPED
GRASS-VENEER

WHITMER'S GRASS-VENEER

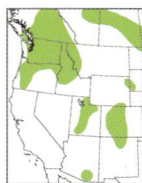

COMMON GRASS-VENEER
Crambus praefectellus

Uncommon
80a-1565 (5355)

TL 15–16 mm Resembles Leach's Grass-veneer, but white central stripe is narrower at base than at midpoint, sharply pointed without fragment, and bordered by a wide brown costa. Inner terminal area is gray to brown, with black dots edged with white. Separable from similar grass-veneers by brown costa and stripe that narrows at base. Separable from Brown-edged Grass-veneer by range and color of inner terminal area. **HOSTS** Grasses and cereal grains.

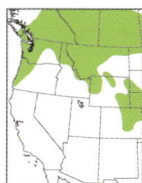

LEACH'S GRASS-VENEER
Crambus leachellus

Uncommon
80a-1567 (5357)

TL 15–16 mm Resembles the smaller Sperry's Grass-veneer, but white stripe usually has a small white fragment (sometimes fused) near point on costal side, and base of white stripe nearly touches costa. Area between point of stripe and ST line is paler tan. ST line is silvery, edged basally with brown, and double at costa. Inner terminal area is brownish gray, with thin black dashes at veins. **HOSTS** Grasses.

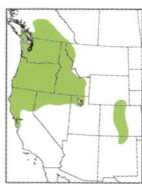

SHADED GRASS-VENEER
Crambus cypridalis

Rare
80a-1568 (5358)

TL 15–19 mm Golden-tan to dark reddish-brown FW has semimetallic grayish shading along inner half, particularly between tan veins in PM area. Broad white stripe has a long point, straightish on inner side and curving or angled on costal side. Area between point of stripe and ST line is pale tan. Separable from similar white-striped grass-veneers by gray shading along inner margin and shape of point of white stripe. **HOSTS** Grasses.

BROWN-EDGED GRASS-VENEER
Crambus rickseckerellus

Rare
80a-1570 (5360)

TL 13–15 mm Golden-brown FW has relatively narrow white stripe, often slightly narrower at base, that evenly tapers to a sharp point. Stripe is bordered by brown costa, slightly darker toward base. Area between point of stripe and ST line is pale tan to whitish. PM area has faint, semimetallic gray shading between veins. Separable from similar white-striped grass-veneers by brown costa, and from Common Grass-veneer by range, and area between point of stripe and ST line. **HOSTS** Grasses.

SPERRY'S GRASS-VENEER
Crambus sperryellus

Common
80a-1580 (5370)

TL 11–15 mm Shiny, dark golden-tan FW has broad, brown-edged, white stripe from base to PM area, tapering to a symmetrical blunt point. Stripe has a narrow strip of brown costa adjacent at base. PM area has soft, metallic gray along veins. Soft-edged whitish patch runs from point of main stripe to outer margin. Separable from similar white-striped grass-veneers by blunt tip to stripe, pale area crossing ST line, and narrow brown costa. **HOSTS** Grasses, including bluegrass.

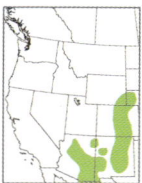

PALE-EDGED GRASS-VENEER
Crambus cyrilellus

Rare
80a-1583 (5373)

TL 12–14 mm Resembles Leach's Grass-veneer, but inner half of FW is shaded darker brown, with distinct silvery stripes in PM area and a pale wash along inner margin. Tip of white streak lacks white fragment. Inner terminal area is gray with white behind black dots nearest the costa. **HOSTS** Grasses.

COMMON GRASS-VENEER

LEACH'S GRASS-VENEER

SHADED GRASS-VENEER

actual size

BROWN-EDGED
GRASS-VENEER

SPERRY'S
GRASS-VENEER

PALE-EDGED
GRASS-VENEER

THYATIRIDS and HOOKTIPS

SUPERFAMILY **Drepanoidea** (85)
FAMILY **Drepanidae**

Two distinctly different subfamilies united by a shared ancestral body structure. The habrosynes and thyatirids are medium-sized, chunky moths that resemble noctuids, with strongly patterned forewings. The thyatirids have a tufted thoracic crest. The hooktips are slimmer moths more similar to geometrids, with broad wings that are strongly falcate at the apex. Adults of both groups will come to lights in small numbers.

LETTERED HABROSYNE

Habrosyne scripta

Common RANGE-WIDE

85-0003 (6235)

TL 20–21 mm Brownish-gray FW has squiggly pattern of brown-edged whitish lines in PM area. White AM line is shallowly wavy, flaring to white patch in outer basal area. Orbicular spot is oblong. **HOSTS** Birch, black raspberry, and purple-flowered raspberry. **NOTE** Glorious Habrosyne (*H. gloriosa*, not shown) is virtually identical, but white AM line has a right-angled bend at midpoint and orbicular spot is round; it is primarily northeastern, with a disjunct population in CO and NM.

TUFTED THYATIRID

Pseudothyatira cymatophoroides

Common RANGE-WIDE

85-0005 (6237)

TL 23–25 mm Gray to white FW has a contrasting black basal patch, dark triple AM line, and a dark patch at inner ST line. Basal area and outer ends of ST area are usually shaded brown. In form "expultrix," black markings are reduced or absent. **HOSTS** Generalist on deciduous shrubs and trees, particularly *Rubus*.

DOGWOOD THYATIRID

Euthyatira pudens

Uncommon RANGE-WIDE

85-0008 (6240)

TL 23–25 mm Gray FW has large pink patches in basal, costal, and apical areas. Tufted thorax has pinkish sides. Individuals of the uncommon form "pennsylvanicus" are mostly gray with ghostlike markings of the typical form. **HOSTS** Dogwoods.

BANDED THYATIRID

Euthyatira semicircularis

Uncommon RANGE-WIDE

85-0009 (6241)

TL 21–24 mm Light gray FW has a pale gray or pinkish basal area, multiple dark AM lines, and a paler patch at outer ST area. Tufted thorax is gray. **HOSTS** Unknown.

RIBBONED LUTESTRING

Ceranemota fasciata

Uncommon RANGE-WIDE

85-0011 (6243)

TL 18–22 mm Gray FW has double, thin, black AM and PM lines bordered by brown. Basal area is pale. Gray thorax is tufted. **HOSTS** Serviceberry and cherry.

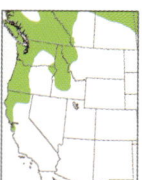

ARCHED HOOKTIP

Drepana arcuata

Common RANGE-WIDE

85-0019 (6251)

TL 15–21 mm Pale orange FW has a bold rusty PM line that curves toward falcate apex. Reniform area is marked with two black, sometimes hollow, spots. **HOSTS** Red and green alder.

222   THYATIRIDS AND HOOKTIPS

form "expultrix"

LETTERED HABROSYNE

TUFTED THYATIRID

form "pennsylvanicus"

DOGWOOD THYATIRID

RIBBONED LUTESTRING

BANDED THYATIRID

ARCHED HOOKTIP

actual size

223

TWO-LINED HOOKTIP
Falcaria bilineata

Uncommon
85-0020 (6252)

RANGE-WIDE

TL 16–21 mm Pale orange FW is marked with slanting, parallel AM and PM lines. Outer margin is irregularly wavy and apex is falcate. Spring brood brindled with brown; summer brood more uniform. Single reniform dot. **HOSTS** Alder, birch, and elm.

TENT CATERPILLAR and LAPPET MOTHS, and TOLYPES

SUPERFAMILY Lasiocampoidea (87)
FAMILY Lasiocampidae

Medium-sized, broad-winged moths with chunky, densely hairy bodies. This family includes the common and familiar tent caterpillar moths, which can be quite abundant in outbreak years. Lappet moths typically rest with the scalloped hindwings projecting from beneath the folded forewings. Tolypes have distinctive blue-black thoracic stripes and rest with forelegs stretched out in front. Adults will come to lights at night.

AMERICAN LAPPET MOTH
Phyllodesma americana

Very Common
87-0003 (7687)

NORTH

SOUTH

TL 15–22 mm Variable. Rusty to reddish-brown (sometimes grayish-brown) FW is usually washed grayish in ST/terminal area. AM, PM, and ST lines are blackish scallops between rusty to brown veins. Reniform spot is marked with a blackish crescent. Scalloped HW typically projects from below costa of FW when wings are folded at rest, a unique characteristic. **HOSTS** Alder, birch, oak, poplar, and rose.

CROWNED LAPPET
Dicogaster coronada

Uncommon
87-0008 (7692)

RANGE-WIDE

TL 33–50 mm Rounded FW is uniformly warm brown with pale speckling. Evenly curving white AM and PM lines and small discal spot are edged with darker brown. Females are larger than males. **HOSTS** Emory oak.

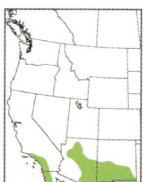

SOOTY LAPPET
Gloveria gargamelle

Uncommon
87-0011 (7695)

RANGE-WIDE

TL 30–45 mm Resembles Arizona Lappet, but FW is more brownish and lacks hoary white shading in PM area. PM line is straight. Females are larger and darker than males. **HOSTS** Oaks; also Apache plume. **NOTE** Males often fly during the afternoon; females are nocturnal.

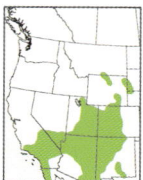

ARIZONA LAPPET
Gloveria arizonensis

Common
87-0012 (7696)

RANGE-WIDE

TL 27–42 mm Hoary gray FW has a prominent white discal spot. Curved AM and jagged PM lines are dark. Jagged ST line is filled basally with hoary white. Females are larger than males. **HOSTS** Juniper, cypress, and pine.

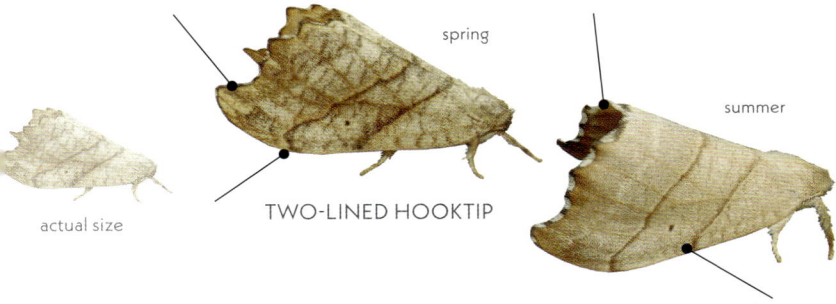

spring

summer

actual size

TWO-LINED HOOKTIP

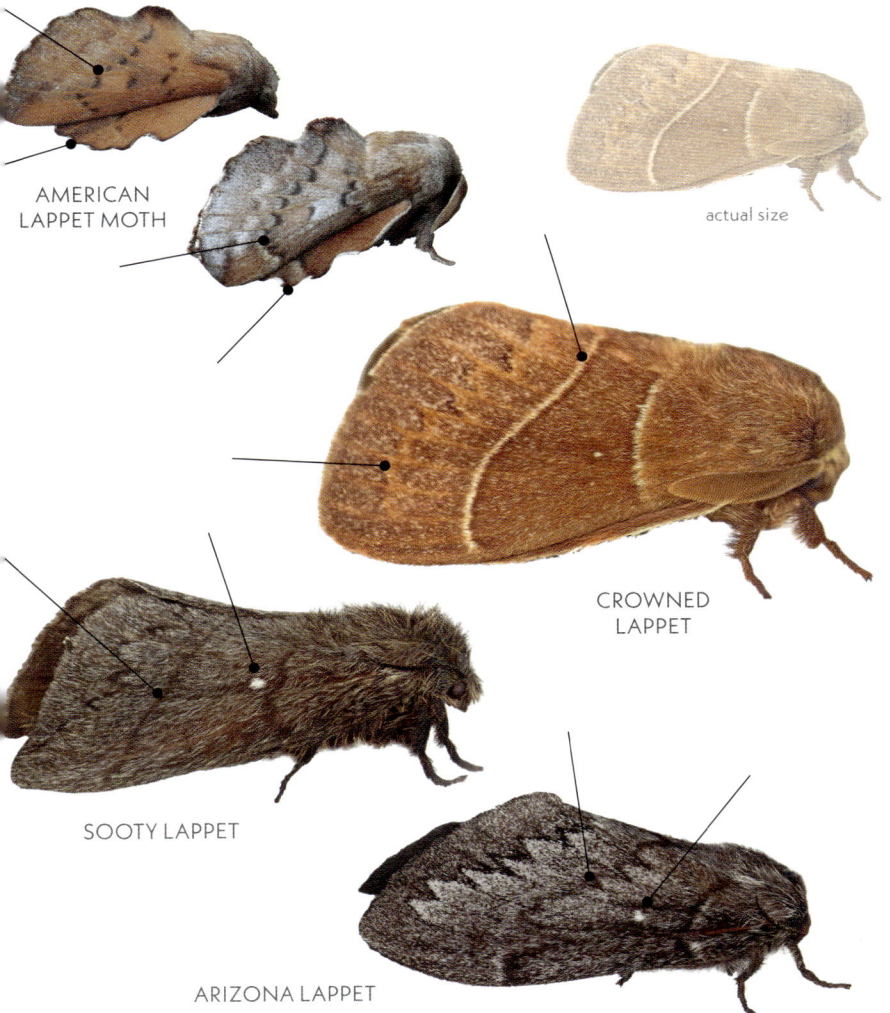

AMERICAN
LAPPET MOTH

actual size

CROWNED
LAPPET

SOOTY LAPPET

ARIZONA LAPPET

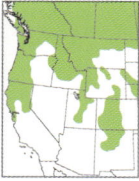

FOREST TENT CATERPILLAR MOTH
Malacosoma disstria

Very Common
87-0014 (7698)

TL 17–21 mm Light golden-brown FW has a slightly (sometimes much) darker median area. Thick AM and PM lines are dark brown; AM line kinks upward near costa. Females are usually larger than males. HOSTS Alder, aspen, basswood, birch, cherry, maple, oak, and other deciduous trees.

PACIFIC TENT CATERPILLAR MOTH
Malacosoma constricta

Very Common
87-0015 (7699)

TL 15–19 mm Resembles Forest Tent, but AM line is straightish and PM line curves to meet costa perpendicularly. Median area is uniformly colored with the rest of the FW. Sexually dimorphic; males are usually a lighter, more yellow brown, occasionally quite pale, while females are a darker reddish brown. HOSTS Garry oak and other oaks.

SONORAN TENT CATERPILLAR MOTH
Malacosoma tigris

Uncommon
87-0016 (7700)

TL 15–20 mm Resembles Forest and Pacific Tents, but AM and PM lines are thinner and meet costa relatively perpendicularly or form a slight apical angle. Sexes are dimorphic; male is pale with concolorous median area, and female is reddish brown with darker median area. HOSTS Shrubby oaks.

EASTERN TENT CATERPILLAR MOTH
Malacosoma americana

Rare
87-0017 (7701)

TL 15–24 mm Resembles Western and Southwestern Tents, but straight, toothless AM and PM lines meet costa without curving. FW is usually warm brown, sometimes slightly hoary, and median area can be shaded lighter. HOSTS Deciduous trees, especially apple, cherry, and crab apple.

WESTERN TENT CATERPILLAR MOTH
Malacosoma californica

Very Common
87-0018 (7702)

TL 15–22 mm Warm brown to yellowish-brown FW has thick, pale AM and PM lines, framing darker brown median area; in rare individuals, lines may be brown, not pale. Southwestern individuals may show lines with points at the veins, often connecting across the darker median area. AM and PM lines curve away from median area at costa. HOSTS Generalist on deciduous trees and shrubs. NOTE By far the most common tent caterpillar moth in our area.

SOUTHWESTERN TENT CATERPILLAR MOTH
Malacosoma incurva

Common
87-0019 (7703)

TL 15–20 mm Resembles Western Tent, but light brown to grayish FW is hoary. HOSTS Fremont cottonwood and other poplars.

GRAY LAPPET
Hypopacha grisea

Rare
87-0020 (7669)

TL 15–17 mm Peppery gray FW has bold, black, wavy, double AM and PM lines that are paler in the middle. Bold black ST line is single and straighter. Gray thorax and forelegs are densely hairy. HOSTS Unknown.

FOREST TENT
CATERPILLAR
MOTH

PACIFIC TENT
CATERPILLAR
MOTH

SONORAN TENT
CATERPILLAR
MOTH

actual size

EASTERN TENT
CATERPILLAR
MOTH

SOUTHWESTERN TENT
CATERPILLAR MOTH

WESTERN TENT
CATERPILLAR
MOTH

GRAY
LAPPET

LARGE TOLYPE
Tolype velleda

Uncommon
87-0021 (7670)

RANGE-WIDE

TL 17–28 mm Resembles Distinct Tolype, but PM line is better defined toward inner margin, and median area is more uniformly shaded. Thorax is usually silvery to all white. ST area usually shaded more evenly, without a clear dark band. Basal half of HW gray, sometimes with diffuse whitish median band. **HOSTS** Deciduous trees, including apple, ash, birch, elm, oak, and plum.

NOTCHED TOLYPE
Tolype glenwoodii

Uncommon
87-0027 (7676)

RANGE-WIDE

TL 25–35 mm Closely resembles Large and Distinct Tolypes, but AM line has a triangular point near inner margin on innermost vein. Thorax is white to light gray, and ST area has a distinct dark band bordering PM line. **HOSTS** Gambel oak; possibly other oaks.

DISTINCT TOLYPE
Tolype distincta

Common
87-0028 (7677)

NORTH & EAST

CALIFORNIA

TL 25–35 mm Densely hairy, light gray to white thorax has a dark gray and blue crest down middle. Median area of FW is light gray, shading darker toward edges. Wavy double PM line is diffuse along most of length. ST area is shaded darker gray, with a distinct dark band bordering PM line. Fringe is usually solid gray, and basal half of HW is shaded grayish. **HOSTS** Pines.

DAY'S TOLYPE
Tolype dayi

Uncommon
87-0030 (7679)

RANGE-WIDE

TL 20–28 mm Resembles Distinct Tolype, but PM line is better defined across its entire length. Thorax is medium gray, with a white border to the raised blue scales. Median and ST areas, and HW, are medium dark gray. **HOSTS** Douglas-fir, ponderosa and lodgepole pine, Engelmann and white spruce, and western hemlock. **NOTE** DNA barcoding suggests Day's Tolype to be a well-defined species restricted to the Pacific Northwest, while Distinct Tolype may consist of multiple cryptic species across the Southwest. Many intermediates may not be identifiable to species where ranges overlap.

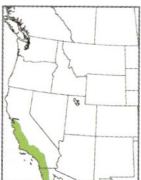

CALIFORNIA TOLYPE
Tolype lowriei

Uncommon
87-0031 (7680)

RANGE-WIDE

TL 20–25 mm Resembles Distinct Tolype, but PM line is very diffuse, with the two lines often blending. Median area is usually pale with shaded edges. Fringe usually has a checkered appearance, and basal half of HW is whitish. **HOSTS** Pines. **NOTE** Both California and Distinct Tolype can be variable, and some intermediates may not be visually identifiable to species.

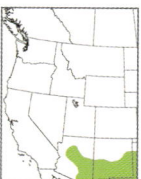

SHORT-CRESTED TOLYPE
Apotolype brevicrista

Uncommon
87-0032 (7681)

RANGE-WIDE

TL 20–25 mm Resembles *Tolype* species, but thoracic crest is much smaller. Wavy double PM line and ST area are roughly evenly spaced. Median area is usually pale. **HOSTS** Mesquite.

LARGE TOLYPE

NOTCHED
TOLYPE

DISTINCT
TOLYPE

DAY'S TOLYPE

actual size

CALIFORNIA
TOLYPE

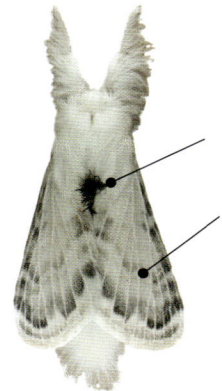

SHORT-CRESTED
TOLYPE

SIERRA MADRE WHITE

Rare

RANGE-WIDE

Artace colaria   87-0034 (7684)

TL 18–26 mm Pure white FW has black dots at the veins along all of the major wing lines. Spaces between veins are less fully scaled, sometimes appearing grayish. Hairy thorax is plain white. Females are substantially larger than males. **HOSTS** Pointleaf manzanita.

APATELODID MOTHS

SUPERFAMILY Bombycoidea (89)
FAMILY Apatelodidae

Medium-sized woodland moths that typically rest with their head down and abdomen and wings raised, often appearing to do a headstand. The ST line of the forewing has a small translucent "window." Adults are nocturnal and will come to lights.

LINED APATELODES

Common RANGE-WIDE

Apatelodes pudefacta   89-0002 (7664)

TL 20–22 mm Triangular FW is soft brown with thin medium brown lines and dark patches at inner basal and outer ST areas, the latter containing a translucent spot. Usually rests with wings held away from body and abdomen curled. **HOSTS** Bigelow's false willow. **NOTE** Typically rests with abdomen curled.

THE SERAPH

Uncommon RANGE-WIDE

Olceclostera seraphica   89-0006 (7667)

TL 20–25 mm Lavender-gray FW has wide, wavy, brownish AM and PM bands. Angular outer edge is scalloped when fresh. Midpoint of ST line has a small translucent window. Typically rests with body and wings held at an angle from resting surface, sometimes perpendicular. **HOSTS** Ash and desert willow.

ROYAL SILKMOTHS

SUPERFAMILY Bombycoidea (89)
FAMILY Saturniidae SUBFAMILY Ceratocampinae

Medium to very large, brightly colored moths with pointed wings. The royal and imperial moths have broad wings that are held slightly spread, while the oakworm and small silkmoths have narrower wings usually held folded at rest. Adults are nocturnal and will come to lights.

SPLENDID ROYAL MOTH

Uncommon RANGE-WIDE

Citheronia splendens   89-0011 (7707)

WS 10–15 cm Large. Gray FW is patterned with orange veins and pale yellow patches along AM and PM lines. Thorax is striped orange and pale yellow. **HOSTS** Desert cotton, pointleaf manzanita, and skunkbush and evergreen sumac. Often within a local population, only a single host species is used.

OSLAR'S IMPERIAL MOTH

Uncommon RANGE-WIDE

Eacles oslari   89-0013 (7705)

WS 11–15 cm Large. Pale yellow (sometimes orange-brown) wings are heavily speckled purplish brown, with thick, wavy, purplish AM and PM lines. Basal and ST areas are frequently shaded purplish. **HOSTS** Emory oak and Mexican blue oak, western soapberry, and border pinyon.

actual size

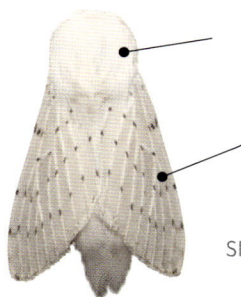

SIERRA MADRE WHITE

LINED APATELODES

THE SERAPH

actual size

SPLENDID ROYAL MOTH

actual size

OSLAR'S
IMPERIAL MOTH

OSLAR'S OAKWORM

Anisota oslari

Uncommon

89-0021 (7722)

TL 25–44 mm Long, narrow, brownish-orange FW has a single prominent white spot. Median area is sometimes shaded more orange, with faint purplish PM line. Thorax is orange. **HOSTS** Sonoran scrub oak, Mexican blue oak, and Emory oak.

HUBBARD'S SMALL SILKMOTH

Syssphinx hubbardi

Very Common

89-0026 (7711)

TL 26–36 mm Hoary gray FW has scalloped AM and PM lines and a small white spot; some individuals show an even smaller second white dot beside the first. HW is rosy pink with a large black spot and narrow gray border. **HOSTS** Wright and catclaw acacia, and honey mesquite.

SIERRA MADRE SMALL SILKMOTH

Syssphinx montana

Uncommon

89-0028 (7713.1)

TL 30–40 mm Sexually dimorphic. Larger females have yellowish to tan FW speckled with brown. Thin brown AM and PM lines border paler median area containing a single white spot. Thorax is yellow to tan. Smaller males are similar, but FW is purplish with an orange-brown median area and two white spots; thorax is orange. HW in both sexes is rosy pink. **HOSTS** Velvetpod mimosa.

PINEMOTHS, BUCKMOTHS, and EYED-SILKMOTHS

SUPERFAMILY Bombycoidea (89)

FAMILY Saturniidae SUBFAMILY Hemileucinae

Medium to large moths with broad wings that are sometimes held tight against the body when at rest. Most *Hemileuca* are boldly patterned in black and white, and often have red, pink, or orange on the body or hindwings. They are typically diurnal, found resting in low vegetation or flying rapidly above it, though some species may occasionally visit lights. The eyed-silkmoths have drab forewings and large, bold eyespots on the hindwings that are thought to be used to startle predators. They are nocturnal and will come to lights.

DORIS' PINEMOTH

Coloradia doris

Rare

89-0032 (7725)

TL 30–40 mm Sexually dimorphic. Resembles Pandora Pinemoth, but PM line is straighter and meets costa at an angle, and semitranslucent HW is largely unmarked. HW is washed pink in male, gray in female. **HOSTS** Ponderosa pine; possibly other pines.

PANDORA PINEMOTH

Coloradia pandora

Very Common

89-0034 (7724)

TL 35–45 mm Hoary brown FW has scalloped dark AM, PM, and ST lines with whitish edge and a single black spot in median area. PM line bends to meet costa perpendicularly. HW is light pink with a gray PM line and median spot. Antennae are yellow. Abdomen is banded with yellow. **HOSTS** Pines.

female

male

OSLAR'S OAKWORM

actual size

HUBBARD'S SMALL SILKMOTH

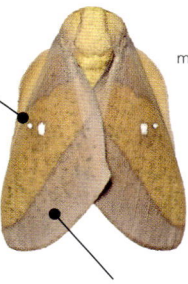

male

female

SIERRA MADRE SMALL SILKMOTH

actual size

DORIS' PINEMOTH

PANDORA PINEMOTH

TRICOLOR BUCKMOTH
Hemileuca tricolor   **Uncommon**   89-0036 (7727)   RANGE-WIDE

TL 25–39 mm Hoary gray FW has thick white AM, PM, and terminal lines and a large yellow median spot outlined in dark gray. Gray thorax is densely hairy, and abdomen is reddish orange to brown. **HOSTS** Legumes, including palo verde, mesquite, and catclaw acacia. **NOTE** Diurnal.

OLIVE SHEEPMOTH
Hemileuca oliviae   **Common**   89-0039 (7729)   RANGE-WIDE

TL 24–38 mm Olive-tan FW has a darker median area, often faintly edged paler, containing a large, dark tan spot. Hairy thorax is tan, and abdomen is reddish to brownish. **HOSTS** Grasses. **NOTE** Diurnal.

NEVADA BUCKMOTH
Hemileuca nevadensis   **Common**   89-0042 (7731)   RANGE-WIDE

TL 30–42 mm Black wings have a wide white median band containing a black-edged yellow spot. Thorax is black with a white collar. Black abdomen has an orange tip in males. **HOSTS** Willow and cottonwood; also coast live oak. **NOTE** Diurnal.

ELECTRA BUCKMOTH
Hemileuca electra   **Uncommon**   89-0044 (7736)   RANGE-WIDE

TL 25–35 mm FW pattern resembles Hera Buckmoth, but all-black markings are much thicker, such that the white areas are reduced to patches, and HW is red with wide black terminal band. Thorax is gray with a white collar and red patches. Abdomen is red above and banded with white and black below. Subspecies *H. e. mohavensis* resembles Nevada Buckmoth, but HW and abdomen are red. **HOSTS** Primarily California buckwheat; also boojum tree. **NOTE** Diurnal.

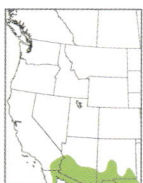

JUNO BUCKMOTH
Hemileuca juno   **Uncommon**   89-0045 (7735)   RANGE-WIDE

TL 29–40 mm Resembles Nevada Buckmoth, but veins in median area are traced with black. Usually has a narrow white dash in basal area. White areas can be very reduced in some individuals. Thorax is hoary gray with a white collar. Abdomen is black, with orange tip in male. **HOSTS** Mesquite. **NOTE** Diurnal.

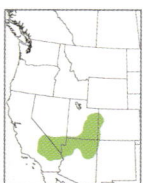

NEUMOEGEN'S BUCKMOTH
Hemileuca neumoegeni   **Rare**   89-0048 (7738)   RANGE-WIDE

TL 24–32 mm Sexually dimorphic. White FW has bold, narrow, wavy, black AM and PM lines. Kidney-shaped reniform spot and teardrop-shaped orbicular spot (bisecting AM line, sometimes absent) are yellow, boldly outlined in black. White thorax has a hoary red dorsal stripe. Abdomen of male is plain orange red, that of female is hoary. **HOSTS** Skunkbush sumac and desert almond. **NOTE** Diurnal.

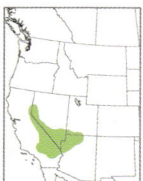

BURNS' BUCKMOTH
Hemileuca burnsi   **Uncommon**   89-0049 (7737)   RANGE-WIDE

TL 24–32 mm Resembles Neumoegen's Buckmoth, but FW lacks orbicular spot and reniform spot is rounder. Thoracic stripe is white bordered thinly with reddish black, and abdomen is striped white and black. **HOSTS** Indigobush and smoke tree. **NOTE** Diurnal.

TRICOLOR
BUCKMOTH

actual size

NEVADA
BUCKMOTH

OLIVE SHEEPMOTH

ELECTRA
BUCKMOTH

JUNO
BUCKMOTH

NEUMOEGEN'S BUCKMOTH

BURNS' BUCKMOTH

CHINATI SHEEPMOTH

Hemileuca chinatiensis   **Rare**   89-0050 (7739)

TL 34–40 mm Yellow-cream FW is thickly marked with black along veins, AM, PM, and terminal lines and discal spot, creating a netted appearance. Abdomen is reddish orange. **HOSTS** Littleleaf and fragrant sumac, catclaw and pink mimosa, agarita, and acacia. **NOTE** Diurnal. Griffin's Sheepmoth (*H. griffini*, not shown), of southern UT and northern AZ, is similar but has narrower black wing markings and a red-tipped black (male) or red-and-black-banded (female) abdomen.

ELEGANT SHEEPMOTH

Hemileuca eglanterina   **Very Common**   89-0052 (7744)

TL 34–45 mm FW pattern resembles Hera Buckmoth, but wings are sunset pink and orange. Black markings can be quite variable, ranging from nearly absent to nearly obscuring the base color, though are most commonly well defined and moderate in extent. Thorax and abdomen orange with black bands. **HOSTS** Generalist on shrubby Rosaceae and Rhamnaceae. **NOTE** Diurnal.

NUTTALL'S SHEEPMOTH

Hemileuca nuttalli   **Uncommon**   89-0053 (7743)

TL 35–42 mm Similar to Hera Buckmoth, but HW is orange, and orange thorax has no white. ST area of FW is sometimes suffused with orange. **HOSTS** Antelope bitterbrush and snowberry. **NOTE** Diurnal.

HERA BUCKMOTH

Hemileuca hera   **Very Common**   89-0054 (7741)

TL 35–50 mm White wings have thick black AM and PM lines and black edging around perimeter. Yellow median spot is edged thickly in black, and veins in ST area are black. Thorax has a dark gray center, white sides, and orange collar. Abdomen is orange with black bands. **HOSTS** Big and sand sagebrush. **NOTE** Diurnal.

IO MOTH

Automeris io   **Uncommon**   89-0055 (7746)

TL 27–45 mm Sexually dimorphic. FW is largely yellow in male, bronzy gray to purplish pink in female. Dark reniform spot appears broken or speckled. Yellow HW has a large blue eyespot boldly outlined with black. **HOSTS** Birch, clover, corn, maple, oak, willow, and many other trees, shrubs, and plants.

CECROPS EYED-SILKMOTH

Automeris cecrops   **Uncommon**   89-0059 (7748)

TL 40–56 mm Similar to Zephyr Eyed-Silkmoth, but FW is a warmer brown, slanting line is yellowish, and FW eyespot is often shaded in slightly darker brown. HW sometimes has pinkish border. **HOSTS** Oak, mountain mahogany, catclaw acacia, Fendler's ceanothus, and several other deciduous shrubs and trees.

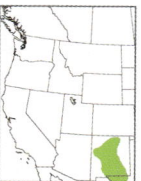

ZEPHYR EYED-SILKMOTH

Automeris zephyria   **Common**   89-0060 (7749)

TL 40–54 mm Pointed FW is chocolate brown with a bold white line running from midpoint of inner margin to apex, and an eyespot consisting of a simple thin black ring. HW is yellow with a brown border and bold, black-edged, blue eyespots. **HOSTS** Oak and mountain mahogany; also willow, rose, plum, and cherry.

CHINATI SHEEPMOTH

ELEGANT
SHEEPMOTH

NUTTALL'S SHEEPMOTH

actual size

HERA
BUCKMOTH

male

IO MOTH

female

CECROPS EYED-SILKMOTH

ZEPHYR EYED-SILKMOTH

TYPICAL SILKMOTHS
SUPERFAMILY Bombycoidea (89)
FAMILY Saturniidae SUBFAMILY Saturniinae

This group contains some of our largest and flashiest moths. Large to very large, with rounded wings that are held spread or folded above the abdomen at rest, and broadly bipectinate antennae in males. The group's common name comes from the caterpillars' strong silk cocoons, often bound within dead leaves. Adults of most species are nocturnal and visit lights in small numbers, but Mendocino Silkmoth and males of Calleta Silkmoth are diurnal.

MENDOCINO SILKMOTH
Calosaturnia mendocino
Rare
89-0064 (7751)
RANGE-WIDE

TL 28–34 mm Warm brown FW has a bold black-ringed eyespot edged thickly with white basally. Apex has a blue-gray spot bordered by reddish-brown and black patches. Orange HW has a bold eyespot and thick black PM band. Reddish thorax has a bold white collar. **HOSTS** Manzanita and madrone. **NOTE** Diurnal; males active mid-afternoon, females in early evening.

MEXICAN AGAPEMA
Agapema anona
Rare
89-0068 (7754.1)
RANGE-WIDE

TL 30–35 mm Grayish-brown FW has wide white AM, PM, and terminal lines and a short red apical dash. Veins in inner median area usually traced with white. Bold round eyespot is dark gray circled by yellow and black and has a white crescent basally; eyespot does not touch PM line. Hairy thorax is mixed brown and dark gray. **HOSTS** Jujube and knifeleaf condalia. **NOTE** Dyar's Agapema (*A. dyari*) is very similar, but eyespot touches PM line. Pale Agapema (*A. platensis*) has a white or pale median area; eyespot does not touch PM line.

ROCKY MOUNTAIN AGAPEMA
Agapema homogena
Uncommon
89-0069 (7756)
RANGE-WIDE

TL 32–35 mm Similar to Mexican Agapema, but AM and PM lines are narrower and often peach orange, red apical dash is smaller, and all veins are traced with peach orange. Thorax is completely dark gray. **HOSTS** Coffeeberry, wax currant, and narrowleaf willow.

POLYPHEMUS MOTH
Antheraea polyphemus
Very Common
89-0070 (7757)
RANGE-WIDE

WS 10–15 cm Warm brown wings have red AM and pink-edged PM lines, warm brown median line, and grayish costa. Transparent eyespots are outlined with yellow and black; HW spots also have blue shading basally. **HOSTS** Ash, birch, grape, hickory, maple, oak, pine, and other woody plants.

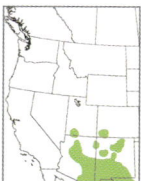

WESTERN POLYPHEMUS MOTH
Antheraea oculea
Common
89-0071 (7757.1)
RANGE-WIDE

WS 14–17 cm Resembles the smaller Polyphemus Moth, but PM line is dark gray, median area is shaded darker, and FW eyespots are more heavily outlined. **HOSTS** Primarily oaks.

MENDOCINO SILKMOTH

MEXICAN AGAPEMA

ROCKY MOUNTAIN AGAPEMA

actual size

POLYPHEMUS MOTH

WESTERN
POLYPHEMUS MOTH

CALLETA SILKMOTH
Eupackardia calleta

Uncommon
89-0078 (7763)

RANGE-WIDE

WS 10–12 cm Dark brown wings have creamy PM and terminal lines. ST area of FW is accented with black-edged blue eyespots, the largest of which is clouded red near apex. All wings are marked with pale wedge-shaped spots in central median area (sometimes reduced or absent, especially on females). **HOSTS** Cenizo, ash, willow, and others. **NOTE** Males are diurnal, active during morning hours.

GLOVER'S SILKMOTH
Hyalophora gloveri

Very Common
89-0084 (7769)

RANGE-WIDE

WS 10–15 cm Resembles Ceanothus Silkmoth, but all eyespots are smaller and less curved and ST area is hoary gray to brownish. **HOSTS** Generalist on deciduous trees and shrubs.

CEANOTHUS SILKMOTH
Hyalophora euryalus

Very Common
89-0085 (7770)

NORTH

SOUTH

WS 10–14 cm Reddish-brown wings have black-edged AM and PM lines. White eyespots are thinly outlined in dark gray; FW spots are comma-shaped, while HW spots are more hooked and elongated, often touching PM line. ST area is hoary brown to pinkish. Northern populations may average overall browner with more grayish ST area. **HOSTS** Ceanothus and coffeeberry; also willow, *Prunus*, manzanita, alder, currant, sumac, and Douglas-fir.

TYPICAL SPHINX MOTHS

SUPERFAMILY Bombycoidea (89)
FAMILY Sphingidae SUBFAMILY Sphinginae

Medium to very large moths with long pointed wings and a chunky body with a tapered abdomen. Most show cryptic patterns of gray or brown that resemble bark. The large caterpillars have a horn-like projection on the posterior end; two species, the Tobacco and Tomato Hornworms (Carolina Sphinx and Five-spotted Hawkmoth), are familiar garden pests. Adults are crepuscular and nocturnal; they will come to lights but can also be found nectaring at flowers at dusk, including at tubular garden flowers such as beebalm and phlox.

PINK-SPOTTED HAWKMOTH
Agrius cingulata

Common
89-0086 (7771)

RANGE-WIDE

TL 55–65 mm Light brown FW appears mottled as a result of thin, black, jagged, double basal, AM, PM, and ST lines. Outer median area is darker brown with a small white discal spot. HW is pale gray with black bands and pink basal patch. Abdomen has five pink spots on each side of brownish midline. **HOSTS** Herbaceous plants and shrubs, including sweet potato and jimsonweed.

GIANT SPHINX
Cocytius antaeus

Uncommon
89-0087 (7772)

RANGE-WIDE

TL 65–90 mm Resembles the smaller Pink-spotted Hawkmoth, but FW is grayer and lacks dark shading in median area. Double AM line is typically broken at midpoint by pale patch. Abdomen is gray with three orange spots at sides of basal segments. **HOSTS** Pond apple. **NOTE** Occurs in our area primarily as a regular stray from Mexico.

CALLETA
SILKMOTH

GLOVER'S SILKMOTH

CEANOTHUS SILKMOTH

actual size

actual size

PINK-SPOTTED
HAWKMOTH

GIANT SPHINX

241

CAROLINA SPHINX
Manduca sexta

Very Common
89-0090 (7775)

TL 55–65 mm Resembles Pink-spotted Hawkmoth, but median area has no shading, and spots along abdomen are orange. HW is gray with black-and-white bands. **HOSTS** Crops, including potato, tobacco, and tomato. **NOTE** Caterpillar is known as Tobacco Hornworm.

FIVE-SPOTTED HAWKMOTH
Manduca quinquemaculatus

Very Common
89-0091 (7776)

TL 50–70 mm Grayish-brown FW has thin, wavy, black AM, PM, and ST lines sloping toward apex, recalling a wood-grain pattern. HW is pale gray with widely separated, jagged, black lines. Abdomen has five or six pairs of black-edged yellow spots. **HOSTS** Crops, including potato, tobacco, and tomato. **NOTE** Caterpillar is known as Tomato Hornworm.

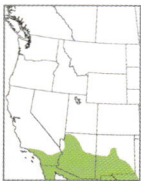

RUSTIC SPHINX
Manduca rustica

Very Common
89-0092 (7778)

TL 50–80 mm FW pattern is similar to that of Pink-spotted Hawkmoth but is much bolder with white base color, strong black lines, and rich brown to chocolate shading in median and basal areas. Abdomen has three pairs of yellow spots near base. Thorax has a mottled white patch. **HOSTS** Fringetree, jasmine, and *Bignonia*.

MUSCOSA SPHINX
Manduca muscosa

Uncommon
89-0094 (7781)

TL 50–65 mm Olive-gray FW has scalloped brownish AM (double), PM (triple), and ST (single) lines. Lowest PM line is shaded below by dark gray. Gray median area has small white discal spot. **HOSTS** Toothleaf goldeneye.

FLORESTAN SPHINX
Manduca florestan

Uncommon
89-0095 (7782)

TL 46–54 mm Resembles Rustic Sphinx, but markings are less bold and shaded areas appear washed out. Basal area of FW is unshaded, and inner half is paler. **HOSTS** Fiddlewood and yellow trumpet flower.

ELM SPHINX
Ceratomia amyntor

Uncommon
89-0102 (7786)

TL 50–60 mm Brown FW has thin black streaks, often set in a dark band running from base to apex. Costa is pale. Parallel thin ST lines are most visible near costa. FW has small white discal spot. HW is brown with dusky bands. Brown thorax is bordered by dark bands. **HOSTS** Elm, birch, basswood, and cherry.

CAROLINA SPHINX

FIVE-SPOTTED HAWKMOTH

RUSTIC SPHINX

actual size

MUSCOSA SPHINX

FLORESTAN SPHINX

ELM SPHINX

WAVED SPHINX

Ceratomia undulosa

Common
89-0103 (7787)

TL 45–60 mm Similar to Pink-spotted Hawkmoth, but median area of FW is a smooth and even gray, so white discal spot is conspicuous, and spots along abdomen are small and whitish gray. Jagged black apical dash points to three or four thin vertical dashes spaced diagonally across PM and median areas. Gray thorax is circled by a thin black ring. **HOSTS** Ash, privet, oak, hawthorn, and fringetree.

ELSA SPHINX

Sagenosoma elsa

Uncommon
89-0109 (7792)

TL 32–40 mm Hoary gray-brown FW has wide pale band along length of costa. Narrow, dark brown PM line and white ST band slant toward apex. Thorax has pale sides and hoary gray back bordered by thin orange and black lines. **HOSTS** Boxthorn and wolfberry.

GREAT ASH SPHINX

Sphinx chersis

Very Common
89-0111 (7802)

TL 50–70 mm Elongated gray FW has a long thin apical dash that points to three or four thin black dashes crossing diagonally through PM and median area. White ST line is sometimes edged thinly with black. HW is dark gray with a white median line. Thorax has two thin black dorsal lines. **HOSTS** Ash, lilac, privet, quaking aspen, and *Prunus* species.

VASHTI SPHINX

Sphinx vashti

Very Common
89-0112 (7803)

TL 35–45 mm Resembles Great Ash Sphinx, but darker gray FW has a broad white basal patch, white ST line is distinctly edged with black, and dorsal lines on thorax are broad. **HOSTS** Snowberry.

ELEGANT SPHINX

Sphinx perelegans

Very Common
89-0114 (7805)

TL 45–55 mm Resembles Vashti Sphinx, but pale basal patch extends to PM area, white ST line has no black edging, and thorax is dark gray in center. **HOSTS** Manzanita and hollyleaf cherry.

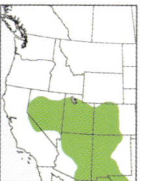

ASELLUS SPHINX

Sphinx asellus

Rare
89-0115 (7806)

TL 45–58 mm Resembles Great Ash Sphinx, but two additional long and thin black dashes are present in median area and near inner margin of FW. ST line is less pronounced, and black dorsal lines are thin and doubled. **HOSTS** Manzanita; possibly other shrubs.

WILD CHERRY SPHINX

Sphinx drupiferarum

Uncommon
89-0123 (7812)

TL 45–60 mm Resembles Elegant Sphinx, but FW is darker gray, and white basal patch extends out to costa and reaches to ST area. Small black discal spot is usually noticeable. Dark thorax is typically light brown at base. **HOSTS** Wild cherry, plum, lilac, hackberry, and apple.

WAVED SPHINX

ELSA SPHINX

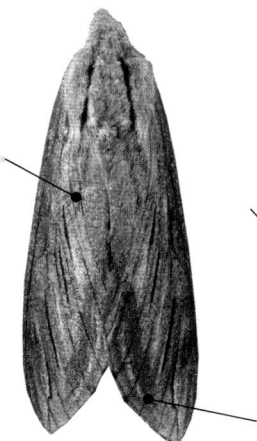

GREAT ASH SPHINX

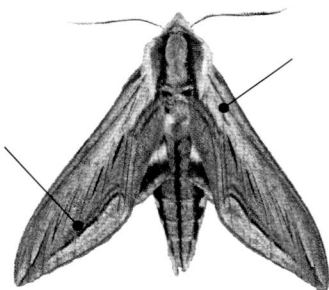

VASHTI SPHINX

actual size

ELEGANT SPHINX

ASELLUS SPHINX

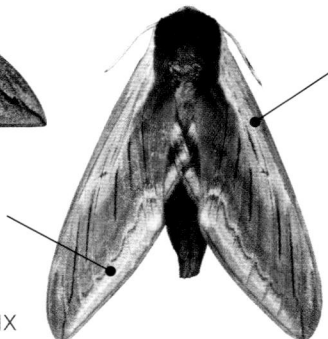

WILD CHERRY SPHINX

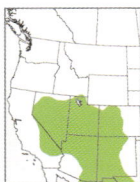

DOLL'S SPHINX

Common

Sphinx dollii
89-0124 (7813)

TL 25–32 mm Elongated, pale gray FW has a series of thick black dashes forming a line diagonally from base to pointed apex. FW is shaded slightly darker gray on inner side of black dashed line. Smoky, diffuse AM and PM lines are often faint bands. Pale thorax has two wide, dark gray bands laterally. **HOSTS** Alligator juniper and other junipers.

SEQUOIA SPHINX

Uncommon

Sphinx sequoiae
89-0125 (7814)

TL 27–37 mm Resembles Great Ash Sphinx but is about half the size. Thin black basal dash on FW connects to white-edged dorsal lines on thorax. ST line is black edged with diffuse pale shading. **HOSTS** Juniper and incense-cedar. Despite the name, not known to feed on sequoia.

ISTAR SPHINX

Rare

Lintneria istar
89-0132 (7799)

TL 55–60 mm Resembles Great Ash Sphinx, but FW has brownish tones and dark grayish shading that runs from the base of the dark dorsal stripes on the thorax, along the inner AM line, thick black median dashes, inner PM line, and then ST line to end at the midpoint of the outer margin. **HOSTS** Mint.

EYED SPHINX MOTHS

SUPERFAMILY Bombycoidea (89)
FAMILY Sphingidae SUBFAMILY Smerinthinae

Medium to large moths with long pointed wings that flare at the wavy outer margin and are typically held level above the body, and a tapered abdomen that is frequently curled upward when at rest. A number of species have hindwings that project from beneath the forewings when folded, and several have blue eyespots on the hindwings. Adults are nocturnal and will visit lights.

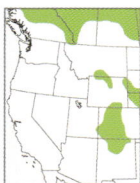

TWIN-SPOTTED SPHINX

Uncommon

Smerinthus jamaicensis
89-0140 (7821)

TL 38–45 mm Resembles One-eyed Sphinx, but shading in median area is reduced to a vertical bar at the point of the AM line. Basal line is pale, median line is usually narrow and dark, and thin PM line is whitish without an accompanying dark dashed line. Apex has a dark brown, white-edged semicircle. Blue eyespot on HW is divided in two by a black bar. **HOSTS** Deciduous trees, including apple, ash, elm, poplar, and birch.

ONE-EYED SPHINX

Common

Smerinthus cerisyi
89-0141 (7822)

TL 45–55 mm Light brownish-gray FW has darker brown shading in inner median area and ST area. White-edged brown AM line is angled at midpoint. White-edged brown basal line is scalloped, and PM line appears as a line of scalloped dashes. Blue eyespot on sunset-colored HW has a black dot in center that does not touch black border. **HOSTS** Poplar and willow.

DOLL'S SPHINX

SEQUOIA SPHINX

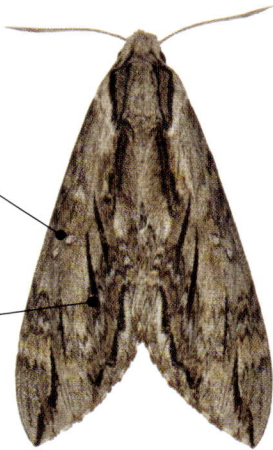

ISTAR SPHINX

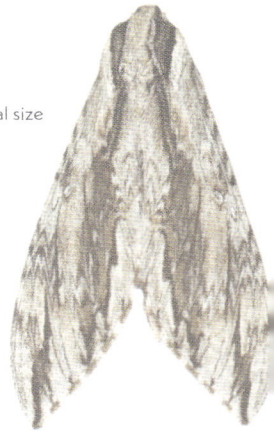

actual size

TWIN-SPOTTED SPHINX

actual size

ONE-EYED SPHINX

WESTERN EYED SPHINX COMPLEX — Very Common
Smerinthus astarte/ophthalmica   89-0141.1/0142 (7822.2/22.1)

TL 40–55 mm Two virtually identical, variable species. They resemble One-eyed Sphinx, but AM line is crisply defined and sharply pointed. Basal line is typically lighter and less curvy, and PM line is a series of straight dashes broken by pale veins. Center of HW eyespot touches black border at each side. Both species are variable in markings and warmth of color, as well as genital morphology, with no consistent visual differences. HOSTS Willow, cottonwood, and quaking aspen. NOTE Drowsy-eyed Sphinx (*S. astarte*) is found in the central states, while Western Eyed Sphinx (*S. ophthalmica*) occurs to the west and north. Best identified by range, though range boundaries in ID and UT are not yet well defined and may overlap; individuals occurring near where illustrated ranges meet require identification through DNA barcoding. One-eyed, Drowsy-eyed, and Western Eyed Sphinxes were once considered conspecific. U.S. populations that were previously identified as Salicet Sphinx (*S. saliceti*, which is a strictly Mexican species) are now considered actually to be Drowsy-eyed Sphinx.

DROWSY-EYED SPHINX

WESTERN EYED SPHINX

DROWSY-EYED SPHINX

WESTERN EYED SPHINX

CALIFORNIA

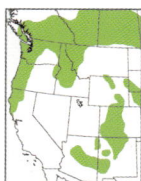

BLINDED SPHINX — Common
Paonias excaecata   89-0144 (7824)

TL 35–50 mm Similar to Twin-spotted Sphinx, but FW averages browner, median area usually has purplish shading, and AM line is slanting instead of V-shaped. Discal spot is a dark dot, and FW apex lacks a dark crescent. Outer margin is strongly scalloped. Blue eyespot on HW has no black dot in center. Apex of HW often sticks out beyond FW costa when moth is at rest. HOSTS Deciduous trees, including basswood, willow, birch, and poplar.

RANGE-WIDE

SMALL-EYED SPHINX — Common
Paonias myops   89-0145 (7825)

TL 32–35 mm Dark FW has purplish tones and yellow-orange patches at apex and anal angle. Thorax has a yellow-orange central stripe bordered by violet bands. HW is yellow with an empty blue eyespot. Yellow-orange apex of HW often sticks out from under FW costa when moth is at rest. HOSTS Deciduous trees, including black cherry, serviceberry, and basswood.

RANGE-WIDE

MODEST SPHINX — Common
Pachysphinx modesta   89-0148 (7828)

TL 45–65 mm Large. Violet-gray FW has a darker median band with small white discal spot. AM line is straightish and edged basally with pale tan. Outer margin is scalloped. Rosy HW has a blue patch bordered by black at anal angle. HOSTS Cottonwood, poplar, and willow. NOTE Also known as Big Poplar Sphinx.

RANGE-WIDE

WESTERN POPLAR SPHINX — Very Common
Pachysphinx occidentalis   89-0149 (7829)

TL 65–75 mm Very similar to Modest Sphinx, but FW is overall browner, median area is lighter, and AM line has a distinct downward-pointing tooth near midpoint. Basal line is irregular and lacks edging. Apex has a small paler patch. Patch at anal angle of HW is faded blue or grayish. HOSTS Cottonwood, poplar, and willow.

NORTH

SOUTH

WESTERN EYED SPHINX COMPLEX

actual size

BLINDED SPHINX

SMALL-EYED SPHINX

MODEST SPHINX

WESTERN POPLAR SPHINX

DIURNAL and STRIPED SPHINX MOTHS

SUPERFAMILY Bombycoidea (89)
FAMILY Sphingidae SUBFAMILY Macroglossinae

Small to large moths with long pointed wings and a tapered abdomen; most are quite colorful or boldly patterned. Some resemble typical sphinx moths but have a squarer outer margin, while others more resemble eyed sphinx moths, with a flared anal angle, but have a stouter abdomen often terminating in an anal tuft. About half of the species are crepuscular or nocturnal and will visit lights. The other half are strictly diurnal and often to be found nectaring at flowers during daytime; most of these are small and may be initially mistaken for a bumblebee or hummingbird.

ELLO SPHINX
Erinnyis ello

Common
89-0154 (7834)

RANGE-WIDE

TL 53–59 mm Sexually dimorphic. Elongated FW of male is streaky grayish brown with a bold dark stripe from base to apex. Male thorax has a dark stripe down center. Female FW and thorax are streaky pale gray, with no bold stripes. Orange HW has a blackish ST band. Abdomen is banded with black and pale gray, with a gray dorsal stripe. **HOSTS** Cassava, poinsettia, and a variety of woody plants, including guava and willow bustic. **NOTE** Known to wander occasionally north of its usual range.

OBSCURE SPHINX
Erinnyis obscura

Uncommon
89-0156 (7837)

RANGE-WIDE

TL 29–37 mm Sexually dimorphic. Both sexes resemble those of the much larger Ello Sphinx, but FW of both sexes has dusky shading along costa at AM and PM lines, as well as at apex. Male has two dark stripes on thorax. Orange HW has an incomplete dusky ST band. Lacks pale bands on abdomen. A darker morph with dark brown shading through most of the FW occurs uncommonly. **HOSTS** Dogbane, papaya, and spurge.

ACHEMON SPHINX
Eumorpha achemon

Very Common
89-0184 (7861)

RANGE-WIDE

TL 45–55 mm Pinkish-brown FW has thin AM and PM lines connecting to bold, dark brown patch at inner median area, and additional dark patches at outsides of ST area. Thorax has dark brown triangles laterally. Pink HW has a brownish ST band with a dark brown dashed ST line. **HOSTS** Grape and peppervine.

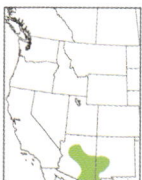

TYPHON SPHINX
Eumorpha typhon

Uncommon
89-0185 (7863)

RANGE-WIDE

TL 30–35 mm Resembles the larger Achemon Sphinx, but outer median area is shaded with dark brown, ST area patches are connected by a thin, brown-edged, white ST line, and PM area and thorax are grayish. HW has a black-edged grayish patch above straight black ST line. **HOSTS** Grape.

VINE SPHINX
Eumorpha vitis

Uncommon
89-0186 (7864)

RANGE-WIDE

TL 42–52 mm Forest-green FW has bold tan stripes from base to apex and across angled AM and PM lines. Three inner veins are traced with tan from central stripe to tan terminal area. Olive thorax has white-edged green triangles laterally. **HOSTS** Grape, treebine, Virginia creeper, primrose-willow, and magnolia.

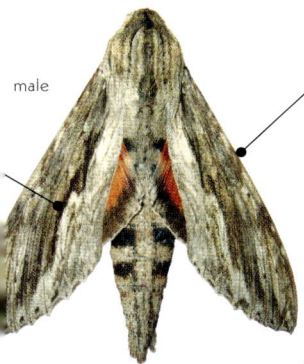

male

female

ELLO SPHINX

female

female

male

male

female

OBSCURE
SPHINX

actual size

ACHEMON SPHINX

TYPHON SPHINX

VINE SPHINX

251

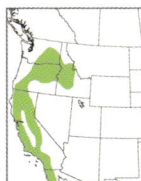

PACIFIC GREEN SPHINX
Proserpinus lucidus

Very Common
89-0195 (7872)

TL 27–32 mm Green FW has double AM and PM lines filled with violet gray that join near inner margin. Tooth at midpoint of AM line creates a heart shape in green median area. HW is orange basally, bordered by rosy red and light tan. **HOSTS** Evening primrose and clarkia. **NOTE** Diurnal visitor to nectar-rich flowers.

JUANITA SPHINX
Proserpinus juanita

Uncommon
89-0197 (7875)

RANGE-WIDE

TL 26–35 mm Very similar to Clark's Day Sphinx but lacks thin white line at collar on thorax, and mark at FW apex is a white dash. Where photos do not clearly show these markings, can generally be identified by range. **HOSTS** Evening primrose, beeblossom, and willowherb. **NOTE** Diurnal visitor to nectar-rich flowers.

CLARK'S DAY SPHINX
Proserpinus clarkiae

Uncommon
89-0198 (7876)

RANGE-WIDE

TL 22–25 mm Green FW has smooth whitish AM and PM lines framing dark green median area with darker green, white-edged spot. ST area is dark green with a black patch at apex. Thorax has two thin white dorsal stripes and a thin white stripe at the back of the collar that creates a pi symbol (π). **HOSTS** Evening primrose and clarkia. **NOTE** Diurnal visitor to nectar-rich flowers.

YELLOW-BANDED DAY SPHINX
Proserpinus flavofasciata

Rare
89-0199 (7877)

RANGE-WIDE

TL 26–28 mm A bumblebee mimic. Hairy thorax is yellow, black FW has a grayish PM band, and black abdomen has a yellow and black anal tuft. HW is orange with a black terminal band. **HOSTS** Fireweed and willowherb. **NOTE** Diurnal visitor to nectar-rich flowers.

TERLOO SPHINX
Proserpinus terlooii

Uncommon
89-0201 (7879)

RANGE-WIDE

TL 25–29 mm Similar to Juanita Sphinx, but median area lacks dark spot, shading in ST area is reduced or nearly absent, and apex has no marking. **HOSTS** Spiderling. **NOTE** Diurnal visitor to nectar-rich flowers.

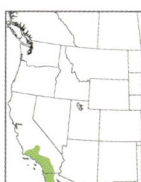

PHAETON PRIMROSE SPHINX
Euproserpinus phaeton

Uncommon
89-0202 (7880)

RANGE-WIDE

TL 19–25 mm Light gray FW has blackish AM and PM lines, a dark discal spot, and dark shading in ST area. HW is white with black ST band. Thorax is hoary gray, and abdomen is black with a white band that is sometimes divided into lateral spots. **HOSTS** Suncup and evening primrose. **NOTE** Diurnal visitor to nectar-rich flowers.

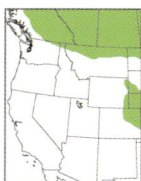

HUMMINGBIRD CLEARWING
Hemaris thysbe

Uncommon
89-0204.3 (7853)

RANGE-WIDE

TL 25–30 mm Resembles Rocky Mountain Clearwing, but olive thorax lacks lateral stripes, base of abdomen is pale tan, dark abdominal band is wine red, and anal tuft has a brownish central patch. Head and thorax are white below, with a thin black line running through eye. **HOSTS** Honeysuckle, snowberry, hawthorn, cherry, and plum. **NOTE** Diurnal visitor to nectar-rich flowers.

PACIFIC GREEN SPHINX

actual size

JUANITA SPHINX

CLARK'S DAY SPHINX

TERLOO SPHINX

YELLOW-BANDED DAY SPHINX

PHAETON PRIMROSE SPHINX

HUMMINGBIRD CLEARWING

SNOWBERRY CLEARWING
Hemaris diffinis

Uncommon
89-0204.5 (7855)

TL 22–30 mm Resembles Rocky Mountain Clearwing, but yellow thoracic stripes extend into black abdominal band, yellow abdominal band is always unbroken, and anal tuft has a solid central yellow patch. **HOSTS** Snowberry, dogbane, and honeysuckle. **NOTE** Diurnal visitor to nectar-rich flowers.

ROCKY MOUNTAIN CLEARWING
Hemaris thetis

Very Common
89-0204.6 (7855.1)

TL 22–30 mm Clear wings have a raspberry-red border and veins. Thorax is olive, sometimes with yellowish stripes laterally. Abdomen is olive basally, with black and yellow bands below; yellow band is often partially bisected by a black dorsal stripe. Anal tuft is completely black or with a thin patch of yellow hairs centrally. **HOSTS** Snowberry, honeysuckle, and related species. **NOTE** Diurnal visitor to nectar-rich flowers.

FALCON SPHINX
Xylophanes falco

Uncommon
89-0210 (7889)

TL 40–45 mm Streamlined moth has distinctly tapered brown FW and abdomen. Multiple thin lines from apex to midpoint of inner margin create a wood-grain look. Anal angle is falcate. Brown thorax has a gray stripe down center. **HOSTS** Firecrackerbush.

SPURGE HAWKMOTH
Hyles euphorbiae

[IN] Very Common
89-0215 (7892)

TL 38–50 mm Resembles Gallium Sphinx but is overall more brownish olive and FW stripe is pinkish and broadly diffuses to costa at AM, PM, and ST areas. **HOSTS** Leafy spurge. **NOTE** Introduced from Europe as biocontrol for the invasive leafy spurge.

GALLIUM SPHINX
Hyles gallii

Very Common
89-0216 (7893)

TL 38–50 mm Brownish to greenish-olive FW has a bold tan stripe running from apex to base of thorax, with two teeth pointing toward costa at AM and PM lines. Thorax has thin white stripes laterally. HW is pink with a black ST band and white patch at anal angle. **HOSTS** Bedstraw, willowherb, woodruff, and godetia.

WHITE-LINED SPHINX
Hyles lineata

Very Common
89-0217 (7894)

TL 35–50 mm Resembles Gallium Sphinx, but FW veins are all traced with white. Olive thorax has six thin, parallel, white lines. Dorsal side of abdomen is banded in black and white with two olive stripes and a thin, central, white stripe running from its base to tip. **HOSTS** Various trees and plants, including apple, elm, evening primrose, grape, tomato, and purslane.

ELEPHANT HAWKMOTH
Deilephila elpenor

[IN] Uncommon
89-0218 (7894.1)

TL 32–40 mm Unmistakable olive-and-pink sphinx with white edging to thorax, inner margin, legs, and antennae. **HOSTS** Fireweed and willowherb; also clarkia, bedstraw, impatiens, and marsh calla. **NOTE** First discovered in 2002 near Vancouver and is slowly spreading into surrounding parts of BC and WA.

SNOWBERRY CLEARWING

ROCKY MOUNTAIN CLEARWING

FALCON SPHINX

SPURGE HAWKMOTH

actual size

GALLIUM SPHINX

WHITE-LINED SPHINX

ELEPHANT
HAWKMOTH

GEOMETER MOTHS

SUPERFAMILY Geometroidea (91a)

Moths in this large group most closely resemble butterflies, with broad flimsy wings and slender bodies. Most species rest with their wings outspread, though a few sometimes or habitually rest with wings folded above the body. The antennae of many species show sexual dimorphism: Males' antennae are feathery, and females' are filiform. The caterpillars are "inchworms," traveling by lifting the posterior end and curling the body to place it immediately behind the front legs, then lifting the front legs to stretch out straight again. Many of the caterpillars are remarkable twig mimics. Adults are almost entirely nocturnal, though a handful of species fly during the day, and some others may be flushed from low vegetation or forest trails. The geometers represent one of our largest superfamilies, with over 1,400 species in North America. The group's taxonomy is continually evolving as more research is completed.

SCOOPWINGS
SUPERFAMILY Geometroidea (91a)
FAMILY Uraniidae

Small flat-winged moths that rest with their forewings spread and hindwings folded tight to the body, creating a distinctive gap in between. They are nocturnal and visit lights in small numbers.

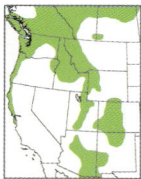

GRAY SCOOPWING
Callizzia amorata
Common
91a-0002 (7650)
RANGE-WIDE

WS 15–22 mm Brindled gray FW has tan-edged brown AM and PM lines that pinch together near inner margin. Midpoint of outer margin has a black-edged triangle, sometimes filled with dark brown. Inner half of HW is folded at rest, creating a distinctive gap between the FW and HW. **HOSTS** Honeysuckle and snowberry.

WAVES
SUPERFAMILY Geometroidea (91a)
FAMILY Geometridae SUBFAMILY Sterrhinae

Small to medium-sized moths that have broad wings in shades of tan, brown, or white, which they hold broadly spread when at rest. Many species may be similar in appearance, and identification often relies on the presence or absence of discal dots, transverse lines, and small patches of shading. Most adults are nocturnal and will visit lights, but many can be found resting among low vegetation during the day. The Chickweed Geometer is commonly active during the day, sometimes found in lawns.

HOLLOW-SPOTTED GEOMETER
Cyclophora dataria
Common
91a-0020 (7135)
RANGE-WIDE

WS 19–21 mm Warm tan to pinkish-brown wings are lightly peppered and slightly falcate at apex. Discal spots are white, bordered (sometimes thickly) with dark brown. Median line is thick and diffuse. AM and PM lines are often faint except for black dots at veins. **HOSTS** Garry oak and tarweed.

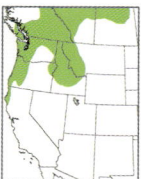

SWEETFERN GEOMETER
Cyclophora pendulinaria
Uncommon
91a-0024 (7139)
RANGE-WIDE

WS 17–24 mm Resembles Hollow-spotted Geometer, but wings are whitish to light gray. Median and ST areas are sometimes shaded slightly darker. Apex is not falcate. **HOSTS** Deciduous trees and herbaceous plants, including alder, beech, blueberry, and sweet fern.

GRAY SCOOPWING

actual size

HOLLOW-SPOTTED
GEOMETER

SWEETFERN GEOMETER

actual size

DWARF TAWNY WAVE
Cyclophora nanaria

Very Common
91a-0025 (7140)

NORTH

SOUTH

WS 16–21 mm Resembles Hollow-spotted Geometer, but wings are mousy brown and heavily speckled. Median band is quite diffuse, and ST area is shaded darker. **HOSTS** Black-eyed Susan, pricklyleaf, and *Coreocarpus*.

CHICKWEED GEOMETER
Haematopis grataria

Common
91a-0026 (7146)

RANGE-WIDE

TL 10-13 mm Yellow wings have bold pink median and ST lines and fringe. FW has pink discal dot. Male has bipectinate antennae. **HOSTS** Chickweed, clover, and other herbaceous plants. **NOTE** Commonly encountered in daytime.

MANY-LINED WAVE
Arcobara multilineata

Uncommon
91a-0033 (7128)

RANGE-WIDE

WS 22–25 mm Light tan wings are pointed at FW apex and HW outer margin. Many narrow, dark brown lines cross in straight lines from inner margin to apex. Median and PM lines are paired and shaded darker. All wings have tiny black discal dot, sometimes indistinct. **HOSTS** Unknown.

NARROW-WINGED WAVE
Euacidalia sericearia

Rare
91a-0037 (7087)

RANGE-WIDE

WS 17–20 mm Dingy gray FW has an indistinct row of black dots along AM and PM lines, with darkest dots at costa. Terminal line is solid and black when fresh. FW apex is slightly pointed. HW is hidden beneath FW when at rest. Abdomen has a narrow white stripe dorsally. **HOSTS** Unknown.

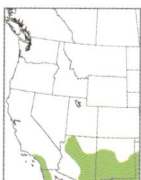

DRAB BROWN WAVE
Lobocleta ossularia

Uncommon
91a-0044 (7094)

RANGE-WIDE

WS 13–19 mm Peppery tan to brown wings have a diffuse, slightly wavy, dark median line. AM, PM, and ST lines usually narrower and more poorly defined. PM line has dark dots at veins. All wings have small black discal dots. **HOSTS** Chickweed, bedstraw, clover, and strawberry.

STRAIGHT-LINED WAVE
Lobocleta plemyraria

Uncommon
91a-0047 (7097)

NORTH

SOUTH

WS 14–20 mm Peppery tan to light yellow FW has pointed apex. Faint, dark brown AM, median, and PM lines are relatively straight and diffuse. Solid terminal line and discal dot are black but sometimes indistinctly marked. **HOSTS** Unknown; possibly sandmat.

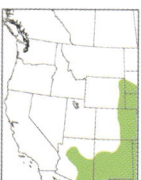

SPECKLED WAVE
Lobocleta peralbata

Common
91a-0050 (7100)

RANGE-WIDE

WS 11–20 mm Lightly peppered white wings have wavy yellowish-brown AM, PM, and ST lines. Tiny black discal dots mark all wings. Some individuals show a terminal line of black dots. **HOSTS** Unknown.

DWARF
TAWNY WAVE

CHICKWEED
GEOMETER

MANY-LINED WAVE

actual size

NARROW-WINGED
WAVE

DRAB BROWN
WAVE

STRAIGHT-LINED WAVE

SPECKLED WAVE

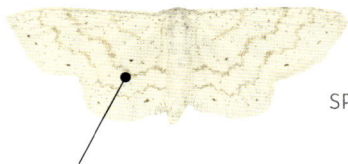

FORTUNATE WAVE
Common RANGE-WIDE

Idaea bonifata 91a-0052 (7102)

WS 13–15 mm Shiny straw-colored wings have incomplete, diffuse, brown AM, median, and PM lines, often darkest at costa. Dark discal spot is usually within median line. ST area is shaded brown. Fringe has a row of dark dots. **HOSTS** Decaying leaves and stored grains.

RED-AND-WHITE WAVE
Uncommon RANGE-WIDE

Idaea basinta 91a-0060 (7110)

WS 10–15 mm Shiny, pale yellow wings have pinkish to maroon basal and PM/ST areas. Thorax and abdomen are reddish. Fringe is yellow when fresh. **HOSTS** Unknown.

RED-BORDERED WAVE
Common RANGE-WIDE

Idaea demissaria 91a-0067 (7114)

WS 16–19 mm Resembles Red-and-white Wave, but reddish areas are bordered by darker AM and PM lines, and all wings have a dark discal spot. Median line is usually present but sometimes indistinct. In some individuals, reddish basal shading extends into median area. **HOSTS** Unknown.

STRAW WAVE
Uncommon RANGE-WIDE

Idaea eremiata 91a-0068 (7115)

WS 14–19 mm Strawberry-tan to yellowish FW resembles Red-bordered Wave without darker reddish shading in basal and ST/terminal areas, but PM line is darker than median or AM lines, which are usually indistinct or even absent. **HOSTS** Plant detritus, including dead leaves.

GEM WAVE
Uncommon RANGE-WIDE

Idaea gemmata 91a-0069 (7116)

WS 12–15 mm Pale tan to yellowish wings have dusky median area that becomes pale at costa. PM line curves basally at costa and toward anal angle at inner margin, with a single broad arc from central wing to inner margin. Discal spots on all wings are warm brown. PM/ST area has smoky shading. Abdomen can be peppery or shaded dusky like median band. **HOSTS** Unknown.

SIGNET WAVE
Uncommon RANGE-WIDE

Idaea occidentaria 91a-0070 (7117)

WS 13–16 mm Resembles Gem Wave, but has stronger shading in basal area of HW, and in ST/terminal area on all wings. Basal area of FW sometimes also shaded dusky. Pale, diffuse ST line of HW mirrors the shape of PM line. Fringe is checkered. **HOSTS** Unknown.

JEWELED WAVE
Uncommon RANGE-WIDE

Idaea asceta 91a-0080 (7117.1)

WS 12–15 mm Resembles Gem Wave, but has dark gray and rusty shading in ST/terminal area of HW. Median band is usually blackish, sometimes faded in inner half of FW, with a diffuse rusty discal spot on all wings. **HOSTS** Unknown.

FORTUNATE WAVE

actual size

RED-AND-WHITE WAVE

RED-BORDERED WAVE

STRAW WAVE

GEM WAVE

SIGNET WAVE

JEWELED WAVE

SINGLE-DOTTED WAVE
Idaea dimidiata

[IN] Very Common
91a-0079 (7126)

WS 18–21 mm Cream-colored FW has diffuse brown patch at anal angle bisected by scalloped pale ST line. Dark brown median line is broader and diffuse. AM and PM lines are thin and indistinct, with small dots at veins along PM line. All wings have small discal spots. **HOSTS** Cow parsley, burnet saxifrage, hedge bedstraw, and dandelion.

ORBED WAVE
Odontoptila obrimo

Uncommon
91a-0082 (7130)

WS 17–24 mm Warm brown wings have broad whitish AM band and thin wavy ST line. Median area is shaded darker basally, with tiny dark discal dot. Terminal line is double. Often rests with FW slightly creased. **HOSTS** Unknown.

DARK-RIBBONED WAVE
Leptostales rubromarginaria

Uncommon
91a-0090 (7179)

WS 17–20 mm Warm brown to orangish wings have wavy dark AM, median, and PM lines with black dots at veins. AM line has white edging basally. AM area is darker brown. Separable from dark Red-bordered Wave by lack of discal dots and even width of PM area. **HOSTS** Unknown; possibly plum.

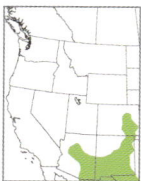

ROYAL WAVE
Scopula plantagenaria

Uncommon
91a-0099 (7154)

WS 18–26 mm Light tan wings are lightly peppery. Brown AM, median, and PM lines are heavily scalloped, curving around dark brown discal spots and forming diffuse dark patches at costa. PM and ST lines have dark brown blotches at midpoint and inner margin, less distinct on HW. **HOSTS** Unknown.

LARGE LACE-BORDER
Scopula limboundata

Uncommon
91a-0104 (7159)

WS 20–31 mm Whitish to pale tan wings have mottled black shading below wavy PM line, heaviest in inner half of FW. In some individuals, this shading may be tan, matching other wing lines. Dark tan AM and median lines are relatively straight with strong angle near costa. All wings have strong black discal dots. FW apex is pointed and often very slightly falcate; HW is angled at midpoint. **HOSTS** Apple, bedstraw, blueberry, cherry, dandelion, sweet pepperbush, and many others.

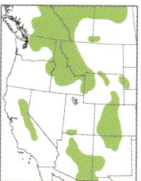

ANGLED WAVE
Scopula ancellata

Rare
91a-0107 (7162)

WS 23–26 mm Resembles tan-shaded Large Lace-border, but median and PM line are darker and nearly straight. FW apex is not falcate, and HW is less strongly angulate. **HOSTS** Unknown.

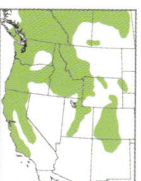

SIMPLE WAVE
Scopula junctaria

Common
91a-0109 (7164)

WS 24–30 mm Resembles unmarked Large Lace-border, but soft median and PM lines are straighter and ST and adterminal lines are often indistinct. Peppery wings nearly always lack black discal spot. Apex of FW is pointed and often slightly falcate. **HOSTS** Deciduous trees and plants, including chickweed, clover, and elm.

SINGLE-DOTTED WAVE

ORBED WAVE

DARK-RIBBONED WAVE

actual size

ROYAL WAVE

LARGE LACE-BORDER

ANGLED WAVE

SIMPLE WAVE

MODEST WAVE
Scopula quinquelinearia
Rare RANGE-WIDE
91a-0110 (7164.1)

WS 24–30 mm Resembles Simple Wave, but PM line is slightly wavy and more defined. Median and PM lines are often darker. Wings commonly have black discal dots, but may be absent on some individuals. ST and adterminal lines usually distinctly visible. **HOSTS** Unknown.

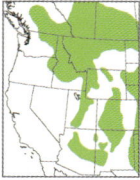

SOFT-LINED WAVE
Scopula inductata
Uncommon RANGE-WIDE
91a-0115 (7169)

WS 17–24 mm Pale tan wings are lightly peppered brown. Brownish AM and diffuse median lines are relatively straight; PM line is slightly wavy. ST area is lightly shaded with wavy pale ST line through middle. Has tiny black discal dots on all wings. Outer margin of HW is rounded. **HOSTS** Aster, clover, dandelion, ragweed, sweetclover, and other herbaceous plants.

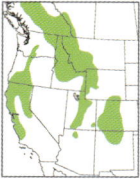

GENTLE WAVE
Scopula luteolata
Uncommon RANGE-WIDE
91a-0116 (7170)

WS 23–25 mm Resembles Soft-lined Wave, but is darker tan, with indistinct curving AM line, PM line slightly darker, and PM area often forming a lighter band. All wings lack discal dots. **HOSTS** Unknown.

CARPETS, PUGS, HYDRIOMENAS, and JESTERS

SUPERFAMILY Geometroidea (91a)
FAMILY Geometridae SUBFAMILY Larentiinae

Small to medium-sized moths with short broad wings that are typically held folded flat beside the body when at rest, though a few species may rest with wings partly spread. The pugs (genus *Eupithecia*) have longish rounded wings that they hold spread at rest, with the hindwing typically tucked beneath the forewing. Patterns are often cryptic, providing camouflage against tree bark during daytime; similar species are often separated by subtle differences in wing markings. The pugs can be particularly difficult to identify visually, and definitive identification can often be made only by genitalia or DNA barcoding. Adults regularly come to lights at night, and a few species can be somewhat abundant.

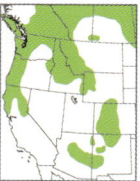

MOTTLED GRAY CARPET
Cladara limitaria
Common RANGE-WIDE
91a-0125 (7637)

TL 14–17 mm Extremely variable. Pale to dark gray FW has a black-shaded PM line that curves widely around oblong black discal spot. Thin black basal line forms a right angle. Some individuals have green or olive bands below basal, AM, and PM lines. Otherwise, weight of lines, darkness of PM shading, and color between lines all occur on a broad spectrum. **HOSTS** Coniferous trees.

POWDERED BIGWING
Lobophora nivigerata
Uncommon RANGE-WIDE
91a-0127 (7640)

TL 13–15 mm Powdery whitish-gray FW has indistinct lines ending as diffuse patches along costa. Basal area often shaded slightly darker. Elongated discal spot is blackish. **HOSTS** Quaking aspen; also balsam poplar, alder, birch, and willow.

MODEST
WAVE

SOFT-LINED WAVE

actual size

GENTLE WAVE

MOTTLED GRAY CARPET

actual size

POWDERED BIGWING

TREBLE-BAR MOTH
Aplocera plagiata

[IN] **Common**
91a-0128 (7627)

TL 23–27 mm Light gray FW has a pointed apex, creating a sharply deltoid appearance. Median and PM lines are triple, shaded slightly inside and dark at costa; triple AM lines are fainter with no costal patch. Apex has a dark brown apical dash. **HOSTS** St. John's wort. **NOTE** Introduced from Europe in 1970s to control St. John's wort.

ORANGE BEGGAR
Eubaphe unicolor

Uncommon
91a-0135 (7444)

RANGE-WIDE

TL 12–14 mm Semitranslucent orange FW has opaque orange central veins. Antennae and legs are contrastingly black. **HOSTS** Violet.

TWELVE-LINED CARPET
Nomenia duodecemlineata

Uncommon
91a-0141 (7426)

RANGE-WIDE

TL 10–13 mm Gray FW has double PM line (or with shadow of a third line). Basal line crosses wing fully. Midpoint veins may be darkly marked at PM line but lack black patches. Appears "broad-shouldered" with pointed apex. Male antennae unipectinate. **HOSTS** Gray oak; probably also other oaks.

FORGOTTEN CARPET
Nomenia obsoleta

Uncommon
91a-0142 (7427)

RANGE-WIDE

TL 10–13 mm Whitish FW has large dark patches at costa and midpoint of PM line and pair of black spots on inner margin at AM and PM lines. Basal line reaches only to midpoint vein. Appears "broad-shouldered" with pointed apex. Male antennae unipectinate. **HOSTS** Unknown. **NOTE** Until very recently, there had been much confusion and misidentification among the *Nomenia* and *Venusia* species. The name *obsoleta* was used for paler Pearsall's, while true *obsoleta* had been lumped in with Twelve-lined. DNA barcoding efforts suggest there may in fact be a number of cryptic *Nomenia* and *Venusia* species in our area as well, requiring further study.

WELSH WAVE
Venusia cambrica

Common
91a-0143 (7425)

RANGE-WIDE

TL 12–14 mm Pale gray to whitish FW has many faint, brownish, wavy lines. Double AM and PM lines are black medially and brownish outside; black lines are thicker at costa. Midpoint of PM line has two black dashes along veins. Terminal line is a series of triangles. **HOSTS** Deciduous trees, including alder, apple, birch, serviceberry, and willow.

PEARSALL'S CARPET
Venusia pearsalli

Common
91a-0145 (7429)

RANGE-WIDE

TL 10–13 mm Gray to whitish FW has triple PM line, particularly evident at inner margin, with narrower "shoulders" than the *Nomenia*, and more rounded apex. Male antennae simple and ciliated. A gray "interior form" in eastern BC has very diffuse, indistinct markings and may constitute a separate species. **HOSTS** Unknown.

TREBLE-BAR MOTH

ORANGE BEGGAR

actual size

TWELVE-LINED
CARPET

FORGOTTEN
CARPET

WELSH
WAVE

PEARSALL'S
CARPET

interior
form

GRAND RIVULET
Common

Martania grandis   91a-0150 (7317)

TL 12–13 mm Brown FW has scalloped double AM and PM lines; medial is white, and outer is tan. Median area is darker brown. ST line is a series of white chevrons with a larger white patch at midpoint. Outer portion of ST area is darker. **HOSTS** Unknown.

BROWN BARK CARPET
Uncommon

RANGE-WIDE

Horisme intestinata   91a-0152 (7445)

WS 21–33 mm Light brown FW has many fine parallel lines that create the appearance of broken bark edge. PM and ST lines are deeply scalloped. Lines stop before costa, creating a paler area parallel to costa. **HOSTS** Clematis.

MOURNFUL PUG
Rare

NORTH

Eupithecia maestosa   91a-0166 (7460)

SOUTH

WS 17–21 mm Tan FW has a slight olive wash when fresh. PM line is edged dark brown at costal end. Dusky brown ST area has two large tan patches in central and costal FW. Thin ST line is white. Abdomen has dark band near base and a line of black-edged white dots dorsally. **HOSTS** Unknown.

EDNA'S PUG
Rare

RANGE-WIDE

Eupithecia edna   91a-0175 (7466)

WS 21–24 mm Pale gray FW has wavy black AM and PM lines that pinch together into black discal spot to connect near costa. Median area is shaded darker on costal side of the connection point, creating a darker triangle. Dark basal area is bounded by thin, black basal line. Pale ST line is sometimes visible. **HOSTS** Unknown.

BROWN PUG
Common

NORTH

Eupithecia unicolor   91a-0181 (7472)

SOUTH

WS 25–28 mm Warm brown FW has faded, dark brown AM and PM lines that bend to form a V beside dark discal spot. ST line is a series of white dots bordered by black shading. **HOSTS** Douglas-fir, spruce, hemlock, fir, and redcedar.

COMMON EUPITHECIA
Very Common

NORTH & EAST

Eupithecia miserulata   91a-0184 (7474)

CALIFORNIA

WS 12–20 mm Gray FW has faint wavy lines that are usually most obvious along costa. Small white spot is present on FW near inner margin of inconspicuous white ST line. Central vein appears as dark dashes in basal and AM areas. ST and terminal areas are shaded very slightly darker. Black discal spot is slightly elongate. Tip of abdomen is often pale or whitish. **HOSTS** Aster, Canadian horseweed, fleabane, grape, oak, willow, and many others.

TAMARACK LOOPER
Uncommon

NORTH & EAST

Eupithecia misturata   91a-0186 (7476)

CALIFORNIA

WS 18–24 mm Resembles Common Eupithecia but lacks white spot on ST line. Thin, sometimes faint, dark median line loops around and touches inner end of elongate black discal spot. PM line has darker black dots at veins. **HOSTS** Generalist on deciduous and coniferous trees and woody shrubs.

GRAND RIVULET

BROWN BARK CARPET

actual size

MOURNFUL PUG

EDNA'S PUG

BROWN PUG

COMMON EUPITHECIA

TAMARACK LOOPER

RED-BANDED PUG
Uncommon · RANGE-WIDE

Eupithecia rotundopuncta 91a-0203 (7496)

WS 17–22 mm FW has reddish-brown (sometimes brownish-gray) AM and PM bands separated from dusky median area by pale, double AM and PM lines. White ST line is relatively straight. ST area has a dusky patch near midpoint, adjacent to a whitish apical dash. Abdomen has a white-edged, dark band at base. **HOSTS** Willow.

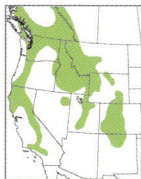

CHALKY PUG
Uncommon · RANGE-WIDE

Eupithecia cretaceata 91a-0236 (7533)

WS 26–34 mm Chalky whitish FW has many wavy tan lines that are darker at costa. AM and PM lines have tiny black dots at veins. ST area is lightly shaded tan behind thick, wavy, white ST line. Abdomen has thin black line at base. **HOSTS** Green false hellebore.

GRAY-LINED PUG
Rare · RANGE-WIDE

Eupithecia behrensata 91a-0238 (7535)

WS 24–28 mm Light gray FW has thin, dark gray, double AM, median, and PM lines that turn upward at costa. ST area is shaded dark gray. Discal spot is black. Abdomen has a darker band at base. **HOSTS** Unknown.

LARCH PUG
Uncommon · RANGE-WIDE

Eupithecia annulata 91a-0245 (7543)

WS 23–29 mm Light tan to brown FW has scalloped white ST line well defined within shaded gray ST/terminal area. Double AM, median, and PM lines are most evident at costa where they form a light brown patch. Black discal spot is elongate. Costal and central veins are often marked black with pale dashes. **HOSTS** Redcedar, spruce, fir, hemlock, Douglas-fir, and pine; also Garry oak.

OLIVE PUG
Uncommon · RANGE-WIDE

Eupithecia olivacea 91a-0247 (7546)

WS 24–30 mm Resembles Larch Pug, but pale AM, median, and PM bands are usually more clearly defined, and ST/terminal area is not noticeably darker than rest of FW. Discal spot is black but sometimes small. FW color variable from pale gray to olive brown. **HOSTS** Douglas-fir.

ZELMIRA PUG
Uncommon · RANGE-WIDE

Eupithecia zelmira 91a-0256 (7555)

WS 22–27 mm Light gray to whitish FW has broad brownish rectangles along costa. Thin median line touches black discal spot. Teeth on deeply scalloped PM line are black, appearing as dashes when line is faint. ST line is double, sometimes faint, and shaded brownish at midpoint and anal angle. Abdomen has dark band at base. **HOSTS** Unknown.

SPRUCE CONE PUG
Rare · RANGE-WIDE

Eupithecia mutata 91a-0278 (7575)

WS 17–22 mm Light tan to light gray FW has reddish-brown bands outward to doubled AM and PM lines. Large, black discal spot is often set in a band of dusky shading. Abdomen has a reddish-brown band at base. In rare individuals, reddish-brown ST band is faded or absent. **HOSTS** Seeds of spruce cones.

RED-BANDED PUG

CHALKY PUG

GRAY-LINED PUG

actual size

LARCH PUG

OLIVE PUG

ZELMIRA PUG

SPRUCE CONE PUG

DUN-STRIPED PUG
Eupithecia gilvipennata

Uncommon
91a-0284 (7581)

RANGE-WIDE

WS 23–27 mm Variable. Dark gray wings often have a broad tan streak running from base to apex. This is sometimes reduced to tan patches on a gray background, and some individuals are entirely gray. Whitish (sometimes indistinct or rarely absent) AM, median, and PM lines are double, with one broad and one narrow line. Thin white ST line is sharply wavy, with a white spot near anal angle. Front of thorax has a narrow tan patch. **HOSTS** Manzanita.

SHARP-WINGED PUG
Eupithecia acutipennis

Uncommon
91a-0288 (7586)

RANGE-WIDE

WS 18–25 mm Narrow and pointed FW is light brown with gray shading alongside costa. White ST line is relatively straight, broader at inner margin, and disappearing before black apical dash, with brown shading basally. Dark streak at inner margin of median line is in line with dark band at base of abdomen. Discal spot is black. **HOSTS** Sagebrush.

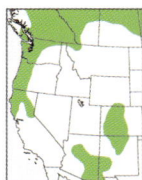

WORMWOOD PUG
Eupithecia absinthiata

Rare
91a-0289 (7586.1)

RANGE-WIDE

WS 22–26 mm Pale gray FW has large black discal spot and black marks along costa at basal, AM, median, and PM lines. Lines are faint except as tiny black dots along veins. White ST line is very thin (often indistinct), with larger white patch near anal angle. Abdomen has dark band at base. **HOSTS** A wide variety of plants, including aster, goldenrod, and wormwood.

YELLOW-SPOTTED PUG
Eupithecia subapicata

Common
91a-0290 (7587)

NORTH

SOUTH

WS 22–28 mm Dusky FW has large, pale yellowish patch near apex and warm brown shading through central area, along inner margin, and outer ST area. PM line is most visible as tan patch at costa, and thin wavy ST line widens to white spot at anal angle. Discal spot is black. **HOSTS** Unknown.

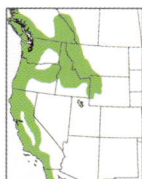

GRAEF'S PUG
Eupithecia graefii

Common
91a-0303 (7600)

RANGE-WIDE

WS 20–26 mm Variable. Brown, tan, gray or dark gray FW has brown discal spot and diffuse brown patches in ST area at midpoint and near apex. Dark AM and PM lines are triple. Abdomen has a dark band at base. In some individuals, ST/terminal area is shaded dark. **HOSTS** Manzanita and madrone.

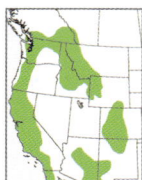

BROWN-PATCHED PUG
Eupithecia nevadata

Uncommon
91a-0304 (7601)

NORTH & EAST

CALIFORNIA

WS 22–26 mm Resembles Tawny Eupithecia, but FW is grayer and brown patch at median area usually reaches inward to discal spot. Black dashes on veins are usually well defined, and pale ST line is less distinct. Thorax is gray to brownish. **HOSTS** Ceanothus, cliffrose, bitterbrush, coffeeberry, deerweed, and island broom.

DUN-STRIPED PUG

SHARP-WINGED PUG

actual size

WORMWOOD PUG

YELLOW-SPOTTED PUG

GRAEF'S PUG

BROWN-PATCHED PUG

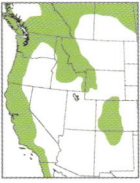

TAWNY EUPITHECIA

Eupithecia ravocostaliata

Common

91a-0308 (7605)

WS 22–25 mm Silvery-white FW has wide brown patches along costa. Inner margin of median area and ST area, and central ST area, are shaded brownish. Central vein and veins in PM area are lightly traced with black dashes. Terminal area is shaded grayish, contrasting slightly with wavy pale ST line. Thorax is pale, and abdomen is brown with a pale tip. HOSTS Willow; also poplar, alder, birch, *Prunus*, cascara, and viburnum.

GRAY-PATCHED PUG

Eupithecia gypsata

Rare

91a-0317 (7611)

WS 16–20 mm Snowy-white FW has large square gray patch at outer median area, smaller gray patch at apex bisected by white ST line, and gray dashes along costa basally. Inner median area is washed lightly brownish. Terminal area has brownish shading. Thorax is white, and abdomen is brown with white tip. HOSTS Unknown.

GREEN PUG

Pasiphila rectangulata

[IN] Common

91a-0331 (7625)

WS 19–23 mm Green wings are broad and rounded. FW has black, slightly wavy AM line and pale double PM line edged in black medially. Median and ST areas are washed dusky. Wavy pale ST line is sometimes indistinct. Green color may fade with age. HOSTS Apple, hawthorn, cherry, pear, and serviceberry.

AUTUMNAL MOTH

Epirrita autumnata

Common

91a-0332 (7433)

TL 18–20 mm Pale gray FW has many faint, scalloped, gray lines. Double AM and PM lines often shaded brownish. Central vein is heavily marked black at AM and PM lines; other veins are marked with black dots at tips of scallop teeth. Terminal line is paired black dots. HOSTS Deciduous and coniferous trees, including ash, birch, maple, oak, and pine.

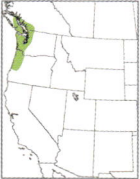

WINTER MOTH

Operophtera brumata

[IN] Common

91a-0335 (7436)

TL 15–17 mm Resembles Bruce Spanworm, but all markings are more indistinct. AM and median lines are straightish and do not angle strongly basally at costa, and median area is usually shaded slightly darker. Central vein is usually dark from base to AM area. Some individuals are weakly marked, with dark vein dashes at ST line and central vein being most prominent. Female has undeveloped wings and is flightless. HOSTS Willow, quaking aspen, paper birch, balsam poplar, and bigleaf maple.

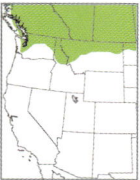

BRUCE SPANWORM

Operophtera bruceata

Uncommon

91a-0336 (7437)

TL 17–19 mm Pale gray to sooty-brown FW has soft scalloped cross-lines marked with darker vertical dashes at veins. Median line has a dark dash across the split central vein. Inner vein is pale or unmarked where it crosses median area. In some individuals, the narrow PM area is shaded darker. Terminal line is single dark dots. Abdomen is gray. Separable from Autumnal Moth by AM and PM lines, central vein, and terminal line. Female has undeveloped wings and is flightless. HOSTS Deciduous trees, including aspen, beech, maple, and willow.

TAWNY EUPITHECIA

GRAY-PATCHED PUG

GREEN PUG

AUTUMNAL MOTH

WINTER MOTH

BRUCE SPANWORM

actual size

WESTERN BRUCE SPANWORM
Operophtera occidentalis

Common
91a-0337 (7438)

RANGE-WIDE

TL 17–19 mm Resembles Bruce Spanworm, but FW is more coarsely hoary and cross-lines are darker and more distinctly marked. Cross-lines are thick where they reach costa, creating a checkered appearance along costa. Abdomen and costa are washed golden brown, sometimes just weakly. Inner vein is usually dark where it crosses median area. Median area is often contrastingly pale. Many individuals have a broad streak of dark shading in central ST/terminal area. A dark form appears sooty with checkered veins and costa, sometimes with weakly visible cross-lines. Female has undeveloped wings and is flightless. Some individuals with weak golden-brown coloring may resemble particularly well-marked Bruce Spanworm but should be identifiable by the costal pattern. **HOSTS** Quaking aspen, maple, willow, and apple.

GRAY BRUCE SPANWORM
Operophtera danbyi

Uncommon
91a-0338 (7439)

RANGE-WIDE

TL 17–19 mm Resembles Bruce Spanworm, but gray FW has wavy pale ST line, and pale AM and PM lines are more tightly paired. Veins are marked with black dashes. Terminal line is single dots. **HOSTS** Ocean spray.

BARBERRY GEOMETER
Rheumaptera meadii

Very Common
91a-0341 (7290)

RANGE-WIDE

WS 36–44 mm Resembles Tissue Moth, but tooth at costal end of PM line is W-shaped and median area has a prominent black discal spot. Median area is usually shaded in inner half, and AM and PM areas are often orangish or brown. **HOSTS** Barberry.

SCALLOP SHELL
Rheumaptera undulata

Uncommon
91a-0342 (7291)

RANGE-WIDE

WS 36–40 mm Grayish to warm brown FW has many thin, tightly spaced, wavy, dark lines. Median area is often shaded slightly darker. Wavy ST line is white. Fringe of HW is usually scalloped when fresh but may appear straight once worn. **HOSTS** Willow, alder, poplar, and blueberry; also spirea, azalea, and rhodora.

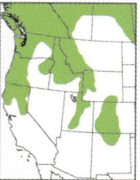

SPEAR-MARKED BLACK COMPLEX
Rheumaptera hastata/subhastata

Uncommon
91a-0344/45 (7293/94)

RANGE-WIDE

WS 30–36 mm Black wings have a broad white PM band with outward-pointing spearhead-like marking at midpoint. The amount of white on the wings is variable, with some individuals showing extensive white markings, and others with just the broad PM band on FW. HW may be entirely black in some individuals. **HOSTS** Deciduous trees and shrubs, including alder, birch, blueberry, poplar, and willow. **NOTE** DNA barcoding data suggests these may actually be a single variable species; more study is needed.

TISSUE MOTH
Triphosa haesitata

Very Common
91a-0347 (7285)

INTERIOR

COASTAL

WS 38–44 mm Grayish to brown FW has multiple scalloped lines. AM and PM bands are darker, sometimes only in outer half. Median area frequently has a diffuse orange patch. Outer margin of all wings is very strongly scalloped. **HOSTS** Coffeeberry; also buckthorn, barberry, hawthorn, oak, and plum.

actual size

WESTERN BRUCE
SPANWORM

GRAY BRUCE
SPANWORM

BARBERRY
GEOMETER

SCALLOP SHELL

SPEAR-MARKED
BLACK COMPLEX

TISSUE MOTH

CALIFORNIA TISSUE MOTH
Triphosa californiata

Uncommon
91a-0348 (7287)

RANGE-WIDE

WS 35–42 mm Resembles Tissue Moth, but veins are speckled black and tan. AM and PM bands are usually less bold. **HOSTS** Hollyleaf redberry.

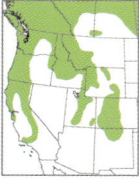

DARK MARBLED CARPET
Dysstroma citrata

Common
91a-0354 (7182)

NORTH

SOUTH

TL 16–20 mm Variable. In western populations, FW has a hoary, dark brown to blackish median area, warm brown AM and PM areas, and a pale yellowish patch at costa of PM area. From costa, PM line drops sharply toward outer margin in a deep V shape. ST line is thin white crescents, sometimes faint. **HOSTS** Generalist on deciduous shrubs and herbaceous plants.

TEN-SPOTTED RHODODENDRON CARPET
Dysstroma sobria

Uncommon
91a-0356 (7184)

RANGE-WIDE

TL 16–20 mm FW has reddish-brown AM and PM bands, and white to hoary gray median area. Costal PM line slopes gently into V shape, and inner PM line is deeply scalloped with long, medial-pointing teeth. Basal and ST areas may be gray or reddish-brown. **HOSTS** Rhododendron; possibly also salal and wintergreen.

MARBLED CARPET
Dysstroma truncata

Rare
91a-0360 (7187)

RANGE-WIDE

TL 15–18 mm Variable. Can resemble Dark Marbled Carpet, but costal PM line is more gently sloping into V shape, and pale yellow patch at costa is smaller; or Ten-spotted Rhododendron Carpet, but basal and ST areas are contrastingly gray or dark brown, AM and PM bands are grayish at costa, and PM area is shaded more heavily along ST line and usually has a narrow yellowish patch at costa. **HOSTS** Alder and willow; also other shrubs and herbaceous plants.

ORANGE-BARRED CARPET
Dysstroma hersiliata

Rare
91a-0363 (7189)

RANGE-WIDE

TL 15–18 mm Variable. FW has hoary black median area (sometimes washed brownish orange in inner half) and a brown to reddish-orange (rarely gray) AM area. PM area may be orange, brown, or gray, but outer portion where ST line creates deeper scallops is nearly always shaded orange or brown. FW always has a short black apical dash, and PM line usually has a pronounced inward tooth near costa. **HOSTS** Currant and gooseberry.

FORMOSA CARPET
Dysstroma formosa

Rare
91a-0365 (7191)

RANGE-WIDE

TL 15–18 mm Variable. Resembles Orange-barred Carpet, but basal edge of AM band is wavy. AM band is gray or only slightly shaded orange, with dark shading along basal line. Outer PM line lacks pronounced inward-pointing tooth. **HOSTS** Currant and gooseberry.

CALIFORNIA TISSUE MOTH

DARK MARBLED CARPET

TEN-SPOTTED RHODODENDRON CARPET

MARBLED CARPET

ORANGE-BARRED CARPET

actual size

FORMOSA CARPET

279

BROWN-BANDED CARPET
Dysstroma brunneata

Uncommon
91a-0368 (7194)

RANGE-WIDE

TL 14–16 mm Dimorphic. FW has hoary black median area, brownish PM area, and either white or brownish AM area. Strongly scalloped PM line crosses the FW in a relatively straight line, with a shallow dip at midpoint. White edging to PM line often becomes a diffuse white patch at costa in white-AM morphs. HOSTS Currant and gooseberry.

BROWN-VEINED CARPET
Dysstroma mancipata

Uncommon
91a-0369 (7195)

RANGE-WIDE

TL 15–18 mm Resembles Dark Marbled Carpet, but costal PM line is gently sloping with a medial-pointing tooth. Hoary median area of FW has white to gray (sometimes dusky brown) central band with dark discal spot, bordered by darker gray bands. Veins in AM and ST areas are reddish brown. HOSTS Possibly woodland star.

CURRANT EULITHIS
Eulithis propulsata

Uncommon
91a-0373 (7199)

RANGE-WIDE

TL 16–19 mm Yellow to tan FW has slightly falcate tip. AM and PM lines are widely double and lightly shaded darker inside. Separable from similar yellow *Eulithis* species (not illustrated) by the scalloped W at the midpoint bend of the PM line. HOSTS Currant, poplar, and willow.

VARIABLE EULITHIS
Eulithis destinata

Uncommon
91a-0379 (7204)

RANGE-WIDE

TL 16–19 mm Variable FW is either entirely shades of gray or brown with a dark median area and warm yellowish AM and PM bands. Patterning resembles *Dysstroma* carpets. Both gray and brown morphs have a dark semicircle on outer margin near apex, ST line as a series of dark crescents, and a prominent three-lobed bulge at the midpoint of the PM line. HOSTS Alder, willow, and subalpine fir.

NORTHWESTERN PHOENIX
Eulithis xylina

Common
91a-0382 (7207)

RANGE-WIDE

TL 15–18 mm Resembles Variable Eulithis, but outer ST area has two sharp black darts. AM and PM bands are edged with white, and AM band is usually lightly shaded dark, creating a filigreed look. HOSTS Alder, willow, and shrubby cinquefoil.

BLACK-BANDED CARPET
Antepirrhoe semiatrata

Uncommon
91a-0385 (7210)

RANGE-WIDE

TL 14–17 mm AM band and dark median area resemble *Dysstroma* carpets, but PM and ST areas are of uniform color. Bulge of PM line turns toward inner margin at a rough right angle. An uncommon dark morph is entirely blackish, with white AM, PM, and ST lines. HOSTS Willowherb.

SMALL PHOENIX
Ecliptopera silaceata

Uncommon
91a-0388 (7213)

RANGE-WIDE

TL 15–18 mm Resembles Northwestern Phoenix, but darker FW has a line of white-edged, black-filled chevrons along ST line. Veins in ST area are brownish. Dark thorax and abdomen are shaded brownish dorsally. HOSTS Impatiens and willowherb.

actual size

BROWN-BANDED
CARPET

BROWN-VEINED
CARPET

CURRANT EULITHIS

VARIABLE EULITHIS

BLACK-BANDED
CARPET

NORTHWESTERN
PHOENIX

SMALL PHOENIX

GEORGE'S CARPET
Plemyria georgii

Common
91a-0392 (7216)

RANGE-WIDE

TL 15–18 mm Pale gray FW has brownish median and basal areas. Sharply toothed PM line meets AM line at inner half of median area, creating a line of circles, with innermost often just a dark spot. Dark basal line is triple. **HOSTS** Deciduous trees, including birch, poplar, and willow.

JUNIPER CARPET
Thera juniperata

[IN] Uncommon
91a-0393 (7217)

RANGE-WIDE

TL 15–18 mm Light gray FW has darker gray basal and median areas and a thin black apical dash. AM, PM, and basal lines are thin and blackish, edged in white. PM line is strongly scalloped toward inner margin, often connecting to AM line. **HOSTS** Juniper.

ROCKY CARPET
Thera otisi

Rare
91a-0395 (7219)

RANGE-WIDE

TL 14–16 mm Hoary gray FW has relatively straight, parallel basal, AM, and PM lines that angle basally at costa. Basal and median areas are shaded slightly darker gray. Median area has a small black discal spot. White ST line is scalloped and bordered above with darker shading. **HOSTS** Juniper.

GUENÉE'S CARPET
Ceratodalia gueneata

Common
91a-0397 (7221)

RANGE-WIDE

WS 23–28 mm Underside of wings is mottled brownish to grayish and has dark median area and white PM line with prominent angle at midpoint. Outer margin of HW is also distinctly angled at midpoint. **HOSTS** Knotweed; possibly also western hemlock. **NOTE** Often rests with wings folded above its body.

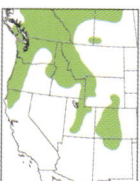

WHITE-STRIPED BLACK
Trichodezia albovittata

Common
91a-0398 (7430)

RANGE-WIDE

WS 20–25 mm Black FW has a distinctive broad white median band. HW is black with a partly white fringe. **HOSTS** Impatiens, willowherb, and meadow-rue. **NOTE** Often seen during daytime along wooded trails and forest edges.

CALIFORNIA BLACK
Trichodezia californiata

Uncommon
91a-0400 (7432)

RANGE-WIDE

WS 20–25 mm Black FW has a bold white PM line and narrow white AM and basal lines. Sometimes shows a small white discal spot and white patch at midpoint of terminal line. Fringe is checkered. **HOSTS** Oxalis; likely others. **NOTE** Often seen during daytime along wooded trails and forest edges.

SOMBER CARPET
Disclisioprocta stellata

Very Common
91a-0401 (7417)

RANGE-WIDE

TL 13–17 mm Dark brown (sometimes warm brown to tan) wings are densely patterned with dark scalloped lines. Basal, AM, PM, and ST lines are thinly edged with white, sometimes appearing fragmented or indistinct. PM line has a W shape at midpoint. Median area is often slightly darker. **HOSTS** Amaranth, American pokeweed, bougainvillea, and common devil's-claw.

GEORGE'S CARPET

JUNIPER CARPET

ROCKY CARPET

actual size

GUENÉE'S CARPET

WHITE-STRIPED
BLACK

CALIFORNIA BLACK

SOMBER CARPET

SHARP-ANGLED CARPET

Euphyia intermediata

Uncommon
91a-0403 (7399)

TL 11–14 mm White FW has dark brown median area and a brown subapical patch on costa bisected by wavy, white ST line. PM line has a tiny sharp tooth projecting into median area near costa, and double white AM lines are wavy. Terminal area is shaded with dusky brown. **HOSTS** Chickweed, elm, impatiens, mustard, and others.

WOODGRAIN RIVULET

"Perizoma" custodiata

Common
91a-0405 (7328)

TL 12–16 mm Brown FW has darker median and basal areas, with dark brown triple basal, AM, and PM lines thinly edged outwardly with white (sometimes nearly absent). AM and PM bands are usually two-toned warm brown. Thin white ST line is slightly scalloped. Pale apical dash is bordered by dark shading.
HOSTS Alkali heath; possibly other sea-heaths. **NOTE** Reassigned to as-yet-unnamed new genus in Euphyiini, Larentiinae.

NEW MEXICO CARPET

Archirhoe neomexicana

Very Common
91a-0406 (7295)

TL 19–22 mm Light brown to grayish FW has many parallel scalloped lines; basal, AM, and PM lines are blackish. AM and PM bands are darker and connected near midpoint. Midpoint of PM line bulges downward. Wavy ST line is white. **HOSTS** Black cherry.

LABRADOR CARPET

Xanthorhoe labradorensis

Uncommon
91a-0411 (7368)

TL 12–14 mm Pale gray FW has reddish-brown median band and basal area. Bold black AM line is nearly straight. Outer third of triple PM line and costal portion of ST area are shaded black. **HOSTS** A variety of plants, including cabbage, hemlock, peppergrass, and radish.

NORTHERN CARPET

Xanthorhoe abrasaria

Rare
91a-0413 (7370)

TL 12–14 mm Gray to light brown FW has darker, narrow AM band and tripled PM lines that are darker in costal half. White ST line is wavy. Median area is slightly paler. **HOSTS** Bedstraw. **NOTE** Often encountered in vegetation during daytime.

WESTERN RED TWIN-SPOT

Xanthorhoe defensaria

Very Common
91a-0429 (7386)

TL 13–16 mm Dimorphic. Resembles Northern Carpet, but triple AM and PM bands are darker and median area is often completely shaded dark. Median area is usually pinched by AM and PM lines in inner half. AM band often has dusky shading, and terminal area has blackish patches at anal angle and inside from pale apical dash. Wavy ST line is white. **HOSTS** Alder, maple, currant and gooseberry, and willow; also other deciduous shrubs and trees.

SHARP-ANGLED CARPET

actual size

WOODGRAIN RIVULET

NEW MEXICO CARPET

LABRADOR CARPET

NORTHERN CARPET

WESTERN RED TWIN-SPOT

RED TWIN-SPOT
Xanthorhoe ferrugata

Common (uncommon in South)
91a-0431 (7388)

RANGE-WIDE

TL 10–13 mm Tricolored FW has blackish (sometimes maroon) median area, reddish-brown AM, basal, and outer ST areas, and light gray inner ST and terminal areas. Indistinct, white, scalloped ST line has a black-shaded patch just in from costa, creating the appearance of twinned spots. **HOSTS** Herbaceous plants, including chickweed and ground ivy.

TOOTHED BROWN CARPET
Xanthorhoe lacustrata

Uncommon
91a-0434 (7390)

RANGE-WIDE

TL 11–14 mm Resembles Sharp-angled Carpet, but AM line is smoothly curving, subapical patch is darkest above ST line, and white PM line lacks tiny tooth near costa. Terminal area is only lightly shaded. **HOSTS** Birch, blackberry, hawthorn, and willow.

CALICO CARPET
Enchoria lacteata

Uncommon
91a-0439 (7403)

RANGE-WIDE

TL 10–12 mm Warm brown FW has broad white (sometimes brownish) median band bordered by brown bands and containing a small black discal dot, and thin white AM and PM lines. AM line is edged medially with black in inner half. Apex has a pale patch beside large diffuse blackish patch in ST area. **HOSTS** Miner's lettuce and other *Claytonia*.

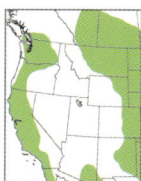

GEM MOTH
Orthonama obstipata

Uncommon
91a-0446 (7414)

NORTH

SOUTH

TL 8–12 mm Sexually dimorphic. Male has tan FW with a dark discal spot. Female has maroon FW with a white-ringed discal spot. Both sexes have a dusky to brownish median band that is sometimes fragmented. Wavy lines are dark, thinly edged with white, sometimes appearing fragmented. **HOSTS** Dock, ragwort, and other herbaceous plants.

BENT-LINE CARPET
Costaconvexa centrostrigaria

Common
91a-0448 (7416)

INLAND

COASTAL

TL 9–12 mm Sexually dimorphic. Gray FW has a blackish AM line and outer PM line, often shaded brown medially. Median area is pale in male and dusky in female and in both sexes contains a black discal dot. Wavy line in ST area has strong black dots along veins. Terminal area usually has dusky shading. **HOSTS** Knotweed, smartweed, and other low plants.

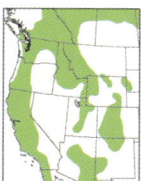

FALCATE BROWN CARPET
Zenophleps lignicolorata

Common
91a-0450 (7406)

NORTH & EAST

CALIFORNIA

TL 13–16 mm Soft brown FW has smooth AM and PM lines shaded darker medially. Faint scalloped lines in median area connect at veins to create a chain appearance. Veins in AM and PM areas are marked with paired black dots. Pale zigzag ST line is often indistinct. Apex is slightly falcate. **HOSTS** Engelmann spruce, limber pine, and other conifers; also bedstraw.

RED TWIN-SPOT

TOOTHED BROWN CARPET

CALICO CARPET

male

GEM MOTH

female

female

BENT-LINE CARPET

male

actual size

FALCATE BROWN CARPET

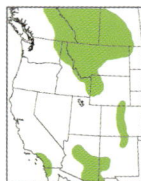

ALPINE BROWN CARPET

Zenophleps alpinata

Rare 91a-0452 (7408)

RANGE-WIDE

TL 10–13 mm Very similar to Falcate Brown Carpet but smaller, and FW apex is pointed but not falcate. Wings average paler brown. AM line is flatter, not noticeably curving where it meets inner margin, and curving less strongly to costa. **HOSTS** Unknown.

OCHRE BROWN CARPET

Zenophleps obscurata

Uncommon 91a-0453 (7409)

RANGE-WIDE

TL 12–14 mm Pale to brown FW has jagged black double basal, AM, and PM bands filled with olive yellow. Discal spot is olive yellow, sometimes with black dot at center. Outer ST area and middle of ST/PM area are shaded diffuse gray. **HOSTS** Unknown.

WHITE-BANDED TOOTHED CARPET

Epirrhoe alternata

Uncommon 91a-0454 (7394)

RANGE-WIDE

TL 10–13 mm Brown FW has broad double PM line, darker median area with black discal spot, and narrow scalloped ST line. Most individuals also have a white double AM line, but this can be reduced or absent. Inner ST area is often washed white. **HOSTS** Bedstraw.

ORANGE-WINGED CARPET

Epirrhoe plebeculata

Common 91a-0455 (7395)

INLAND

COASTAL

TL 12–13 mm Warm brown FW has double AM and PM lines, with medial line white and outer line tan. Median area is grayish, with darker brown AM and PM borders and strong black discal spot. HW is orange with a dashed black terminal line. **HOSTS** Bedstraw.

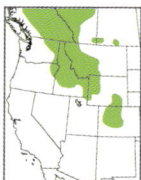

SPERRY'S TOOTHED CARPET

Epirrhoe sperryi

Rare 91a-0456 (7396)

RANGE-WIDE

TL 12–14 mm Resembles Spear-marked Black but lacks spearhead projections at midpoints of PM line. **HOSTS** Unknown, but likely bedstraw. **NOTE** Diurnal.

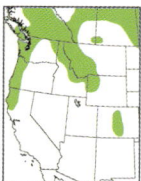

WHITE-RIBBONED CARPET

Mesoleuca ruficillata

Uncommon 91a-0458 (7307)

RANGE-WIDE

TL 15–17 mm Dark blue-gray FW has broad, pure white median area with a small dark discal dot. Scalloped white lines in PM and ST area are backed by light bluish gray in middle. Terminal band is warm brown. **HOSTS** Birch and blackberry.

WESTERN WHITE-RIBBONED CARPET

Mesoleuca gratulata

Very Common 91a-0459 (7308)

RANGE-WIDE

TL 14–16 mm Dark gray FW has broad, pure white median band bordered by olive bands, containing large dark discal spot. White ST line bleeds into terminal area along inner half. Apex has an olive patch bisected by white ST line. **HOSTS** *Rubus* and beaked hazelnut.

ALPINE BROWN CARPET

OCHRE BROWN
CARPET

WHITE-BANDED
TOOTHED CARPET

ORANGE-WINGED CARPET

SPERRY'S TOOTHED CARPET

actual size

WHITE-RIBBONED CARPET

WESTERN WHITE-RIBBONED
CARPET

DOUBLE-BANDED CARPET

Common

Spargania magnoliata   91a-0463 (7312)

TL 15–17 mm   Pale gray FW has many parallel wavy lines. AM and PM bands are paler, and median area is slightly darker. AM band often has teeth pointing into median area. Dark discal spot is a thick rectangle. HOSTS Willowherb and evening primrose.

WHITE-BANDED CARPET

Uncommon

Spargania luctuata   91a-0464 (7313)

TL 16–18 mm   Blackish FW has broad, white, double, scalloped PM band. Median area is often darker black. HW is usually black but uncommonly may have a white PM band. HOSTS Fireweed.

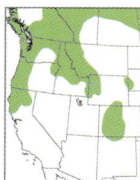

VARIABLE CARPET

Common

Anticlea vasiliata   91a-0465 (7329)

TL 14–17 mm   Variable. Brown FW has triple AM line with uppermost line bold black. Scalloped black PM line has bold black section costal to W shape at midpoint. Thin white ST line has small white spot at midpoint. Median area may be brown or pale, and AM and PM bands may be brown or purplish gray. HOSTS Raspberry and Carolina rose.

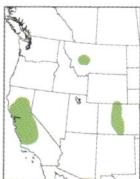

PURGED HYDRIOMENA

Rare

Hydriomena expurgata   91a-0471 (7224)

TL 16–19 mm   Resembles Coastal Hydriomena, but basal line is bold and usually evenly curved to meet inner margin at a shallow angle. Dark AM band is broader and less wavy. PM line lacks tooth beneath discal dash. HOSTS Unknown.

FURIOUS HYDRIOMENA

Rare

Hydriomena irata   91a-0475 (7228)

TL 17–19 mm   Resembles Coastal Hydriomena, but basal line is less steep and has a small bend at midpoint. Black borders to median area are strongly scalloped, with teeth often meeting to create circles in inner half. Basal, AM, and PM bands are usually brownish, and inner median area is often shaded lightly to heavily brownish. HOSTS Spruce, hemlock, and Douglas-fir; occasionally other conifers.

COASTAL HYDRIOMENA

Uncommon

Hydriomena marinata   91a-0478 (7231)

TL 16–19 mm   Resembles Oak Winter Highflier but is less variable in color, ranging from pale to dark grayish brown. Central band of median area is always paler. Straightish black basal line meets inner margin at a steep angle. Median line is straightish, and PM line is straight near inner margin and has a pronounced tooth at costal end pointing toward small black discal dash. Black apical dashes are usually pronounced. HOSTS Sitka spruce, Douglas-fir, and western hemlock; probably also mountain hemlock.

DOUBLE-BANDED CARPET

WHITE-BANDED CARPET

VARIABLE
CARPET

PURGED
HYDRIOMENA

FURIOUS HYDRIOMENA

COASTAL HYDRIOMENA

actual size

EDEN HYDRIOMENA
Hydriomena edenata   **Rare**   91a-0479 (7232)

NORTH

SOUTH

TL 16–20 mm Resembles Coastal Hydriomena, but basal line usually connects to black streak along inner margin. A dark dash connects basal and median lines at midpoint. Bold apical dashes reach PM line on costal side of tooth. **HOSTS** Unknown; probably Douglas-fir.

JULY HIGHFLYER
Hydriomena furcata   **Uncommon**   91a-0504 (7257)

RANGE-WIDE

TL 16–19 mm Resembles Oak Winter Highflier, but basal line on FW is not as steep, and PM band is straight across wing, turns sharply up, then straightens to meet costa. Commonly shows a pale patch at midpoint of ST band. **HOSTS** Willow; also poplars, alder, birch, and *Prunus*.

FIVE-BANDED HYDRIOMENA
Hydriomena quinquefasciata   **Rare**   91a-0505 (7258)

RANGE-WIDE

TL 16–19 mm Resembles July Highflyer, but blackish ST band substantially narrows at inner margin and costa. Basal, median, and PM lines are distinctly narrower than AM and ST bands. **HOSTS** Willow.

WHITE-BANDED HYDRIOMENA
Hydriomena albifasciata   **Uncommon**   91a-0508 (7261)

RANGE-WIDE

TL 15–19 mm Extremely variable. FW usually olive green to dark gray, sometimes pale gray. AM band appears as a long, thick, black line in inner half of FW. PM line is incomplete, often absent from inner half of FW. ST line has a white mark or patch near midpoint. Median area is often white but may be dark or uniform with the rest of the wing color. **HOSTS** Coast live oak; also reported from maple.

BEAUTIFUL HYDRIOMENA
Hydriomena speciosata   **Uncommon**   91a-0510 (7263)

RANGE-WIDE

TL 18–20 mm Boldly marked hydriomena has alternating gray to olive-green and blackish bands on FW. Median area is shaded partly to fully white, frequently in irregular patches that do not appear to match up with the usual FW lines. Squiggly green to gray PM line has a white-edged W shape near midpoint. **HOSTS** Grand fir, Douglas-fir, and western hemlock; also spruce and pine.

OAK WINTER HIGHFLIER
Hydriomena nubilofasciata   **Very Common**   91a-0523 (7276)

RANGE-WIDE

TL 15–17 mm Extremely variable. FW can be greenish gray, brownish, or olive. Wide basal, AM, PM, and ST bands are usually a contrasting color. Median area is often lighter, sometimes darker, occasionally with a white or brownish patch in outer half. However, across all color morphs, basal line is strongly angled, terminal area is dark, and AM and PM lines are roughly straight or slightly curved through inner half, then angle sharply upward to meet costa. **HOSTS** Oaks, including Garry and coast live oak. **NOTE** By far our most common hydriomena.

EDEN HYDRIOMENA

JULY HIGHFLYER

FIVE-BANDED
HYDRIOMENA

WHITE-BANDED
HYDRIOMENA

BEAUTIFUL HYDRIOMENA

actual size

OAK WINTER
HIGHFLIER

MANZANITA HYDRIOMENA

Uncommon

Hydriomena manzanita 91a-0524 (7277)

TL 18–21 mm Gray FW is much narrower than on our other hydriomenas. Bands may be dark gray or barely visible. PM and ST lines are pale gray, and veins are blackish. AM and PM areas are shaded lightly with brown. HOSTS Madrone and manzanita.

DARK-PATCHED JESTER

Rare

Stamnodes blackmorei 91a-0547 (7334)

TL 14–18 mm Resembles White-patched Jester, but pale patch at apex of FW underside is creamy and mottled with brown, and white costal dashes at AM and median lines are more crisply defined. HOSTS Unknown.

WHITE-PATCHED JESTER

Uncommon

Stamnodes albiapicata 91a-0548 (7335)

TL 16–19 mm Underside of wings is mottled brown. FW has large white apical patch and white dash at costa of PM line. HW usually has diffuse brown patches near costa and inner margin of PM area. HOSTS Baby blue eyes, fiesta flower, and caterpillar scorpionweed. NOTE Rests with wings folded above body. Sometimes encountered in vegetation during daytime.

BROWN JESTER

Uncommon

Stamnodes affiliata 91a-0550 (7337)

TL 15–19 mm Underside of wings is mottled warm brown. Diffuse, dark PM band is broken at midpoint to form two sections, sometimes bordered medially by a thin white line. Median area of HW is slightly paler. HOSTS Unknown. NOTE Rests with wings folded above body. Sometimes encountered in vegetation during daytime.

CHESTNUT JESTER

Uncommon

Stamnodes coenonymphata 91a-0554 (7341)

TL 15–19 mm Underside of wings is warm brown to grayish and lightly mottled. Median line on HW forms a V that is shaded dark brown basally, with tiny white discal dot. FW has thick white PM line, visible as a bold rectangle at costa when wings are folded. HOSTS Unknown. NOTE Rests with wings folded above body. Sometimes encountered in vegetation during daytime.

FESTIVE JESTER

Uncommon

Stamnodes formosata 91a-0560 (7347)

TL 13–15 mm Underside of wings is bright peach tan with reddish striations, darker on basal side of bold white PM band of HW. HW has white costal patch at base and large white discal spot. Fringe is checkered white and dark brown. HOSTS Unknown. NOTE Rests with wings folded above body. Sometimes encountered in vegetation during daytime.

MANZANITA
HYDRIOMENA

DARK-PATCHED
JESTER

actual size

WHITE-PATCHED JESTER

BROWN JESTER

CHESTNUT JESTER

FESTIVE JESTER

TOPAZ JESTER
Stamnodes topazata   **Rare**   91a-0562 (7349)

TL 12–13 mm Underside of HW is white, mottled with light brown, with a dark, sharply V-shaped median band and blackish patch at apex. Underside of FW is warm brown, with thick white PM band and white costal marks at AM and basal lines. Fringe is checkered black and white. Upper side of all wings is usually orange. **HOSTS** Unknown. **NOTE** Rests with wings folded above body. Sometimes encountered in vegetation during daytime.

FORMAL JESTER
Stamnodes modocata   **Rare**   91a-0564 (7351)

TL 15–19 mm Underside of wings is mottled gray, with scalloped, black, V-shaped median line, usually bordered outwardly with white. Small, black-bordered, white discal spot is indistinct. Solid terminal line is black. FW has white-bordered black PM line with dark patch apically. **HOSTS** Unknown. **NOTE** Rests with wings folded above body. Sometimes encountered in vegetation during daytime.

ORANGE JESTER
Stamnodes seiferti   **Uncommon**   91a-0566 (7353)

TL 17–20 mm. Underside of wings is mottled gray, usually with a peach cast. HW can appear mostly unmarked or may have a darker V-shaped median band bordered outwardly with white. Underside of FW has white patch along costa at PM line. Upper side of wings is orange; FW has thick curving median bar and wide black apex. **HOSTS** Mountain mahogany and Apache plume. **NOTE** Rests with wings folded above body. Sometimes encountered in vegetation during daytime.

PAINTED JESTER
Stamnodes deceptiva   **Uncommon**   91a-0568 (7355)

TL 13–15 mm Underside of wings is boldly marked with cream-colored lines and black-rimmed, dark gray patches. Center of median band is diffusely orange, and base of wing is pinkish. Inner portion of FW (usually hidden when wings folded) is orange. Upper side of wings is orange with black patches in the pattern of the underside. **HOSTS** Unknown. **NOTE** Rests with wings folded above body. Sometimes encountered in vegetation during daytime.

MARBLED JESTER
Stamnodes marmorata   **Uncommon**   91a-0573 (7363)

TL 17–20 mm Underside of wings is dark brown, with warm brown shading along veins and white striations throughout. HW has bold white median line with V-shaped dip at midpoint, a broken white ST band, and two white discal spots. FW has white PM and ST lines. Fringe is checkered white and dark brown. Upper side of wings is soft brown. **HOSTS** Unknown. **NOTE** Rests with wings folded above body. Sometimes encountered in vegetation during daytime.

TOPAZ JESTER

FORMAL JESTER

ORANGE JESTER

PAINTED JESTER

MARBLED JESTER

actual size

CHECKERED JESTER
Stamnodes tessellata

Uncommon
91a-0574 (7364)

TL 16–19 mm Underside of wings is brownish orange, with bold, black-and-white, checkered pattern along costa, inner margin, and median vein. Fringe is checkered black and white. Upper side is orange, with checkered fringe and costa. **HOSTS** Unknown. **NOTE** Rests with wings folded above body. Sometimes encountered in vegetation during daytime.

SMOOTH JESTER
Stamnoctenis pearsalli

Uncommon
91a-0578 (7357)

TL 17–19 mm Underside of wings is reddish brown. HW has narrow reddish median line that is solid at inner half and reduced to darkened veins in costal half. FW has paler area along costa from dark median dash to apex. **HOSTS** Unknown.

GRAY RIVULET
"Perizoma" curvilinea

Common
91a-0583 (7324)

TL 12–16 mm Light gray FW has dark gray AM and PM bands bordered by narrow scalloped lines. Scalloped ST line is white. Veins in PM and ST areas are marked with black dots. Terminal line is paired black dots at veins. **HOSTS** Unknown. **NOTE** Reassigned to as-yet-unnamed new genus in Larentiinae; unplaced to tribe.

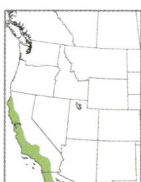

SPOTTED RIVULET
"Perizoma" costiguttata

Common
91a-0584 (7325)

TL 11–13 mm Light gray FW has dark patches along costa at basal, AM, and PM lines. AM and PM lines are thin and double; lower of PM lines is toothed at veins. Indistinct wavy ST line is white, with darker shading at midpoint. **HOSTS** Ocean spray. **NOTE** Reassigned to as-yet-unnamed new genus in Larentiinae; unplaced to tribe.

TWO-BANDED RIVULET
"Perizoma" epictata

Common
91a-0585 (7326)

TL 11–13 mm Light brown FW has dark, scalloped, triple AM and PM lines; medial lines are lighter than outer line, and inner spaces are sometimes shaded dark. Basal line is also dark. AM and ST areas are often warm brown. Pale brown patch at apex is bordered by a dark wavy apical dash. **HOSTS** Unknown. **NOTE** Reassigned to as-yet-unnamed new genus in Larentiinae; unplaced to tribe. Often encountered in vegetation during daytime.

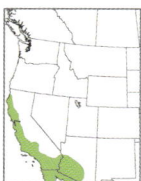

BEADED CARPET
"Euphyia" implicata

Uncommon
91a-0587 (7400)

TL 16–18 mm Warm brown FW has many closely spaced scalloped brown or blackish lines with teeth pointing toward median area. Pale whitish or brown median area is narrow and resembles a necklace of beads. AM and PM lines are whitish and edged medially with black. **HOSTS** Sand verbena. **NOTE** Reassigned to as-yet-unnamed new genus in Larentiinae; unplaced to tribe.

CHECKERED JESTER

SMOOTH JESTER

GRAY RIVULET

SPOTTED RIVULET

actual size

TWO-BANDED RIVULET

BEADED CARPET

299

INFANTS

SUPERFAMILY Geometroidea (91a)

FAMILY Geometridae SUBFAMILY Archiearinae

Small diurnal moths that fly on warm days in early spring, sometimes before the snow has fully melted. Usually found along forest edges or tracks, where the flash of the orange hindwings while in flight or resting on the ground may initially call to mind small butterflies.

THE INFANT
Archiearis infans

Uncommon
91a-0596 (6256)

RANGE-WIDE

TL 15–17 mm Dark brown FW has white patches at outer median and ST areas. HW is orange with black at inner margin and highly visible when moth is flying. **HOSTS** Birch; also willow and poplar. **NOTE** Diurnal.

EMERALDS

SUPERFAMILY Geometroidea (91a)

FAMILY Geometridae SUBFAMILY Geometrinae

Small eye-catchingly green moths with broad rounded wings that are held spread at rest. Many species are superficially similar, and identification often relies on fringe and abdominal markings. Caterpillars of *Synchlora* species "decorate" themselves with bits of their host flowers for camouflage. Adults of virtually all species are nocturnal and will visit lights, though the Day Emerald flies during daytime in grassy woodland openings.

BANK'S EMERALD
Chlorosea banksaria

Uncommon
91a-0601 (7013)

RANGE-WIDE

WS 30–42 mm Pale green wings are mottled with white, with a thin white costa and fringe, and straight white PM line. Abdomen has yellowish spots encircled with reddish color. Forehead and antennae are white. **HOSTS** Blueblossom and ocean spray.

MARGARET'S EMERALD
Chlorosea margaretaria

Uncommon
91a-0602 (7014)

RANGE-WIDE

WS 29–38 mm Resembles Bank's Emerald, but white PM line angles more toward apex, and costa and fringe are more boldly edged with white. FW shape is slightly narrower and more pointed, and green color often has a bluish tinge. **HOSTS** Unknown.

LOVELY EMERALD
Nemoria pulcherrima

Common
91a-0604 (7016)

RANGE-WIDE

WS 25–34 mm Dimorphic. Green form resembles Bank's Emerald, but white PM line is thinner and less distinct, costa has a reddish tinge, fringe is checkered with dark red spots, and wings have a small black discal spot. Brown form has thick, diffuse, black AM and PM lines, a dark fringe, and small black discal dot. **HOSTS** Oak catkins.

SINGLE-LINED EMERALD
Nemoria unitaria

Uncommon
91a-0606 (7018)

RANGE-WIDE

WS 25–34 mm Smooth green wings have thin white AM and PM lines that connect near inner margin of HW to form a loop. Spots on abdomen are white. **HOSTS** Unknown; has been reared on gooseberry.

THE INFANT

actual size

BANK'S EMERALD

MARGARET'S EMERALD

actual size

LOVELY EMERALD

SINGLE-LINED
EMERALD

ARIZONA EMERALD
Nemoria arizonaria **Rare** 91a-0608 (7021)

WS 30–35 mm Heavily mottled green wings have strong, straight AM and PM lines and thick yellowish edging along costa, fringe, and collar of thorax. Brownish abdomen has dark-edged white spots. A second, summer brood is smaller, with lighter coloration and fainter markings. **HOSTS** Oak catkins and leaves.

ROSY EMERALD
Nemoria pistaciaria **Uncommon** 91a-0614 (7027)

RANGE-WIDE

WS 25–32 mm Resembles Lovely Emerald, but fringe is solid pinkish, rather than checkered, costa has a pinkish tinge, and abdomen is plain green or just with small white marks. Discal dots can vary from faint to bold black spots. **HOSTS** Uncertain; probably ceanothus and oak.

COLUMBIAN EMERALD
Nemoria darwiniata **Common** 91a-0622 (7035)

NORTH

SOUTH

WS 27–36 mm Light green wings have narrow, wavy, white AM line (sometimes indistinct) and relatively straight PM line. PM line takes a sharp turn at midpoint of HW to meet inner margin. Abdomen has a chestnut-bordered white spot at base separated by green from a pair of spots at middle. Discal dots are usually brown. Costa is white, and fringe may appear pale or have reddish spots along apical half. **HOSTS** Generalist on deciduous trees and shrubs.

ARDENT EMERALD
Nemoria zelotes **Rare** 91a-0623 (7036)

RANGE-WIDE

WS 25–35 mm Resembles Columbian Emerald, but fringe is white with a thin red terminal line. FW has narrow, curving, white AM line and PM line, with small teeth at thin white veins. **HOSTS** Unknown.

SLANTED EMERALD
Nemoria obliqua **Uncommon** 91a-0624 (7037)

RANGE-WIDE

WS 24–32 mm Green FW has bold, white AM and PM lines that converge toward inner margin. Fringe is pinkish with reddish dots and costa is pinkish. Abdomen has three white dots circled in red. **HOSTS** Skunkbush sumac.

PINK-MARGINED GREEN
Nemoria leptalea **Common** 91a-0628 (7041)

RANGE-WIDE

WS 25–35 mm Resembles Columbian Emerald, but abdomen is plain green or with small pinkish spots at middle segments. Costa may be yellowish or peach. Fringe is pinkish, without any distinct spots at apex. Discal dots are usually indistinctly marked. **HOSTS** California buckwheat and toyon.

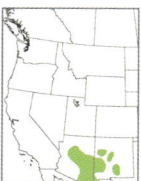

FESTIVE EMERALD
Nemoria festaria **Uncommon** 91a-0631 (7044)

RANGE-WIDE

WS 22–30 mm Resembles Columbian Emerald, but abdomen has four red-bordered white spots (second may be tinted greenish), with no green spaces. Fringe is slightly pinkish, with reddish dots at veins. Wings lack dark discal spots. **HOSTS** Unknown.

ARIZONA EMERALD

actual size

ROSY EMERALD

COLUMBIAN EMERALD

ARDENT EMERALD

SLANTED EMERALD

PINK-MARGINED GREEN

FESTIVE EMERALD

YELLOW-FRINGED EMERALD
Uncommon

RANGE-WIDE

Nemoria glaucomarginaria   91a-0637 (7049)

WS 28–38 mm Resembles Columbian Emerald, but abdominal spots are larger and bordered with a lighter shade of red, fringe is yellowish with reddish dots at veins (sometimes plain yellowish with just reddish tip at apex), and HW has a thin white discal dash. FW lacks any discal markings. **HOSTS** Manzanita.

ILLUSTRATED EMERALD
Very Common   RANGE-WIDE

Dichorda illustraria   91a-0643 (7055)

WS 28–44 mm Green to yellowish-green wings have bold white PM lines that slant from FW apex to HW anal angle when wings are held open. Slanting AM line is present on FW only. Costa is white with black speckles. Veins are yellowish. Tiny black discal dots mark all wings. **HOSTS** Laurel and skunkbush sumac; possibly also *Ribes*.

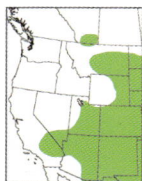

HONEST EMERALD
Uncommon   RANGE-WIDE

Dichorda rectaria   91a-0644 (7056)

WS 27–30 mm Resembles Illustrated Emerald, but speckles on costa are often a lighter brown. PM line on HW terminates higher up the inner margin. Discal spots are usually tiny or absent altogether. **HOSTS** Skunkbush sumac.

WAVY-LINED EMERALD
Very Common   RANGE-WIDE

Synchlora aerata   91a-0645 (7058)

WS 15–24 mm Green to yellow-green wings have slightly wavy white AM and PM lines and (often) white veins. Costa is white. Terminal line is white dots or scallops, and fringe is pale green. All wings have a white discal dash, sometimes indistinct. Abdomen has a narrow white dorsal stripe running entire length. **HOSTS** Generalist on deciduous trees, shrubs, and herbaceous plants.

SOUTHERN EMERALD
Common   RANGE-WIDE

Synchlora frondaria   91a-0646 (7059)

WS 16–20 mm Resembles Wavy-lined Emerald, but AM and PM lines are more dentate. Faint white veins are only sometimes obvious in median and ST areas. **HOSTS** Herbaceous plants, including blackberry, chrysanthemum, and Spanish needle.

OBLIQUE-STRIPED EMERALD
Uncommon   RANGE-WIDE

Synchlora bistriaria   91a-0653 (7065)

WS 23–35 mm Green FW has bold white AM and PM lines gently curving across wing. HW pale green to whitish, with green shading at anal angle. Often rests with wings held tented over abdomen. **HOSTS** Flowers and buds of sunflower, goldenrod, and other composites; occasionally also rose.

WHITE-WINGED EMERALD
Uncommon   RANGE-WIDE

Synchlora faseolaria   91a-0655 (7067)

WS 21–25 mm Green FW has faint, thin white PM line that fades toward costa. AM line is absent. Costa is brownish. HW is white with green shading at anal angle. **HOSTS** California sagebrush.

ILLUSTRATED EMERALD

YELLOW-FRINGED EMERALD

actual size

HONEST EMERALD

WAVY-LINED EMERALD

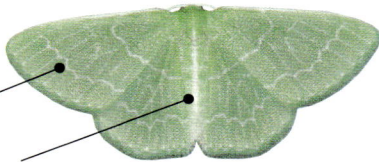

SOUTHERN EMERALD

OBLIQUE-STRIPED
EMERALD

WHITE-WINGED
EMERALD

305

PHOENIX EMERALD
Dichordophora phoenix **Rare** 91a-0657 (7057)

WS 25–33 mm Resembles Honest Emerald, but costa is maroon, lightly speckled with white, and bordered by diffuse yellow green. PM line curves very slightly basally at costa. HOSTS Skunkbush sumac.

BLACKBERRY LOOPER
Chlorochlamys chloroleucaria **Uncommon** 91a-0661 (7071)

WS 16–20 mm Resembles Thick-lined Ivory Emerald, but color is denser and often forms pale striations, particularly on HW. PM line is straight and even. HOSTS Generalist on many herbacous plants, shrubs, and small trees, including blackberry and raspberry.

WESTERN IVORY EMERALD
Chlorochlamys triangularis **Uncommon** 91a-0662 (7072)

WS 16–22 mm Resembles Blackberry Looper, but FW is proportionally slightly longer, and AM and PM lines are often thinner. Easily separable by range; Blackberry Looper does not occur west of the Rocky Mountains. HOSTS Unknown.

THICK-LINED IVORY EMERALD
Chlorochlamys appellaria **Common** 91a-0663 (7073)

WS 15–20 mm Wings are ivory, with heavy green or brownish-red speckling, creating a densely peppery appearance. Ivory AM (absent on HW) and PM lines are thick and slightly uneven. Median area is sometimes shaded darker at lines, and costa is ivory. Thorax and abdomen have a white dorsal line. Discal dots are absent. HOSTS Buckwheat, rabbitbrush, and marsh baccharis.

THIN-LINED IVORY EMERALD
Chlorochlamys phyllinaria **Common** 91a-0664 (7074)

WS 16–23 mm Resembles Thick-lined Ivory Emerald, but AM and PM lines are thinner and wavy. Color is denser and often forms pale striations, particularly on HW. HOSTS Unknown.

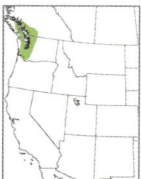

COMMON EMERALD
Hemithea aestivaria **[IN] Very Common** 91a-0673 (7083)

WS 26–38 mm Dark blue-green wings are pointed at FW apex and HW outer margin. Thin white AM and PM lines are wavy, often edged medially with darker shading. Terminal line is red, and fringe is checkered red and white. HOSTS Generalist on deciduous trees and shrubs.

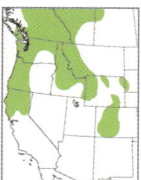

DAY EMERALD
Mesothea incertata **Rare** 91a-0675 (7085)

WS 18–22 mm Yellow-green to blue-green wings have faint wavy AM and PM lines; AM line is often indistinct or absent. Thorax and abdomen plain green. HOSTS Generalist on deciduous trees and shrubs, including birch, willow, and Rosaceae. NOTE Diurnal.

RANGE-WIDE

PHOENIX EMERALD

BLACKBERRY LOOPER

WESTERN IVORY EMERALD

THICK-LINED IVORY
EMERALD

THIN-LINED IVORY EMERALD

COMMON EMERALD

DAY EMERALD

actual size

ANGLES and GRANITES

SUPERFAMILY Geometroidea (91a)

FAMILY Geometridae SUBFAMILY Ennominae TRIBE Macariini

Small to medium-sized moths with broad triangular wings that are usually held flat beside the body or slightly spread. Most species are shades of white, tan, or gray; many have dark costal bars and often ST patches, while other groups have black, wavy transverse lines. These groups, and the *Digrammia* particularly, contain many species that may share overlapping phenotypic traits, making visual identification to species difficult for many individuals. Given here are the most stereotypical traits for each species, but not every individual within a given species will necessarily match all ID marks, and ID should be made, if possible, through consideration of the whole. Adults are nocturnal and will regularly visit lights.

RED-WINGED WAVE
Dasyfidonia avuncularia

Uncommon

91a-0685 (6426)

RANGE-WIDE

TL 14–17 mm Striated, warm to dark brown wing has paler ST area and white patch at costa of ST line. Narrow cross-lines are black and slightly wavy; PM line has a pronounced bulge near costa, pointing toward outer margin. HW is orange, with strong black median and PM lines. **HOSTS** Bitter cherry and other *Prunus*.

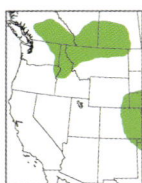

BROWN-BORDERED GEOMETER
Eumacaria madopata

Uncommon

91a-0691 (6272)

RANGE-WIDE

WS 20–25 mm Light grayish-brown wings have relatively straight, crisp, brown AM and median lines that angle sharply basally at costa. Brown PM line is curving. Area below PM line is mottled brown with pale veins and paler apical patch. FW apex and midpoint of HW outer margin are pointed. **HOSTS** Apple, cherry, and plum.

DOT-LINED ANGLE
Psamatodes abydata

Common

91a-0697 (6332)

RANGE-WIDE

WS 22–27 mm Lightly speckled, pale brown wings have a wide mottled gray band in ST area, often with blackish patches on FW in central and costal areas. Scalloped brown AM, median, and PM bands are often broken or indistinct. Dashed terminal line is black. Angulate HW has strong black discal spot. Head and shoulders are orangish. **HOSTS** Sweet acacia, Mexican palo verde, *Sesbania*, and soybean.

BLACK-BANDED ORANGE
Macaria truncataria

Uncommon

91a-0700 (6321)

RANGE-WIDE

WS 12–20 mm Brindled pale to dark orange wings have thick, yellow-edged, black basal, AM, median, and PM bands. **HOSTS** Bearberry and leatherleaf. **NOTE** Diurnal; often found flitting among low vegetation in bogs.

PLAIN ANGLE
Macaria marcescaria

Uncommon

91a-0702 (6323)

RANGE-WIDE

TL 15–17 mm Pale tan to brownish wings look heavily worn or faded even when fresh. AM line is thin and indistinct; PM line is slightly darker, with a small dark dot at midpoint in line with dark, angled discal dash. ST area is shaded slightly darker. Underside of HW has most prominent markings: a darkish brown band in PM area and small black discal dot. FW apex is falcate, with a concavity in apical half of outer margin; HW is lightly scalloped and angled at midpoint. Palpi are long, creating a snouted appearance. Often rests with wings folded above abdomen. **HOSTS** Coyote brush.

RED-WINGED WAVE

BROWN-BORDERED GEOMETER

actual size

DOT-LINED ANGLE

BLACK-BANDED ORANGE

PLAIN ANGLE

FAWN ANGLE

Uncommon

Macaria metanemaria　　91a-0703 (6325)

TL 13–15 mm　Soft brown, occasionally peppery, FW has dark curving AM and straight PM lines bordered medially with pale yellow. ST area is uncommonly shaded darker or with darkish patch at midpoint. Concavity in apical half of outer margin is dark brown. HW is sharply angled at midpoint. **HOSTS** Desert broom, coyote brush, and false goldenaster.

RANNOCH LOOPER

Uncommon

Macaria brunneata　　91a-0704 (6286)

TL 12–15 mm　Yellowish to tan wings are often peppery. Diffuse (sometimes indistinct) brownish AM, median, and PM lines terminate with brown triangle at costa; a fourth costal mark is present at otherwise-absent ST line. Often rests with wings held above abdomen; underside of wings shows the same markings as upper side. **HOSTS** Primarily blueberry but also recorded feeding on aspen, birch, and buffalo berry.

YELLOW-WINGED ANGLE

Rare

Macaria amboflava　　91a-0708 (6284)

TL 13–16 mm　Peppery yellow wings have a large brown blotch at inner margin of PM line, small dark triangles along costa at basal, PM, and ST lines, and a brown-ringed gray discal spot. ST line is a row of diffuse brown blotches, occasionally indistinct. Costa is sometimes paler yellow. Terminal line is a series of dark brown triangles, and fringe is brown. **HOSTS** Wild licorice; possibly also bearberry.

SPLIT-LINED ANGLE

Uncommon

Macaria bitactata　　91a-0720 (6304)

TL 13–17 mm　Light gray FW has a distinctive black median line that begins as a thick bar at costa, slanting toward outer margin, sharply angles basally near midpoint, then turns downward again to meet inner margin. Costa has three other black marks at AM, PM, and ST lines. AM line may be thinly present or visible as a row of black dots. PM line is indistinct, often appearing as black dashes at costa and inner margin and black dots at veins in between. **HOSTS** Currants and gooseberries.

REFINED ANGLE

Uncommon

Macaria colata　　91a-0726 (6308)

TL 13–15 mm　Light gray to whitish FW is striated with black, sometimes heavily. Black, curving PM and ST lines are usually best defined on inner half and shaded between with dark brown. AM and median lines are diffuse, sometimes indistinct; AM line is usually lightly shaded brown basally. Costa has four bold dashes at lines; that at ST line is usually edged apically with whitish. **HOSTS** Big sagebrush and bitterbrush.

DISPATCHED ANGLE

Uncommon

Macaria occiduaria　　91a-0727 (6279)

TL 15–18 mm　Yellow to light tan FW has an indistinct, sinuous PM line shaded diffusely below with brown. Costa has large dark triangles at AM, median, and PM lines and a small rectangle at ST line. AM and median lines are usually indistinct but may be well marked in dark brown. Apical half of yellow fringe is dark brown. Females have reduced FW and rarely fly. **HOSTS** Generalist on deciduous trees and shrubs.

FAWN ANGLE

RANNOCH LOOPER

YELLOW-WINGED ANGLE

SPLIT-LINED ANGLE

REFINED ANGLE

actual size

DISPATCHED ANGLE

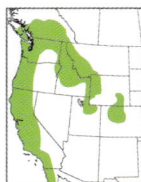

LORQUIN'S ANGLE
Macaria lorquinaria

Uncommon
91a-0734 (6324)

RANGE-WIDE

TL 13–15 mm Light tan to grayish-brown FW has curving, yellowish, brown-edged AM and PM lines that do not reach costa. ST area is lightly shaded, with two dark spots near costa and third at midpoint. Dark discal dash is slanting. Apex of FW is falcate, and HW is slightly scalloped when fresh. Palpi are long, creating a snouted appearance. HOSTS Alder, birch, and willow.

MAPLE ANGLE
Macaria plumosata

Rare
91a-0736 (6296)

RANGE-WIDE

TL 13–15 mm Pale yellowish to tan FW has tiny dark striations. Sinuous PM line is bordered by dark shading in ST area and a large, diffuse, dark patch at midpoint. Diffuse AM and median lines are curving and darken as they reach costa; dark discal spot is often incorporated into median line. Terminal line is a row of distinct black dots. HOSTS Maple.

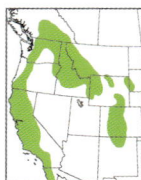

DUSKY ANGLE
Macaria quadrilinearia

Rare
91a-0738 (6288)

RANGE-WIDE

TL 13–18 mm Peppery, pale gray to brownish FW has a dusky patch where diffuse, parallel PM and ST lines meet inner margin. Central terminal area and fringe is shaded dusky, often with a warm brown patch adjacent in ST area. All lines are diffuse and indistinct except for four dark spots where they reach costa. Elongate discal spot is dark. HOSTS Ceanothus.

BUFF-BANDED ANGLE
Macaria guenearia

Uncommon
91a-0740 (6301)

RANGE-WIDE

TL 14–16 mm Peppery purplish-gray FW has a bold yellow ST line edged with warm brown below and, frequently, with a dark-edged tooth at midpoint. Costa is usually also yellow. Black discal spot and terminal line of small black dots are usually the only other markings. HOSTS Unknown.

YELLOW-BANDED ANGLE
Macaria austrinata

Uncommon
91a-0741 (6301.1)

RANGE-WIDE

TL 12–17 mm Resembles Buff-banded Angle, but yellow ST line is edged basally with blackish shading, yellow costa is reduced or absent, and diffuse, indistinct AM and median lines terminate at dark patches along costa. HOSTS Redberry buckthorn.

DECEPTIVE ANGLE
Macaria deceptrix

Rare
91a-0742 (6312)

RANGE-WIDE

TL 11–14 mm White FW has a diffuse black band in ST area, with brown patches at midpoint and sometimes near costa. Black AM and PM lines are thin and wavy; AM line occasionally has black and brown shading basally that mirrors ST band. A dark patch at costa of median line runs into the black discal spot. FW is usually peppery, sometimes quite heavily, so the wing looks gray. HOSTS Knifeleaf condalia.

LORQUIN'S ANGLE

MAPLE ANGLE

DUSKY ANGLE

BUFF-BANDED ANGLE

YELLOW-BANDED ANGLE

actual size

DECEPTIVE ANGLE

GRAY-BANDED ANGLE

Macaria pallipennata

Uncommon
91a-0750 (6317)

RANGE-WIDE

TL 11–15 mm White FW is lightly striated with brown and has diffuse, dark brown AM and PM lines. Diffuse mousy-brown shading fills most of lower half of wing, darkest below PM line and costally from short black subapical dash, and cut through by wide, wavy, white ST line. **HOSTS** Unknown.

BIRCH ANGLE

Macaria ulsterata

Uncommon
91a-0751 (6330)

RANGE-WIDE

TL 13–15 mm Whitish to pale gray FW has four brown to blackish patches along costa, with the largest a rectangle at ST line. Midpoint of ST area has a large, circular, black patch fragmented by white veins. Pale brownish AM, median, and PM lines are faint and often indistinct. Head is orange, and abdomen has paired black dots on each segment. Apex of FW is falcate, with a black-fringed concavity just to the inside. HW is strongly pointed at midpoint. **HOSTS** Birch and alder.

COMMON ANGLE

Macaria aemulataria

Uncommon
91a-0752 (6326)

RANGE-WIDE

TL 11–14 mm Resembles Birch Angle, but FW is usually grayer, with slightly darker shading in ST area, especially on HW. Costal patches are smaller and usually warmer brown, connecting to lines that are usually slightly darker and more visible. **HOSTS** Maple.

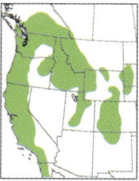

ADONIS ANGLE

Macaria adonis

Common
91a-0760 (6338)

RANGE-WIDE

TL 14–17 mm Striated gray to purplish-gray wings have wide, diffuse, warm brown ST band edged basally with diffuse tan. ST area is sometimes also shaded brown. AM, median, and PM lines are indistinct, widening to strong black patches at costa; PM line has thick black dash at midpoint. Head is orange. Apical half of outer margin of FW is very slightly concave, and HW is strongly angled at midpoint. **HOSTS** Pines, including ponderosa and lodgepole pine.

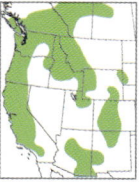

PALE-MARKED ANGLE

Macaria signaria

Very Common
91a-0767 (6344)

RANGE-WIDE

TL 15–19 mm Pale gray FW has tiny dark striations. Thick wavy AM, median, and PM lines are gray; costal patches are only slightly darker than lines. ST area has brown costal patch and diffuse black patch at midpoint indistinctly fractured by white veins. Area below PM line is shaded darker gray, with blurry white ST line curving from anal angle to apex. **HOSTS** Coniferous trees.

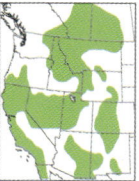

CALIFORNIAN GRANITE

Digrammia californiaria

Uncommon
91a-0774 (6380)

RANGE-WIDE

TL 12–15 mm Peppery brownish-gray FW has three bold black marks along costa at AM, median, and PM lines. AM and PM marks are rectangles perpendicular to the costa, while median mark is usually more triangular and angles downward toward PM area. PM line is primarily small black spots at the veins (sometimes indistinct or backed with a faint dusky band) and an elongate dash near midpoint with a dark smudge (sometimes two dark streaks) immediately below. Terminal line is dusky dots but often indistinct or absent, and fringe is pale. **HOSTS** Big deervetch and other legumes.

BIRCH ANGLE

GRAY-BANDED ANGLE

COMMON ANGLE

ADONIS ANGLE

PALE-MARKED ANGLE

CALIFORNIAN GRANITE

actual size

CREOSOTE MOTH
Digrammia colorata

Very Common
91a-0775 (6381)

RANGE-WIDE

TL 10–16 mm Striated, pale gray to brownish FW has dusky shading below PM line, often with blackish patches at costa and midpoint of ST area, and an indistinct pale ST line. Costa has four dark patches that can be black to faded gray. AM, median, and PM lines are often faintly visible as pale tan to reddish lines. Terminal line is dusky dashes, and fringe is checkered. HOSTS Creosote bush.

PEARL GRANITE
Digrammia pervolata

Rare
91a-0776 (6383)

NORTH

SOUTH

TL 13–16 mm Resembles Creosote Moth but ST/terminal area is not noticeably shaded darker and ST line is indistinct. Dark triangle at costa of PM line points at a small, dark dot. AM and PM lines are usually weakly present. Terminal line is dusky dots, and fringe is pale or faintly checkered. HOSTS Birchleaf mountain mahogany.

UBIQUITOUS GRANITE
Digrammia ubiquitata

Uncommon
91a-0780 (6374.1)

INLAND

COASTAL

TL 12–16 mm Resembles Toothed Granite, but costal markings are thicker, with mark at PM line often longer and curving. Central PM line has a dark black section without a blackish patch below. Area below PM line is often slightly shaded darker, with a diffuse pale ST line. AM, median, and PM lines sometimes visible as faint brown shadows. Terminal line is absent. HOSTS Spreading buckwheat and buckbrush.

TOOTHED GRANITE
Digrammia denticulata

Uncommon
91a-0781 (6373)

RANGE-WIDE

TL 13–16 mm Resembles California Granite, but AM and PM lines are present as diffuse (sometimes indistinct) bands behind the dark vein spots, and ovate dash and shading at midpoint of PM line are usually reduced or indistinct. Area below PM line is shaded darker, with a pale wavy ST band dividing it. When fresh, fringe is checkered dark gray, with a pale line running through the center, creating a scalloped effect with the dark, dashed terminal line. HOSTS Unknown.

PAINTED GRANITE
Digrammia pictipennata

Uncommon
91a-0782 (6372)

NORTH

SOUTH

TL 11–13 mm Resembles Toothed Granite, but terminal area lacks shading, and shading of ST area overlaps PM line, which is visible as a pale thin line. Outer ST line has a shaded patch at costa. When fresh, fringe is checkered gray, with a pale line running through center, creating look of double line with black-dashed terminal line. HW is peppery but otherwise largely unmarked. HOSTS Narrowleaf goldenbush.

TAWNY-VEINED GRANITE
Digrammia nubiculata

Rare
91a-0784 (6371)

RANGE-WIDE

TL 11–15 mm Striated white to pale gray FW has broad dark ST band, edged basally with thin white and black PM line, and often with blackish patch at midpoint. AM and median lines are curving and often somewhat indistinct. Costa lacks dark patches. Black terminal line is edged with white along fringe. Many individuals have tawny veins. HOSTS Rabbitbrush.

CREOSOTE MOTH

PEARL GRANITE

UBIQUITOUS GRANITE

TOOTHED GRANITE

actual size

PAINTED GRANITE

TAWNY-VEINED GRANITE

SHADED GRANITE
Digrammia curvata

Uncommon
91a-0786 (6370)

TL 11–14 mm Resembles Curve-lined Angle, but thick AM and PM lines end at midpoint of FW. Median line is shaded darker along inner half and curves upward to meet end of AM line. ST area is shaded gray to costa, contrasting with broad, pale ST line. **HOSTS** Rubber rabbitbrush.

THREE-PATCHED GRANITE
Digrammia triviata

Uncommon
91a-0788 (6385)

TL 12–17 mm Resembles California Granite, but black costal patch at PM line pinches into a thin U-shaped tail that trails into PM line. Lacks both ovate dash in PM line and associated shaded patch. Area below PM line is shaded slightly darker, without pale ST line. Uncommonly, some individuals may have noticeable thin AM and/or PM lines. **HOSTS** Juniper.

CURVE-LINED ANGLE
Digrammia continuata

Common
91a-0789 (6362)

TL 11–13 mm Lightly striated gray FW has bold black PM line that is thick and straight in inner half, then narrows as it curves down and to costa. Uniformly thick AM line is relatively straight but curves slightly basally at costa. AM and PM lines are often narrowly edged with diffuse shading outwardly. Median line is usually indistinct, curving in a gentle S shape to costa. Median area is usually paler, and diffuse ST line is pale. **HOSTS** Whitecedar, redcedar, and Guadalupe cypress; possibly also juniper.

STRANGE ANGLE
Digrammia imparilata

Rare
91a-0793 (6366.1)

TL 15–18 mm Resembles Curve-lined Angle, but straightish AM and PM lines are thin or broken at costal end. Diffuse shading outside from AM and PM lines is often indistinct. Soft median line is relatively straight, embedded with an indistinct black discal dash. **HOSTS** Cypress.

DOWNWARD ANGLE
Digrammia excurvata

Uncommon
91a-0794 (6363)

TL 15–17 mm Resembles Curve-lined Angle, but AM and median lines angle apically at costa and bulge in PM line is shallower. Bold AM and PM lines are usually narrower. Indistinct median line curves evenly downward to meet costa pointing apically. Separable from Broad-lined Angle by slightly more curving PM line and lack of white wash in central median area. **HOSTS** California juniper and other juniper.

FADED ANGLE
Digrammia pallorata

Uncommon
91a-0795 (6363.1)

TL 12–17 mm Resembles Strange Angle, but FW is whiter and usually more coarsely striated, with thinner AM and PM lines; AM line curves apically at inner margin. Costa has a patch of dark shading at ST area. **HOSTS** One-seed and redberry junipers.

SHADED GRANITE

THREE-PATCHED GRANITE

CURVE-LINED ANGLE

actual size

STRANGE ANGLE

DOWNWARD ANGLE

FADED ANGLE

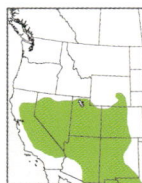

SMOKY ANGLE
Rare

Digrammia cinereola   91a-0796 (6363.2)

TL 14–17 mm Resembles Faded Angle, but striations are finer, AM and PM line are smoother, median line is faint or absent, and ST area usually lacks any shading. Median area can appear slightly paler. **HOSTS** Juniper.

BROAD-LINED ANGLE
Uncommon

Digrammia atrofasciata   91a-0798 (6368)

TL 13–15 mm Olive to brown FW has bold black PM line that is relatively straight, and median area that is usually washed paler centrally, with darker shading at costa and inner margin. PM line is shaded below with a diffuse black band, which shifts before costa to border indistinct pale ST line. **HOSTS** Juniper.

HUMBLE GRANITE
Common

Digrammia muscariata   91a-0800 (6377)

TL 15–17 mm Peppery, pale gray FW has smoothly curving, narrow, gray AM and PM lines that slightly widen and curve basally at costa. Rarely, PM line will have darker spot at midpoint. Median line is diffuse and most distinct at costa and inner margin. Gray discal dash is angled. **HOSTS** Oaks.

DECORATED GRANITE
Rare

Digrammia decorata   91a-0809 (6389)

TL 12–17 mm Peppery, pale gray FW has strong black PM line that makes an S-bend at midpoint and ends before reaching costa. AM line ends at midpoint, and median line is indistinct or absent. ST area is darkly shaded beneath PM line only. Sometimes AM and PM lines are diffuse brownish. **HOSTS** Willow.

BROWN-LINED GRANITE
Rare

Digrammia yavapai   91a-0813 (6393)

TL 14–18 mm Peppery, pale gray FW has warm brown AM, median (sometimes indistinct), and PM lines, with black dots at veins. PM line curves lower at center of wing and is connected by narrow thread to slightly higher, dark costal patch; line is usually lightly shaded outwardly, with a large diffuse gray patch at midpoint. Discal spot is a dark ring with pale center. **HOSTS** New Mexico locust.

VERMILION GRANITE
Uncommon

Digrammia subminiata   91a-0816 (6399)

TL 11–13 mm Gray FW has thick, curving, black AM and PM lines that end just inside from costa; PM line sometimes also has dark fragment at costa. Costa, costal end of AM and PM lines, and veins are all shaded with reddish (fainter orange brown when worn). ST area is shaded blackish. Discal spot has dark outline and pale center. Some individuals are pale; others lack thick black AM and PM lines. **HOSTS** Willow.

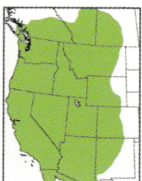

DARK-BORDERED GRANITE
Very Common

Digrammia neptaria   91a-0817 (6396)

TL 11–16 mm Pale gray FW has curving, yellow-edged, warm brown AM and PM lines. PM line is bordered below by a blackish line that blends into diffuse blackish shading. Discal spots have dark outline and pale center. **HOSTS** Poplar and willow.

actual size

SMOKY ANGLE

BROAD-LINED ANGLE

HUMBLE GRANITE

DECORATED GRANITE

BROWN-LINED GRANITE

VERMILION GRANITE

DARK-BORDERED GRANITE

321

PALE-LINED ANGLE

Digrammia irrorata

Common

91a-0818 (6395)

NORTH

SOUTH

TL 11–15 mm Peppery, light brown FW has straight, pale yellow AM and PM lines edged outwardly with dark brown. Blotchy, dark brown shading fills ST area. Discal spot has dark outline and pale center. Veins are sometimes traced with pale yellow. **HOSTS** Poplar and willow.

RUSTY-EDGED ANGLE

Frederickia nigricomma

Uncommon

91a-0824 (6407)

RANGE-WIDE

TL 11–15 mm Peppery, light gray FW has wavy, light brown AM, median, and PM lines that terminate with dark patch at costa. Fragmented black patch at midpoint of PM line is usually a horizontal dash with two vertical dashes below. Area below PM line is shaded darker gray. Sometimes rests with wings raised; underside of both wings peppery, with dark, diffuse median and ST lines and warm brown shading below PM line. **HOSTS** Sweet acacia.

SIGNATE LOOPER

Frederickia s-signata

Common

91a-0832 (6414)

RANGE-WIDE

TL 12–15 mm Striated, pale gray FW has bold, black, S-shaped PM line that ends in central wing and is edged in pale color above; barcoded specimens on average show curve of PM line turning slightly to meet inner margin perpendicularly. ST area is shaded dark gray, with a diffuse white ST line below. Costa has dark spots at AM, median, and PM lines. AM and median lines are usually faint warm brown; AM line occasionally blackish in inner half. **HOSTS** Mesquite.

MESQUITE LOOPER

Frederickia cyda

Uncommon

91a-0833 (6415)

NORTH

SOUTH

TL 12–15 mm Virtually identical to Signate Looper, and individuals in areas of range overlap may be identifiable only by dissection or barcode. However, barcoded specimens on average have bold PM line curving smoothly to meet inner margin at a slight angle (versus straight in Signate), and pale edging to PM line may be on average thicker. **HOSTS** Mesquite.

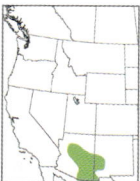

BUCKSKIN LOOPER

Frederickia hypaethrata

Uncommon

91a-0835 (6417)

RANGE-WIDE

TL 13–16 mm FW has tan median area sharply contrasting with dark brown basal and ST areas. AM and PM lines are evenly curving. Costa shows diffuse pale apical patch and diffuse dark median dash. An indistinct white ST line is sometimes present. **HOSTS** Mesquite.

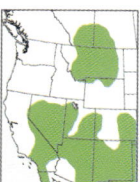

GREEN BROOMWEED LOOPER

Narraga fimetaria

Common

91a-0836 (6420)

RANGE-WIDE

TL 6–8 mm Small moth, atypical for this tribe, that typically rests with wings folded and held above thorax. Underside of wings is boldly marked with broad, dark-edged, white AM and PM bands (sometimes broken) and line of white spots in ST area. Upper side of wings largely brown, with pale patch at outer PM line and checkered white fringe. **HOSTS** Prairie broomweed. **NOTE** Often seen at flowers during daytime.

ANGLES

PALE-LINED ANGLE

RUSTY-EDGED ANGLE

SIGNATE LOOPER

SIGNATE LOOPER

MESQUITE LOOPER

actual size

MESQUITE LOOPER

BUCKSKIN LOOPER

GREEN BROOMWEED LOOPER

323

GRAYS and ALLIES
SUPERFAMILY Geometroidea (91a)
FAMILY Geometridae SUBFAMILY Ennominae TRIBE Boarmiini

Medium-sized moths with long rounded forewings and broad hindwings in shades of gray or occasionally brown, often peppery. All or most transverse lines are usually present, with variable amounts of adjacent shading. Most individuals hold their wings spread when at rest, but the cold-season species of this group fold their wings flat. Adults are nocturnal and will visit lights, sometimes in abundance.

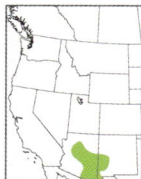

SHARP-TOOTHED GRAY
Glena grisearia

Uncommon
91a-0841 (6445)

RANGE-WIDE

WS 32–42 mm Peppery, pale gray wings have strongly toothed black PM line and three black patches along costa. AM and PM lines are indistinct. Terminal line is black dots. **HOSTS** Oak.

STRAIGHT-LINE GRAY
Glena nigricaria

Uncommon
91a-0844 (6448)

RANGE-WIDE

WS 30–40 mm Gray FW has thin, toothed, relatively straight, black PM line and AM line that curves strongly basally at costa. Thicker (sometimes wide) median line is diffuse and often nears or touches PM line near inner margin. PM and median lines also cross HW, but AM line does not. Area below PM line shaded slightly darker, with thin, white, wavy ST line (often indistinct). Often has darker shading at inner half of PM line that curves across ST area toward apex. Terminal line is dotted. Abdomen is gray, with dark bands. Separable from Common Gray by AM line on HW, shape of PM line, and abdomen. **HOSTS** Douglas-fir and pine, especially ponderosa pine.

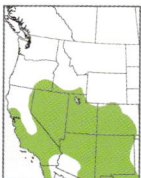

FIVE-LINED GRAY
Glena quinquelinearia

Common
91a-0849 (6453)

RANGE-WIDE

WS 26–36 mm Elongate gray FW has straight black PM line running from middle of inner margin toward apex, ending at ST area. Black AM and gray PM lines run parallel and end in center of wing. Gray ST area has diffuse, pale, wavy ST line, often indistinct. **HOSTS** Unknown.

TWIN-DOTTED GRAY
Stenoporpia anastomosaria

Rare
91a-0859 (6462)

RANGE-WIDE

WS 38–42 mm Resembles Straight-line Gray, but PM line curves basally at costa and has a basal-pointing tooth near inner margin, white ST line is series of strong scallops often edged basally with white shading, dark shading basal to AM line makes line appear doubled, and, usually, dark shading along inner PM line curves across ST area to subapical outer margin (sometimes indistinct except in terminal area). Abdomen is gray with paired black dots. Separable from Common Gray by AM line on HW, ST line, and abdomen. **HOSTS** Unknown.

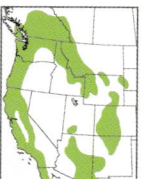

ORANGE-WASHED GRAY
Stenoporpia pulmonaria

Uncommon
91a-0860 (6463)

RANGE-WIDE

WS 34–38 mm Pale to gray FW has a light, warm brown wash along inner margin and apex. Thin black PM line is straight in inner half and curves deeply near costa and can be smooth or toothed. AM and PM lines edged outwardly with dark shading. Narrow white ST line is wavy. **HOSTS** Douglas-fir, hemlock, fir, Sitka spruce, and pine.

SHARP-TOOTHED GRAY

actual size

STRAIGHT-LINE GRAY

FIVE-LINED GRAY

TWIN-DOTTED GRAY

ORANGE-WASHED GRAY

RUSSET GRAY
Stenoporpia macdunnoughi   **Rare**

91a-0867 (6470)

WS 38–46 mm Pale gray wings have curving black AM and PM lines broadly edged with diffuse, warm brown shading. Veins and inner margin are usually washed rusty. Pale median area sometimes has dark discal spot. **HOSTS** Unknown.

RUSTY-PATCHED GRAY
Stenoporpia glaucomarginaria   **Rare**

91a-0869 (6472)

WS 34–42 mm Dark gray FW has strong, warm brown patches at inner margin and central and subapical ST area. Black AM and PM lines usually have inward-pointing tooth at central vein and are shaded outwardly with darker gray. White ST line is relatively straight. **HOSTS** Unknown.

EXALTED GRAY
Stenoporpia excelsaria   **Rare**

91a-0871 (6474)

WS 38–44 mm Peppery, pale gray FW has straight, black, toothed PM line that tapers out before costa, thin blackish median line that curves sharply basally at costa, dark shading along curving AM line that makes line appear doubled, and pale, slightly wavy ST line. Gently scalloped outer margin has thin black terminal line. Abdomen has dark bands. Separable from Straight-line Gray by median and terminal lines, and from Twin-dotted Gray and Common Gray by abdomen pattern and PM line. **HOSTS** Douglas-fir, and ponderosa and lodgepole pine.

GOLDEN NARROWWING
Tornos benjamini   **Rare**

91a-0881 (6483)

WS 22–32 mm Elongate perpendicular wings resemble those of pugs. Dark brown FW has speckly tan costa (clearest at apex), basal area, and central ST area. Large discal spot is black. PM line is a row of often-indistinct dots. Thorax and abdomen are tan. **HOSTS** Unknown.

BROWN NARROWWING
Tornos erectarius   **Uncommon**

91a-0882 (6484)

WS 28–32 mm Mottled brown FW has thin, smoothly curving, black AM and PM lines. Median area often shaded darker and containing large black discal spot. Indistinct white ST line is strongly zigzagged. **HOSTS** Unknown.

HOARY NARROWWING
Glaucina eroraria   **Uncommon**

91a-0885 (6488)

WS 28–36 mm Hoary gray FW has irregularly scalloped, dark gray AM and PM lines that are closer together at inner margin than at costa. Small dark discal dash is close to PM line. HW is hoary along inner margin but lacks any distinct markings. **HOSTS** Unknown.

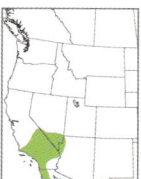

STREAKED NARROWWING
Glaucina macdunnoughi   **Rare**

91a-0889 (6492)

WS 24–30 mm Gray FW has toothed black PM line (sometimes indistinct) that often spikes sharply basally at inner margin. Area below PM line is usually shaded slightly darker and streaked with black interveinal dashes. ST line is absent. Markings can be somewhat variable in extent and strength. **HOSTS** Unknown.

RUSSET GRAY

RUSTY-PATCHED
GRAY

actual size

EXALTED GRAY

GOLDEN NARROWWING

BROWN NARROWWING

HOARY NARROWWING

STREAKED NARROWWING

327

INTERRUPTED NARROWWING COMPLEX   Uncommon

RANGE-WIDE

Glaucina epiphysaria/interruptaria/gonia/platia/nephos
91a-0890/899/901/902/903 (6493/6502/6504/6505/6506)

WS 23–32 mm   Smooth gray FW has thread-thin black AM and PM lines. AM line has two sharp medial-pointing teeth, and PM line has one, which touch in inner wing to pinch median area into two rounded sections, often separating them entirely. Sometimes has whitish, scalloped ST line. The species in this group are all very similar and not easily identified to species visually. *G. epiphysaria, G. gonia,* and *G. platia* are largely restricted to CA, while *G. nephos* ranges from AZ to WY. Interrupted Narrowwing, *G. interruptaria*, is the most widespread, occurring from CA to CO and western TX. **HOSTS** Unknown for most. *G. epiphysaria* feeds on desert ironwood and California buckwheat.

SCALLOPED NARROWWING   Uncommon

RANGE-WIDE

Glaucina eupetheciaria   91a-0893 (6496)

WS 28–36 mm   Resembles Hoary Narrowwing, but thin black PM line is more evenly scalloped, and HW usually has a thin, black, scalloped line at inner margin of PM line. **HOSTS** Unknown.

OCOTILLO MOTH   Uncommon

RANGE-WIDE

Eubarnesia ritaria   91a-0918 (6520)

TL 10–14 mm   Gray FW has dark blackish AM, median, and PM lines spaced closely together in middle of wing. PM line is lightly bordered by warm brown below, and inner margin of basal area has a brown patch. Median and terminal areas have light gray shading. Abdomen has thick white band at base. Often rests with wings held more folded than other grays. **HOSTS** Ocotillo.

LINEAR GRAY   Uncommon

RANGE-WIDE

Pterotaea lamiaria   91a-0953 (6553)

WS 28–32 mm   Pale gray to brownish wings have straight or slightly curving black AM and PM lines terminating in central wing and edged outwardly with cool brown. Median area is shaded darker gray, usually with large black discal spot. Dark blackish apical dash cuts through scalloped white ST line. ST area has whitish patch at inner margin and dark gray to blackish patch in central FW. HW lacks discal spot. **HOSTS** Chamise.

CROSS-LINED GRAY   Rare

RANGE-WIDE

Pterotaea cariosa   91a-0968 (6568)

WS 35–40 mm   Speckled, pale brownish-gray FW has thick, black, curving AM and PM lines that pinch together near inner margin, creating appearance of crossed lines. HW has thin black PM line. Scalloped outer margin has solid black terminal line with dots at veins. **HOSTS** Unknown.

BROWN-WINGED GRAY   Uncommon

RANGE-WIDE

Iridopsis dataria   91a-0973 (6573)

WS 30–40 mm   Brown to pale FW has thin black AM and PM lines edged basally with warm brown. AM line is evenly curving; PM line has two basal-pointing curves in inner half. Indistinct white ST line is scalloped. Pale discal spot has thin dark edge. Abdomen has dark band near base. **HOSTS** Unknown.

INTERRUPTED NARROWWING
COMPLEX

actual size

SCALLOPED NARROWWING

OCOTILLO MOTH

LINEAR GRAY

CROSS-LINED GRAY

BROWN-WINGED GRAY

MOUNTAIN-MAHOGANY LOOPER

Iridopsis clivinaria

Uncommon

91a-0975 (6575)

RANGE-WIDE

WS 45–50 mm Resembles Linear Gray, but gray to pale gray FW lacks dark apical dash or discal spot. Black PM line slopes apically from inner margin with two shallow, broad waves, then turns quickly upward and ends at fourth vein from costa. Slightly wavy AM line ends at midpoint of FW. AM and PM lines (sometimes fractured or faint) shaded outwardly with medium to warm brown. Indistinct median line parallels PM line. ST area has blackish patch below costal end of PM line. Pale ST line is scalloped, and terminal area is shaded darker with tan streaks along veins. HW has straight black PM line and distinct median line; discal spot is usually faint or absent. **HOSTS** Manzanita, mountain mahogany, rose, mock orange, cliffrose, and bitterbrush.

OBLIQUE LOOPER

Iridopsis obliquaria

Uncommon

91a-0977 (6577)

RANGE-WIDE

WS 28–38 mm Resembles larger Mountain-Mahogany Looper, but shading along black AM and PM lines is brown to brownish gray, and indistinct median line is equidistant to both AM and PM lines, or curves to touch both. Blackish shading in outer ST area runs into apical dash. Pale ST line is often edged below with diffuse, dark gray scallops. PM line on HW is thin. **HOSTS** Western soapberry and catclaw mimosa. **NOTE** Occurs east and north of coastal southern CA.

FRAGILE GRAY

Iridopsis fragilaria

Uncommon

91a-0985 (6585)

RANGE-WIDE

WS 27–40 mm Lightly peppered gray FW has thin, wavy, black PM line that is most distinct as doubled curves at inner margin and midpoint; sometimes also has a dark section with a tooth just inside from costa. Midpoint of ST area has a diffuse darkish patch. HW has relatively level, thin, black PM line and indistinct dark-rimmed pale discal spot. **HOSTS** Generalist on deciduous trees and shrubs.

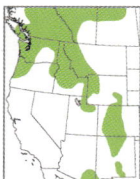

BENT-LINE GRAY

Iridopsis larvaria

Uncommon

91a-0989 (6588)

RANGE-WIDE

WS 26–36 mm Light gray FW has narrow, black, evenly curving AM line and PM line that has a large basal-pointed curve at inner margin, blunt downward point costal from midpoint, and angles strongly basally to meet costa. AM and PM lines broadly bordered outwardly with yellow orange. Zigzag ST line is pale, with dark patch below blunt tooth of PM line. HW has curving PM line with a single sharp downward tooth and narrow to wide black median line. All wings have dark-rimmed pale discal spots. **HOSTS** Alder, apple, birch, cherry, hawthorn, poplar, willow, and many other plants.

TARNISHED GRAY

Iridopsis emasculatum

Very Common

91a-0990 (6589)

RANGE-WIDE

WS 30–36 mm AM and PM lines resemble those of Bent-line Gray but shallower, with stronger brown shading outwardly. Wings are overall more peppery gray. Discal spots are more indistinct and often incorporated into diffuse gray median line on FW. Wavy pale ST line is bordered above with dusky gray shading on HW. Median line on HW has blackish shading. **HOSTS** Alder, maple, willow, and blueberry.

MOUNTAIN-MAHOGANY LOOPER

actual size

OBLIQUE LOOPER

FRAGILE GRAY

BENT-LINE GRAY

TARNISHED GRAY

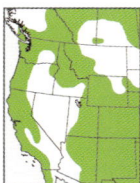

COMMON GRAY
Anavitrinella pampinaria

Very Common
91a-0991 (6590)

RANGE-WIDE

WS 23–34 mm Light gray FW has thin black AM and PM lines that curve strongly basally near costa. Thin median line (sometimes indistinct) nearly touches PM line near inner margin. All lines present on HW (AM may sometimes be indistinct or placed high and hidden by FW when at rest). AM and PM lines are sometimes edged outwardly with indistinct brown shading. Dark shading fills inner half of median and PM lines, then curves across ST area to costa; shading sometimes indistinct and most obvious near outer margin. Separable from all similar grays except Gulf Coast Gray by pale band at base of abdomen. **HOSTS** Apple, ash, clover, cotton, fir, tamarack, maple, poplar, and many others.

GULF COAST GRAY
Anavitrinella atristrigaria

Rare
91a-0992 (6591)

NORTH

SOUTH

WS 24–28 mm Resembles Common Gray, but thick black AM line crosses abdomen above pale abdominal band to form a solid line from costa to costa, inner PM line is thicker and bows toward AM line, and dark shading from central median area to subapical outer margin creates the appearance of a straight blackish line across wings when at rest. Basal area and area below PM line are usually lightly shaded brown. **HOSTS** Unknown.

SMALL ENGRAILED
Ectropis crepuscularia

Common
91a-0997 (6597)

RANGE-WIDE

WS 26–37 mm Peppery gray FW has thin, relatively straight, black PM line that has small teeth at veins. Midpoint of lightly shaded ST area adjacent to PM line has two strong black teeth and another double-toothed blackish patch against wavy pale ST line. AM and median lines are often indistinct other than dark patch at costa. Terminal line is small dots. Base of abdomen has two pairs of black dots, often large enough to blur together into a black band. Separable from Sharp-toothed Gray by shading along AM and PM lines and shape of PM at costa, and from other grays by abdomen. **HOSTS** Apple, birch, fir, hemlock, tamarack, oak, poplar, spruce, willow, and many others.

PORCELAIN GRAY
Protoboarmia porcelaria

Uncommon
91a-0998 (6598)

RANGE-WIDE

WS 26–35 mm Resembles Small Engrailed, but PM line curves basally at costa, costal patches are larger and darker, and PM line on HW bends strongly toward anal angle at inner margin. Area below PM line is usually shaded slightly darker, with wavy white ST line through middle. Abdomen is gray, sometimes with a thin white band on first segment (rarely entire segment pale) and paired diffuse dark patches, and thorax has dark stripes laterally, indistinct when worn. Separable from Sharp-toothed Gray by shaded ST/terminal area and thorax. **HOSTS** Fir, elm, hemlock, poplar, willow, and many other trees.

CINNAMON GRAY
Mericisca gracea

Rare
91a-1004 (6605)

RANGE-WIDE

WS 32–44 mm Pale gray FW has reddish-brown shading in basal area and below curvy black PM line. Broad median band is diffuse and heavily shaded with reddish brown. Terminal area is lightly shaded gray, with darker patch subapically. Terminal line is solid black with wider spots interveinally. Easily separable from Russet Gray by median band. **HOSTS** Unknown.

COMMON GRAY

GULF COAST GRAY

SMALL ENGRAILED

actual size

PORCELAIN GRAY

CINNAMON GRAY

BRINDLED GRAY
Uncommon
Parapheromia cassinoi 91a-1007 (6608)

WS 26–30 mm Brindled, dark gray FW has wide, diffuse, white ST line bordered basally with dark shading. Black AM, median, and PM lines are narrow and smoothly curving. Terminal area, inner margin, and central median are all washed with reddish brown. HW is paler. Thorax has large black patch, and abdomen is warm brown, which separates this species from all similar grays. HOSTS Unknown.

BELTED DELTA
Uncommon
Prionomelia spododea 91a-1012 (6613)

RANGE-WIDE

TL 14–18 mm Brown FW has thick, straight, black median line; slightly curved, narrow, black PM line; and double AM line. Whitish scalloped ST line is sometimes indistinct. Area below PM line is lightly speckled, with dusky shading slightly inward from apex. HOSTS Unknown.

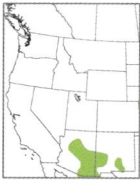

GREEN DELTA
Rare
Tracheops bolteri 91a-1014 (6615)

RANGE-WIDE

TL 13–18 mm Pale FW has highly scalloped, thin, black AM and PM lines edged outwardly with white and bordered by grayish green. Thin black median line is strongly wavy and bordered below with blackish shading. Black-edged white ST line is wavy and broken near midpoint. Thorax is gray green, and abdomen is light orange. HOSTS Unknown.

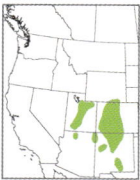

STREAKED DELTA
Rare
Carphoides incopriarius 91a-1024 (6624)

RANGE-WIDE

TL 15–17 mm Gray to tan FW has diffuse white stripes down center and inner margin of wing, overlaid with thin black streaks along veins in ST area; three thicker black streaks progress diagonally from ST to basal area in central FW. Abdomen is light tan. HOSTS Unknown.

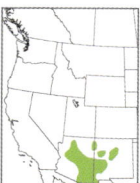

SMOKY DELTA
Rare
Galenara lixaria 91a-1030 (6630)

RANGE-WIDE

TL 18–25 mm Smoky-gray FW has thick black median line and diffuse paler patches in median area. Thin scalloped AM line curves sharply basally at costa, and PM line curves gently around short black discal dash. Indistinct pale ST line is scalloped and bordered basally with diffuse pale shading. Abdomen is light tan. HOSTS Unknown.

ENIGMATIC DELTA
Uncommon
Vinemina opacaria 91a-1035 (6635)

RANGE-WIDE

TL 14–19 mm Light gray FW has blackish median band that dips outwardly at midpoint. A diffuse vertical band connects median line to outer margin. Thin, scalloped, black PM line is most visible at costa. Indistinct pale ST line has dark edging at costa. All lines are bordered by diffuse, cool brown shading. HOSTS Unknown.

WESTERN CARPET
Very Common
Melanolophia imitata 91a-1038 (6618)

RANGE-WIDE

WS 34–46 mm Peppery pale to brownish FW has thin, scalloped, dark PM line that slightly thickens at inner margin and near midpoint, sometimes edged below with darker shading. Scalloped AM and diffuse median lines are often indistinct. ST line is irregularly spaced black spots with whitish edge below. HOSTS Douglas-fir, western hemlock, and western redcedar; also fir, spruce, larch, and pine.

BRINDLED GRAY

BELTED DELTA

GREEN DELTA

STREAKED DELTA

SMOKY DELTA

actual size

ENIGMATIC DELTA

WESTERN CARPET

SULPHUR MOTH
Hesperumia sulphuraria

Very Common
91a-1046 (6431)

TL 16–19 mm Pale yellow to yellow-orange FW has large, blurry-edged, brown discal spot, sometimes filled with paler brown. Curving AM and PM lines are a row of brown spots (sometimes indistinct), with a brown patch at costa. Fringe is usually yellow. Some rare individuals show variations, such as shaded median area, solid AM and PM lines, or lack of discal spot. **HOSTS** Various trees and shrubs, including cherry and snowberry. **NOTE** Often seen in vegetation during daytime.

SMOKY-EYED SULPHUR MOTH
Hesperumia fumosaria

Uncommon
91a-1047 (6432)

TL 16–22 mm Resembles Sulphur Moth, but fringe is dark brown and FW lacks dark patch at costa of AM and PM lines. Discal spot rarely has a pale center. **HOSTS** Generalist on Rosaceae, Rhamnaceae, and Ericaceae.

BROAD-WINGED GRAY
Hesperumia latipennis

Rare
91a-1048 (6433)

WS 34–36 mm White to gray wings are heavily peppered with black. Thin, dark gray PM line has sharp black teeth at veins. Thin AM and diffuse median lines are often indistinct. Strong black terminal line is wavy with scalloped outer margin and helps separate from Small Engrailed. ST area sometimes shaded darker with wavy pale ST line. **HOSTS** Cascara, ocean spray, common snowberry, elder, and other deciduous shrubs and trees.

FOUR-BARRED GRAY
Aethalura intertexta

Uncommon
91a-1050 (6570)

WS 21–25 mm Peppery gray wings have curving, dark gray AM, median, and PM lines that are darkest as spots at veins and widening at costa. Jagged white ST line is shaded basally with gray. **HOSTS** Alder and birch.

BROWN-LINED LOOPER
Neoalcis californiaria

Very Common
91a-1051 (6435)

WS 30–36 mm Variable. Pale whitish to dark gray wings have thin black AM on FW that curves strongly basally at costa, and nearly straight PM line has two shallow, downward waves. Usually has some to a lot of diffuse blackish shading outwardly along AM and PM lines, as well as along wavy pale ST line. Outer margin is scalloped with thin black terminal line. Inner margin sometimes has reddish-brown wash. **HOSTS** Douglas-fir, hemlock, western redcedar, red fir, coast redwood, and other conifers; also ceanothus.

PEPPERED MOTH
Biston betularia

Very Common
91a-1060 (6640)

WS 40–45 mm Chunky, with a distinctively long FW profile. Heavily peppered whitish or light gray wings have wavy black AM and PM lines; PM line has large outward bulge toward costa. Blackish median line is broad and diffuse; AM line often also bordered basally with diffuse black shading. Melanic individuals are sooty black. **HOSTS** Alder, birch, cherry, dogwood, elm, oak, tamarack, willow, and many other trees.

SULPHUR MOTH

actual size

SMOKY-EYED
SULPHUR MOTH

BROAD-WINGED GRAY

FOUR-BARRED GRAY

BROWN-LINED LOOPER

PEPPERED MOTH

337

SILVERY-GRAY GEOMETER
Biston sinuaria

Uncommon
91a-1069 (6650)

RANGE-WIDE

WS 40–58 mm Resembles Peppered Moth, but silvery-gray wings are less peppery. Median line is usually faint to absent. Thin black AM and PM lines disappear just inside from costa, then reappear at costa, and are usually edged outwardly with darker gray. Sinuous PM line has large basal-pointing curve at inner margin, then angles toward outer margin. HOSTS Mountain mahogany, manzanita, oak, laurel sumac, chamise, and other woody shrubs.

LINDEN LOOPER
Erannis tiliaria

Uncommon
91a-1070 (6665)

RANGE-WIDE

TL 16–21 mm Peppery, warm brown FW has straight, dark brown to black PM line with bulge near costa. Wavy AM line curves up at costa. Median area is paler, with dark discal spot. Female is wingless and heavily spotted dorsally. HOSTS Ash, birch, cherry, elm, maple, oak, poplar, and other deciduous trees. NOTE Can experience outbreaks in some regions, where it may be abundant.

VANCOUVER LOOPER
Erannis vancouverensis

Common
91a-1071 (6665.1)

RANGE-WIDE

TL 18–26 mm Peppery pale FW has broad blackish shading bordering thin black AM and PM lines. White ST line is edged basally with blackish shading. Median area has dark discal spot. Warm brown often suffuses veins and areas of dark shading. Female is wingless and heavily spotted dorsally. HOSTS Maple, oak, apple, willow, and birch; also western hemlock and western white pine.

WALNUT SPANWORM
Phigalia plumogeraria

Common
91a-1079 (6661)

RANGE-WIDE

TL 19–25 mm Hoary, light gray FW has thin, straight, dark PM line with small blackish teeth near midpoint and costa. Dark ST line is scalloped. AM and median lines and dark discal spot are sometimes indistinct. Antennae are broadly feathery. Female is wingless. HOSTS Oak, maple, willow, walnut, apple, and plum; also ceanothus and arbutus.

WINTER CANKERWORM
Paleacrita longiciliata

Rare
91a-1082 (6664)

RANGE-WIDE

TL 13–18 mm Narrow, hoary, pale gray FW has thin black veins. Costa and inner margin have dark marks representing wing lines. Terminal area is paler with diffuse interveinal smudges. Antennae are broadly feathery. Female is wingless. HOSTS Chamise.

STOUT SPANWORM
Lycia ursaria

Rare
91a-1083 (6651)

RANGE-WIDE

TL 20–27 mm Distinctively chunky moth with a furry thorax and hoary gray FW crossed by dark gray AM, median, and PM lines. Pale, diffuse ST line is sometimes indistinct. Female has small wings and does not fly. HOSTS Deciduous trees and woody plants, including alder, ash, birch, cherry, cranberry, maple, and poplar.

SILVERY-GRAY GEOMETER

LINDEN LOOPER

VANCOUVER LOOPER

actual size

WALNUT SPANWORM

WINTER CANKERWORM

STOUT SPANWORM

ASSORTED GEOMETERS

SUPERFAMILY Geometroidea (91a)

FAMILY Geometridae SUBFAMILY Ennominae TRIBES Caberini, Alsophilini, Campaeini, Epirranthini, Diptychini, Baptini, Anagogini, Gnophini, and Nacophorini

This large assemblage contains varied moths from several tribes, many of which have only one or two member genera. All have the broad wings and narrow body typical of geometers. Most sit with wings spread, but several fold their wings flat and a few rest with wings closed above their body. Adults of most species are nocturnal and regularly visit lights; a small number, such as the spring moths, are diurnal, and some species may be flushed from forest trails during daytime. Note: Some of the moths at the beginning and end of this section are presented slightly out of taxonomic order with the preceding and following sections, to facilitate guide organization.

EIGHT-LINED GEOMETER
Taeniogramma octolineata
Uncommon
91a-0678 (6423)
RANGE-WIDE

WS 20–27 mm Peppery tan to gray FW has densely speckly, dark AM, median, PM, and ST lines that curve slightly basally at costa and terminate at wider dark patch. Apical half of pale FW fringe is dark. **HOSTS** Unknown; possibly oak or manzanita.

VIRGIN MOTH
Protitame virginalis
Rare
91a-0681 (6270)
RANGE-WIDE

WS 20–26 mm Satiny white wings are peppered with brown scales, especially along costa of FW. Some individuals have speckly brown AM, median, and PM lines, or just the PM line. Brown terminal line, if present, is dashed. Outer margin of HW is rounded. **HOSTS** Poplar and willow.

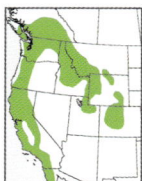

FRECKLED VIRGIN MOTH
Protitame subalbaria
Uncommon
91a-0682 (6266)
RANGE-WIDE

WS 20–28 mm Resembles Yellow-dusted Cream, but is more darkly speckled, median and PM lines are closer together, and wings have solid brown terminal line. **HOSTS** Aspen and willow.

YELLOW-DUSTED CREAM
Cabera erythemaria
Uncommon
91a-1087 (6677)
RANGE-WIDE

WS 21–30 mm Lightly speckled, creamy to pale yellowish wings have evenly-spaced, wavy tan AM, median, and PM lines, sometimes indistinct. **HOSTS** Willow; also aspen and poplar.

DELICATE WAVE
Eudrepanulatrix rectifascia
Common
91a-1091 (6681)
NORTH

SOUTH

TL 13–15 mm Speckled, pale pinkish-tan FW has straight, slightly diffuse, dusky PM line and tiny dark discal dot. Fringe is slightly rosy when fresh, bordered by terminal line of small black dots. **HOSTS** Ceanothus.

EIGHT-LINED GEOMETER

VIRGIN MOTH

FRECKLED VIRGIN MOTH

actual size

YELLOW-DUSTED CREAM

DELICATE WAVE

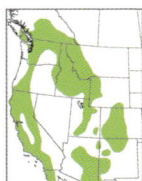

SPURRED WAVE

Common

Drepanulatrix unicalcararia

91a-1092 (6682)

TL 15–17 mm Speckled or brindled, pale tan to brown FW has pale yellow ST line that bends at midpoint so inner half is higher. ST line is sometimes bordered narrowly along inner half by fragment of warm brown PM line, and usually below with rows of two to four diffuse black dots at costal and inner ends. FW has dark black discal spot and sometimes shows indistinct dusky AM line. **HOSTS** Ceanothus.

CONVERGENT WAVE

Uncommon

Drepanulatrix hulstii

91a-1093 (6683)

TL 14–19 mm Speckled, light brown FW has indistinct, slightly wavy, dark brown AM and PM lines. Broadly diffuse median band crosses dark discal spot and meets PM line at inner margin. ST line is a series of three to six black dots usually edged below with white. **HOSTS** Ceanothus.

PARALLEL WAVE

Uncommon

Drepanulatrix bifilata

91a-1094 (6684)

TL 12–15 mm Plain or lightly speckled, pale brown FW has straight brown AM and PM lines that angle slightly basally at costa. Terminal area is usually shaded dusky, sometimes with diffuse dusky spots along ST line. **HOSTS** Possibly oak and mountain mahogany.

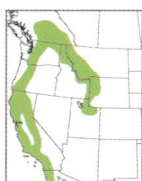

QUADRATE WAVE

Uncommon

Drepanulatrix quadraria

91a-1095 (6685)

TL 13–16 mm Resembles Spurred Wave, but ST line lacks pale yellow band and FW has straight, dark brown PM line, often darkest at costa. Terminal area often shaded darker. **HOSTS** Unknown; likely ceanothus.

ORANGE-BANDED WAVE

Uncommon

Drepanulatrix foeminaria

91a-1096 (6686)

TL 14–16 mm Heavily speckled, pale brown to dusky FW has thin ST line of often-connected white dots usually bordered above with dusky shading, and small, white-rimmed, black discal spot surrounded by dusky shading. Slightly wavy dark AM and PM lines are usually indistinct. Terminal area sometimes shaded dusky, and ST area shaded slightly orangish. **HOSTS** Ceanothus.

RED-LINED WAVE

Uncommon

Drepanulatrix carnearia

91a-1099 (6688)

TL 12–14 mm Plain or lightly speckled, peach FW has straight brown AM, median, and PM lines; median line often passes through or against white-rimmed black discal spot. Area below PM line is shaded purplish in inner half and orangish toward costa. **HOSTS** Unknown; likely ceanothus.

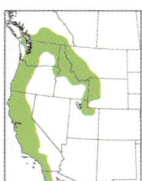

FALCATE WAVE

Uncommon

Drepanulatrix falcataria

91a-1100 (6689)

TL 12–17 mm Resembles Convergent Wave, but FW is more coarsely speckled and median line does not touch or approach PM line. FW apex is falcate, and black discal spot is not rimmed white. HW is pale and lightly speckled. **HOSTS** Ceanothus.

SPURRED WAVE

CONVERGENT WAVE

PARALLEL WAVE

QUADRATE
WAVE

ORANGE-BANDED WAVE

RED-LINED WAVE

FALCATE WAVE

actual size

343

SECONDARY WAVE

Drepanulatrix secundaria

Uncommon
91a-1101 (6690)

TL 12–14 mm Resembles Falcate Wave, but warm to orange-brown FW is more finely speckled, apex is not falcate, and black-edged white chevrons of ST line are more crisply defined, often shaded basally with warm brown. Dark discal spot has no or only narrow white ring. Contrastingly paler HW is lightly speckled, particularly toward inner margin. HOSTS Ceanothus.

PEBBLED WAVE

Drepanulatrix monicaria

Common
91a-1103 (6692)

TL 14–18 mm Resembles Secondary Wave, but FW is brown or reddish brown, with ST line usually weakly defined. In some individuals, median and/or ST area is shaded dusky. HW is pale brownish and minimally contrasting with FW, with moderate, evenly distributed speckling. HOSTS Ceanothus.

DESPERATE WAVE

Ixala desperaria

Rare
91a-1106 (6695)

TL 16–18 mm Heavily brindled, light brown to gray wings have thin, wavy, dark AM, median, and PM lines. FW has round blackish discal spot, and HW has large white discal spot edged in dark brown. Thin blackish ST line is edged with white below, shaded brown basally, and strongly zigzagged, with a distinct W near midpoint. HOSTS Fendler's ceanothus.

BICOLORED NUMIA

Numia bicoloraria

Common
91a-1110 (6700)

TL 10–12 mm Mint- to yellow-green (occasionally light brown) FW has pale brown costa, often with dark patches at median, PM, and ST lines. Dark-edged white discal spot is sometimes present. Light orange HW has a reddish median line. Antennae are white. HOSTS Unknown.

FALL CANKERWORM

Alsophila pometaria

Uncommon
91a-1128 (6258)

TL 14–16 mm Hoary gray FW has scalloped white AM and PM lines shaded darker medially, with short black dashes at veins. PM line is sharply offset at costa. Female lacks wings and is flightless. HOSTS Apple, basswood, elm, maple, oak, and other trees and shrubs.

PALE BEAUTY

Campaea perlata

Very Common
91a-1130 (6796)

WS 28–51 mm Pale green wings have straight, white-edged, dark green AM and PM lines that angle sharply basally near costa. Median area is shaded slightly darker. HOSTS Deciduous trees, including alder, birch, elm, oak, and willow.

DOG-FACE GEOMETER

Spodolepis danbyi

Uncommon
91a-1132 (6799.1)

TL 23–25 mm Speckled gray FW has thin black AM and PM lines that are toothed at the veins. Straight AM line angles basally at costa, while PM line dips strongly toward outer margin in central area. Large black discal spots have thin white centers. ST area usually has brown shading at costa and inner margin. FW pattern may fancifully resemble a bulldog's face. HOSTS Unknown.

SECONDARY WAVE

PEBBLED WAVE

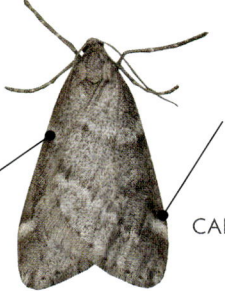

DESPERATE WAVE

FALL
CANKERWORM

BICOLORED
NUMIA

PALE BEAUTY

DOG-FACE
GEOMETER

actual size

345

BLACK-DASHED GEOMETER
Philedia punctomacularia

Uncommon

91a-1133 (6802)

RANGE-WIDE

TL 17–19 mm Pale gray FW is finely peppered with black. Faint gray AM and PM lines are marked with thick, elongate, white-edged, black dashes at veins. Veins are marked black in an inverted V at discal spot. **HOSTS** Bracken, brake, and other ferns.

TAYLOR'S THALLOPHAGA
Thallophaga taylorata

Uncommon

91a-1143 (6808)

RANGE-WIDE

TL 16–19 mm Resembles Northern Thallophaga, but black dots along AM and PM lines are usually reduced or nearly absent, and median band is narrower and fades toward costa. **HOSTS** Western sword fern.

NORTHERN THALLOPHAGA
Thallophaga hyperborea

Common

91a-1144 (6809)

RANGE-WIDE

TL 19–21 mm Lightly speckled, warm brown FW has white-edged black dots at pale veins along AM and PM lines. Brown median band is diffuse. FW apex is slightly falcate. **HOSTS** Western hemlock; also other conifers and sometimes alder.

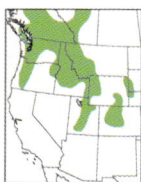

BLUISH SPRING MOTH
Lomographa semiclarata

Uncommon

91a-1146 (6666)

RANGE-WIDE

TL 10–12 mm Peppery silvery-gray FW has diffuse brown PM and ST bands. Often rests with wings held over abdomen. Underside of FW has blackish fringe, large black discal spots, and black PM and ST lines that are thick at costa and taper to small dots on HW. **HOSTS** Cherry, chokecherry, hawthorn, ninebark, and other members of rose family. **NOTE** Often seen in dirt clearings and trails during daytime.

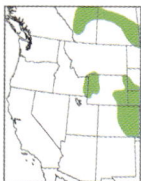

WHITE SPRING MOTH
Lomographa vestaliata

Uncommon

91a-1147 (6667)

RANGE-WIDE

TL 8–12 mm Small, satiny, white moth is virtually unmarked except for pale yellowish brown along costa of underside of FW. **HOSTS** Generalist on many deciduous trees and shrubs, including *Prunus*, hawthorn, oak, viburnum, maple, and others.

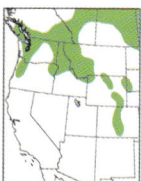

RUDDY METARRANTHIS
Metarranthis duaria

Rare

91a-1156 (6822)

RANGE-WIDE

WS 35–40 mm Speckled, pale brown wings have curving dark AM and PM lines that are shaded medially with warm brown, and strong, dark brown discal spots. FW apex is falcate with thin dark apical dash. Some individuals have dark patches in central ST area. **HOSTS** Birch, blueberry, and cherry in captivity.

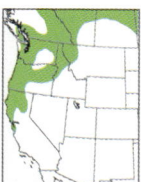

BROWN-TIPPED THORN
Selenia alciphearia

Common

91a-1173 (6817)

RANGE-WIDE

TL 20–24 mm Wings are closed tightly above abdomen at rest. Underside of brindled HW is golden brown, with straight, dark brown median line and dark-edged white PM line. White crescent discal spot touches median line. Upper surface of wings is brindled golden brown, with darker brown AM, median, and PM lines and speckled white costa. **HOSTS** Deciduous trees and woody plants, including alder, birch, cherry, currant, hickory, and maple.

ASSORTED GEOMETERS

BLACK-DASHED GEOMETER

TAYLOR'S THALLOPHAGA

NORTHERN THALLOPHAGA

BLUISH SPRING
MOTH

WHITE SPRING
MOTH

RUDDY METARRANTHIS

BROWN-TIPPED
THORN

actual size

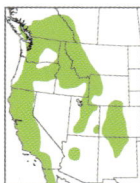

BORDERED FAWN MOTH

Common

Sericosema juturnaria   91a-1175 (6672)

TL 18–22 mm Speckled, pale reddish-tan FW has broad, diffuse, curving, dusky PM line. Area below PM line slightly darker tan, with dusky shading at midpoint connecting PM line to outer margin. Often rests with wings folded above body; underside of wings similar, but PM line of HW is absent centrally. **HOSTS** Ceanothus and false-buckthorn. **NOTE** Often disturbed from vegetation during daytime, alighting with wings folded above body.

FRIENDLY PROBOLE

Uncommon

Probole amicaria   91a-1183 (6838)

WS 23–34 mm Variable. Golden wings are brindled with brown veins. Thin brown PM line dips sharply in central FW, often nearly reaching pointed outer margin. Area below PM line is usually shaded darker (occasionally entirely purplish brown), with small blackish patches in ST area. **HOSTS** Primarily dogwood. **NOTE** Taxonomic work on this genus is ongoing. Some authors believe there to be a single species (*P. amicaria*), while others suggest there may be two or three species based on wing pattern, genitalia, and host-plant preferences.

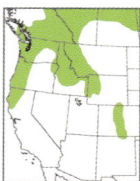

STRAIGHT-LINED PLAGODIS

Uncommon

Plagodis phlogosaria   91a-1187 (6842)

TL 11–16 mm Seasonally dimorphic. FW of spring brood is straw-colored with straight, dark brown AM and PM lines shaded medially with dark brown. Summer brood is yellow orange, usually shaded dark purple in basal and inner ST areas. FW sometimes has dark discal spots. **HOSTS** Alder, apple, basswood, birch, cherry, oak, poplar, and other trees.

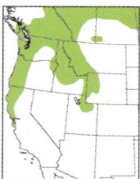

AMERICAN BARRED UMBER

Uncommon

Plagodis pulveraria   91a-1188 (6836)

TL 13–16 mm Speckled, light brown FW has straight AM line and toothed PM line that has a rounded bulge pointing medially in inner half. Median area is uniformly darker brown. **HOSTS** Alder, birch, cherry, fir, hemlock, spruce, willow, and others.

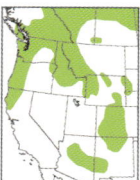

PALE METANEMA

Uncommon

Metanema inatomaria   91a-1191 (6819)

WS 26–36 mm Purplish-gray wings have pale veins and straight, rust-edged, yellowish lines. Midpoint of outer margin is sharply pointed, and a dark crescent curves between midpoint and apex. Large dark discal spot on FW is often bisected by pale veins. **HOSTS** Poplar and willow.

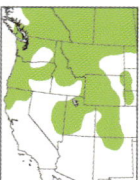

JOHNSON'S EUCHLAENA

Uncommon

Euchlaena johnsonaria   91a-1209 (6729)

WS 26–37 mm Smooth or lightly speckled, light brown wings have a scalloped outer margin and darker brown ST/terminal area, usually with blackish shading along outer and inner ST line. Thin straight AM and PM lines are darker brown; AM line angles sharply basally to meet costa. Diffuse brown median band crosses all wings. FW has pale triangular apical spot usually edged with diffuse blackish shading. Veins are thinly traced with brown. Black discal spots may or may not be present. **HOSTS** Ash, birch, elm, hawthorn, willow, and others.

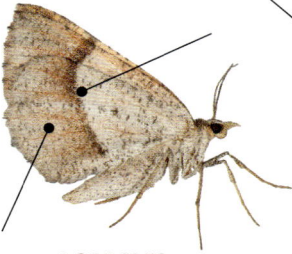

BORDERED
FAWN MOTH

spring

FRIENDLY
PROBOLE

summer

STRAIGHT-LINED
PLAGODIS

AMERICAN
BARRED UMBER

PALE METANEMA

actual size

JOHNSON'S
EUCHLAENA

349

SOFT EUCHLAENA

Euchlaena mollisaria

Rare

91a-1210 (6730)

RANGE-WIDE

WS 36–46 mm Resembles Johnson's Euchlaena, but wings are paler and finely speckled and brown veins are more conspicuous. PM lines have a very thin second line below (sometimes indistinct) that bows outwardly across central wing. Black shading along ST line is usually reduced or often absent. All wings usually have an indistinct black discal dot, sometimes faint on HW. **HOSTS** Unknown; possibly false-buckthorn, willow, raspberry, and/or current. **NOTE** DNA barcoding data suggests Johnson's and Soft Euchlaenas may actually be a complex of overlapping species, but more study is needed.

SCRUB EUCHLAENA

Euchlaena madusaria

Rare

91a-1211 (6731)

RANGE-WIDE

WS 26–37 mm Resembles Soft Euchlaena, but outer margin of FW is not scalloped and FW usually lacks distinct median band. Midpoint of ST area often has a pale patch. All wings have a distinct black discal dot. **HOSTS** Unknown, but reported on buffalo berry and conifers in North.

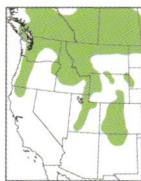

MOTTLED EUCHLAENA

Euchlaena tigrinaria

Uncommon

91a-1217 (6737)

RANGE-WIDE

WS 33–41 mm Lightly speckled, brown to rusty FW is usually heavily mottled with large blackish speckles. Median area is paler and pale subapical patch is edged outwardly with black. Thin reddish-brown AM and PM lines are bordered inwardly with pale yellow. Area below PM line shaded darker, sometimes only diffusely immediately below line. Some individuals may have heavy blackish blotches along median and ST lines. **HOSTS** White birch; also oak, poplar, and willow.

WILLOW BEAUTY

Phaeoura cristifera

Uncommon

91a-1252 (6764)

RANGE-WIDE

WS 36–58 mm Sexually dimorphic. Males have lightly speckled, grayish-brown wings with thick black AM and PM lines and a black-outlined discal spot. AM line bends sharply to meet costa perpendicularly. Brown shading in ST area forms a diffuse band. HW has mostly straightish PM line, sometimes with weak points. Thorax is gray dorsally, and base of abdomen has a broad pale band. Basal area lacks blackish basal dash. Females are similar but are white, with dark speckles in basal and ST areas, and median area has a strong white discal spot. PM line on HW is often more strongly pointed. **HOSTS** Willow, oak, and manzanita.

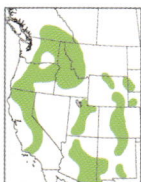

PINE BEAUTY

Phaeoura mexicanaria

Uncommon

91a-1254 (6766)

RANGE-WIDE

WS 44–60 mm Lightly speckled gray (rarely brown) wings have thick, unevenly scalloped, black AM and PM lines bordering a dusky median area. PM line has a double-toothed projection at midpoint and bows strongly basally in inner half. ST area below bowed section of PM line and near costa is an even golden brown; some individuals are also washed brown in basal and median areas. Outer bulge of PM line is bluntly pointed. Dark discal spot is indistinct. HW has similar double-toothed PM line. Posterior of thorax has a small brown patch, and base of abdomen has a narrow pale band. **HOSTS** Ponderosa pine.

SOFT
EUCHLAENA

SCRUB
EUCHLAENA

actual size

MOTTLED
EUCHLAENA

female

male

male

WILLOW BEAUTY

PINE BEAUTY

ENIGMATIC BEAUTY
Phaeoura perfidaria

Uncommon
91a-1256 (6768)

WS 36–55 mm Sexually dimorphic. Male resembles male Willow Beauty, but FW is grayish to brownish gray, with weak brown shading along indistinct whitish ST line, and outer bulge of AM line is shallow and bluntly pointed. Black-outlined white discal spots are elongate. Basal area has black basal dash. PM line on HW is more curving with a deeper point. Thorax is entirely gray, and abdomen lacks pale band. Females are gray, strongly washed with reddish brown along costa and veins and in basal and ST areas. Discal spots are often indistinct, and bulge of AM line may be rounded. Thorax is dusky, with a brown central dorsal patch, and abdomen is reddish brown, with a pale band at base. **HOSTS** Unknown. **NOTE** Rugged Beauty (*P. belua*, not shown) of southern AZ and CA is similar to male but browner, with rounded bulges to AM line; thorax is gray, with a weak brown dorsal patch at posterior, and base of abdomen has a narrow pale band.

ANTIQUE BEAUTY
Phaeoura aetha

Uncommon
91a-1258 (6770)

WS 45–55 mm Resembles Pine Beauty, but FW is light brown to tan, and golden-brown patch below PM line shades darker toward outer margin. Outer bulge of AM line is rounded; AM line may meet costa at an angle or perpendicularly. Brown-washed basal area has a diffuse blackish basal dash. HW usually has a diffuse median line. Thorax has brown central dorsal patch, and base of abdomen a broad pale band. **HOSTS** Unknown.

DISTANT WAVE
Holochroa dissociarius

Uncommon
91a-1261 (6773)

WS 32–44 mm Elongate brownish-gray wings have a slightly darker median area containing a round white discal spot. Thin black AM and PM lines are edged outwardly with white. PM line has an outward-pointing tooth near midpoint. Typically rests with abdomen curled upward. **HOSTS** Possibly juniper.

SECLUDED WAVE
Aethaloida packardaria

Uncommon
91a-1262 (6774)

WS 26–30 mm Resembles Distant Wave, but wings are slightly broader, outer margin is scalloped, and PM line lacks large tooth at midpoint. Often has a paler brown or gray patch at inner portion of ST area. **HOSTS** Ceanothus, oak, manzanita, chamise, redshanks, and sweet gale.

DYAR'S LOOPER
Gabriola dyari

Common
91a-1279 (6781)

TL 14–18 mm Speckled grayish-brown FW has thick black AM and PM lines bordering a paler median area. Basal area and inner portion of ST line have large, diffuse, white patches. **HOSTS** Western hemlock and Douglas-fir; also other conifers.

male

ENIGMATIC BEAUTY

female

ANTIQUE BEAUTY

DISTANT WAVE

SECLUDED WAVE

actual size

DYAR'S LOOPER

PEROS

SUPERFAMILY Geometroidea (91a)
FAMILY **Geometridae** SUBFAMILY **Ennominae** TRIBE Odontoperini

Small moths that have the distinctive habit of rolling the costal margin of their broad triangular wings. All species are brown or brownish gray, with a scalloped outer margin; small differences in markings often separate similar species. Adults are nocturnal and will visit lights.

MESKE'S PERO
Pero meskaria
Uncommon
91a-1231 (6747)

NORTH

SOUTH

TL 17–19 mm Light brown or gray FW is lightly speckled, with darker median area containing large white to tan discal spot. PM line has outward bulge near inner margin, then continues relatively straight to costa. PM line is edged below with white, and adjacent ST area is usually shaded dusky. Basal area has a warm brown wash. **HOSTS** Clematis.

STRAIGHT-LINED PERO
Pero radiosaria
Uncommon
91a-1233 (6749)

RANGE-WIDE

TL 14–18 mm Brown FW has straight, dark brown AM and PM lines edged outwardly with white. Median area is slightly darker. Discal spot is a brown-edged white dot but is often indistinct. **HOSTS** Unknown; possibly willow and honeysuckle.

YELLOW-BROWN PERO
Pero flavisaria
Rare
91a-1236 (6751)

RANGE-WIDE

TL 17–21 mm Resembles Meske's Pero, but yellow orange to yellow brown FW has gray shading below PM line only in costal half, no brown wash in basal area, and discal spot is small and often not obvious. **HOSTS** Unknown.

HONEST PERO
Pero honestaria
Uncommon
91a-1241 (6753)

RANGE-WIDE

TL 17–18 mm Resembles Mizon Pero, but finely speckled FW usually lacks discal spot, PM line lacks tooth near inner margin, and basal area is similar in color to median area. AM line is often indistinct. FW is dark gray (male) or violet brown (female). **HOSTS** Unclear because of potential confusion with other *Pero* species.

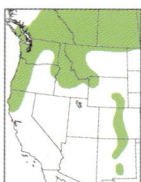

MORRISON'S PERO
Pero morrisonaria
Common
91a-1242 (6755)

RANGE-WIDE

TL 17–20 mm Resembles Mizon Pero, but brown FW is strongly mottled with heavy speckles, particularly in basal and median areas, and pale ST line is wavy. ST area below inward bow of PM line is lightly shaded with dusky scales. **HOSTS** Fir, tamarack, pine, and spruce.

MIZON PERO
Pero mizon
Common
91a-1244 (6757)

RANGE-WIDE

TL 22–26 mm Median area is darker purplish brown, shaded darker along PM line. AM line often becomes indistinct in inner half. PM line bows inward at midpoint, where it is usually shaded below by a diffuse blackish patch. Pale ST line is smoothly curving. Small white discal spot is often obscured by costal fold. **HOSTS** Generalist on deciduous trees and shrubs and needled conifer trees.

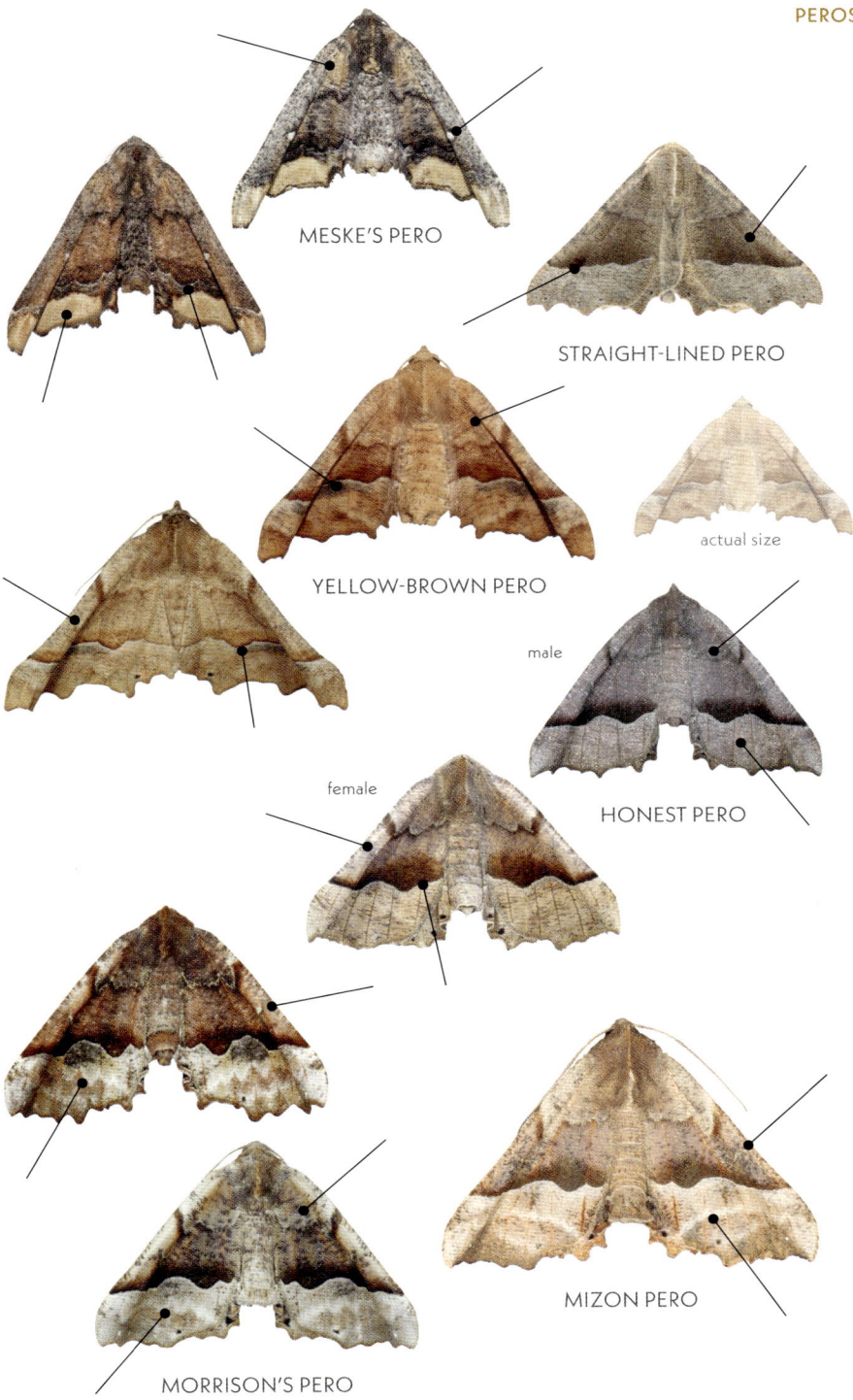

MESKE'S PERO

STRAIGHT-LINED PERO

YELLOW-BROWN PERO

actual size

male

female

HONEST PERO

MORRISON'S PERO

MIZON PERO

MCDUNNOUGH'S PERO

Pero macdunnoughi

Uncommon 91a-1245 (6758)

RANGE-WIDE

TL 20–23 mm Resembles Mizon Pero, but FW usually has reddish-brown shading at costa in upper median area, veins are lightly traced with brown, and pale ST line is indistinct. **HOSTS** California buckwheat, California sagebrush, and redberry buckthorn.

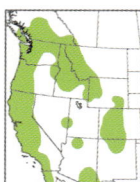

BEHR'S PERO

Pero behrensaria

Common 91a-1248 (6760)

RANGE-WIDE

TL 17–19 mm Resembles Morrison's Pero, but FW brownish gray FW has a warm brown median area bordered by blackish AM and PM lines. AM line is bold and relatively straight across FW. ST area below bow in PM line lacks dusky patch. White discal spot is usually better defined, and terminal area usually has two or three white dots. **HOSTS** Douglas-fir and other needled conifers.

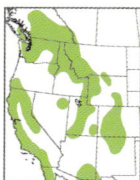

WESTERN PERO

Pero occidentalis

Uncommon 91a-1250 (6761)

RANGE-WIDE

TL 17–20 mm Resembles Honest Pero, but FW has a small white discal spot, and usually also shows two or three small white dots in terminal area. PM line has a small tooth near inner margin. Thin, black AM line is wavy. FW often plain, sometimes speckled in basal area or rarely throughout. **HOSTS** Lodgepole pine, ceanothus, Canadian buffalo-berry, and mock azalea.

BROAD-WINGED GEOMETERS

SUPERFAMILY Geometroidea (91a)

FAMILY Geometridae SUBFAMILY Ennominae TRIBE Ennomini

The largest tribe within the Ennominae, this group is varied in appearance, but moths generally have broad triangular forewings with a convex or pointed outer margin. They largely rest with wings flat, either spread or folded, with the exception of *Sicya*, which curve the wings tent-like over the body, and the thorns, which hold them folded above. Adults are nocturnal and regularly visit lights.

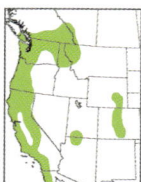

BROWN-EDGED YELLOW

Sicya crocearia

Rare 91a-1284 (6911)

RANGE-WIDE

TL 14–18 mm Resembles Sharp-lined Yellow, but brown area below PM line extends closer to apex, and basal area has a brown patch at inner margin. Costa and apical end of terminal line are brownish. **HOSTS** Unknown; likely ceanothus.

SHARP-LINED YELLOW

Sicya macularia

Common 91a-1285 (6912)

RANGE-WIDE

TL 14–18 mm Lemon-yellow FW has large cinnamon-brown patch in inner ST/terminal area. AM and PM lines may be either solid brown line or row of loose, diffuse, brown to blackish spots. Costa and apical end of terminal line are yellow. Fringe is checkered when very fresh. Some individuals may have light speckling, and black discal and basal dots may be present or absent. Rests with wings in a rounded tent. **HOSTS** Trees and shrubs, including alder, blueberry, poplar, and willow.

MCDUNNOUGH'S PERO

BEHR'S PERO

actual size

WESTERN PERO

actual size

BROWN-EDGED
YELLOW

SHARP-LINED YELLOW

357

CINNAMON YELLOW
Sicya morsicaria

Uncommon
91a-1288 (6915)

TL 14–18 mm Dimorphic. In more common morph, cinnamon-brown FW is speckled, with scalloped white AM and PM lines edged medially with black shading. Less commonly, white AM and PM lines are straight, and black shading and FW speckling are light or absent. Middle segments of legs are speckled gray in both morphs. Most individuals have small black discal dot. HOSTS Mistletoe; also possibly oak.

FALCATE EUSARCA
Eusarca falcata

Uncommon
91a-1293 (6927)

WS 38–44 mm Lightly speckled, light brown FW has angled AM line and straight, dark-edged, pale yellow PM line that does not reach costa. Veins are thinly traced with yellow. Tiny discal dot is blackish. Some individuals have black blotches in outer ST area. Outer margin of all wings is slightly pointed at midpoint, and FW apex is falcate. HOSTS Unknown.

RECTANGULAR EUSARCA
Eusarca geniculata

Rare
91a-1304 (6939)

WS 28–36 mm Resembles Falcate Eusarca, but PM line is evenly straight and angles sharply basally to meet costa. Typically rests with wings held less wide than Falcate Eusarca, so the AM and PM lines have a rectangular appearance. Falcate apex is shaded dusky. HOSTS Unknown.

CONFUSED EUSARCA
Eusarca confusaria

Uncommon
91a-1306 (6941)

WS 29–41 mm Resembles Falcate Eusarca, but ranges do not overlap. Outer margin is rounded, and apex is not usually falcate. AM line is rounded, and pale portion of PM line is often reduced or nearly absent. Some individuals have blotchy dark shading along part or whole of ST line. HOSTS Goldenrod, asters, and other Asteraceae.

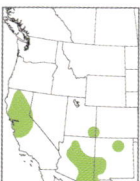

FADED THORN
Neoterpes ephelidaria

Uncommon
91a-1318 (6859)

TL 17–19 mm Pale yellow FW has thick, slightly wavy, brown PM line that angles toward apex, often very faded. AM line is often broken or indistinct, most visible at costa. Small brown discal spot may be indistinct. HOSTS Southwestern and flatbud prickly poppy.

CANARY THORN
Neoterpes trianguliferata

Common
91a-1319 (6860)

TL 15–21 mm Straw- to lemon-yellow FW has bold, elongate, brown triangles on costa at AM and PM lines. Many individuals also have a thick brown line at inner margin of PM line and, occasionally, additional brown marks along costa. Midpoint of outer margin is pointed, and apex is slightly falcate. HOSTS Currant and gooseberry.

CINNAMON
YELLOW

FALCATE EUSARCA

RECTANGULAR EUSARCA

CONFUSED EUSARCA

FADED THORN

CANARY THORN

actual size

YELLOW THORN

Common

Neoterpes edwardsata

91a-1320 (6861)

RANGE-WIDE

TL 14–18 mm Bright yellow, speckled FW has wavy, brown, slanting PM line, usually bordering purplish ST/terminal area. AM line is sometimes visible as blotches at costa, and large discal dots are dark or faded. Resembles female Scallop-lined Geometer, but FW is strongly speckled with brown and AM and PM lines are thicker, with blackish filling. Pale HW is unmarked except speckles. **HOSTS** California poppy; also cultivated bush poppy and Coulter's matilija poppy.

SMOOTH-LINED GIRDLE

Rare

Plataea calcaria

91a-1322 (6921)

RANGE-WIDE

TL 11–19 mm Sexually dimorphic. FW of male is grayish brown, with darker brown median area and darker shading along white ST line. White PM line is S-shaped, connecting with elongate white discal dash near midpoint and joining with AM line near inner margin. FW of female is uniformly grayish to yellowish brown, with relatively straight white AM and PM lines that do not connect to discal spot or each other. Outer margin in both sexes is strongly scalloped. **HOSTS** Unknown; likely sagebrush.

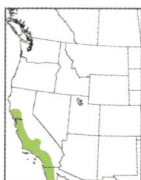

SHARP-TOOTHED GIRDLE

Uncommon

Plataea personaria

91a-1323 (6922)

RANGE-WIDE

TL 13–19 mm Resembles Sagebrush Girdle, but PM line has multiple sharp, inward-pointing teeth and discal spot is dark. ST area is shaded darker than terminal area. Outer margin is slightly pointed at midpoint, and apex is falcate. **HOSTS** California sagebrush. **NOTE** Paler individuals are commonly misidentified as California Girdle (*P. californiaria*, not shown). True California Girdle is relatively rare, found only along the coast from San Francisco to Santa Barbara, CA, and most resembles Hoary Girdle; Hoary occurs more inland, and California has a more distinct ST line.

LARGE GIRDLE

Uncommon

Plataea ursaria

91a-1325 (6924)

RANGE-WIDE

TL 19–24 mm Nearly identical to smaller Sharp-toothed Girdle, but ST line has a small tooth near midpoint. Best identified by range and elevation; Large Girdle is found at higher elevations in the Sierra Nevada and Transverse ranges, while Sharp-toothed Girdle is found at lower elevations in the Coast and Peninsular ranges. **HOSTS** Unknown.

HOARY GIRDLE

Uncommon

Plataea diva

91a-1326 (6925)

RANGE-WIDE

TL 13–21 mm Hoary gray FW has sloping dark AM and PM lines that are similar in pattern to Sharp-toothed Girdle but lack white edging. White discal spots are small and dark-ringed, and ST line is indistinct. Veins are rarely marked. Midpoint of outer margin is rounded, but apex is falcate. **HOSTS** Unknown; likely sagebrush.

YELLOW THORN

actual size

male

male

SMOOTH-LINED
GIRDLE

female

SHARP-TOOTHED
GIRDLE

LARGE GIRDLE

HOARY GIRDLE

SAGEBRUSH GIRDLE
Plataea trilinearia

Common
91a-1327 (6926)

RANGE-WIDE

TL 16–24 mm Sandy-brown FW has sharply sloping white AM and PM lines that have a single, large, medially pointing tooth in inner half, then connect near inner margin. Median area is shaded slightly darker. Elongate median spots and ST line are white. Veins are usually traced with whitish color. HOSTS Unknown; likely sagebrush and possibly goldenbush.

BOLD-LINED PHERNE
Pherne placeraria

Uncommon
91a-1330 (6949)

RANGE-WIDE

TL 15–18 mm Heavily speckled, warm to golden-brown FW has thin, dark reddish-brown, straight AM and gently S-shaped PM lines that are boldly edged outwardly with warm white to tan and indistinctly medially with warm brown shading. PM line is usually roughly parallel to AM line in inner half then bows outward in costal half. Median area is slightly darker and more densely speckled, and sometimes contains a hint of a dark discal spot. Speckling denser in terminal area, and base of fringe is slightly darker than AM/PM lines. HOSTS Unknown; possibly sage. NOTE DNA barcoding data suggests Bold-lined and Thin-lined Pherne may be conspecific; further study is needed.

BROWN-LINED PHERNE
Pherne parallelia

Rare
91a-1331 (6950)

RANGE-WIDE

TL 15–18 mm Golden-brown FW is lightly speckled evenly from base to outer margin. Pale-edged olive-brown AM and PM lines are usually closer together than in other phernes and appear nearly parallel, sometimes narrowly shaded darker medially. Sometimes shows indistinct, elongate, dark discal dash. Base of fringe is the same color as AM/PM lines, and terminal area sometimes has weak olive shading in apical half. HOSTS Unknown; possibly sage.

SHADED PHERNE
Pherne sperryi

Uncommon
91a-1332 (6951)

RANGE-WIDE

TL 14–17 mm Pale FW is lightly dusted with golden-brown speckles. Straight AM line and S-shaped PM line are thin or unmarked, and sharply defined by olive-brown to brown shading that diffuses broadly into wide median area. Fringe does not contrast with FW. HOSTS Unknown; possibly sage.

THIN-LINED PHERNE
Pherne subpunctata

Uncommon
91a-1333 (6952)

RANGE-WIDE

TL 15–18 mm Heavily speckled violet- to brownish-gray, sometimes warm brown, FW has thin, darker brownish-gray, straight AM and gently S-shaped PM lines narrowly bordered outwardly with warm white. AM line is sometimes weak or indistinct. Median area is often slightly darker. Speckling becomes denser toward outer margin. Base of fringe is darker than AM/PM lines, often contrastingly orange brown, and outer fringe is white when fresh. HOSTS Sage.

THICK-LINED GEOMETER
Eriplatymetra coloradaria

Uncommon
91a-1334 (6852)

RANGE-WIDE

TL 16–19 mm Lightly speckled, pale yellow FW has thick, curving, dusky AM and PM lines with blackish edge. Midpoint of outer margin is slightly angled. HOSTS Unknown.

actual size

SAGEBRUSH GIRDLE

BOLD-LINED
PHERNE

BROWN-LINED
PHERNE

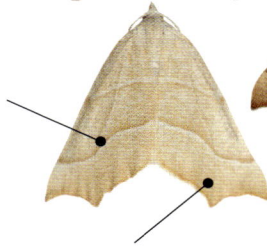

SHADED
PHERNE

THIN-LINED
PHERNE

THICK-LINED GEOMETER

FALCATE GEOMETER

Rare

RANGE-WIDE

Eriplatymetra grotearia   91a-1336 (6854)

TL 18–20 mm Speckled, light brown FW has wavy, brown PM line that crosses FW straight, then angles basally at costa. AM line is visible as brown dashes at costa. Area below PM line is shaded slightly darker. Outer margin is falcate at apex and strongly pointed at midpoint. **HOSTS** Unknown.

SCALLOP-LINED GEOMETER

Rare

RANGE-WIDE

Lychnosea helveolaria   91a-1337 (6857)

TL 15–19 mm Sexually dimorphic. Faintly speckled FW is pale (male) to bright (female) yellow, with brown to purplish basal and ST/terminal areas. Straight PM line is scalloped, and AM line angles basally near costa. Males resemble Falcate Geometer, but outer margin is more rounded and apex is not falcate. Pale HW bears a faint shadow of FW pattern. **HOSTS** Unknown.

GRAY SPRUCE LOOPER

Common

RANGE-WIDE

Caripeta divisata   91a-1339 (6863)

TL 13–19 mm Brown to grayish-brown FW is lightly brindled with thin, dark, irregular AM and PM lines thickly bordered with white. Darker median area contains large white discal spots. **HOSTS** Fir, hemlock, tamarack, pine, and spruce.

RED GIRDLE MOTH

Common

RANGE-WIDE

Caripeta aequaliaria   91a-1341 (6865)

TL 18–21 mm Resembles Gray Spruce Looper, but FW is reddish brown and AM and PM lines are thinner. Veins are often traced with pinkish color, and terminal area has bold cream interveinal dashes. **HOSTS** Douglas-fir, pine, and western hemlock.

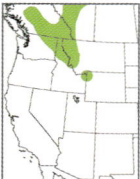

BROWN PINE LOOPER

Rare

RANGE-WIDE

Caripeta angustiorata   91a-1344 (6867)

TL 14–16 mm Warm brown FW has thin, brown, sharply pointed AM and uneven but straightish PM lines thickly bordered white. Broad ST line and basal dash are speckled gray, and large discal spot is white. **HOSTS** Pine; also fir, spruce, and tamarack.

SNOWY MOUNTAIN MOTH

Rare

RANGE-WIDE

Snowia montanaria   91a-1353 (6875)

TL 17–19 mm Brindled golden-brown FW has blackish AM and PM lines sharply angled in costal half. Long black basal dash bisects AM line, and point of AM line reaches PM line. Veins are pinkish, and terminal area has speckled gray patches. **HOSTS** Unknown.

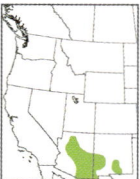

BESMA THORN

Uncommon

RANGE-WIDE

Destutia flumenata   91a-1356 (6880)

WS 22–25 mm Heavily speckled, straw-colored FW has thin, gently wavy, brown AM and PM lines and tiny black discal dots. Veins are thinly traced with brown. **HOSTS** Unknown.

FALCATE GEOMETER

SCALLOP-LINED
GEOMETER

GRAY SPRUCE
LOOPER

RED GIRDLE MOTH

BROWN PINE LOOPER

SNOWY MOUNTAIN MOTH

actual size

BESMA THORN

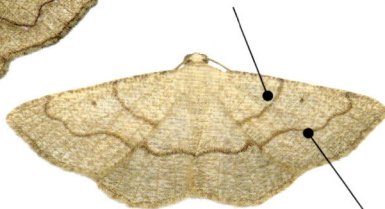

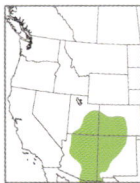

STRAIGHT-LINED THORN
Destutia excelsa

Uncommon
91a-1359 (6883)

TL 13–17 mm Tan FW has cream-edged, curving, brown AM and PM lines and small brown discal spot. Costa is lightly peppered with black. Outer margin is pointed at midpoint, and apex is falcate. Resembles some *Tetracis* species, but HW outer margin is not as pointed and PM line is straighter. **HOSTS** Oak.

OAK BESMA
Besma quercivoraria

Rare
91a-1361 (6885)

WS 27–41 mm Variable and sexually dimorphic. Female is speckled tan, with straight AM and PM lines and dark discal dot. Veins are traced with brown. Male is similar but has dark shading below PM line and sometimes basal to AM line, and warm brown edging along PM line. In some individuals, AM and PM lines may connect near inner margin. **HOSTS** Oak, willow, maple, and other trees.

HEMLOCK LOOPER
Lambdina fiscellaria

Very Common
91a-1366 (6888)

WS 29–45 mm Peppery, semitranslucent, grayish-brown wings have yellow-edged dark AM and PM lines. PM line is angled at midpoint on all wings. Median area has a dark discal spot and is sometimes shaded darker. **HOSTS** Hemlock, fir, spruce, maple, oak, and many other trees.

HOARY CONIFER LOOPER
Nepytia umbrosaria

Uncommon
91a-1386 (6899)

TL 17–21 mm Hoary gray FW has darker median area bordered by scalloped AM and PM lines. Veins are dark and broken below PM line by diffuse pale ST line. Thin elongate discal spot is sometimes indistinct. **HOSTS** Fir, Douglas-fir, and western hemlock; also other needled conifers.

BLACK-VEINED CONIFER LOOPER
Nepytia swetti

Uncommon
91a-1392 (6905)

TL 19–21 mm Lightly peppered gray FW has thin, indistinct, scalloped AM and PM lines. All veins slightly darker; central veins are black from AM line to outer margin, darkest in median area, and often backed by dusky shading. Curving black discal spot sometimes appears joined to central vein. **HOSTS** Unknown, but likely needled conifers.

PHANTOM HEMLOCK LOOPER
Nepytia phantasmaria

Very Common
91a-1394 (6907)

TL 17–20 mm White to pale gray FW has bold, black, strongly scalloped AM and PM lines and black discal spot. Veins are often traced with black, and fringe is checkered. Front of head and shoulders are orange. **HOSTS** Western hemlock and Douglas-fir; also other conifers.

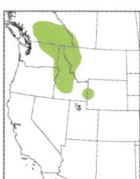

WESTERN FALSE HEMLOCK LOOPER
Nepytia freemani

Uncommon
91a-1397 (6910)

TL 11–15 mm Peppery gray FW has thin, gray, scalloped AM and PM lines and indistinct discal spot. Veins are sometimes indistinctly darker. Separable from *Sabulodes* species by rounded PM line that curves up at costa. **HOSTS** Douglas-fir.

male

male

STRAIGHT-LINED
THORN

female

OAK BESMA

HEMLOCK LOOPER

HOARY CONIFER LOOPER

BLACK-VEINED
CONIFER LOOPER

actual size

WESTERN FALSE
HEMLOCK LOOPER

PHANTOM HEMLOCK
LOOPER

VARIABLE DALMATION
Eucaterva variaria

Uncommon
91a-1398 (6918)

TL 16–18 mm White FW has variable amount of black speckling through central wing; some individuals may be mostly white and others mostly black aside from a speckled white costa, while most individuals will fall somewhere in between. Uncommonly, some individuals may have thick, curving, blackish AM and PM lines and large discal spot that may be indistinct behind the black speckling. Rare individuals may be uniformly speckled gray, with indistinct AM and PM lines. **HOSTS** Desert willow.

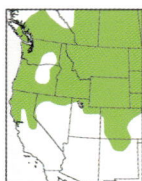

MAPLE SPANWORM
Ennomos magnaria

Common
91a-1405 (6797)

TL 24–30 mm Golden-orange wings have scalloped outer margin and falcate apex and are peppered with thick, dark brown speckles. Some individuals have curving, dark brown AM and PM lines and/or dark brown shading below PM line. Typically rests with wings held slightly aloft. **HOSTS** Alder, basswood, maple, oak, poplar, willow, and other trees.

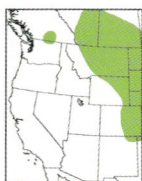

WHITE SLANT-LINE
Tetracis cachexiata

Uncommon
91a-1410 (6964)

TL 17–25 mm White FW has a bold, warm brown PM line that slants from midpoint of inner margin to apex. Midpoint of outer margin is slightly angled. **HOSTS** Alder, ash, birch, cherry, maple, spruce, willow, and many other trees.

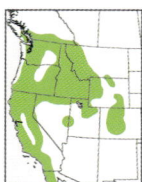

ELEGANT THORN
Tetracis cervinaria

Very Common
91a-1411 (6956)

WS 38–46 mm Variable. Fawn- to orange-brown wings have long elegant points at midpoint and falcate FW apex. AM line curves to meet costa, while PM line is straight across all wings; lines may be maroon or pale. Median area is sometimes shaded darker and contains a small black discal dot. Veins on fawn-brown individuals may be pale tan. Filiform antennae are white. **HOSTS** Bitter cherry and chokecherry; possibly also alder, madrone, manzanita, bitterbrush, ceanothus, cascara, and willow. **NOTE** Rests with wings spread more than other *Tetracis*.

SHADED THORN
Tetracis pallulata

Uncommon
91a-1414 (6955)

TL 18–24 mm Resembles October Thorn, but thick black AM and PM lines are boldly edged outwardly with pale yellowish color, and dark speckling is much stronger. Median area is slightly darker, and dark speckling forms a shaded band in ST area. **HOSTS** White and grand fir, Engelmann and Sitka spruce, western hemlock, and Douglas-fir.

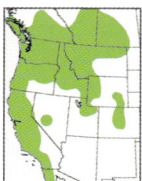

OCTOBER THORN
Tetracis jubararia

Very Common
91a-1416 (6954)

TL 17–26 mm Lightly striated, pale to warm brown (rarely blackish) FW has thin, blackish, rounded AM and gently S-shaped PM lines sometimes with faint pale yellowish edge outwardly. FW apex is slightly falcate, and outer margin of all wings is pointed at midpoint. Hairy thorax is orange. Both sexes have filiform antennae. **HOSTS** Alder, birch, dogwood, poplar, and *Ribes*.

VARIABLE DALMATION

MAPLE SPANWORM

actual size

WHITE SLANT-LINE

ELEGANT THORN

SHADED THORN

OCTOBER
THORN

369

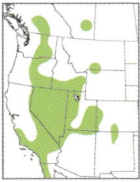

SPECKLED THORN

Rare

Tetracis formosa 91a-1419 (6958)

TL 18–22 mm Peppery gray FW has a paler costa and terminal area. Thick, black, white-edged AM line forms a long loop with basal dash. White-edged black PM line is scalloped or wavy, while black ST line is straight. Median area is shaded darker with white-traced veins. Both sexes have filiform antennae. **HOSTS** Desert peach; possibly others.

CALIFORNIA THORN

Uncommon

Tetracis hirsutaria 91a-1420 (6960)

TL 14–17 mm Resembles October Thorn, but AM and PM lines have dark dots (often edged with white dots) at veins; dots may be tiny or indistinct in some individuals. Inner margin of HW has dark median line. Males have bipectinate antennae. **HOSTS** Ceanothus, mountain mahogany, bitter cherry, and chaparral currant.

WHITE-DOTTED THORN

Uncommon

Prochoerodes truxaliata 91a-1436 (6977)

TL 18–22 mm Yellowish-brown FW has diffuse brown AM, PM, and ST lines, typically with white-edged black dots at veins. HW lacks median line. Both sexes have filiform antennae. **HOSTS** Coyote brush.

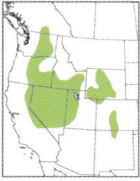

PEPPERED THORN

Rare

Prochoerodes amplicineraria 91a-1437 (6978)

TL 15–18 mm Hoary gray-brown FW has darker median area and brown ST area. Diffuse dusky AM, PM, and ST lines are edged with white. Veins in ST area often slightly blackish. **HOSTS** Unknown.

WAVY-LINED THORN

Uncommon

Prochoerodes forficaria 91a-1439 (6981)

WS 36–42 mm Lightly speckled, grayish to warm brown FW has dark wavy AM and gently curved PM lines edged medially with diffuse, warm brown shading. White-edged PM line slants toward apex, then turns sharply basally at costa. ST area sometimes has dusky shading. Discal spot is black. Midpoint of outer margin is usually only slightly pointed. **HOSTS** Coffeeberry, trailing blackberry, ash, and willow.

OMNIVOROUS LOOPER

Very Common

Sabulodes aegrotata 91a-1447 (6995)

WS 34–52 mm Lightly speckled, light brown wings have diffuse dusky median band and indistinct AM and PM lines dotted with black at veins. Median and PM lines touch at inner margin. PM area is sometimes shaded darker. Outer margin is slightly pointed at midpoint. **HOSTS** Broad generalist. May be an occasional pest on fruit trees.

WHITE-STRIPED GIRDLE

Uncommon

Sabulodes niveostriata 91a-1452 (7000)

TL 19–24 mm Rusty-orange FW is boldly striped with white. Stripes in central FW form an inverted V shape. Terminal area has white stripes at veins. **HOSTS** Likely pine.

SPECKLED THORN

actual size

CALIFORNIA THORN

WHITE-DOTTED THORN

PEPPERED THORN

WAVY-LINED THORN

OMNIVOROUS LOOPER

WHITE-STRIPED GIRDLE

DARK-BANDED GIRDLE

Uncommon

RANGE-WIDE

Sabulodes spoliata   91a-1455 (7003)

TL 16–19 mm Warm brown FW has darker median area. Dark brown V-shaped AM and wavy PM lines are edged outwardly with white. ST/terminal area is broad. **HOSTS** Conifers, likely including pine and Douglas-fir.

GOLDEN GIRDLE

Uncommon

NORTH

Sabulodes edwardsata   91a-1456 (7004)

SOUTH

TL 19–24 mm Resembles Dark-banded Girdle, but FW is more golden brown and AM line has a bulge at midpoint that resembles a dripping paint droplet. ST/terminal area is narrower. **HOSTS** Pine and Douglas-fir; also western hemlock, Sitka spruce, and other needled conifers.

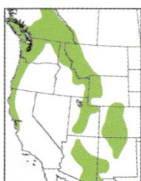

VARIABLE GIRDLE

Uncommon

RANGE-WIDE

Sabulodes venata   91a-1457 (7005)

TL 16–19 mm Peppery gray to white FW resembles Western False Hemlock Looper, but AM line has a bulge that touches elongate discal dash, and midpoint of PM line has a small medial bulge. AM and PM lines are often thick, and median area is usually shaded blackish. Terminal line and fringe are dotted. **HOSTS** Douglas-fir, western hemlock, and fir.

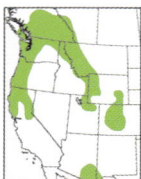

GRAY GIRDLE

Rare

RANGE-WIDE

Sabulodes griseata   91a-1458 (7006)

TL 16–18 mm Resembles Variable Girdle, but bulge in AM line is not as pronounced. Elongate blackish discal dash is distinct. PM line on HW is often indistinct. Terminal line and fringe are plain or weakly dotted. **HOSTS** Conifers, likely including Douglas-fir and western hemlock.

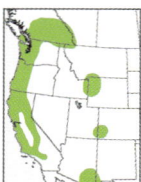

PACKARD'S GIRDLE

Common

RANGE-WIDE

Sabulodes packardata   91a-1459 (7007)

TL 16–18 mm Very similar to Gray Girdle, but AM and PM lines are usually darker and more deeply scalloped, especially near costa. Terminal line and fringe are strongly dotted. **HOSTS** Douglas-fir, grand fir, and western hemlock; also mountain hemlock, Sitka spruce, Pacific silver fir, and western redcedar.

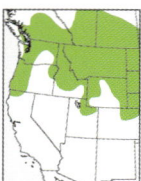

HORNED SPANWORM

Common

RANGE-WIDE

Nematocampa resistaria   91a-1461 (7010)

WS 28–32 mm Yellow-tan (male) to whitish (female) wings have a network of wavy brown AM, median, and PM lines crossed by thin brown veins and backed by short brown striations. Area below PM line is extensively shaded brownish except costal half of FW. In rare individuals, shading is absent and ST area resembles upper half of wing. Northwestern individuals are larger than eastern. **HOSTS** Alder, ash, birch, maple, oak, strawberry, and many other trees, shrubs, and herbaceous plants.

SPURRED SPANWORM

Uncommon

RANGE-WIDE

Nematocampa brehmeata   91a-1462 (7011)

WS 28–32 mm Resembles Horned Spanworm, but dark shading in male usually reduced, with a diffuse border and/or pale patches. Female lacks dark shading. Some individuals may be best identified by range. **HOSTS** Unknown; possibly willow or California buckeye.

DARK-BANDED
GIRDLE

GOLDEN GIRDLE

VARIABLE GIRDLE

GRAY GIRDLE

PACKARD'S GIRDLE

actual size

HORNED SPANWORM

SPURRED SPANWORM

NOCTUID MOTHS

SUPERFAMILY Noctuoidea (93)

This is our largest superfamily of moths, with more than 3,700 species in North America. It contains most of the macromoths, predominantly medium to large, stout-bodied species, though there is a great deal of variability in size, shape, and posture, and some groups resemble geometers or micromoths. It is considered to be the most evolutionarily recent moth lineage; a common characteristic not shared by other lepidopterans is the presence of specialized organs that allow the moths to detect the calls of bats and avoid predation by them. Adults are mostly nocturnal, though some, such as the wasp moths and flower moths, are often encountered at flowers, and others, like the litter moths, may be flushed from low vegetation or forest trails.

PROMINENTS
SUPERFAMILY Noctuoidea (93)
FAMILY Notodontidae

Medium-sized moths with long, broadly rounded wings that are usually held tent-like or curved against the body, though a few species more commonly rest with wings flat. Bodies are stout, and most species have a hairy thorax, sometimes with a small thoracic crest. Adults are nocturnal and regularly visit lights, though California Oak Moth is primarily diurnal.

SIGMOID PROMINENT
Clostera albosigma

Uncommon
93-0003 (7895)

RANGE-WIDE

TL 16–20 mm Gray-brown to warm brown FW has chestnut-brown patch in costal ST area bordered by bold white dash at outer PM line. Inner PM, median, and AM lines are straight, thin, and pale grayish white; FW also shows a brownish-white fourth line in a shallow diagonal across PM area. Front of thorax and head is dark blackish brown. ST line is indistinct. HOSTS Aspen, poplar, and willow.

BRUCE'S CHOCOLATE-TIP
Clostera brucei

Uncommon
93-0008 (7900)

RANGE-WIDE

TL 14–17 mm Resembles the larger Sigmoid Prominent, but AM, median, PM, and fourth line are wavy, and area between PM and fourth line is shaded dark brown. ST line is a series of black dashes. HOSTS Quaking aspen.

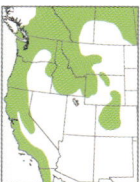

APICAL PROMINENT
Clostera apicalis

Uncommon
93-0009 (7901)

RANGE-WIDE

TL 17–19 mm Resembles Sigmoid Prominent, but diagonal fourth line is distinct and runs from inner margin of PM line to costal end of AM line, and ST patch is reddish, often with rusty spots, surrounded by reddish-brown shading that diffuses across PM line, and bordered by gray patch in central ST area. ST line is a series of black dashes. HOSTS Willow and poplar.

ORNATE PROMINENT
Clostera ornata

Uncommon
93-0009.1 (7901.1)

RANGE-WIDE

TL 17–19 mm Very similar to Apical Prominent, but reddish ST area patch is reduced and surrounding shading is dark brown, reaching outer margin and ending sharply at central vein, so entire FW apical area appears darkened. Costa is distinctly convex at PM area. HOSTS Likely willow and poplar.

actual size

SIGMOID PROMINENT

BRUCE'S CHOCOLATE-TIP

APICAL PROMINENT

ORNATE PROMINENT

RUSSET PROMINENT
Clostera inornata

Uncommon
93-0005 (7897)

RANGE-WIDE

TL 22–24 mm FW line pattern resembles that of Apical Prominent, but entire FW is suffused with rusty brown, becoming grayish along inner margin. Large reniform spot and central ST area are light gray. Thorax usually lacks dark front. **HOSTS** Cottonwood and willow.

BLACK-RIMMED PROMINENT
Pheosia rimosa

Uncommon
93-0012 (7922)

RANGE-WIDE

TL 25–32 mm Distinctive FW has a thick white arc from base to apex, bordered by warm brown and edged along inner margin and costa with dark brown. Dark patch along costa is narrow and has black streaks in outer ST area; FW sometimes has grayish patch at median area. PM line is present only as white dash at costa. Inner terminal area has white streaks at veins, with innermost reaching nearly to hooked white AM line. Inner basal area is golden brown. **HOSTS** Poplar and willow.

BROWN-RIMMED PROMINENT
Pheosia californica

Uncommon
93-0012.1 (7923)

RANGE-WIDE

TL 25–32 mm Dimorphic. Dark form has hoary gray FW, with warm brown shading along inner margin and a scalloped white PM line (sometimes indistinct). AM line is present as small, curved, white spot at inner margin. Outer ST area has thick black streaks between veins and a diffuse white patch at apex. Inner basal area is golden brown. Pale form resembles Black-rimmed Prominent, but hoary gray median area diffuses into white arc, black along inner margin is narrower, costa has a bold white square at PM line, and white streaks in terminal area are reduced. **HOSTS** Willow, cottonwood, and quaking aspen.

ELEGANT PROMINENT
Pheosidea elegans

Uncommon
93-0013 (7924)

RANGE-WIDE

TL 27–32 mm Gray FW has a tawny patch in inner basal area edged thinly with white. Outer ST area has two black streaks. Inner margin has hairy tufts that project upward when wings are folded. **HOSTS** Poplar.

COMMON GLUPHISIA
Gluphisia septentrionis

Common
93-0019 (7931)

RANGE-WIDE

TL 14–17 mm Variable. Hoary gray FW has tawny patches in basal and inner median area. Dark gray, wavy AM line is straight; gently scalloped PM line is often indistinct. ST line may be dark gray or indistinct. Median and/or ST area sometimes shaded dusky, rarely completely tan. Rare individuals are very pale gray. HW is plain gray or lightly marked. Thorax is densely hairy. **HOSTS** Poplar.

BANDED GLUPHISIA
Gluphisia severa

Uncommon
93-0023 (7935)

RANGE-WIDE

TL 19–21 mm Resembles the smaller Common Gluphisia, but ST line is edged with tan shading, AM line is slightly toothed, and HW has dark dashes at anal angle. Some individuals have a distinct tan reniform spot. **HOSTS** Willow and poplar.

RUSSET
PROMINENT

BLACK-RIMMED PROMINENT

BROWN-RIMMED PROMINENT

ELEGANT PROMINENT

actual size

COMMON GLUPHISIA

BANDED GLUPHISIA

GRAY FURCULA
Furcula cinerea

Rare
93-0025 (7937)

RANGE-WIDE

TL 17–22 mm Light gray FW has darker, hoary, gray AM and ST bands bordered by orange dots, and a row of black dots just adjacent. Soft median line is thin and sometimes indistinct. Hairy gray thorax has two transverse orange and black lines. **HOSTS** Aspen, poplar, and willow.

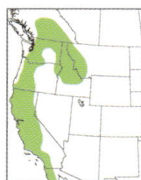

ASHY FURCULA
Furcula cinereoides

Uncommon
93-0025.1 (7937.1)

RANGE-WIDE

TL 17–22 mm Resembles Gray Furcula, but ranges do not overlap. FW is pale gray, sometimes almost white, and thoracic markings are bold black and orange. FW markings may be variably crisp to indistinct, and median line may or may not be visible. **HOSTS** Aspen, poplar, and willow.

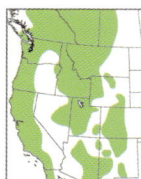

ZIGZAG FURCULA
Furcula scolopendrina

Common
93-0028 (7940)

RANGE-WIDE

TL 19–22 mm Resembles Gray Furcula, but FW is whitish, dark median line is double and strongly scalloped, and broad ST band extends only to midpoint of wing. Orange dots along bands are often small. Thorax is mostly or entirely dark hoary gray mixed with orange. **HOSTS** Willow, birch, and cottonwood.

MODEST FURCULA
Furcula modesta

Rare
93-0029 (7941)

RANGE-WIDE

TL 18–23 mm Resembles Zigzag Furcula, but AM and ST bands are darker and lack orange dots at edges, AM band is broader, sometimes pinched in middle, and median line is faded or indistinct. Thorax is dark bluish gray. **HOSTS** Poplar and willow.

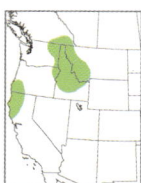

SCALLOPED DATANA
Datana californica

Rare
93-0033.1 (7902.1)

RANGE-WIDE

TL 22–26 mm Resembles Spotted Datana, but spots are usually indistinct or almost absent and outer margin is more strongly scalloped. Apical dash may be indistinct. **HOSTS** Trees and shrubs, including apple, oak, birch, and willow.

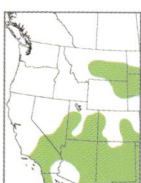

SPOTTED DATANA
Datana perspicua

Common
93-0039 (7908)

RANGE-WIDE

TL 25–30 mm Light brown FW has five thin brown lines: AM, median, and PM lines are dark, while two lines in PM area are fainter. Dark orbicular and reniform spots are round; median line crosses through latter. Long brown apical dash is curving. Outer margin may be slightly wavy. **HOSTS** Sumac and smoke tree.

LARGE DATANA
Datana perfusa

Uncommon
93-0045.1 (7914)

RANGE-WIDE

TL 30–34 mm FW is light brown to brownish orange, with indistinctly darker shading in terminal area. Front of thorax is light orange. Faint reniform spot is smaller and more indistinct than in Spotted Datana. Most lines are indistinct or nearly absent; curving AM and PM lines are usually the most prominent. Veins are sometimes traced darker. **HOSTS** Oak.

GRAY FURCULA

ASHY FURCULA

ZIGZAG FURCULA

actual size

MODEST FURCULA

SCALLOPED DATANA

SPOTTED DATANA

LARGE DATANA

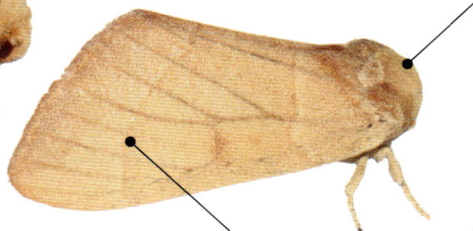

WHITE-DOTTED PROMINENT
Nadata gibbosa **Very Common** 93-0046 (7915) RANGE-WIDE

TL 20–30 mm Tawny FW has pale-edged, straight AM and PM lines. Median area has two small, round, white spots, often bordered basally by a paler patch. Outer margin is scalloped and often falcate. Thorax has pointed crest. **HOSTS** Oak; also birch, cherry, maple, and others.

OREGON PROMINENT
Nadata oregonensis **Uncommon** 93-0047 (7916) RANGE-WIDE

TL 22–26 mm Resembles White-dotted Prominent, but white dots are oblong and PM line curves up to meet costa. Veins are often slightly darker brown. **HOSTS** Unknown.

ORANGE-TAILED PROMINENT
Cargida pyrrha **Uncommon** 93-0052 (7964) RANGE-WIDE

TL 18–26 mm Hoary gray FW has sharply toothed black AM and PM lines and indistinct black dashes at apex. Abdomen is orange. Thorax is densely hairy. **HOSTS** Torrey wolfberry; also lotebush.

MIMOSA PROMINENT
Afilia oslari **Uncommon** 93-0055 (7962) RANGE-WIDE

TL 17–19 mm Peppery, pale gray FW has strong double AM line and jagged black ST line. Basal and PM lines are double, with basal line black and lower line brown. Reniform spot is a long black dash. Thorax is crested with a brownish center, and abdomen is rusty orange. **HOSTS** Graham's mimosa.

OLIVE-PATCHED PROMINENT
Astiptodonta wymola **Uncommon** 93-0062 (7970) RANGE-WIDE

TL 15–17 mm Hoary gray FW is olive in inner half, with veins strongly traced with black. Apex has an angled dusky smudge, often adjacent to a diffuse, pale gray patch. **HOSTS** Unknown.

BLACK-CRESCENT PROMINENT
Heterocampa averna **Uncommon** 93-0083 (7991) RANGE-WIDE

TL 22–29 mm Olive-gray FW has a grayish costa and wavy, double, black AM and PM lines. ST area has two angled sets of black-shaded dashes; the inner one connects to dark shading in PM area that curves to meet dark reniform dash, creating a dusky crescent shape. **HOSTS** Emory oak.

SMILING PROMINENT
Cecrita lunata **Uncommon** 93-0087.15 (7993) RANGE-WIDE

TL 23–25 mm Brownish-gray FW has a thick, curving, black reniform dash and diffuse rusty shading in basal area. Veins are indistinctly traced with grayish color; ST area contains dusky interveinal dashes. Some individuals show indistinct double AM line. **HOSTS** Unknown; likely quaking aspen.

WHITE-DOTTED
PROMINENT

OREGON PROMINENT

ORANGE-TAILED
PROMINENT

actual size

MIMOSA PROMINENT

OLIVE-PATCHED PROMINENT

BLACK-CRESCENT PROMINENT

SMILING PROMINENT

STEEL-GRAY PROMINENT
Theroa zethus

Uncommon
93-0093 (8000)

RANGE-WIDE

TL 17–18 mm Steel-gray FW has brown-edged tan reniform spot and dotted ST line. AM and PM lines are present as thin pale dashes at veins. Thorax is crested. **HOSTS** Flowering spurge and spotted spurge.

WHITE-LINED PROMINENT
Ursia noctuiformis

Uncommon
93-0095 (8002)

RANGE-WIDE

TL 12–14 mm Hoary gray FW has narrow white PM line running from inner margin to black-centered tan reniform spot. Long dashes below PM line are black in ST area and white in terminal area; central dash is often more prominent than reniform spot. **HOSTS** Unknown.

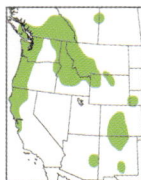

MORNING-GLORY PROMINENT
Schizura ipomaeae

Common
93-0098 (8005)

RANGE-WIDE

TL 20–25 mm Dimorphic. Pale form has light brown FW, with scalloped AM and PM lines, and black-centered tan reniform spot. Central median area and outer half of basal area are shaded dark brown, and inner half of basal area is tan. Fringe is checkered, and apex has a pale patch. Dark form is similar, but central wing is dark brown from base to outer margin. **HOSTS** Basswood, beech, birch, maple, oak, and many other woody plants.

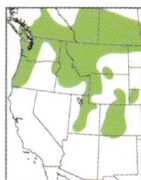

UNICORN PROMINENT
Coelodasys unicornis

Common
93-0100 (8007)

RANGE-WIDE

TL 18–25 mm FW variable in color from grayish green to warm brown with a greenish cast. Scalloped AM and PM lines are double, and reniform spot is a long black dash. Outer median area and apex are pale, and central ST/terminal area is warm brown. ST line is a series of black dashes edged below with white dashes, sometimes visible only on either side of brown patch. **HOSTS** Hickory, maple, oak, willow, and other deciduous trees.

RED-HUMPED CATERPILLAR MOTH
Oedemasia concinna

Uncommon
93-0103 (8010)

RANGE-WIDE

TL 17–21 mm Tan FW is shaded maroon along inner margin, grayish along apical half of costa, and warmer brown in central basal area. Small black discal dot is only strong marking. ST area has black patches at inner margin and costa, sometimes with indistinct interveinal brown patches in central ST area. **HOSTS** Apple, blueberry, elm, hickory, maple, oak, and many other woody plants.

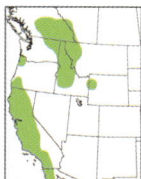

WESTERN RED-HUMPED CATERPILLAR MOTH
Oedemasia salicis

Common
93-0103.1 (8010.1)

NORTH

SOUTH

TL 17–21 mm Resembles Red-humped Caterpillar Moth, but central FW is more straw-colored and has diffuse brown PM line. Reliably separated by range. **HOSTS** Presumably similar to those of Red-humped Caterpillar Moth.

STEEL-GRAY PROMINENT

WHITE-LINED PROMINENT

MORNING-GLORY
PROMINENT

UNICORN PROMINENT

RED-HUMPED
CATERPILLAR MOTH

actual size

WESTERN RED-HUMPED
CATERPILLAR MOTH

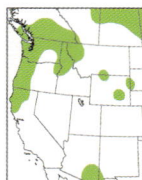

RED-WASHED PROMINENT
Oedemasia semirufescens

Uncommon
93-0105 (8012)

TL 24–25 mm Light brown to dark brown FW has a thin reniform dash, with a small black spot at inner end and bordered outwardly by dusky shading, and a whitish apical patch. Veins are thinly traced with dark brown. Rest of wing is variable and often patchy, usually shaded darker in inner basal area, but generally shows inner half of indistinct double scalloped PM line. HOSTS Poplar and willow; also alder, birch, oak, and rose.

PALE PROMINENT
Ianassa pallida

Uncommon
93-0107 (8014)

TL 18–26 mm Light, warm brown FW is broadly grayish along costa, with thick black reniform dash shaded outwardly with darker brown. Outer ST area usually shows two black dashes. Often has reddish shading at shoulder. HOSTS Maple, willow, poplar, and apple.

HOLLOW-SPOTTED PROMINENT
Pseudhapigia brunnea

Uncommon
93-0120 (8029)

TL 28–32 mm Fawn to reddish-brown FW has distinctive white spots: circular orbicular spot is solid, and kidney-shaped reniform spot is hollow. Dark ST line is a series of disconnected scallops with a white crescent at costa. Pale PM line is smoothly curving. HOSTS Desert kidneywood.

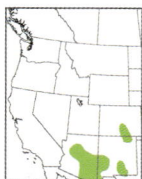

WAVY SYMMERISTA
Symmerista zacualpana

Rare
93-0131 (7955)

TL 18–22 mm Gray FW has long, wavy-edged, white patches adjacent to gray costa from apex to median area, bordered inwardly with dark reddish brown shading. Black ST line is strongly scalloped. Front of thorax is tan. HOSTS Oak. NOTE Suave Symmerista (*S. suavis*, not shown), also of AZ but rare, is nearly identical but white patch lacks the tooth near ST line.

GRAY-STREAKED PROMINENT
Notela jaliscana

Uncommon
93-0136 (7960)

TL 18–20 mm Dimorphic. Hoary gray FW has a zigzag double AM and PM line, sometimes most noticeable at inner margin, and light brownish wash along costa and inner median area. Veins are thinly traced with black. Costal end of ST area has dusky interveinal dashes. Some individuals have a thick, dark gray, fragmented streak running from basal area to outer margin. HOSTS Desert kidneywood.

CALIFORNIA OAK MOTH
Phryganidia californica

Very Common
93-0138 (8031)

TL 16–20 mm Dusky brownish FW has dark veins and a diffuse tan patch (sometimes indistinct) in PM area. Sides of dusky thorax are dull orange. HOSTS Coast live oak, chinquapin, stone oak, and other oaks and Fagaceae. NOTE Diurnal.

RED-WASHED
PROMINENT

PALE PROMINENT

HOLLOW-SPOTTED PROMINENT

actual size

WAVY SYMMERISTA

GRAY-STREAKED
PROMINENT

CALIFORNIA OAK MOTH

TUSSOCK MOTHS
SUPERFAMILY Noctuoidea (93)
FAMILY Erebidae SUBFAMILY Lymantriinae

Medium-sized moths with broadly rounded wings that are held flat, and densely hairy forelegs that are extended in front of the head at rest. The common name of the group comes from the caterpillars, which in many species bear mounded tufts (tussocks) of hairs along their back. Females lack wings and are flightless. Adults are nocturnal and regularly visit lights.

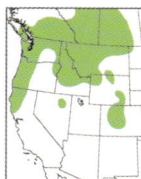

VARIABLE TUSSOCK MOTH
Dasychira vagans

Uncommon
93-0146 (8294)

RANGE-WIDE

TL 18–26 mm Peppery gray FW has blackish wavy AM and PM lines that are slightly thicker toward costa, and paler median area that is often washed whitish along costa. Large reniform spot is pale and thinly outlined in black. ST area is usually unmarked but may show a small white spot at inner margin or diffuse whitish ST line. HOSTS Deciduous trees, including apple, birch, poplar, and willow.

GRIZZLED TUSSOCK MOTH
Dasychira grisefacta

Uncommon
93-0158 (8306)

RANGE-WIDE

TL 18–21 mm Hoary gray FW has small white spot at inner margin of ST area and diffuse white ST line from costa to midpoint. Reniform spot is sometimes paler but often indistinct. Dusky AM and PM lines are thin, toothed, and usually faint. HOSTS Cottonwood, willow, and oak.

RUSTY TUSSOCK MOTH
Orgyia antiqua

Very Common
93-0160 (8308)

RANGE-WIDE

TL 12–16 mm Warm brown to rusty-brown FW has bold white spot at inner margin of ST area and dusky shading above and below diffuse dusky AM and PM lines. Lightly traced, kidney-shaped reniform spot is faint. HOSTS Deciduous trees and shrubs, including alder, cherry, maple, pine, and willow.

WESTERN TUSSOCK MOTH
Orgyia vetusta

Very Common
93-0161 (8309)

RANGE-WIDE

TL 13–15 mm Warm brown to hoary gray-brown FW has a bold white spot at inner margin and dusky patch at costa of ST area, as well as kidney-shaped, dark-centered, tan reniform spot that is sometimes large and diffuse. AM and PM lines are jagged black, sometimes indistinct; AM is slightly offset at midpoint, and PM curves widely around reniform spot. Median area is often slightly grayish. HOSTS Coast live oak, ragweed, lupine, and saltbush.

DOUGLAS-FIR TUSSOCK MOTH
Orgyia pseudotsugata

Very Common
93-0164 (8312)

RANGE-WIDE

TL 14–17 mm Resembles Western Tussock Moth, but AM line is straight, and white spot has a dusky patch above and indistinct dusky ST line that connects to dark patch at costa of ST area. HOSTS Douglas-fir, spruce, fir, and hemlock.

WHITE SATIN MOTH
Leucoma salicis

[IN] Very Common
93-0170 (8319)

RANGE-WIDE

TL 22–27 mm Satiny white FW is completely unmarked. White abdomen has blackish bands at each segment. Superficially similar to Virginian Tiger Moth, but legs are banded with black and white. Antennae are broadly bipectinate. HOSTS Poplar and willow.

VARIABLE TUSSOCK
MOTH

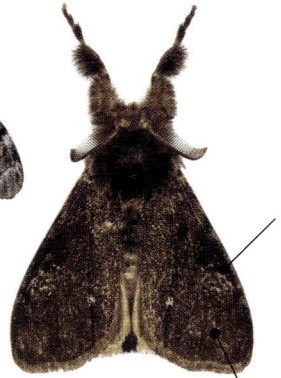

GRIZZLED
TUSSOCK MOTH

RUSTY TUSSOCK
MOTH

WESTERN TUSSOCK
MOTH

actual size

DOUGLAS-FIR TUSSOCK
MOTH

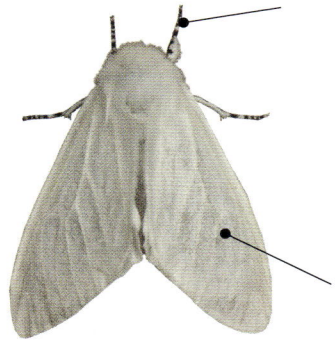

WHITE SATIN MOTH

LICHEN MOTHS

SUPERFAMILY Noctuoidea (93)

FAMILY Erebidae SUBFAMILY Arctiinae TRIBE Lithosiini

Small moths that are often boldly colored in black and orange. Most rest with wings folded flat and overlapping, though *Cisthene* hold wings tented. For all species where larval habits are known, caterpillars feed on algae and the algal component of lichens. Adults are nocturnal and regularly visit lights, though the *Lycomorpha* and *Ptychoglene* are often seen at flowers during daytime.

ARID EUDESMIA
Eudesmia arida

Uncommon
93-0174 (8096)

RANGE-WIDE

TL 13–16 mm Long wings are boldly banded with orange across AM and ST areas. Basal, PM, and terminal areas are crisp black. Thorax and abdomen are orange. **HOSTS** Lichen on rocks or walls.

CRESCENT-MARKED LICHEN MOTH
Cisthene liberomacula

Uncommon
93-0179 (8062)

RANGE-WIDE

TL 9–12 mm Dusky FW has two white crescent-shaped patches at inner margin and costa of median band. Basal area, head, and thorax are gray. HW is pale orange, with wide gray band along costa. Abdomen is light orange, with dusky shading. Some individuals have reduced white patches and mostly gray FW. Form "basijuncta" (not shown) has a white basal streak parallel to inner margin and orange stripes laterally on thorax. **HOSTS** Lichen on coast live oak and other oaks.

WHITE-PATCHED LICHEN MOTH
Cisthene deserta

Uncommon
93-0180 (8063)

RANGE-WIDE

TL 10–11 mm Hoary gray FW has large white patch at inner median area and small white spot at outer median area, edged thinly with black. Inner basal area is white. Thorax and head are gray. HW is pale yellow orange with a black terminal band. **HOSTS** Lichen and algae.

THREE-SPOTTED LICHEN MOTH
Cisthene faustinula

Uncommon
93-0181 (8064)

RANGE-WIDE

TL 10–12 mm Pale gray to dusky FW has thick white to tan median band pinched twice to form three bulges or, sometimes, discrete spots. Inner basal area is white to tan, sometimes extending a short distance along inner margin. Head and thorax are white to tan, with a dusky dorsal patch. HW is white to pale gray, with almost no terminal border. **HOSTS** Lichen and algae.

GRAY LICHEN MOTH
Cisthene dorsimacula

Rare
93-0182 (8065)

RANGE-WIDE

TL 10–12 mm Dusky gray FW variably resembles White-patched and Three-spotted Lichen Moths, but basal area is entirely gray or with just a thin sliver of white along inner margin. Head and thorax are gray. HW is pale peach orange. **HOSTS** Lichen and algae.

actual size

ARID EUDESMIA

CRESCENT-MARKED LICHEN MOTH

WHITE-PATCHED LICHEN MOTH

THREE-SPOTTED LICHEN MOTH

GRAY LICHEN MOTH

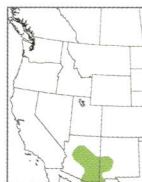

SCHWARZ'S LICHEN MOTH

Common

Cisthene schwarziorum   93-0183.1 (8066)

TL 8–10 mm Blackish FW has an orange median band that is concave on basal side; in some individuals, it may be so narrowed that the band is fragmented. Orange patch at base of inner margin tapers to a point and does not usually meet median band. Reddish HW has a black terminal band. Orange thorax has a black dorsal patch. Top of head is orange. Abdomen is reddish. HOSTS Lichen and algae. NOTE Previously considered a western subspecies of Thin-banded Lichen Moth (*C. tenuifascia*, not shown), which is very similar but has a thinner median band and does not occur west of central NM.

ROSY LICHEN MOTH

Uncommon

Cisthene perrosea   93-0186 (8069)

TL 10–12 mm Resembles Schwarz's Lichen Moth, but pinkish orange along inner margin is uniformly narrow and connects to narrow median band. HW and abdomen are lightish red. HOSTS Lichen and algae.

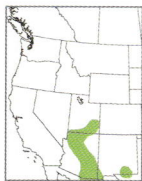

ANGEL LICHEN MOTH

Uncommon

Cisthene angelus   93-0187 (8070)

TL 11–13 mm Resembles Schwarz's Lichen Moth, but orange along inner margin is uniformly broad and connects to median band. Thorax lacks black dorsal patch. HOSTS Lichen and algae.

BARNES' LICHEN MOTH

Uncommon

Cisthene barnesii   93-0191 (8074)

TL 9–13 mm Resembles Rosy Lichen Moth, but markings are orange and narrower, median band is usually fragmented or sometimes absent altogether, and orange along inner margin is slightly tapered toward median band. Black dorsal patch is broad, and orange portions are reduced to lateral lines. Top of head is gray. HW and abdomen are orange. HOSTS Lichen and algae.

JUANITA'S LICHEN MOTH

Uncommon

Cisthene juanita   93-0193 (8076)

TL 10–12 mm Blackish FW has a narrow reddish-orange stripe along inner margin and a short narrow stripe along costa at median area. Black thorax has narrow reddish-orange lateral stripes. Head is black. HW and abdomen are rosy orange. HOSTS Lichen and algae. NOTE Uncommonly found at mule fat flowers during daytime.

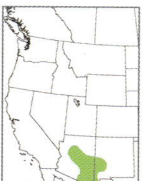

MARTIN'S LICHEN MOTH

Uncommon

Cisthene martini   93-0195 (8078)

TL 10–12 mm Resembles Schwarz's Lichen Moth, but markings are red to reddish orange, often with a yellow-orange tinge at the edges, and stripe along inner margin is wider and connects to median band. Thorax has black dorsal spot, and top of head is reddish orange. HW and abdomen are rosy red. HOSTS Lichen and algae.

SCHWARZ'S LICHEN MOTH

ROSY LICHEN MOTH

ANGEL LICHEN MOTH

BARNES' LICHEN MOTH

JUANITA'S
LICHEN MOTH

actual size

MARTIN'S
LICHEN MOTH

ROYAL LICHEN MOTH
Lycomorpha regulus   **Rare**

93-0198 (8084)

TL 14–16 mm Bright orange FW has a black border along outer margin and continuing part-way up costa. Thorax is orange. HW is scarlet to red orange on costal half and broadly black along outer and inner margins. The extent of black is variable, from one- to two-thirds of the HW; individuals with greater extents of black have commonly been misidentified online as Shining Lichen Moth, but are readily separable by the color of the thorax. **HOSTS** Lichen of genus *Parmelia*. **NOTE** Often seen at flowers during daytime.

SHINING LICHEN MOTH
Lycomorpha fulgens   **Rare**

93-0199 (8085)

TL 14–16 mm Resembles Royal Lichen Moth, but central thorax is black, FW is crimson orange, and inner margin of FW has a very thin, black edge. HW is almost entirely black, with a narrow red stripe along costa that doesn't quite reach apex. **HOSTS** Lichen. **NOTE** Often seen at flowers during daytime.

SPLENDID LICHEN MOTH
Lycomorpha splendens   **Rare**

93-0200 (8086)

TL 13–16 mm Crisp black FW is unmarked. Black thorax has small red patch laterally at base of each wing. HW is orange red, with black fringe. **HOSTS** Lichen. **NOTE** Unlike other *Lycomorpha*, this species is largely nocturnal.

BLACK-AND-YELLOW LICHEN MOTH
Lycomorpha pholus   **Uncommon**

93-0201 (8087)

TL 13–17 mm Bicolored FW is yellow orange (rarely reddish) in basal half and black in outer half. Head and thorax are black. **HOSTS** Lichen. **NOTE** Resembles net-winged beetles in the genus *Calopteron*. Often seen at flowers during daytime.

SCARLET-WINGED LICHEN MOTH
Hypoprepia miniata   **Uncommon**

93-0204 (8089)

TL 15–21 mm Dark gray FW has scarlet border and inverted Y-shaped stripe along central veins. Red thorax sometimes has a black dorsal spot. Some individuals may be orange or yellow orange, but color is uniform across wing. **HOSTS** Lichen and blue-green algae.

PAINTED LICHEN MOTH
Hypoprepia fucosa   **Uncommon**

93-0205 (8090)

TL 14–18 mm Resembles Scarlet-winged Lichen Moth, but FW coloring is yellow orange, shifting to rosy red in ST/terminal area. Rosy thorax has a gray dorsal spot. **HOSTS** Lichen, algae, and moss.

CADAVER LICHEN MOTH
Hypoprepia cadaverosa   **Rare**

93-0206 (8091)

TL 14–18 mm Resembles Painted Lichen Moth but lacks rosy patch on FW, and thorax is solid yellow orange. **HOSTS** Lichen.

ROYAL LICHEN
MOTH

SHINING LICHEN MOTH

SPLENDID LICHEN MOTH

BLACK-AND-YELLOW
LICHEN MOTH

actual size

SCARLET-WINGED
LICHEN MOTH

PAINTED LICHEN MOTH

CADAVER LICHEN MOTH

PINK-LINED LICHEN MOTH
Rare RANGE-WIDE

Hypoprepia inculta   93-0207 (8092)

TL 11–16 mm Dusky FW has a hint of a pinkish wash. Central vein is very thinly traced with pink, terminating above a small pale pinkish spot. Dusky thorax has narrow pinkish lines laterally, connecting to pink FW vein. **HOSTS** Lichen.

CRIMSON LICHEN MOTH
Uncommon RANGE-WIDE

Ptychoglene coccinea   93-0208 (8079)

TL 10–14 mm Bright reddish-orange FW is black in ST/ terminal area; PM edge is slanting such that, when wings are folded, black areas form the shape of a triangle. Head and thorax are reddish orange. **HOSTS** Lichen. **NOTE** Often seen at flowers during daytime.

VERMILION LICHEN MOTH
Uncommon RANGE-WIDE

Ptychoglene phrada   93-0209 (8080)

TL 10–14 mm Resembles Crimson Lichen Moth, but FW is scarlet red and head and thorax are black. HW is nearly entirely black, with narrow red inner margin. Separable from Shining Lichen Moth by fully black thorax, and teeth on black FW border. **HOSTS** Lichen. **NOTE** Often seen at flowers during daytime.

PEPPERED LICHEN MOTH
Uncommon RANGE-WIDE

Bruceia hubbardi   93-0214 (8095)

TL 11–14 mm Straw-colored FW has blackish zigzag AM and PM lines, with peppery edges. Median and basal areas are peppered with blackish scales, except at pale reniform spot. Pale thorax is shaded dusky. **HOSTS** Lichen.

LITTLE SHADED LICHEN MOTH
Common RANGE-WIDE

Clemensia umbrata   93-0215.1 (8098.2)

TL 10–13 mm Peppery white FW has jagged brownish-gray AM, median, and PM lines; costal third of AM line is straight and blackish. Median area contains small black orbicular dot and bold black reniform chevron. Terminal line is checkered. **HOSTS** Lichen.

BICOLORED MOTH
Uncommon RANGE-WIDE

Manulea bicolor   93-0217 (8043)

TL 14–16 mm Lead-gray FW has yellow-orange costal stripe. Gray thorax has yellow-orange collar. Resembles slightly larger Ancient Tussock Moth but typically rests with wings overlapping and back of head is gray, collar is yellow orange, fringe is gray, and abdomen is gray, with an orange tip. Separable from Yellow-collared Scape Moth by costa and abdomen markings. **HOSTS** Coniferous trees and the lichen that grows on them.

PINK-LINED LICHEN
MOTH

CRIMSON LICHEN
MOTH

VERMILION LICHEN
MOTH

PEPPERED LICHEN
MOTH

LITTLE SHADED
LICHEN MOTH

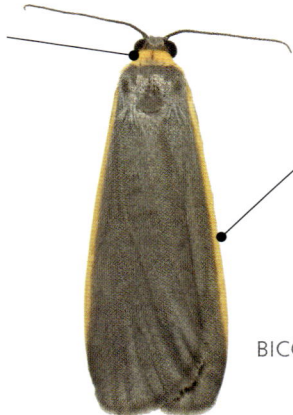

actual size

BICOLORED MOTH

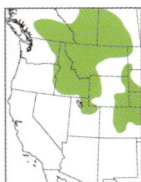

PEARLY-WINGED LICHEN MOTH

Crambidia casta

Uncommon
93-0225 (8051)

TL 16–20 mm Satiny white to silvery-gray FW is completely unmarked. Typically rests with wings curled around body and may superficially resemble grass-veneers, but outer margin is more rounded and head lacks long facial palps. **HOSTS** Lichen.

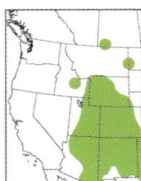

YELLOW-HEADED LICHEN MOTH

Crambidia cephalica

Uncommon
93-0227 (8053)

TL 11–16 mm Resembles larger Pearly-winged Lichen Moth, but front of head is yellow orange and legs are darker. **HOSTS** Lichen.

TIGER MOTHS   SUPERFAMILY Noctuoidea (93)

FAMILY Erebidae SUBFAMILY Arctiinae TRIBE Arctiini

A large and varied group of predominantly medium-sized moths, most with long, somewhat broad wings. Forewings are typically white or boldly patterned with black, and the hindwings and/or abdomen are often contrastingly pinkish red or orange. Most species will come to lights at night, but a number of species are also commonly encountered during daytime, particularly the ctenuchas, *Gnophaela*, and wasp moths at flowers, and the Cinnabar Moth, virbias, and cycnias among low vegetation.

ARGE TIGER MOTH

Apantesis arge

Uncommon
93-0240 (8199)

TL 19–25 mm FW has thick pinkish white stripes along veins and cross-lines, creating the effect of small black shards on a pale wing. HW is pale pinkish or peach, with small black spots. Thorax has thin stripes. Wings and thorax are often suffused strongly pink. **HOSTS** Herbaceous plants, as well as prickly-pear, grape, and sunflower.

VIRGIN TIGER MOTH

Apantesis virgo

Rare
93-0244 (8197)

TL 24–37 mm Resembles the smaller Parthenice Tiger Moth, but central vein is thick, and median line angles toward anal angle. PM line is more strongly V-shaped and points of W-shaped ST line usually touch outer margin. HW is pinkish orange to rosy, and has additional black spots in AM and median areas. **HOSTS** Bedstraw, clover, plantain, and other herbaceous plants.

PARTHENICE TIGER MOTH

Apantesis parthenice

Uncommon
93-0246 (8196)

TL 18–28 mm Black FW has pale yellowish stripes along veins and cross-lines; central vein is always thinly marked. Median line is perpendicular to costa, and may or may not reach vertical stripe. One or both points of W-shaped ST line touch outer margin. HW is orangish pink, with variable large black terminal band spots but none in AM or median area. **HOSTS** Dandelion, ironweed, thistle, and other herbaceous plants.

PEARLY-WINGED LICHEN MOTH

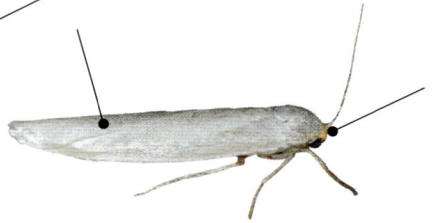

actual size

YELLOW-HEADED LICHEN MOTH

ARGE TIGER MOTH

VIRGIN
TIGER MOTH

actual size

PARTHENICE
TIGER MOTH

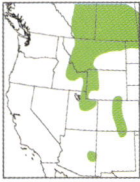

LITTLE VIRGIN TIGER MOTH

Apantesis virguncula   **Rare**   93-0247 (8175)   RANGE-WIDE

TL 19–25 mm Resembles Virgin Tiger Moth but lacks median and PM lines; rarely, median line may be present as short discal bar. Points of W-shaped ST line do not touch outer margin. HW is orange, with large black spots along terminal band to anal angle, sometimes fused into a solid band. **HOSTS** Herbaceous plants, including dandelion, knotweed, and plantain.

PHYLLIRA TIGER MOTH

Apantesis phyllira   **Uncommon**   93-0242 (8194)   RANGE-WIDE

TL 17–21 mm Dimorphic. Common morph resembles Figured Tiger Moth, but median line is perpendicular to inner margin, and pale costa usually has a bulge at AM line. Pale edging to costa and inner margin reaches nearly to outer margin. PM line usually connects to pale costa. Less commonly, some individuals have pale veins and resemble Parthenice Moth, but are separable by median line, costal bulge, and fringe color. HW is reddish pink with a narrow, black terminal band and black spots in ST area. **HOSTS** Corn, lupine, tobacco, and other herbaceous plants.

FIGURED TIGER MOTH

Apantesis figurata   **Uncommon**   93-0253 (8188)   RANGE-WIDE

TL 18–22 mm Extremely variable. Black FW has inverted, pale yellowish F shape created by median and PM lines connected to vertical stripe. W-shaped ST band sometimes present. Costa and inner margin commonly have pale edging at base, but may be entirely black. Fringe may be black or pale. Thorax is pale, with three black stripes, and two black spots at collar. Abdomen is orange to pink dorsally and black below (rarely pale below). HW is reddish to pink (rarely yellow or black), with a wide black ST/terminal area. Females resemble males. **HOSTS** Alfalfa, plantain, and other herbaceous plants. **NOTE** Where range overlaps with Lettered Tiger Moth, individuals with an ST band, pale edging to costa, inner margin, or fringe, pale underside to abdomen, a black HW, or that are observed in the spring or autumn, can be reliably identified as Figured Tiger Moth. Some summer males with reduced markings may not be reliably identifiable visually.

LETTERED TIGER MOTH

Apantesis f-pallida   **Uncommon**   93-0254 (8189)   RANGE-WIDE

TL 16–19 mm Variable. Resembles Figured Tiger Moth, but always lacks W-shaped ST line, and costa, inner margin, and fringe are always black. Abdomen is uniformly pinkish red above and always black below. Females have all-black thorax and FW, sometimes with faint traces of pale FW markings. Reddish (rarely yellow in CO) HW has thick black terminal band. **HOSTS** Generalist on herbaceous plants. **NOTE** Females can always be reliably identified, but where range overlaps with Figured Tiger Moth, most males are probably not reliably identifiable visually.

WILLIAMS' TIGER MOTH

Apantesis williamsii   **Common**   93-0264 (8186)   RANGE-WIDE

TL 16–18 mm Resembles Figured Tiger Moth, but markings are reduced and often fragmented. Median line is small or sometimes absent. Inner margin is thinly edged pale from base to anal angle. Salmon-pink or yellow HW has a broad black terminal line and spots in median area. **HOSTS** Herbaceous plants, including ragwort.

LITTLE VIRGIN TIGER MOTH

PHYLLIRA TIGER MOTH

FIGURED TIGER MOTH

actual size

LETTERED TIGER MOTH

WILLIAMS' TIGER MOTH

IMMACULATE TIGER MOTH

Apantesis incorrupta

Common

93-0255 (8180)

RANGE-WIDE

TL 18–21 mm Resembles Mexican Tiger Moth, but FW has a thick pale basal line from costa to vertical stripe. Points of ST line touch fringe. Collar usually has two black dots, but these may be absent in some individuals; rarely, thorax may be entirely black. HW is rosy orange to rosy red, with three spaced-out black spots along ST band. **HOSTS** Generalist on herbaceous plants.

LITTLE BEAR TIGER MOTH

Apantesis ursina

Uncommon

93-0257 (8186.2)

RANGE-WIDE

TL 16–20 mm FW resembles Immaculate Tiger Moth, but ranges do not overlap. Lines are usually thinner. Points of ST line touch fringe. HW has larger black spots along terminal line. Some individuals have an incomplete median line. Thorax is usually black, rarely striped. **HOSTS** Likely a generalist on herbaceous plants.

NEVADA TIGER MOTH

Apantesis nevadensis

Common

93-0258 (8179)

RANGE-WIDE

TL 15–20 mm Variable. Resembles Immaculate Tiger Moth, but lines are narrower and often incomplete or fragmented, particularly AM line. Points of ST line do not reach fringe. HW is usually rosy red, rarely pale, with a terminal band of black spots that can be large, small, or nearly absent. Thorax is either striped, with spots on collar, or entirely black. **HOSTS** Big sagebrush and groundcherry; possibly other herbaceous plants.

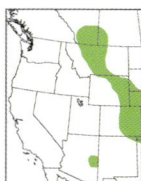

BLAKE'S TIGER MOTH

Apantesis blakei

Rare

93-0269 (8185)

RANGE-WIDE

TL 13–17 mm Resembles Nevada Tiger Moth, but FW lacks basal line (except occasionally a pale bulge at costa) and inner basal area has a thin pale dash along vein between inner margin and vertical stripe. HW is orange to pale orange, with terminal band of thick connected spots. Thorax is striped, and collar has spots. **HOSTS** Generalist on herbaceous plants.

ORNATE TIGER MOTH

Apantesis ornata

Very Common

93-0271 (8177)

NORTH

SOUTH

TL 19–23 mm Variable. Black FW has thin pale edging along outside of FW. A single vertical stripe runs from base to anal angle. Thick AM, median, and PM lines run from costa and may or may not cross vertical stripe to inner margin; AM line is sometimes fragmented. ST line is a strong W shape; costal point does not usually reach pale fringe. Veins are often marked in the same color as the cross-lines. HW is yellow orange to reddish orange, with black median and terminal bands broken into large round spots. Thorax is striped, and collar has two black dots. **HOSTS** Generalist on herbaceous plants. **NOTE** Our most common *Apantesis* tiger moth.

MEXICAN TIGER MOTH

Apantesis proxima

Very Common

93-0276 (8181)

INLAND

CALIFORNIA

TL 16–22 mm Resembles Ornate Tiger Moth, but stripes are much thicker and slightly paler, usually reaching inner margin, and points of W-shaped ST line touch thick pale fringe. HW is either white and largely unmarked except for rosy wash along inner margin, or rosy red with terminal band of crowded black spots. Collar is slightly yellower than thorax and lacks black spots. **HOSTS** Generalist on herbaceous plants.

IMMACULATE
TIGER MOTH

actual size

LITTLE BEAR
TIGER MOTH

NEVADA TIGER MOTH

BLAKE'S TIGER MOTH

ORNATE
TIGER MOTH

MEXICAN TIGER
MOTH

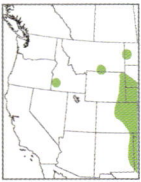

CARLOTTA'S TIGER MOTH
Apantesis carlotta

Uncommon
93-0281 (8171.1)

TL 16–22 mm Black FW has thick, peach-orange, vertical stripe from base to anal angle, and V-shaped PM line and W-shaped ST line that join to form an X in lower part of wing; ST line may be reduced or absent in some individuals. All lines connect to thick pale border, except at costa below PM line where it is black. HW is peach orange, with black costa and well-defined black spots along terminal band. Thorax is striped, but collar has no spots. **HOSTS** Wild lettuce; likely other herbaceous plants.

WOOD TIGER MOTH
Arctia plantaginis

Common
93-0283 (8127)

TL 17–20 mm Variable. Black FW has bold white stripe from base to anal angle, angled white PM band, and large white median spot. Some individuals show reduced markings, while others have a bold, white, V-shaped ST line. HW may be white or orange, with bold black terminal band. Collar and abdomen are orange. **HOSTS** Generalist on herbaceous and woody plants. **NOTE** Diurnal; low, quick fliers.

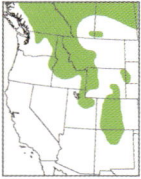

ST. LAWRENCE TIGER MOTH
Arctia parthenos

Common
93-0288 (8162)

TL 27–33 mm Brown FW has large white spots in fragmented rows at AM, median, PM, and ST lines, as well as a short white basal dash. Yellow-orange HW has a black basal area and PM band. Thorax is brown, with white stripes laterally and a red collar. **HOSTS** Deciduous trees and plants, including alder, birch, lettuce, and willow.

RANCHMAN'S TIGER MOTH
Arctia virginalis

Very Common
93-0289 (8163)

TL 27–33 mm Black FW is boldly white or yellowish spotted along AM, median, PM, and ST lines. HW is either yellow orange with black AM and PM bands and median spot, or black with yellow-orange median (often hidden behind FW) and ST bands. Black thorax has white epaulettes and an orange dorsal spot. Head and either entire or just tip of abdomen are orange. **HOSTS** Generalist on herbaceous and woody plants. **NOTE** Diurnal; fast and erratic fliers.

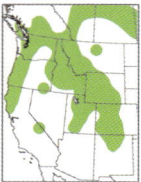

GREAT TIGER MOTH
Arctia caja

Very Common
93-0290 (8166)

TL 28–37 mm Brown FW has bold white basal, PM, and ST lines; PM and ST lines connect at midpoint to form an X. AM and median lines are present as white patches at costa. Orange HW has large blue-black spots along AM and PM bands. Thorax is reddish brown, with red collar, and abdomen is orange. **HOSTS** Deciduous trees, including alder, cherry, poplar, and willow.

VESTAL TIGER MOTH
Spilosoma vestalis

Very Common
93-0312 (8135)

TL 18–26 mm White FW has small black dots forming curving PM line and sparsely across AM and median areas. Base of forelegs is pink, and distal segments are solid black on inside and plain white on outside. White abdomen has black spots dorsally and laterally. **HOSTS** Generalist on herbaceous and woody plants.

CARLOTTA'S
TIGER MOTH

WOOD
TIGER MOTH

ST. LAWRENCE
TIGER MOTH

RANCHMAN'S
TIGER MOTH

actual size

VESTAL TIGER MOTH

GREAT TIGER MOTH

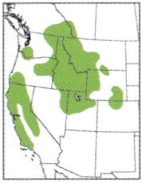

WANDERING TIGER MOTH
Spilosoma vagans   **Very Common**   93-0313 (8138)   RANGE-WIDE

TL 15–19 mm   Variable. Light brown to tan FW usually has faded, black, curving AM and PM lines and reniform spot, and indistinct median line. Some individuals may be heavily mottled dusky; rare individuals are pinkish tan. HW is usually dusky, with a pale fringe, but may rarely be pale, with fragmented dusky PM band and median spot. Thorax is densely hairy. **HOSTS** Lupine.

BROWN TIGER MOTH
Spilosoma pteridis   **Rare**   93-0314 (8139)   RANGE-WIDE

TL 15–19 mm   Brown to reddish-brown FW has an indistinct dusky reniform spot and dusky patch at inner margin of PM line. Some individuals have thick, dusky, curving AM and PM lines. Hairy thorax is yellow brown to reddish orange. **HOSTS** Fireweed; also other herbaceous plants.

VIRGINIAN TIGER MOTH
Spilosoma virginica   **Very Common**   93-0316 (8137)   RANGE-WIDE

TL 17–26 mm   Resembles Vestal Tiger Moth, but FW is mostly white, with just one or two black dots, forelegs are orange at base and striped black distally, and abdomen has orange patches alongside black dorsal spots. **HOSTS** Birch, cabbage, maple, tobacco, walnut, willow, and many other trees, shrubs, and herbaceous plants.

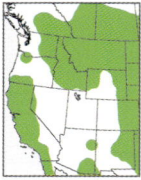

SALT MARSH MOTH
Estigmene acrea   **Very Common**   93-0317 (8131)   RANGE-WIDE

TL 24–35 mm   Sexually dimorphic. White FW has five or six rectangular black spots along costa at five cross-lines and normally also basal area, and variably across wings at those lines; fringe is orangish (male) or white (female). HW is yellow orange (male) or white (female), with variable black spots along outer margin. Thorax is white; abdomen is orange dorsally and (in males) ventrally, with black spots. Legs are orange at base and striped distally. Males have narrowly bipectinate antennae. **HOSTS** Apple, cabbage, corn, potato, tobacco, and other trees and plants.

BARRED TIGER MOTH
Estigmene albida   **Uncommon**   93-0318 (8132)   RANGE-WIDE

TL 21–27 mm   Resembles female Salt Marsh Moth but averages smaller in size and black spots on abdomen are wide bars. Costa of FW nearly always has five or fewer black patches, lacking the sixth, most basal patch often seen in Salt Marsh Moth. FW fringe is white in both sexes. Uncommonly, some individuals show large squarish black spots along lines. Males have narrowly bipectinate antennae. **HOSTS** Scarlet spiderling and white sweetclover; likely also others.

FALL WEBWORM
Hyphantria cunea   **Very Common**   93-0319 (8140)   RANGE-WIDE

TL 14–19 mm   Variable. Often resembles Vestal Tiger Moth, but base of forelegs is orange, legs are solid black on inside and striped outside, and abdomen is completely white. Many individuals have squarish black spots along cross-lines, sometimes quite heavily. Rarely, FW is so densely spotted that it is almost gray. Thorax is usually white, occasionally with small dusky spots on more heavily marked individuals. **HOSTS** Ash, hickory, maple, oak, walnut, and many other woody plants.

actual size

WANDERING
TIGER MOTH

BROWN TIGER MOTH

VIRGINIAN
TIGER MOTH

SALT MARSH
MOTH

male

female

FALL
WEBWORM

BARRED TIGER MOTH

405

MANY-SPOTTED TIGER MOTH

Common

Hypsocompe permaculata

93-0321 (8144)

TL 19–22 mm White FW has multiple rows of dense black spots. Thorax has black central stripe and black spots laterally. Abdomen is orange dorsally, with black segment bars, sometimes joined as a wide black dorsal stripe. Legs are white at base and entirely black distally. HOSTS Knotweed, chickweed, and other herbaceous plants.

GRAY-SPOTTED TIGER MOTH

Uncommon

Stictocompe suffusa

93-0325 (8149)

TL 21–27 mm White FW has lines of large silvery-gray spots edged thinly with black. Thorax is similarly marked, with a central dorsal stripe split into a heart shape at anterior end. HW is white, with sparse, black-edged, gray spots. Abdomen is orange, with black-edged gray spots dorsally, and legs are banded with silvery gray and orange. HOSTS Generalist on herbaceous and woody plants.

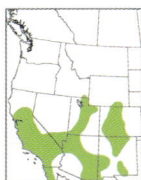

PAINTED TIGER MOTH

Very Common

Arachnis picta

93-0327 (8152)

TL 21–27 mm Resembles Gray-spotted Tiger Moth, but silvery-gray FW markings are more sinuous, thorax has two straight, parallel dorsal lines, and HW and abdomen are pink. HOSTS Generalist on herbaceous and woody plants.

AULAEAN TIGER MOTH

Rare

Arachnis aulaea

93-0330 (8155)

TL 22–28 mm Resembles Painted Tiger Moth, but gray FW markings are darker and more fragmented, appearing as densely packed, elongate spots. Terminal area has black streaks between veins. HW is black, with narrow pink median and terminal bands. Abdomen is dark gray, with a pink base and small orange spots laterally. Legs are mostly gray, with a pink base. HOSTS Generalist on herbaceous and woody plants.

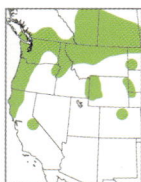

RUBY TIGER MOTH

Uncommon

Phragmatobia fuliginosa

93-0332 (8156)

TL 15–18 mm Reddish-brown FW has a single black reniform spot. Terminal area is sometimes indistinctly dusky. Brown thorax is densely hairy, and abdomen is dull orange, with black spots. Forelegs are pink at base. HOSTS Herbaceous plants, including dock, goldenrod, ironweed, plantain, and sunflower.

ISABELLA TIGER MOTH

Very Common

Pyrrharctia isabella

93-0335 (8129)

TL 24–33 mm Sexually dimorphic. Tan FW has faded, light brown AM, median, and PM lines and dusky reniform spot, apical dash, and spots along AM line. HW is rosy (female) or tan (male, not shown), with sparse black spots. Legs are black, and orange abdomen has black spots dorsally. HOSTS Aster, dandelion, grass, lettuce, meadowsweet, and other herbaceous plants. NOTE Larva is the familiar Woolly Bear Caterpillar.

MANY-SPOTTED
TIGER MOTH

GRAY-SPOTTED TIGER MOTH

PAINTED TIGER MOTH

AULAEAN TIGER MOTH

RUBY TIGER MOTH

ISABELLA
TIGER MOTH

female

actual size

CALIFORNIA TIGER MOTH

Common

Leptarctia californiae

93-0336 (8126)

TL 15–18 mm Highly variable. Hoary, dark gray FW has dusky AM and PM bands, short white basal dash, and white dashes on costa at PM and ST lines. Many individuals may also have white at inner ST area; some have white along part or entire AM, PM, and/or ST lines; and, rarely, white AM and PM bands might connect to form white median area. HW may be yellow, orange, or rosy, with a variably fragmented black terminal band and, rarely, median band, often also with black along inner margin; rare individuals may have mostly black HW with white median band. Dark gray thorax has thin white lines laterally connected to basal dash on FW. **HOSTS** Generalist on trees and herbaceous plants.

GRAY VIRBIA

Uncommon

Virbia costata

93-0336.52 (8115)

TL 10–14 mm Sexually dimorphic. Grayish-pink to pinkish-tan FW has narrow pink costa and small pink reniform spot. Female is typically darker than male. HW is pink, unmarked in male and with a wide black terminal band in female. **HOSTS** Likely a generalist on herbaceous plants.

SHOWY VIRBIA

Uncommon

Virbia ostenta

93-0336.54 (8116)

TL 15–19 mm Dusky orange FW has a pinkish-orange costa but is otherwise unmarked. Pink HW has bold black terminal band that turns up along inner margin to base. Thorax and head are dusky orange, and abdomen is black. **HOSTS** Likely a generalist on herbaceous plants.

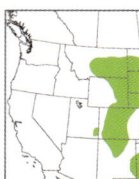

ORANGE VIRBIA

Uncommon

Virbia aurantiaca

93-0336.57 (8121)

TL 10–14 mm Variable. Pinkish- to brownish-orange FW usually has an indistinct dusky discal spot and shading along ST line; markings may be quite faint. Some individuals have whitish spots in median area. HW of male is orange to rosy orange, with faded gray markings; female HW is rosy red with blackish terminal band and discal spot, sometimes strongly pinched in middle or just barely broken. **HOSTS** Corn, dandelion, pigweed, plantain, and other herbaceous plants and crops. **NOTE** Sometimes encountered in daytime.

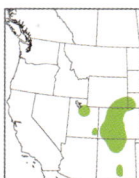

FRAGILE VIRBIA

Rare

Virbia fragilis

93-0336.6 (8125)

TL 13–15 mm Resembles male Orange Virbia, but markings on brownish-tan FW are usually light or absent, and HW is yellow orange and unmarked. Some individuals have a rough-edged white spot in inner median area and/or dusky shading at PM and ST lines. Occasionally, HW will show faded grayish terminal band and discal spot. **HOSTS** Likely a generalist on herbaceous plants. **NOTE** Sometimes encountered in daytime.

RUSTY VIRBIA

Rare

Virbia ferruginosa

93-0336.65 (8123)

TL 13–16 mm Resembles Orange Virbia, but brownish-orange to rusty-brown FW usually has darker ST band and discal spot, and HW is rosy orange to orange, with terminal band cleanly split into two, sometimes largely reduced. Some females may show white spots on FW. **HOSTS** Generalist on herbaceous and woody plants.

CALIFORNIA TIGER MOTH

GRAY VIRBIA

actual size

SHOWY VIRBIA

ORANGE VIRBIA

male

FRAGILE VIRBIA

female

RUSTY VIRBIA

409

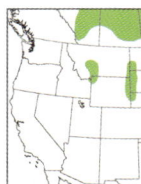

LECONTE'S HAPLOA
Haploa lecontei

Uncommon
93-0345 (8111)

TL 19–26 mm Variable. White FW has a brown to blackish border, often with large teeth at AM, PM, and ST lines that may sometimes connect to divide white areas into large spots. A thick dark line crosses from inner PM area to apex. White thorax has a dark dorsal stripe, and head is orange. HOSTS Apple, thoroughwort, willow, and other trees and herbaceous plants.

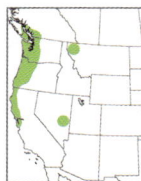

CINNABAR MOTH
Tyria jacobaeae

[IN] Very Common
93-0347 (8113)

TL 18–21 mm Dusky black FW has two large red to fuchsia spots in terminal area and long stripes at inner margin and beside costa. HW is red to fuchsia with black fringe. HOSTS Tansy ragwort. NOTE Intentionally introduced from Europe as a biological control for tansy ragwort. Often seen in vegetation during daytime.

WILD FORGET-ME-NOT MOTH
Gnophaela latipennis

Very Common
93-0353 (8034)

TL 26–28 mm Black FW has three large, rounded, white to yellowish spots in median area and two to four smaller white to yellowish spots in outer ST area. Abdomen is metallic blue, and black collar is orange laterally. HOSTS California stickseed, hound's-tongue, bluebell, forget-me-not, and related species. NOTE Often seen visiting flowers during daytime.

MOUNTAIN BLUEBELL MOTH
Gnophaela discreta

Uncommon
93-0355 (8036)

TL 26–29 mm Nearly identical to Wild Forget-me-not Moth, but innermost spot of median area is longer and pointed at basal end. Also resembles Police Car Moth, but white median-area markings are reduced and basal edge of most costal spot is squared off more. Reliably separable from both by range. HOSTS Bluebell. NOTE Often seen visiting flowers during daytime.

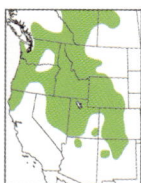

POLICE CAR MOTH
Gnophaela vermiculata

Very Common
93-0356 (8037)

TL 22–26 mm Black FW has three large white spots in median area that are longer and more pointed than those of similar species. Outer ST area has two to four smaller white spots. Abdomen is metallic blue, and black collar is orange laterally. HOSTS Bluebell, puccoon, and stickseed. NOTE Often seen visiting flowers during daytime.

NORTHERN GIANT FLAG MOTH
Dysschema howardi

Uncommon
93-0358 (8040)

TL 40–50 mm Black FW has bold white stripe from base to thick zigzag PM line; center of PM line has large tooth that reaches outer margin. Narrow white dash of outer median line angles to meet intersection of stripe and PM line. HW is white (male) or yellow orange (female), with an orange terminal band containing black-edged blue spots; occasionally, black edging is wide enough to obscure orange band. Abdomen is orange with black dorsal stripe. HOSTS Brickellbush, goldeneye, and sotol.

TIGER MOTHS

LECONTE'S
HAPLOA

actual size

CINNABAR MOTH

WILD FORGET-ME-NOT MOTH

MOUNTAIN
BLUEBELL MOTH

POLICE CAR
MOTH

NORTHERN GIANT
FLAG MOTH

411

LEAST HYPOCRISIAS
Hypocrisias minima

Uncommon
93-0359 (8201)

RANGE-WIDE

TL 15–16 mm Peppery tan FW has dusky shading in median and terminal areas. Basal, PM, and ST lines are rows of brown-edged pale spots; AM and median lines are present as pale spots at costa. Hairy thorax is tan with a brown central stripe. **HOSTS** Toothleaf goldeneye.

DAVIS' TUSSOCK MOTH
Halysidota davisii

Uncommon
93-0362 (8205)

RANGE-WIDE

TL 26–34 mm Peach-tan FW has black-edged yellow-orange spots along costa and, sometimes faded, along ST line and outer margin. Reniform spot is two black-edged yellow-orange bars. Basal area and sometimes inner median line have black-edged pale spots. Tan thorax has orange dorsal stripe edged laterally with turquoise. **HOSTS** Mulberry, little leaf sumac, Emory oak, and hackberry.

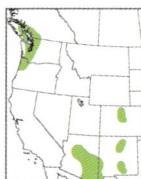

ROSY AEMILIA
Lophocampa roseata

Common
93-0365 (8206)

RANGE-WIDE

TL 18–20 mm Reddish-orange FW has white AM, PM, and ST lines crossed by darker reddish veins. Median line usually a single white patch at costa. Reddish-orange thorax has four thin white stripes. **HOSTS** Pine, Douglas-fir, and maple.

GREAT TUSSOCK MOTH
Lophocampa ingens

Uncommon
93-0367 (8208)

RANGE-WIDE

TL 23–30 mm Resembles the smaller Silver-spotted Tussock Moth, but brown FW is darker and white spots are not noticeably edged. Terminal line a row of small white spots. Thick dark stripes on pale thorax angle together to form a V shape, and collar has solid brown bar. Abdomen is orange. **HOSTS** Pinyon and ponderosa pine.

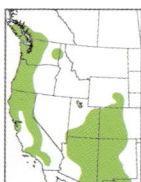

SILVER-SPOTTED TUSSOCK MOTH
Lophocampa argentata

Very Common
93-0368 (8209)

RANGE-WIDE

TL 18–24 mm Hoary brown FW has rows of brown-edged pale spots at PM and ST lines, and loosely ordered spots at AM and median lines. White spots at outer margin on fringe only. Tan thorax has paired, wide, brown stripes with a central brown spot and two small brown spots on collar. Abdomen is light orange. **HOSTS** Pine, fir, Douglas-fir, and other needled conifers.

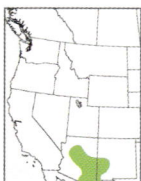

SPECKLED TUSSOCK MOTH
Lophocampa mixta

Uncommon
93-0371 (8212)

RANGE-WIDE

TL 18–20 mm Tan FW is heavily speckled with brown and crossed by rows of brown-edged tan spots at basal, AM, median, PM, and ST lines. A thin, dark, brown line crosses wing from costa at basal line to anal angle. Veins are traced with brown. Tan thorax has V-shaped, thin, brown lines. **HOSTS** Silverleaf oak, evergreen sumac, and pointleaf manzanita; possibly others.

LEAST
HYPOCRISIAS

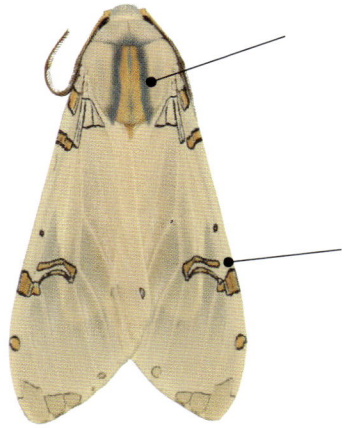

DAVIS' TUSSOCK
MOTH

ROSY AEMILIA

actual size

GREAT TUSSOCK MOTH

SILVER-SPOTTED TUSSOCK MOTH

SPECKLED TUSSOCK
MOTH

MODEST TUSSOCK MOTH
Lophocampa pura

Uncommon
93-0372 (8213)

TL 23–27 mm Resembles Isabella Tiger Moth, but cross-lines are fainter, black apical dash is absent, legs are tan, and abdomen is pale orange without black dorsal spots. HOSTS Pointleaf manzanita; possibly others.

SPOTTED TUSSOCK MOTH
Lophocampa maculata

Very Common
93-0373 (8214)

TL 20–28 mm Warm brown FW has rows of brown-edged tan spots at AM, PM, and ST lines, the latter merging with pale outer margin sometimes to form a row of brown spots. Median line usually a single pale spot at costa. Tan thorax has brown epaulettes. Abdomen is pale orange. HOSTS Deciduous trees, including birch, maple, oak, poplar, and willow.

STRIPED GLASSY-WING
Pseudohemihyalea ambigua

Uncommon
93-0380 (8218)

TL 21–26 mm FW is boldly striped with orange along veins and satiny white between; white areas are sometimes translucent. Rusty thorax is striped darker orange, and abdomen is rosy. HOSTS Pines.

EDWARDS' GLASSY-WING
Pseudohemihyalea edwardsii

Very Common
93-0381 (8222)

TL 29–32 mm Translucent tan FW has brown basal, AM, median, PM, and terminal bands that are often most visible at costa. Abdomen is rosy to orange, with black dorsal spots on posterior segments. Tan thorax has a black dot at base of each FW. HOSTS Oaks, including coast live oak, canyon live oak, and Emory oak.

FRECKLED GLASSY-WING
Pseudohemihyalea labecula

Uncommon
93-0382 (8221)

TL 23–31 mm Translucent brown FW has tan AM, median, PM, and ST lines that are sometimes narrowed and often most visible at costa. Rosy-orange abdomen has variably small to large black dorsal spots on posterior segments. Brownish-gray thorax has bluish-black commas laterally. Separable from Edwards' Glassy-wing by thorax and ST line. HOSTS Oaks; probably also other species.

MOTTLED GLASSY-WING
Pseudohemihyalea splendens

Rare
93-0383 (8220)

TL 28–30 mm Resembles Freckled Glass-wing, but has a thin blackish crescent in reniform area, and hoary gray thorax has thin black stripes laterally, with small black dot at outside of each. Abdomen is pale rosy, with variably small to large black dorsal spots on posterior segments. HOSTS Oaks.

ARIZONA TUSSOCK MOTH
Carales arizonensis

Uncommon
93-0386 (8226)

TL 26–30 mm Resembles some *Lophocampa*, but gray FW is distinctive. All cross-lines are chains of gray-edged pale spots. Costa is checkered dark and pale gray. Gray thorax and collar have eight black dots. Striped legs are orange at base. HOSTS Bigtooth maple.

MODEST
TUSSOCK MOTH

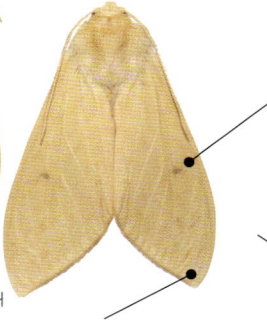

SPOTTED
TUSSOCK MOTH

STRIPED
GLASSY-WING

EDWARDS' GLASSY-WING

FRECKLED GLASSY-WING

actual size

MOTTLED GLASSY-WING

ARIZONA TUSSOCK
MOTH

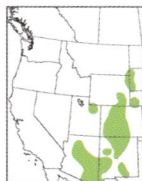

GROTE'S BERTHOLDIA
Bertholdia trigona

Uncommon
93-0390 (8258)

RANGE-WIDE

TL 19–21 mm Speckled rosy-brown FW is shaded gray in ST/terminal area. Large, semitranslucent, pale yellow patch along pink costa is edged with black and contains a handful of randomly placed black dots. Inner basal area has one or two pale yellow spots. **HOSTS** Early instars feed on lichen, moss, and algae on citrus and mountain mahogany. Later instars consume herbaceous plants.

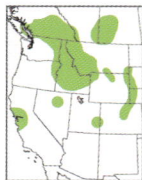

DELICATE CYCNIA
Cycnia tenera

Uncommon
93-0404 (8230)

RANGE-WIDE

TL 16–22 mm White FW has yellow-orange costa that extends to ST area. Yellow orange of collar bleeds into upper thorax. Abdomen is pale yellow orange with black dorsal spots. **HOSTS** Dogbane and milkweed.

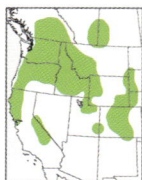

OREGON CYCNIA
Cycnia oregonensis

Uncommon
93-0405 (8231)

RANGE-WIDE

TL 16–22 mm Resembles Delicate Cycnia, but white to light gray FW lacks colored costal streak and veins are often paler than the ground color of FW. Collar and thorax are concolorous with the FW; only the head is yellow orange. Abdomen is yellow orange dorsally and has large black dorsal spots. **HOSTS** Dogbane.

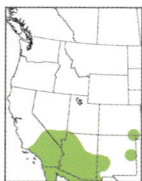

DUSKY TUSSOCK MOTH
Euchaetes zella

Uncommon
93-0406 (8232)

RANGE-WIDE

TL 13–16 mm Dusky brown to blackish FW has small white reniform spot. Abdomen is rosy pink, with black dorsal spots. Thorax is black, with pink edging. **HOSTS** Milkweed and climbing milkweed.

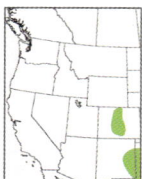

MILKWEED TUSSOCK MOTH
Euchaetes egle

Uncommon
93-0412 (8238)

RANGE-WIDE

TL 16–22 mm Silvery-gray FW is unmarked; veins may be slightly paler, and some individuals may show hint of darker PM line. Abdomen is yellow orange, with black dorsal spots. Thorax is silvery gray; some individuals may show yellow orange at back of head. **HOSTS** Milkweed and dogbane.

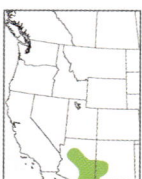

ANCIENT TUSSOCK MOTH
Euchaetes antica

Uncommon
93-0416 (8242)

RANGE-WIDE

TL 15–22 mm Resembles Bicolored Moth, but outer margin of FW is more steeply angled and moth tends to sit with wings held side by side, rather than overlapping. Back of head is orange, collar is gray, and FW fringe is white when fresh. Abdomen is rosy with black bars. Separable from Yellow-collared Scape Moth by width of costa, collar markings, and color of abdomen. **HOSTS** Milkweed and related plants.

GROTE'S BERTHOLDIA

DELICATE CYCNIA

actual size

OREGON CYCNIA

DUSKY TUSSOCK
MOTH

MILKWEED TUSSOCK MOTH

ANCIENT TUSSOCK MOTH

FRECKLED TIGER MOTH

Rare

RANGE-WIDE

Ectypia bivittata   93-0420 (8247)

TL 23–25 mm   White FW is patterned with short black dashes. White thorax has black lateral lines edged with orange. Abdomen has yellow dorsal surface and lines of black dorsal and lateral dots. Black legs have wide white bands. **HOSTS** Unknown.

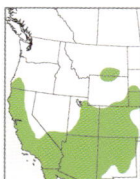

CLIO TIGER MOTH

Common

RANGE-WIDE

Ectypia clio   93-0422 (8249)

TL 19–21 mm   White FW has long black streaks along inner margin, and another from base to central outer margin that joins two shorter black dashes in outer ST/terminal area. White thorax has black dorsal stripe and two black lateral stripes edged with yellow orange. Collar is yellow orange, and abdomen is yellow orange with black dorsal and lateral spots. **HOSTS** Milkweed.

MOUSEY PYGARCTIA

Uncommon

RANGE-WIDE

Pygarctia murina   93-0423 (8250)

TL 15–18 mm   Pale to medium gray FW is unmarked. Gray thorax has rosy posterior margin, and back of head is rosy. Abdomen is rosy pink, with small black dorsal spots. **HOSTS** Spurge, twinevine, caltrop, and menodora; possibly also others.

ROSE-HEADED PYGARCTIA

Common

RANGE-WIDE

Pygarctia roseicapitis   93-0427 (8253)

TL 16–19 mm   White FW and thorax are unmarked. Front of head and base of forelegs are rosy, and abdomen is rosy pink with black dorsal spots. **HOSTS** Hyssop spurge and other spurges.

SPRAGUE'S PYGARCTIA

Rare

RANGE-WIDE

Pygarctia spraguei   93-0429 (8254)

TL 16–19 mm   Grayish-white FW has yellow-orange to pinkish costa and inner margin. Grayish-white thorax has lateral lines that come together and connect to concolorous inner margin. Posterior edge of collar and head are similarly colored. Abdomen is concolorous with inner margin and marked with black dorsal spots. **HOSTS** Unknown; possibly spurge.

SAFFRON-SPOTTED WASP MOTH

Uncommon

RANGE-WIDE

Phoenicoprocta hampsonii   93-0468 (8285)

TL 17–22 mm   Long, narrow, black FW is checkered with thick yellow-orange AM and PM bands and typically held away from body while at rest. Metallic blue abdomen has large orange spots laterally and orange dorsal spot at tip. Black thorax has large orange spots laterally. **HOSTS** Unknown.

actual size

FRECKLED
TIGER MOTH

CLIO TIGER
MOTH

MOUSEY
PYGARCTIA

ROSE-HEADED
PYGARCTIA

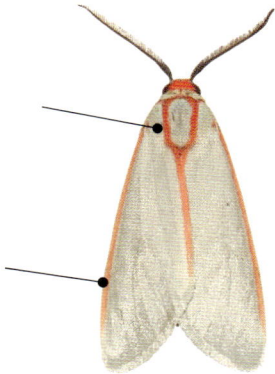

SPRAGUE'S
PYGARCTIA

SAFFRON-SPOTTED
WASP MOTH

419

VEINED CTENUCHA
Ctenucha venosa

Uncommon
93-0433 (8260)

TL 16–18 mm Dark gray FW has bold yellow streaks along central vein, forking at PM area, and adjacent vein costally, as well as costa and parallel to inner margin. Fringe at apex and anal angle is white. Metallic blue thorax has yellow lateral stripes that connect to FW streaks. Head is bright red. **HOSTS** Grasses. **NOTE** Sometimes seen visiting flowers during daytime.

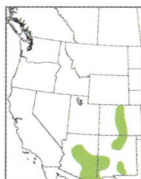

THIN-LINED CTENUCHA
Ctenucha cressonana

Common
93-0434 (8261)

TL 22–25 mm Resembles smaller Veined Ctenucha, but FW streaks are thinner and paler yellow, and short stripe in outer ST/terminal area is usually indistinct. Costa is often red or orange, contrasting with FW stripes. Entire FW fringe is white when fresh. **HOSTS** Grasses. **NOTE** Commonly seen visiting flowers during daytime.

VIRGINIA CTENUCHA
Ctenucha virginica

Very common
93-0435 (8262)

TL 25–27 mm Broad, dark grayish-brown FW has metallic blue sheen at base. Fringe is partly white. Abdomen and thorax are metallic blue, contrasting with orange head and sides of collar. **HOSTS** Grasses, iris, and sedge. **NOTE** Commonly seen visiting flowers during daytime.

WHITE-MARGINED CTENUCHA
Ctenucha multifaria

Uncommon
93-0436 (8263)

TL 23–25 mm Resembles Red-shouldered Ctenucha, but costa and anal angle are also thinly edged with white. Red edging on metallic blue collar does not have central red point. Some individuals may have orange or reddish-orange head and thoracic stripes. **HOSTS** Grasses. **NOTE** Commonly seen visiting flowers during daytime.

RED-SHOULDERED CTENUCHA
Ctenucha rubroscapus

Very Common
93-0437 (8264)

TL 20–22 mm Velvety black FW has white tip at apex. Metallic blue thorax has wide scarlet lateral stripes. Red edging to collar has red point at center. Head is scarlet, and abdomen is metallic blue. **HOSTS** Grasses. **NOTE** Commonly seen visiting flowers during daytime.

BROWN CTENUCHA
Ctenucha brunnea

Uncommon
93-0438 (8265)

TL 20–22 mm Resembles White-margined Ctenucha, but FW is golden brown with black veins and entire fringe is white. **HOSTS** Grasses. **NOTE** Commonly seen visiting flowers during daytime.

YELLOW-COLLARED SCAPE MOTH
Cisseps fulvicollis

Common
93-0440 (8267)

TL 16–20 mm Resembles Bicolored Moth, but yellowish edge along costa is much thinner, FW apex is more elongate, and abdomen is entirely metallic blue black. Thorax and head are larger relative to wings, and yellow-orange collar is wider. Antennae are thick and often held in front of the head. Separable from Ancient Tussock Moth by costa and collar markings, as well as abdomen color. **HOSTS** Grass and sedge. **NOTE** Commonly seen visiting flowers during daytime.

VEINED CTENUCHA

THIN-LINED
CTENUCHA

VIRGINIA CTENUCHA

actual size

WHITE-MARGINED
CTENUCHA

RED-SHOULDERED
CTENUCHA

BROWN CTENUCHA

YELLOW-COLLARED
SCAPE MOTH

LITTER MOTHS

SUPERFAMILY **Noctuoidea (93)**
FAMILY **Erebidae** SUBFAMILY **Herminiinae**

Small to medium-sized moths that rest with wings folded in a broad triangle and usually have curving palps that create a snouted appearance. Most are shades of brown and tan, with darker brown or black markings. Caterpillars of most species are detritivores, feeding on fallen leaves and other decaying plant matter. Adults are nocturnal and regularly come to lights, but many species can also be flushed from leaf litter and forest trails during daytime. Adults will also commonly come to sugar baits.

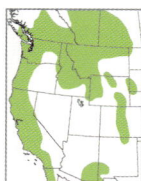

AMERICAN IDIA
Idia americalis

Common
93-0469 (8322)

RANGE-WIDE

TL 13–14 mm Peppery gray FW has scalloped black AM, PM, and ST lines that are edged in white and end at broad dusky square at costa. Central FW has a cinnamon wash, usually most visible running through orbicular and reniform spots to outer margin, but sometimes strong across entire central wing. **HOSTS** Dead leaves and lichens.

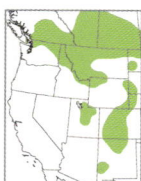

COMMON IDIA
Idia aemula

Common
93-0471 (8323)

RANGE-WIDE

TL 11–16 mm Resembles American Idia, but hoary gray FW lacks large rectangles at costa and cinnamon wash. Reniform and orbicular spots are contrastingly tan, rarely blackish. Blackish median band is often broad and dark. **HOSTS** Presumably leaf litter.

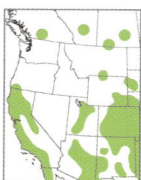

GLOSSY BLACK IDIA
Idia lubricalis

Uncommon
93-0482 (8334)

RANGE-WIDE

TL 18–21 mm Dusky to blackish FW has scalloped white AM, PM, and ST lines. Pale oval reniform spot has a dark central dash; orbicular dot is pale. **HOSTS** Fungi and lichens.

WESTERN IDIA
Idia occidentalis

Uncommon
93-0483 (8334.1)

RANGE-WIDE

TL 15–19 mm Resembles Glossy Black Idia, but FW is lighter brown with a diffuse, dark brown median band and hoary shading in ST area. Reniform spot is pale dash, and orbicular spot is pale dot; both often indistinct. **HOSTS** Lichen, grasses, dead wood, and leaf litter.

WAVY-LINED FAN-FOOT
Zanclognatha jacchusalis

Rare
93-0500 (8353)

RANGE-WIDE

TL 17–18 mm Tan FW has wavy, curving, brown AM and PM lines and straight white ST line. ST and terminal areas shaded slightly darker. Reniform spot is a brown (sometimes blackish) crescent. **HOSTS** Probably dead leaves.

AMERICAN IDIA

actual size

COMMON IDIA

GLOSSY BLACK IDIA

WESTERN IDIA

WAVY-LINED FAN-FOOT

MORBID OWLET
Chytolita morbidalis

Uncommon
93-0502 (8355)

TL 16–19 mm Resembles Wavy-lined Fan-Foot, but FW is more peppery, PM line is less wavy, and ST line is a series of white-edged black dots. Large reniform spot is brown or sometimes blackish. **HOSTS** Dead leaves.

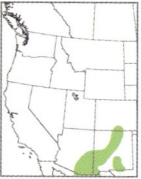

ORANGE-SPOTTED TETANOLITA
Tetanolita palligera

Common
93-0517 (8367)

TL 11–14 mm Grayish-brown FW has wavy, dark AM and PM lines that start straightish then curve sharply to meet costa. Pale wavy ST line is indistinct within diffuse, dark brown band. Orbicular and reniform spots are orangey tan, the latter with a brownish lower half and crossed behind by diffuse, dark brown median line. **HOSTS** Dead leaves.

BENT-WINGED OWLET
Bleptina caradrinalis

Uncommon
93-0520 (8370)

TL 12–17 mm Resembles Wavy-lined Fan-Foot, but FW is more peppery, PM line is less wavy, and ST line is a series of white-edged black dots. Large reniform spot is brown or sometimes blackish. **HOSTS** Dead leaves.

PEPPERED RENIA
Renia hutsoni

Uncommon
93-0534 (8383)

TL 14–17 mm Peppery brown FW has small, dull orange orbicular and reniform spots. Slightly wavy AM and PM lines, and diffuse dark median band, are dusky. Indistinct ST line is edged by patchy, diffuse, dusky ST shading. **HOSTS** Unknown; likely dead leaves.

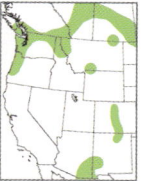

DARK-SPOTTED PALTHIS
Palthis angulalis

Uncommon
93-0551 (8397)

TL 12–18 mm Light brown FW is usually held slightly creased along costa when at rest. Oblique brown median band stops short of costa. Thin AM and PM lines are slightly wavy. Reniform spot is a slanting brown crescent. Male has long upturned labial palps, while female has tufted forelegs. **HOSTS** Coniferous and deciduous trees, including ash, alder, maple, spruce, and willow.

DIMORPHIC SNOUT
Hypena bijugalis

Uncommon
93-0564 (8443)

TL 14–17 mm Sexually dimorphic. Female has rich brown upper FW and grayish-white lower FW, separated by a straight PM line with outward bulge at midpoint. Basal half of inner margin has golden-brown patch edged with white, and apex has blackish dash. Male has sooty-black FW with a square white patch where indistinct white PM line meets inner margin. Both sexes have small black discal dot and ST line of black dots edged thinly with white. **HOSTS** Dogwood.

MORBID OWLET

ORANGE-SPOTTED
TETANOLITA

BENT-WINGED OWLET

actual size

PEPPERED RENIA

male

male

DARK-SPOTTED
PALTHIS

male

female

female

male

DIMORPHIC
SNOUT

425

MOTTLED SNOUT

Hypena palparia

Rare

93-0565 (8444)

RANGE-WIDE

TL 14–18 mm Resembles female Dimorphic Snout, but PM line is scalloped, basal area is paler than median area, and gray to brownish ST/terminal area is heavily mottled. Basal patch along inner margin is grayish. Males average darker than females. **HOSTS** Eastern hornbeam, ironwood, and hazel.

WHITE-LINED SNOUT

Hypena abalienalis

Uncommon

93-0566 (8445)

RANGE-WIDE

TL 14–18 mm Sexually dimorphic. Warm brown (female) to dusky brown (male) FW has curving, double, white to tan PM line, thin white ST line, and pale veins in ST area that create a lacy effect. AM line of female is white, that of male is indistinct. **HOSTS** Slippery elm.

HOP VINE SNOUT

Hypena humuli

Uncommon

93-0584 (8461)

RANGE-WIDE

TL 14–20 mm Light to dark brown FW has a dusky diamond-shaped patch in outer median area, bounded by zigzag brown AM and PM lines that fade before reaching inner margin. Indistinct apical dash divides pale outer ST area and shaded inner terminal area; the latter has dusky veins. **HOSTS** Common hop and stinging nettle.

CALIFORNIA CLOVERWORM

Hypena californica

Common

93-0585 (8462)

INLAND

COASTAL

TL 15–17 mm Tan to brown FW has scalloped brown AM and PM lines. A dusky bar connects dark-centered gray orbicular spot and grayish reniform spot containing two black dots. ST line a series of indistinct dusky dots bordered by brown-shaded ST area. Blackish apical dash is shaded dusky along inner side. **HOSTS** Stinging nettle.

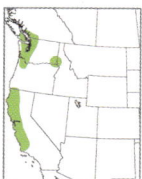

DECORATED CLOVERWORM

Hypena decorata

Rare

93-0586 (8463)

RANGE-WIDE

TL 14–16 mm Resembles California Cloverworm, but orbicular and reniform spots are darker, black line connecting them is thin, and pale area outside of apical dash has two white spots. Inner PM line has a dark brown to black bulge. Rare individuals have a mostly white FW, with just outer median area, costa, and ST/terminal area brown. **HOSTS** Stinging nettle.

GREEN CLOVERWORM

Hypena scabra

Uncommon

93-0588 (8465)

NORTH

SOUTH

TL 14–18 mm Variable. Brown to gray FW has wavy median line that extends straight across wing, becoming blackish in inner half. A triangular patch of darker shading covers inner ST and terminal areas from apex to inner margin. Some individuals have darker outer median patch, extending past AM line into basal area. Other individuals may have blackish vertical dashes in inner and central median area, or they may be unmarked but for black discal dot. **HOSTS** Generalist on herbaceous and woody plants. **NOTE** Northernmost populations cannot survive cold winters and are replenished each year by southern migrants.

male

MOTTLED
SNOUT

female

female

male

WHITE-LINED
SNOUT

HOP VINE SNOUT

CALIFORNIA
CLOVERWORM

actual size

DECORATED
CLOVERWORM

GREEN
CLOVERWORM

427

ASSORTED OWLETS

SUPERFAMILY **Noctuoidea (93)**

FAMILY **Erebidae** SUBFAMILIES Rivulinae, Scoliopteryginae, Calpinae, Hypocalinae, Scolecocampinae, Boletobiinae, and Toxocampinae

A diverse assemblage of small to medium-sized moths representing several subfamilies, some containing only a few genera. This catchall category contains a little bit of everything: moths that rest with wings flat and others that rest with them tented against the body; some with long palps and some with short; many of plain appearance and some with bold or bright colors. Adults are largely nocturnal and regularly come to lights, but a few species may also be flushed from low vegetation during daytime.

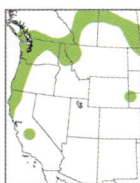

SPOTTED GRASS MOTH
Rivula propinqualis

Uncommon
93-0592 (8404)

RANGE-WIDE

TL 10–11 mm Straw-colored FW has a purplish-gray reniform spot with two black dots. AM and PM lines are warm brown and are parallel near inner margin. **HOSTS** Grass.

CURVE-LINED TAN
Oxycilla tripla

Uncommon
93-0595 (8405)

RANGE-WIDE

TL 12–14 mm Peppery, pale tan FW has roughly parallel, golden-brown AM and PM lines. Terminal line is thin brown dashes. Head and collar are golden brown. **HOSTS** Ticktrefoil.

DUSTY-WINGED TAN
Zelicodes linearis

Uncommon
93-0600 (8410)

RANGE-WIDE

TL 10–11 mm Peppery, tan to grayish FW has wavy brown AM line, curving brown PM line edged indistinctly with pale tan below, and indistinct pale ST line often edged basally with darker shading. Terminal line is dark brown dashes. **HOSTS** Unknown.

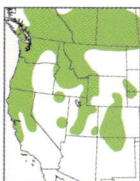

HERALD MOTH
Scoliopteryx libatrix

Common
93-0601 (8555)

RANGE-WIDE

TL 21–23 mm Warm to grayish-brown FW has wide fiery-orange stripe through basal and median areas. Single AM and double PM lines are whitish. Outer margin is strongly scalloped. **HOSTS** Poplar and willow. **NOTE** Often visits sugar bait.

MOONSEED MOTH
Plusiodonta compressipalpis

Rare
93-0622 (8534)

RANGE-WIDE

TL 14–18 mm Fawn-colored FW has brown, curving AM and wavy PM lines edged with lilac. Basal and inner ST areas and large reniform spot are pale with gold edging. Inner margin has large median tuft and small anal tuft that project upward from back when at rest. **HOSTS** Moonseed and snailseed.

SPOTTED GRASS MOTH

CURVE-LINED TAN

DUSTY-WINGED TAN

HERALD MOTH

MOONSEED MOTH

actual size

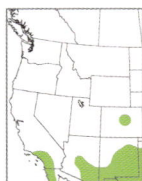

HYPOCALA MOTH
Hypocala andremona

Uncommon
93-0627 (8642)

TL 22–26 mm Peppery, warm brown to gray FW has dark brown double ST line from inner margin, ending abruptly at midpoint. Other cross-lines are thin and wavy but usually indistinct. Reniform spot may be dark brown or indistinct. HW is orange, with black PM and terminal bands. HOSTS Persimmon.

BLACK MOON MOTH
Scolecocampa atriluna

Rare
93-0638 (8515)

RANGE-WIDE

TL 22–24 mm Lightly peppered, pale tan FW has bold, black (rarely hollow), T-shaped reniform spot, small black orbicular and basal dots, and zigzag PM line often most visible as black patches at costa and inner margin. HOSTS Unknown, but eastern sister species, Deadwood Borer (*S. liburna*, not shown), feeds inside rotting logs.

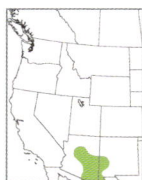

SOFT-LINED TAN
Pseudorgyia russula

Uncommon
93-0648 (8513)

RANGE-WIDE

TL 13–16 mm Light brown FW has indistinct wavy PM and ST lines usually defined by dusky shading basally, rarely appearing as dark lines with pale edging. AM line may or may not be present. Indistinct reniform spot is pale; orbicular spot is a small dark dot. HOSTS Unknown.

THREE-LINED OWLET
Phobolosia anfracta

Uncommon
93-0654 (8439)

NORTH

SOUTH

TL 11–15 mm Gray FW has wavy grayish-brown AM and PM lines edged on both sides with pale gray. Reniform spot is a black crescent. Apical half of costa is striped white and grayish brown. Black dashed terminal line ends with black spot at pointed apex. HOSTS Unknown.

LONG-PALPED MYCTEROPHORA
Mycterophora longipalpata

Rare
93-0676 (8415)

RANGE-WIDE

WS 22–28 mm Resembles Blushing Mycterophora, but wings lack pinkish tinge and have black orbicular and reniform spots. Lower half of wings often shaded dusky, so pale edges to PM and ST lines more evident. Very long palps create a snouted appearance. Some individuals have pale costa. Males have bipectinate antennae; those of females are filiform. HOSTS Unknown.

BLUSHING MYCTEROPHORA
Mycterophora rubricans

Rare
93-0678 (8417)

RANGE-WIDE

WS 23–25 mm Light pinkish-brown FW has wavy, dark brown AM and PM and brown median and ST lines that angle basally to meet costa and somewhat blend in with peppery background. Small black orbicular spot is often indistinct, and reniform spot is absent. HOSTS Unknown.

MISERABLE FUNGUS MOTH
Metalectra miserulata

Rare
93-0686 (8506)

RANGE-WIDE

TL 11–12 mm Resembles Common Idia, but reniform spot is a blackish crescent with rusty shading below; orbicular spot is a black dot. Costa has three or four white dots below PM line. HOSTS Fungus.

HYPOCALA MOTH

BLACK MOON MOTH

SOFT-LINED TAN

THREE-LINED OWLET

actual size

female

LONG-PALPED
MYCTEROPHORA

male

female

BLUSHING MYCTEROPHORA

MISERABLE
FUNGUS MOTH

BICOLORED FUNGUS MOTH
Metalectra edilis
Uncommon
93-0687 (8507)

TL 11–12 mm Warm brown to tan FW has dark brown basal and median bands, sometimes with indistinct, thin, pale line in center, and dark brown patch at costa of ST area. Thin wavy AM and PM lines are dark brown. Orbicular and reniform spots are blackish. **HOSTS** Fungus.

IVORY-EDGED MOTH
"Oruza" albocostata
Rare
93-0692.6 (9026)

WS 22–25 mm Peppery brown FW has creamy-white costa that connects to white stripe across front of thorax. Diffuse, dark brown AM and PM lines are edged outwardly with whitish. ST and terminal lines are rows of blackish dots. **HOSTS** Unknown. **NOTE** Recently reassigned to a new as-yet-unnamed genus.

EVERLASTING EUBLEMMA
Eublemma minima
Common
93-0693 (9076)

TL 8–9 mm Whitish FW has broad olive-brown bands in median and ST areas and thin olive-brown PM line. ST line is a row of black spots, sometimes indistinct except for larger spots near costa and inner margin. Reniform spot is two tiny black dots. Inner margin has a tuft of scales at median band that stick up when wings are folded. **HOSTS** Cudweed.

STRAIGHT-LINED SEED MOTH
Eublemma recta
Uncommon
93-0695 (9078)

TL 8–10 mm Warm brown to tan FW has oblique white median line that slants to apex, shaded basally with brown. Curving white PM line connects to outer median line. ST area has a row of tiny black dots on veins. Thin reddish (sometimes olive-tan) ST line has a black-edged spot at apex. **HOSTS** Morning glory and bindweed.

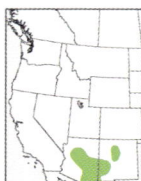

RUSTY HEMEROPLANIS
Hemeroplanis punitalis
Rare
93-0701 (8468)

TL 13–15 mm Tan to brownish-orange FW has gently wavy, yellow-edged, brown AM and PM lines and solid (sometimes hollow), dark brown, kidney-shaped reniform spot. ST line is small black dots edged with white. **HOSTS** Unknown.

BLACK-MARKED HEMEROPLANIS
Hemeroplanis historialis
Common
93-0705 (8472)

TL 12–14 mm Variable. Gray to grayish-brown FW has blackish squares on costa at ends of fragmented thin brown (sometimes indistinct) AM and PM lines. Kidney-shaped reniform spot is black or hollow and dusky. ST/terminal area usually shaded brown, and ST line is indistinct, white-edged, black dots. **HOSTS** Unknown; possibly legumes.

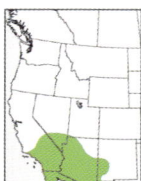

ANVIL-WING MOTH
Hemeroplanis incusalis
Very Common
93-0707 (8474)

TL 12–14 mm Variable. Tan FW has thick, dark brown, slanting AM and curving PM lines and warm brown ST/terminal area. Pale ST line is often indistinct. Hollow reniform spot is indistinct. Median area is sometimes shaded darker. **HOSTS** Mesquite and other legumes.

BICOLORED FUNGUS MOTH

IVORY-EDGED MOTH

EVERLASTING
EUBLEMMA

STRAIGHT-LINED
SEED MOTH

RUSTY HEMEROPLANIS

actual size

BLACK-MARKED
HEMEROPLANIS

BOLD-LINED
HEMEROPLANIS

PALE-LINED HEMEROPLANIS
Hemeroplanis rectalis

Uncommon
93-0709 (8475.1)

TL 12–14 mm Tan to reddish-brown FW has thin, yellow-edged, brown, straight AM and gently wavy PM lines that end with thick brown spot at costa. Kidney-shaped reniform spot usually hollow and indistinct but sometimes black. ST line a row of white-edged black dots, often indistinct. **HOSTS** Unknown.

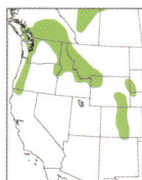

SIX-SPOTTED GRAY
Spargaloma sexpunctata

Rare
93-0715 (8479)

RANGE-WIDE

TL 15–18 mm Gray FW has dusky brown median band and dusky triangular subapical patch tipped with three short black dashes. Tiny orbicular dot is black. Scalloped AM and PM lines are inconspicuous. **HOSTS** Dogbane.

SCALLOPED GRAY
Nychioptera noctuidalis

Uncommon
93-0721 (8485)

RANGE-WIDE

TL 9–11 mm Gray FW has thin, scalloped, blackish AM and PM lines; AM line is straightish, and curved PM line ends with dark square at costa. Reniform spot is dusky. Outer ST area has indistinct darker shading. Terminal line is blackish dashes. **HOSTS** Unknown.

FRAMED ISOGONA
Isogona segura

Rare
93-0737 (8496)

RANGE-WIDE

TL 13–15 mm Tan FW has a sharply V-shaped, brown-edged, yellowish PM line that connects to yellowish apical dash to frame dark brown subapical patch. Brown-edged yellowish AM line is zigzagged. Large kidney-shaped reniform spot has yellowish outline, and thin veins are pale tan. Head and collar are dark brown. **HOSTS** Unknown.

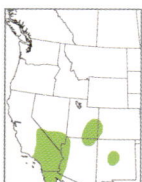

ASTRONAUT OWLET
Allerastria albiciliatus

Uncommon
93-0744 (9020)

RANGE-WIDE

TL 10–12 mm Variable. Ivory FW has dark gray AM and ST bands bordered by brownish-orange AM and PM lines. Median line a gray square at costa. Some individuals are rusty to reddish brown instead of ivory, and some of these have dusky brown bands instead of gray. Rare individuals are mostly reddish brown with indistinct markings. **HOSTS** Unknown.

BROKEN-SPOTTED OWLET
Lygephila victoria

Uncommon
93-0753 (8563)

RANGE-WIDE

TL 19–23 mm Lightly striated, lilac-gray to reddish-brown FW has distinctive T-shaped black (rarely tan) reniform spot shaded dusky below and to costa, with two to five adjacent small black (or tan) spots. Tiny orbicular spot is white, sometimes indistinct. Outer ST area shaded dusky, and costa has dusky marks at AM and often median and PM lines. Head and collar dark brown. **HOSTS** Snowberry.

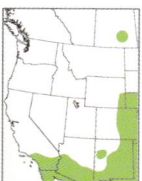

LEVANT BLACKNECK
Tathorhynchus exsiccata

[IN] Uncommon
93-0754 (8466)

RANGE-WIDE

TL 14–16 mm Narrow grayish-brown FW has dark streak behind pale orbicular and two-dotted reniform spots. Veins are often noticeably pale. ST area is shaded dark brown. Dark brown basal dashes meet lateral stripes on thorax. Collar is dark brown. **HOSTS** Alfalfa and Spanish broom.

PALE-LINED
HEMEROPLANIS

SIX-SPOTTED GRAY

SCALLOPED GRAY

FRAMED ISOGONA

actual size

ASTRONAUT OWLET

LEVANT
BLACKNECK

BROKEN-SPOTTED
OWLET

WITCHES
SUPERFAMILY Noctuoidea (93)
FAMILY Erebidae SUBFAMILY Erebinae TRIBE Thermesiini

These impressive moths are among the largest of North American species, and the only member of their genus. They are resident in Mexico and the southernmost parts of the United States but are nomadic and wander widely in summer; individuals have been recorded as far north as Canada and Alaska. They are nocturnal and come to both lights and sugar bait but are also commonly found sheltering in protected places, such as in garages, beneath eaves, or under bridges during daytime.

BLACK WITCH
Ascalapha odorata

Very Common
93-0759 (8649)

NORTH

SOUTH

WS 110–150 mm Unmistakable. Sexually dimorphic. Brown FW has large comma-shaped, metallic blue, reniform eyespot. ST line is black-edged yellow spots, and PM line is black along inner half, bordered by rusty and metallic blue. HW has paired blue-edged eyespots at anal angle. Female has whitish median band that forms a straight line across wings when at rest. **HOSTS** Leguminous trees, including cassia and mesquite.

UNDERWINGS
SUPERFAMILY Noctuoidea (93)
FAMILY Erebidae SUBFAMILY Erebinae TRIBE Catocalini

The large genus *Catocala* contains more than a hundred species in North America. They are medium to large moths with broad wings that are held flat over the body when at rest. The group's common name comes from the bright pink or orange hindwings of most species, which are used to confuse potential predators when flashed during escape. The forewings are gray or brown and cryptically patterned; identification of similar species often relies on subtle differences in lines or shading. Adults are largely nocturnal and will come to lights; some will also visit sugar baits.

PENITENT UNDERWING
Catocala piatrix

Uncommon
93-0762 (8771)

RANGE-WIDE

TL 36–44 mm Grayish-brown FW has relatively smooth, double AM line that angles strongly basally to costa. Adjacent AM area is pale gray, passing through pale claviform spot, with dusky gray median area slanting in parallel below. Reniform spot brown, edged with black, and ringed in gray. Thin black PM line has very long W beneath reniform spot, with black shading below angling to apex, and strong black vertical section parallel to inner margin. Tooth of inner PM line connects to claviform spot. ST area is brownish. Orange HW has broad terminal and median bands and lightly checkered peach fringe. **HOSTS** Pecan, hickory, and walnut.

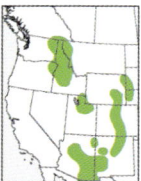

BRIDE UNDERWING
Catocala neogama

Uncommon
93-0790 (8798)

RANGE-WIDE

TL 37–44 mm Resembles Penitent Underwing, but black basal dash is thicker, AM line is slightly scalloped with basal-pointing tooth at outer side of basal dash, and pale area beneath AM line is less pronounced. Claviform spot usually has some brown shading but is sometimes pale, and reniform spot is brown. Terminal area has light shading below black vertical bar of PM line. HW is rosy orange to orange, with broader black bands. **HOSTS** Walnut and butternut.

female

male

female

BLACK WITCH

actual size

PENITENT
UNDERWING

actual size

BRIDE
UNDERWING

AHOLIBAH UNDERWING

Catocala aholibah

Common

93-0791 (8800)

RANGE-WIDE

TL 30–40 mm Gray FW has brown basal and ST areas. Gray-lined reniform spot is brown, and claviform spot is white. Jagged double AM line has large tooth projecting medially nearly to or touching claviform spot. Zigzag ST line is whitish. Pink HW has black median and terminal bands, the latter with black scallops bulging into white fringe. **HOSTS** Oaks, including Garry, Gambel, and bur oak.

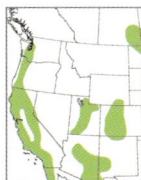

ILIA UNDERWING

Catocala ilia

Uncommon

93-0792 (8801)

RANGE-WIDE

TL 34–45 mm Hoary gray to brownish FW has tan reniform spot edged with black and rimmed with white; white sometimes fills entire spot. Wavy double AM line is relatively straight across wing and strongly suffused dusky. Thin black PM line has long W below reniform spot, but teeth in inner half are moderate in length. Black basal dash bordered outwardly with small white spot. Claviform spot often pale but sometimes indistinct. Median area sometimes paler. Rosy to orange HW has black median band, terminal band that is wider toward apex, and checkered white fringe. **HOSTS** Oak.

WHITE UNDERWING

Catocala relicta

Common

93-0795 (8803)

RANGE-WIDE

TL 37–47 mm FW is white or light gray with dark gray median band that passes behind circular, gray-edged, white claviform spot. Pale wavy AM, PM, and ST lines are edged with dark gray. HW is black, with white median band and fringe. Some individuals have dusky or even blackish FW. **HOSTS** Poplar and willow. **NOTE** One of our most distinctive underwings.

ONCE-MARRIED UNDERWING

Catocala unijuga

Uncommon

93-0797 (8805)

RANGE-WIDE

TL 37–47 mm Hoary gray FW has wavy black AM and PM lines; PM line has very shallow W shape. White zigzag ST line is bordered by gray shading. Claviform spot is slightly pale, and hollow reniform spot is gray; AM area above spots often has a small whitish patch. Terminal line is black crescents. Rosy to rosy-orange HW has black median band and broad black terminal band, with black scallops bulging into white fringe. **HOSTS** Poplar and willow.

MOTHER UNDERWING

Catocala parta

Rare

93-0798 (8806)

RANGE-WIDE

TL 38–43 mm Resembles Bride Underwing, but pale claviform spot is more squarish, AM line has flatter tooth on outer side of basal dash, and thorax has distinctive black lateral lines framing a brownish patch. ST line is strongly zigzagged. Double-lined reniform spot is sometimes filled with brown but often hidden within dusky median band. HW is rosy to rosy orange, with narrow black median and broad terminal band, the latter with black scallops bulging into white fringe. **HOSTS** Poplar and willow.

AHOLIBAH UNDERWING

ILIA
UNDERWING

WHITE UNDERWING

ONCE-MARRIED
UNDERWING

MOTHER
UNDERWING

actual size

439

IRENE'S UNDERWING

Uncommon

Catocala irene   93-0799 (8807)

TL 33–36 mm Resembles Penitent Underwing, with slanting pale AM area, but AM line is wavy. Dark brown reniform spot is often indistinct in blackish median band. Pale gray ST line is sharply zigzagged. Points of W shape in PM line are shorter and blunt. Inner ST area is brown, sometimes strongly so, and FW often has a brownish wash. Rosy HW has narrow black median and broad terminal band, the latter with black scallops bulging into white fringe. **HOSTS** Poplar and willow.

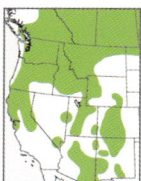

WESTERN UNDERWING

Uncommon

Catocala californica   93-0803 (8814)

TL 33–39 mm Resembles Once-married Underwing, but black AM and PM lines are thinly edged with tan, and pale spots in median area above and below reniform spot, and that in center of ST area, are tan, contrasting with whitish-gray ST line. Claviform spot is often tan but sometimes shaded darker, connected by thin stem to PM line. Reniform spot often obscured by dusky median band. On some individuals, most FW markings indistinct, with just pale ST line obvious. Rosy to rosy-orange HW has narrow median band and broad black terminal band that has black scallops bulging into white fringe. **HOSTS** Poplar and willow.

BRISEIS UNDERWING

Uncommon

Catocala briseis   93-0804 (8817)

TL 32–37 mm Dark gray FW has diffuse white patches above and below reniform spot, and white ST area often has patches of light brown. Thin black PM line cuts through white area, emphasizing W shape. Zigzag ST line is whitish gray and stepped higher at costa. Terminal line is black dashes, with a white dot beneath each. Claviform spot sometimes pale, but reniform spot and AM line are usually dark gray and often difficult to see. Some individuals have very reduced white areas but show brown in ST area. **HOSTS** Poplar and willow.

GROTE'S UNDERWING

Rare

Catocala grotiana   93-0805 (8818)

TL 33–39 mm Resembles Briseis Underwing but averages darker with more white, AM line is more evenly scalloped, and W shape of PM line is not much more pronounced than scallops of the rest of the line. AM line often has white edging. ST area may be dark brown to gray. Pale markings on FW may sometimes be overall more tan than white. Rosy to rosy-orange HW has medium-width median and terminal lines, the latter with black scallops bulging into white fringe. **HOSTS** Poplar and willow.

SEMIRELICT UNDERWING

Uncommon

Catocala semirelicta   93-0806 (8821)

TL 33–39 mm Resembles Once-married Underwing, but lower line of double AM line is darker than upper, zigzag pale ST line is more regular with stronger gray shading, and black crescents of terminal line have white dots beneath. Reniform spot often obscured by dark median-band shading. Black HW median band is often narrower than that of Once-married and ends before inner margin. **HOSTS** Poplar and willow.

IRENE'S
UNDERWING

WESTERN
UNDERWING

BRISEIS
UNDERWING

actual size

GROTE'S
UNDERWING

SEMIRELICT
UNDERWING

441

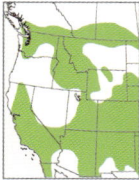

JOINED UNDERWING
Catocala junctura

Common

93-0809 (8829)

RANGE-WIDE

TL 34–43 mm Hoary gray to tan FW has wavy double AM line with paler spot at inner margin. Reniform spot is outlined with double black lines and filled with brown or gray and edged below and costally by dusky median band. Thin black PM line has a shallow W shape below reniform spot and two long narrow bulges adjacent to inner margin. Pale zigzag ST line is roughly straight along inner half. Pink to reddish-orange HW has black median and terminal bands, the latter with scallops bulging into white fringe. FW of some individuals is indistinctly marked, with ST line being most visible marking. **HOSTS** Poplar and willow.

SWEETHEART UNDERWING
Catocala amatrix

Uncommon

93-0815 (8834)

RANGE-WIDE

TL 37–44 mm Brownish-gray FW has broad band of dusky shading from base to apex, interrupted by claviform spot. Thin, jagged, black AM line slants strongly basally to costa; AM line bulge near inner margin has mini M shape at peak. Thin black PM line has moderately long W shape beneath reniform spot and is not thickened at vertical section at inner margin. Claviform spot is connected to PM line by wide neck, and PM line between this and vertical section has a mini W shape. In rare individuals, dusky band is reduced or absent. Resembles heavily marked individuals of Bride Underwing, but dusky band passes through median area, and AM and PM lines have mini M/W shapes near inner margin. **HOSTS** Poplar.

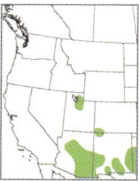

DESDEMONA UNDERWING
Catocala desdemona

Uncommon

93-0817 (8835.1)

RANGE-WIDE

TL 29–37 mm Brown FW has wavy double AM line with one black and one brown of pair. Claviform spot is pale, and brown reniform spot is usually obscured by dark median band. Terminal area has dusky patch below W shape of PM line. Black vertical bar projection at inner PM line is short. ST area is shaded brown. Orange HW has moderately narrow black median and terminal bands and black-checkered, bright orange fringe. Some individuals are shaded chestnut brown below AM line, with a contrastingly pale basal area. **HOSTS** Mexican blue oak and other oaks.

DELICATE UNDERWING
Catocala chelidonia

Rare

93-0825 (8838)

RANGE-WIDE

TL 23–26 mm Gray to brownish-gray FW has slanting, thin, black AM line, with a thicker area directly above pale round claviform spot, creating an angry-eyebrows appearance. Thin black PM line is mostly straight except for large W-shaped section. Gray reniform spot is indistinct. Orange HW has lumpy, curved median band and terminal band that is wider at apex, with concave section near midpoint. Fringe is checkered gray with white at apex. **HOSTS** Sonoran scrub oak.

VERRILL'S UNDERWING
Catocala verrilliana

Uncommon

93-0838 (8852)

RANGE-WIDE

TL 20–25 mm Gray FW has slightly scalloped, sloping AM line shaded dusky, with small tooth at midpoint and larger tooth near inner margin, and short black basal dash. Triangular claviform spot is outlined in black and usually filled with dusky (sometimes pale) shading. Gray reniform spot is smaller and lacks double ring. Area between spots and AM line is pale. Brown-shaded PM line has large W shape and bold vertical line at inner margin; area between is evenly scalloped and relatively flat. Rosy-orange HW has black terminal band that is wider toward apex, as well as C-shaped median band and checkered peach fringe. **HOSTS** Oak.

JOINED
UNDERWING

SWEETHEART
UNDERWING

DESDEMONA
UNDERWING

DELICATE UNDERWING

VERRILL'S UNDERWING

actual size

443

PASSIONATE UNDERWING

Rare

Catocala violenta   93-0839 (8853)

TL 25–29 mm Resembles Verrill's Underwing, but straightish AM line lacks tooth at midpoint, pale claviform spot is connected by a thread to tooth in PM line, and ST area is broadly brown. Usually has dusky basal dash. Rosy-orange HW has black terminal band that is wider toward apex, as well as crescent-shaped median band, with checkered peach-white fringe. HOSTS Gambel oak.

ULTRONIA UNDERWING

Uncommon

RANGE-WIDE

Catocala ultronia   93-0841 (8857)

TL 25–33 mm Brown to gray FW has dark brown patch in inner basal area, often extending down inner margin. Subapical patch is pale brown at apex, with warm brown in outer ST area and grayish shading on inside of apical dash. Reniform spot is a faint pale kidney shape shaded slightly darker inside, and claviform spot is often indistinct. Veins below PM line are often strongly traced with dark brown. Rosy-orange HW has medium-width black median and terminal bands, as well as checkered grayish fringe with white section at apex. Some individuals can be relatively dark brown through central wing, reducing contrast of basal and subapical spots. HOSTS Cherry and plum; also apple and hawthorn.

GRAPHICS, SOMBERWINGS, and ALLIES

SUPERFAMILY Noctuoidea (93)
FAMILY Erebidae SUBFAMILY Erebinae
TRIBES Melipotini and Euclidiini

Medium-sized moths with narrow shoulders and a broad straight outer margin. The graphics are boldly marked with contrasting bands of light and dark brown and, usually, a large pale reniform spot. The loopers are more modestly marked, excepting the somberwings, which have bold blackish triangles. Adults will come to lights at night, but most are also readily flushed from vegetation during daytime, and Little Arches and the somberwings are entirely diurnal.

INDISCRETE CISSUSA

Uncommon

NORTH

Cissusa indiscreta   93-0866 (8594)

SOUTH

TL 16–18 mm Tan to reddish-brown FW has black reniform dash edged with white above and shaded diffusely black below. Pale ST line has small black dots, often with two or three larger black spots toward costa. Thin brown AM line (often indistinct) is narrowly edged with pale tan. Thin brown median line (often indistinct) connects to reniform dash. Sometimes shows a paler patch below reniform dash. HOSTS Oak.

VIGOROUS CISSUSA

Uncommon

RANGE-WIDE

Cissusa valens   93-0867 (8596)

TL 22–26 mm Grayish to warm brown FW has thin brown PM line edged with pale tan that bulges medially in inner half and downward-pointing tooth in outer half; area immediately below tooth is shaded dark brown. Thin brown AM and median lines are gently curving, and reniform dash has dark spots at each end. Some individuals have brown median area, brown shading below median line, extended dark brown shading below PM line, and/or a brown ST area. HOSTS Oak.

PASSIONATE UNDERWING

actual size

ULTRONIA
UNDERWING

INDISCRETE CISSUSA

actual size

VIGOROUS
CISSUSA

445

LITOCALA MOTH
Litocala sexsignata

TL 14–16 mm Hoary gray FW has strongly curving, thin, black PM line that has a long median loop in inner half, then curves around broad grayish-white patch, with downward-pointing tooth at midpoint of patch. PM line and scalloped AM line are bordered outwardly with light brown shading. Wavy pale ST line has two black spots near costa. HW is black, with three creamy-white spots and white fringe. **HOSTS** Oak, chinquapin, and manzanita.

PERPENDICULAR GRAPHIC
Melipotis perpendicularis

Uncommon
93-0869 (8598)

RANGE-WIDE

TL 20–22 mm Sexually dimorphic. Gray to brownish FW has a thick, slanting, pale AM band connected by a thick stem to a large pale spot that is framed by sinuous, smooth PM line. Inner PM area is roughly square. Pale ST line has a basal-pointing tooth near midpoint backed by dark brown shading. Basal area of male is dark; that of female matches AM band. Some individuals are paler through entire inner half of FW, with just square, dark brown patches along costa at median and ST areas, and brown terminal band. Occasionally may have a strong reddish wash across basal and terminal area or throughout. **HOSTS** Hopbush.

INDOMITABLE GRAPHIC
Melipotis indomita

Very Common
93-0871 (8600)

RANGE-WIDE

TL 21–26 mm Sexually dimorphic. Resembles Perpendicular Graphic, but large pale spot is toothed and inner PM line is rounded. Basal area has black spots at costa. Basal area of male is dark; that of female is pale. **HOSTS** Mesquite. **NOTE** May occasionally wander north of its regular range.

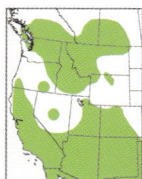

MERRY GRAPHIC
Melipotis jucunda

Very Common
93-0878 (8607)

NORTH

SOUTH

TL 21–26 mm Sexually dimorphic and variable. Male has gray FW with thin black AM line and meandering PM line that has a sharp tooth in inner half, then curves around a large pale area to costa. Dark patch along costa at median band meets indistinct reniform spot. Inner margin usually has a dusky streak from base to inner ST area, and dusky apical dash reaches PM line. Veins thinly traced with black, lending a streaked appearance to FW. Some individuals have rusty brown in ST and inner median areas. Female lacks dusky streak along inner margin and has indistinct AM and PM lines; most obvious markings are usually veins, apical dash, and often pale bulge in PM area. **HOSTS** Catclaw blackbead, acacia, oak, and willow.

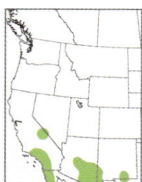

NOVEL GRAPHIC
Melipotis novanda

Uncommon
93-0880 (8609)

RANGE-WIDE

TL 21–25 mm Sexually dimorphic. Male has brown FW with a pale brown, slanted square in median area, connected at one corner to large pale patch with brown reniform spot at top and pale teeth along lower edge. Wavy double AM line is blackish and flares open at inner margin. Brown ST line is edged with black wedges above. Female is similar, but entire inner FW is washed brown, so only markings along costal side are clear. **HOSTS** Catclaw acacia.

LITOCALA MOTH

male

female

PERPENDICULAR
GRAPHIC

male

INDOMITABLE
GRAPHIC

male

female

male

male

male

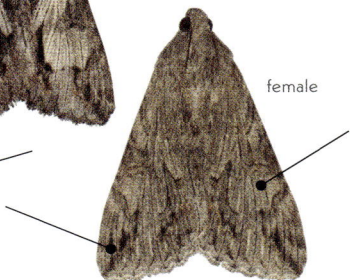

female

MERRY
GRAPHIC

male

female

actual size

NOVEL GRAPHIC

447

ROYAL POINCIANA GRAPHIC

Melipotis acontioides

Common

93-0881 (8610)

TL 22–27 mm Sexually dimorphic. Male has brown to brownish-gray FW with long, brown-edged, black line at inner PM line that curves to meet inner margin perpendicularly; occasionally, PM line may also cross wing to costa. Central FW from base to PM line is washed brownish, while costa and inner margin are gray to dusky, sometimes nearly black. ST area has veins traced with black and small black marks at apical dash. Female is more uniformly gray with black-traced veins and thin, vertical, black, S-shaped section of PM line that does not curve to touch inner margin. **HOSTS** Royal poinciana, and blue and Mexican palo verde.

ASHY GRAPHIC

Forsebia cinis

Very Common

93-0884 (8613)

TL 15–17 mm Sexually dimorphic. Male resembles Indomitable Graphic, but large, pale grayish patch below golden-brown AM band is diffuse and connected directly to band, rather than by a stem. Inner PM line is partially obscured by pale spot. Brown ST line is zigzagged. Female has a speckled peach wash across median area, uneven ST line is brownish orange, and ST area is gray. **HOSTS** Palo verde and other legumes.

DEDUCED GRAPHIC

Bulia deducta

Very Common

93-0885 (8614)

TL 16–20 mm Sexually dimorphic. Male resembles Indomitable Graphic, but brown to pale AM band is horizontal, and brown to pale PM area patch below is kidney-shaped without teeth. Upper basal area and thorax are grayish. Some individuals have grayish AM band, dusky median band, and rusty shading through PM and ST areas. Females are plain grayish, with a dusky patch at costal median area, pale ST line, and small black apical spot. HW has a yellow-orange spot in terminal area not found in *Melipotis*. **HOSTS** Mesquite. **NOTE** In southern CA, AZ, and western TX, Deduced Graphic overlaps with the identical Similar Graphic (*B. similaris*, not shown); ID can be made only by genital dissection or DNA analysis.

WONDERFUL GRAPHIC

Drasteria mirifica

Uncommon

93-0888 (8616)

TL 18–22 mm Resembles Streaked Graphic, but tan thorax has two brown stripes laterally, and upper half of AM band is usually light brown to tan. Pale-edged ST line is warm brown. ST area at costa has diffuse grayish patch. **HOSTS** Snow buckwheat.

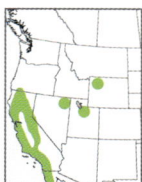

SCRUPULOUS GRAPHIC

Drasteria scrupulosa

Uncommon

93-0894 (8621)

TL 20–23 mm Speckled lilac-gray FW has reddish-purple to dusky purple shading below median line, lightening gradually to pale gray terminal area. Reddish to dusky, scalloped AM line is edged with pale yellow. Pale yellow PM line meanders through shading of lower wing, rounded in inner half, touching pale-outlined reniform spot, and toothed at outer half. Pale wavy ST line has reddish to dusky edge. **HOSTS** Unknown.

male

female

ROYAL POINCIANA
GRAPHIC

male

female

ASHY GRAPHIC

male

female

male

DEDUCED GRAPHIC

actual size

WONDERFUL GRAPHIC

SCRUPULOUS
GRAPHIC

INEPT GRAPHIC
Drasteria inepta

Uncommon
93-0895 (8622)

RANGE-WIDE

TL 18–22 mm Resembles noctuid arches and quakers. FW is variably gray, tan, or reddish orange, with indistinct, wavy, double AM line and narrowly spaced, indistinct, wavy, dark PM and pale ST lines bordering slightly darker (sometimes dusky) ST area. Dark median line sometimes visible at costa. Reniform spot is dusky or often indistinct and sometimes edged by small pale spots. Terminal line a row of dusky dots. **HOSTS** Unknown.

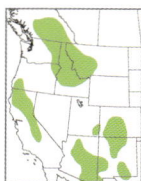

SANDY GRAPHIC
Drasteria sabulosa

Uncommon
93-0896 (8623)

RANGE-WIDE

TL 17–20 mm Hoary gray FW has strongly wavy, dusky, double AM line and meandering black PM line that forms a blocky or curling wave shape in inner half. Median area is paler gray. ST line is indistinct except for black wedges near costa. Some individuals have a tan median area, with white-outlined spiky reniform spot, and pale terminal area; these individuals are separable from other similar graphics by wavy AM line, indistinct or absent ST line, and thorax with brown transverse striping. **HOSTS** Unknown.

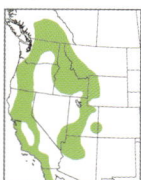

OCHRE GRAPHIC
Drasteria ochracea

Common
93-0898 (8626)

NORTH

SOUTH

TL 21–23 mm Brown to brownish-gray FW has curving, square-ended, dusky reniform spot and veins thinly traced with black. Curving black AM line is faintly doubled and visible only at costa. Outer ST area is dusky with strong black dashes. **HOSTS** Elder.

HANDSOME GRAPHIC
Drasteria edwardsii

Common
93-0899 (8627)

RANGE-WIDE

TL 17–20 mm Grayish-brown FW has curving pale AM line. Meandering pale PM line frames toothed pale PM area patch in outer half; inner half of line turns to touch pale-outlined, dark brown reniform spot and forms a triangular, dark brown ST area. ST line is indistinct with black wedges at costa. **HOSTS** Possibly sumac.

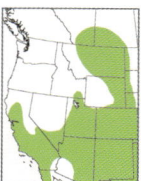

PALER GRAPHIC
Drasteria pallescens

Very Common
93-0900 (8628)

RANGE-WIDE

TL 16–20 mm Resembles Deduced Graphic, but AM line is slanting, inner PM area is triangular, and terminal area is grayish, with gray ST line bearing two black wedges at costa. Upper basal area and thorax are cinnamon. **HOSTS** Unknown.

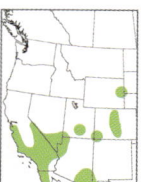

SMOKY ARCHES
Drasteria fumosa

Uncommon
93-0901 (8629)

RANGE-WIDE

TL 19–22 mm Resembles Paler Graphic, but AM line has a rounded tooth near inner margin, costal tooth of pale PM area patch is narrower and sharper, and thorax is pale gray to tan, with two short dark dashes at collar. AM band is often entirely cinnamon brown. **HOSTS** Unknown.

actual size

INEPT GRAPHIC

SANDY GRAPHIC

OCHRE GRAPHIC

HANDSOME GRAPHIC

PALER GRAPHIC

SMOKY ARCHES

DIVERGENT GRAPHIC
Drasteria divergens

Common

93-0902 (8630)

TL 17–23 mm FW has chocolate-brown basal and ST area and sloping tan-and-brown AM area. Double AM line is scalloped, with a large bulge near inner margin. Pale PM area beneath chocolate reniform spot is toothed below. ST area has black wedges near costa. HOSTS Possibly elder.

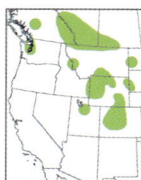

LITTLE ARCHES
Drasteria petricola

Uncommon

93-0905 (8631)

TL 14–16 mm Resembles Divergent Graphic, but lower edge of double AM line has small or absent bulges at inner margin, ST line is smoothly curving across central FW, and dark brown terminal line is straight. HOSTS Sweetvetch. NOTE Diurnal; adults may be seen at flowers during daytime.

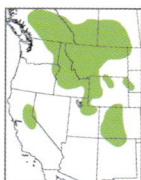

NORTHERN ARCHES
Drasteria hudsonica

Uncommon

93-0906 (8632)

TL 17–19 mm FW has chestnut AM band with yellowish edge, and dusky blackish, slightly wavy AM line and ST area. Whitish PM area patch has two very small teeth below and grayish reniform spot above. Inner PM line is slightly bulbous. Scalloped ST line is pale gray, with small black wedges above at costa. Upper basal area, thorax, and terminal area are hoary gray. Some individuals have entirely gray AM band. HOSTS Canadian buffalo-berry.

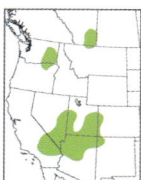

PERPLEXED ARCHES
Drasteria perplexa

Rare

93-0909 (8635)

TL 15–18 mm Resembles Shadowy Arches, but AM band is wider, with straight, pale AM line. Straight, indistinct ST line edging gray terminal area touches inner corner of gray PM area patch. HOSTS Unknown.

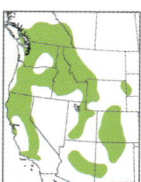

SHADOWY ARCHES
Drasteria adumbrata

Common

93-0910 (8636)

TL 15–18 mm FW has chocolate-brown ST area and double AM line that bulges medially in inner half. Large, light brown PM area patch has a tooth connecting to the median line. Thorax, upper basal area, AM band, and terminal band are light brown. HOSTS Bearberry; probably also *Vaccinium*.

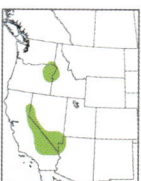

STRETCHED GRAPHIC
Drasteria stretchii

Uncommon

93-0911 (8637)

TL 15–18 mm Resembles Shadowy Arches, but PM patch is not usually connected to median line, and inner AM line lacks a distinct bulge. Medial edge of pale PM patch is often traced with white. HOSTS Unknown.

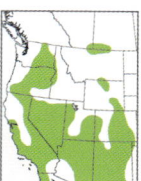

STREAKED GRAPHIC
Drasteria howlandii

Common

93-0913 (8639)

TL 17–19 mm Resembles Shadowy Arches, but PM area patch has pale streaks in inner half, AM line is more evenly wavy, and ST line is warm brown. Terminal area is shaded dusky gray. Separable from Wonderful Graphic by lack of stripes on thorax, gray AM band, and chocolate-brown ST line. HOSTS Buckwheat.

DIVERGENT GRAPHIC

LITTLE ARCHES

NORTHERN ARCHES

PERPLEXED ARCHES

actual size

SHADOWY ARCHES

STRETCHED GRAPHIC

STREAKED
GRAPHIC

STRAIGHT-BANDED WEDGE MOTH

Callistege intercalaris

Uncommon
93-0918 (8740)

TL 14–16 mm Tan to grayish FW has bold pattern of black triangles in central wing, separated by pale median band and central vein. Pale PM line curves at inner margin to connect to median band. ST area and line are layered with warm brown, black, dusky brown, and pale yellowish lines. HOSTS Unknown.

BROKEN-BANDED WEDGE MOTH

Callistege diagonalis

Uncommon
93-0919 (8741)

TL 14–16 mm Similar to Straight-banded Wedge Moth, but median band is disjointed. FW is overall grayer. HOSTS Unknown.

CERULEAN LOOPER

Caenurgina caerulea

Rare
93-0922 (8736)

TL 15–17 mm Resembles Clover Looper, but AM and PM bands lack brown shading inside. Upper outline of AM band is often thickened. AM band touches inner margin but does not connect to PM band. HOSTS Sweet pea, vetchling, and vetch.

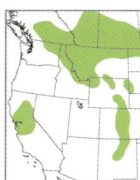

CLOVER LOOPER

Caenurgina crassiuscula

Common
93-0923 (8738)

TL 17–22 mm Overlaps greatly with Forage Looper, and some individuals cannot be visually identified to species with confidence. Pale brown to grayish-brown FW has broad AM and PM bands outlined in thin brown and shaded with brown gradient; basal edge of AM band is edged with pale tan to orange. Bands are variable in shape and extent, and may or may not connect near inner margin in both species. In Clover Looper, discal spot is usually (but not always) absent; when present, it is usually medium-brown and soft-edged. Dark spots at costal end of ST line are usually blackish and clearly separate. HOSTS Legumes and grass.

FORAGE LOOPER

Caenurgina erechtea

Very Common
93-0924 (8739)

TL 17–23 mm Overlaps greatly with Clover Looper, and some individuals cannot be visually identified to species with confidence. AM and PM bands are variable in shape and extent, and may or may not connect near inner margin in both species. In Forage Looper, discal dot is usually (but not always) present, and typically dark and distinct. Dark spots at costal end of ST line are usually slightly merged together; in some individuals, they may be brown or nearly absent. Some individuals have indistinct or faint AM and PM markings, a FW phenotype not generally seen in Clover Looper. HOSTS Clover and grass.

actual size

BROKEN-BANDED
WEDGE MOTH

STRAIGHT-BANDED
WEDGE MOTH

CERULEAN
LOOPER

CLOVER LOOPER

FORAGE
LOOPER

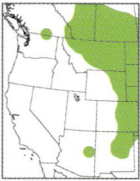

TOOTHED SOMBERWING

Euclidia cuspidea

Uncommon

93-0929 (8731)

TL 17–19 mm Nearly identical to Burnt Somberwing and many individuals may be best identified by range. Light gray to brown FW has a dark brown AM band with a large triangular tooth, and a blackish basal spot. PM line has a sharp, dark brown triangle that points toward dark subapical patch. Dark terminal line is zigzagged. Other features, such as the shape of the lower edge of the AM band, are variable and overlap with Burnt Somberwing. **HOSTS** Clover and other legumes.

BURNT SOMBERWING

Euclidia ardita

Uncommon

93-0930 (8732)

RANGE-WIDE

TL 15–19 mm Nearly identical to Toothed Somberwing, and many individuals may best be identified by range. Triangle on AM band averages more rounded and terminal line is usually absent or shallowly scalloped. **HOSTS** Deerweed.

BROWN-SPECKLED OWLET

Ptichodis ovalis

Rare

93-0936 (8753)

RANGE-WIDE

TL 15–18 mm Brown-speckled gray FW has smooth, thin, yellow-edged, brown AM and PM lines and thin brown reniform dash. ST line is row of indistinct dusky dots, with yellowish-brown veins in terminal area below. **HOSTS** Unknown.

PALE-LINED LOOPER

Caenurgia togataria

Common

93-0939 (8734)

RANGE-WIDE

TL 18–22 mm Tan to brownish-gray FW has similar pattern to Cerulean Looper, but lines are pale and often indistinct. Pale ST line is straight and well defined, sometimes accented with dark dots between veins. **HOSTS** Probably legumes.

ZALES and ALLIES

SUPERFAMILY Noctuoidea (93)

FAMILY Erebidae SUBFAMILY Erebinae TRIBE Omopterini; and SUBFAMILY Eulepidotinae

Small to large moths with long triangular wings that are usually held folded flat or slightly spread. These moths are brown or brownish gray, and most have complex patterns that provide camouflage against tree bark. Some of the zales can be quite variable within species and show subtle differences between species, which can make identification more challenging. Moonlight Azeta and Black-patched Panopoda belong to a separate subfamily but are similar in appearance and habit. Adults are nocturnal and will come to lights, and a number of species will also visit sugar bait.

GRAY-WINGED OWLET

Lesmone griseipennis

Uncommon

93-0973 (8654)

RANGE-WIDE

WS 30–38 mm Sexually dimorphic. Dark brownish-gray FW has thin, dark brown AM and PM lines; AM line is curving, and PM line is straight, rounding (female) or angling (male) basally to meet costa, and edged thinly with pale tan below. Pale ST line is wavy, sometimes indistinct. Indistinct reniform spot has small white dots at outer edges. Costa has tiny white dots between PM line and apex. Males have dark brown shading basal to PM and ST lines; females do not. **HOSTS** Unknown.

TOOTHED SOMBERWING

BURNT SOMBERWING

BROWN-SPECKLED OWLET

PALE-LINED
LOOPER

actual size

male

female

actual size

male

GRAY-WINGED OWLET

457

THE MIME
Heteranassa mima

Very Common
93-0982 (8659)

TL 14–18 mm Mottled brown to gray FW has bold black PM line that curves around lower edge of kidney-shaped reniform spot. AM line is double with small bulge medially at inner margin. Reniform spot is sometimes partially or entirely white. **HOSTS** Mesquite and acacia.

SPOTTED TOXONPRUCHA
Toxonprucha pardalis

Rare
93-0996 (8670)

WS 25–29 mm Brown to grayish wings have multiple wavy dusky lines in median area between pale-edged dark AM and PM lines. ST area is lilac gray, and thick ST line is warm brown. Apex of FW has blackish eyespot. Thin black terminal line is scalloped. **HOSTS** Catclaw acacia.

WINGED TOXONPRUCHA
Toxonprucha volucris

Uncommon
93-0998 (8672)

WS 25–29 mm Grayish-brown FW has wavy brown AM and PM lines and wavy double median line that is blackish at costa and inner margin of HW. Indistinct jagged ST line is edged above with dusky shading. Reniform spot sometimes has white spots along upper edge. Thin black terminal line is scalloped. **HOSTS** Catclaw acacia.

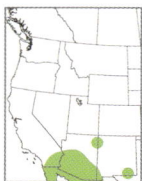

SUDDEN TOXONPRUCHA
Toxonprucha repentis

Rare
93-0999 (8673)

WS 25–29 mm Brown FW has bold black zigzag AM line and PM line that is zigzagged in inner half, touches dusky double median line, then curves widely to meet costa in a rounded W shape. PM area within W shape usually has a thin or bold white reniform zigzag. Pale ST line is often an indistinct row of chevrons, with a dusky patch at costa. **HOSTS** Catclaw acacia.

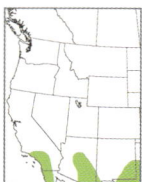

CRUEL TOXONPRUCHA
Toxonprucha crudelis

Common
93-1001 (8674)

WS 20–24 mm Brown to dusky wings have diffuse, dusky, double median band and thin, blackish, zigzag AM and PM lines. PM line is broadly edged with tan and on FW curves in a shallow W shape around indistinct, pale, reniform zigzag. ST line is a row of pale dots with a pale-edged blackish spot at costa. **HOSTS** Catclaw acacia.

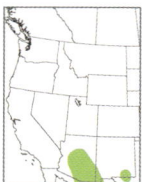

SHADOWED ZALEOPS
Zaleops umbrina

Uncommon
93-1004 (8677)

WS 30–33 mm Dimorphic. Light form has brown FW with disjointed, dark brown, double AM band that crosses thorax, and dark brown PM area bounded below by wavy black PM line that runs from inner median line to apex and is split into Y shape at costa by large brown to pale tan reniform patch. HW has a dark brown ST band. Dark form is chocolate brown from AM line to ST line, with a very broad tan costa. **HOSTS** Catclaw acacia.

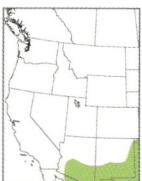

REDDISH MATIGRAMMA
Matigramma rubrosuffusa

Rare
93-1009 (8680)

WS 34–40 mm Light brown to grayish FW has indistinct, dusky, double AM and median lines and broad, faint, reddish washes along veins in central and inner wing. Indistinct reniform spot sometimes has white dots at corners. Pale ST line has dusky edge above. **HOSTS** Scrub oak and whitethorn acacia.

actual size

THE MIME

SPOTTED TOXONPRUCHA

WINGED
TOXONPRUCHA

SUDDEN
TOXONPRUCHA

CRUEL
TOXONPRUCHA

SHADOWED ZALEOPS

REDDISH MATIGRAMMA

BROWNISH MATIGRAMMA

Matigramma emmilta

Uncommon
93-1011 (8680.2)

RANGE-WIDE

WS 34–40 mm Resembles Reddish Matigramma, but AM and median lines are darker and more crisply defined, particularly at inner margin. Terminal area is lighter, with brown along veins. Brown-filled dusky reniform spot is thick and clearly visible. **HOSTS** Rabbitbrush.

LUNATE ZALE

Zale lunata

Very Common
93-1023 (8689)

NORTH

SOUTH

WS 40–55 mm Highly variable. Brown, tan, or dusky FW always has tiny white dots at inside of dusky reniform spot. Double PM line bends sharply at midpoint, then flares open at costa; inner half is edged below with dusky shading. Double AM line is often indistinct, but when visible, section above tiny black discal dot is straight. Median area is crossed by many faint wavy lines. Some individuals have a tan central wing and dusky costa; others are uniformly brown. Some individuals have grayish or white AM and/or terminal bands or just a thin silvery edging to lower PM line. **HOSTS** Various deciduous trees and woody plants, including apple, cherry, oak, plum, and willow.

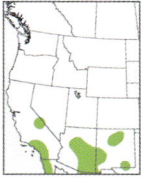

DESERT PEAKS ZALE

Zale insuda

Uncommon
93-1031 (8696)

RANGE-WIDE

WS 31–48 mm Sexually dimorphic. Resembles Lunate and Colorful Zales, but PM line lacks dark shading at costa, and ST area is uniformly dusky or without shading. Reniform spot is concolorous with FW or sometimes narrowly tan, and it lacks white dots. AM line is usually distinct, and section above black discal dot is toothed. Females are darker than males. **HOSTS** Unknown.

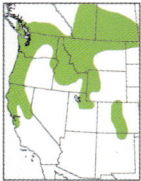

COLORFUL ZALE

Zale minerea

Uncommon
93-1032 (8697)

RANGE-WIDE

WS 37–50 mm Highly variable. Resembles Lunate Zale, but dusky reniform spot is typically well defined and does not usually have white dots, dusky subapical patch on costa has thick, black lower edging, and median area lacks cross-lines or has only double median line. AM line is often visible; section above black discal dot is relatively straight. **HOSTS** Chestnut, hawthorn, oak, and probably others.

DUSKY ZALE

Zale termina

Uncommon
93-1047 (8712)

RANGE-WIDE

WS 32–38 mm Gray FW has double AM and PM lines filled faintly with brown; PM line flares at costa. Lower edge of indistinct kidney-shaped reniform spot is edged with white. Double median line sometimes visible on paler individuals. **HOSTS** Oak and chinquapin.

SOUTHWESTERN ZALE

Zale colorado

Rare
93-1051 (8715)

RANGE-WIDE

WS 35–42 mm Variable. FW has dusky basal area and double AM and PM lines with dusky or dark brown fill. Dusky reniform spot is not edged with white. Median area is variably tan, brown, dusky, or grayish or a gradient combination. Terminal area is usually grayish but sometimes dusky. **HOSTS** Oak.

**BROWNISH
MATIGRAMMA**

LUNATE ZALE

female

male

actual size

DESERT PEAKS ZALE

COLORFUL ZALE

DUSKY ZALE

**SOUTHWESTERN
ZALE**

461

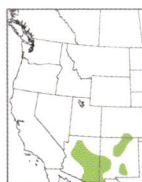

LOCUST UNDERWING
Euparthenos nubilis

Uncommon
93-1055 (8719)

TL 30–37 mm Resembles *Catocala* underwings. Gray to grayish-brown FW has darker basal area with light gray half-circle at inner margin. AM band and rounded PM area patch are marbled light gray to whitish. Thin, wavy, pale ST line cuts through blackish apical dash. Orange HW has thick black basal, AM, PM, and terminal bands and scalloped orange fringe. **HOSTS** Primarily black locust.

MOONLIGHT AZETA
Azeta schausi

Rare
93-1079 (8576)

RANGE-WIDE

WS 38–42 mm Warm brown to purplish-red FW has kidney-shaped, olive reniform spot, often with adjacent pale area at costa; small, round, olive orbicular spot sometimes also present. Wavy AM, median, and PM lines are thin and light brown on FW; only blackish PM line present on HW. ST line is a row of black dots, sometimes edged with olive. **HOSTS** Sonoran indigo.

BLACK-PATCHED PANOPODA
Panopoda rigida

Uncommon
93-1092 (8590)

RANGE-WIDE

WS 39–45 mm Orange-brown FW has brown, straight PM and scalloped ST lines, shaded faintly purplish between. Many individuals have a large black spot at inner margin between PM and ST lines and, often, black dots at tips of ST line scallops. Orbicular spot is small and brown; kidney-shaped reniform spot is a brown outline. AM line is often indistinct. **HOSTS** Unknown.

MARATHYSSAS and PAECTES
SUPERFAMILY Noctuoidea (93)
FAMILY Euteliidae SUBFAMILY Euteliinae

Small distinctive moths that often rest with the abdomen curled above the wings. Marathyssas tightly roll their wings and hold them at an angle from the body against the substrate. Paectes have long broad wings that they do not roll. Adults are nocturnal and will regularly come to lights.

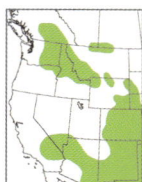

DARK MARATHYSSA
Marathyssa inficita

Uncommon
93-1103 (8955)

RANGE-WIDE

TL 14–17 mm Rests with wings tightly rolled and abdomen raised. Gray to brownish FW has broad grayish median area and rusty bands in basal and ST areas. AM and PM lines are blackish, sometimes indistinct. Base of abdomen has a narrow black band with grayish border. **HOSTS** Sumac and poison ivy.

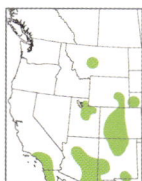

BARRENS PAECTES
Paectes abrostolella

Uncommon
93-1108 (8959.1)

RANGE-WIDE

TL 12–14 mm Gray FW has strongly curving, double, black AM line around pale gray to tan basal patch. Double, black PM line is evenly curving in inner half and sharply pointed near costa. Brownish to gray ST area has paler apical patch and two thin black dashes below PM line point. Small pale orbicular spot is often indistinct or absent. **HOSTS** Sumac.

LOCUST UNDERWING

MOONLIGHT
AZETA

actual size

BLACK-PATCHED
PANOPODA

DARK
MARATHYSSA

actual size

BARRENS PAECTES

463

HIEROGLYPHIC MOTH

SUPERFAMILY Noctuoidea (93)
FAMILY Nolidae SUBFAMILY Diphtherinae

This monogeneric subfamily contains a single species, the Hieroglyphic Moth, so named for the complex, bold patterning on the forewings. It is primarily an eastern species; but also found in the mountains south of Tucson, AZ, where it can be common. The species is nocturnal and will occasionally come to lights.

HIEROGLYPHIC MOTH

Diphthera festiva   **Uncommon**   RANGE-WIDE
93-1120.5 (8560)

TL 20–24 mm Unmistakable. Yellow-orange FW has metallic blue, double PM line and three rows of black spots in ST/terminal area. Blue-black costal streak flares inward at median line. Thorax has three thick black stripes. **HOSTS** Chocolate weed, pyramid flower, soybean, clover, mimosa, and mallow.

NOLAS and MIDGETS

SUPERFAMILY Noctuoidea (93)
FAMILY Nolidae SUBFAMILIES Nolinae and Chloephorinae

Small gray moths with understated markings. The meganolas and nola have pointed wings that form a notched triangle when folded at rest. The garella and midgets are more rectangular in appearance. Adults are nocturnal and will come to lights.

CONFUSED MEGANOLA

Meganola minuscula   **Uncommon**   RANGE-WIDE
93-1121 (8983)

TL 8–12 mm Light gray FW has double, black PM line with dots at veins creating teeth (sometimes visible only as dots), which bulges around grayish white reniform spot. Costa has dark C-shaped lines at basal, median, and ST areas, often filled with gray shading. Dusky AM line is faint or indistinct. **HOSTS** Oak. **NOTE** The rare Dusky Meganola (*M. fuscula*, not shown) is similar but FW is darker gray and costal patches are reduced or absent, reniform spot is gray and indistinct and lacks dusky shading above, and AM and PM lines are solid and smoothly curving, without notable teeth. It is found in coastal CA, south and central AZ, and north-central NM.

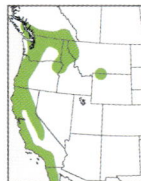

CEANOTHUS NOLA

Nola minna   **Common**   RANGE-WIDE
93-1133 (8993)

TL 11–12 mm Speckled, pale gray FW has black costa in basal area. Outer half of black AM line is thickened. PM line is black dots, sometimes most visible at inner margin. ST line is often indistinct, but sometimes black. Long palps project in front of head. **HOSTS** Alder, oak, ceanothus, false-buckthorn, willow, and tobacco.

SMALL GARELLA

Garella nilotica   **[IN] Uncommon**   RANGE-WIDE
93-1141 (8974)

TL 8–9 mm Highly variable. FW nearly always has a slanting, reddish to brownish patch in outer median/PM area, with a pale spot at inner corner, and a curving, black basal line. Thin, wavy, black AM line is sometimes indistinct. ST area is shaded dusky along thin white ST line. Often has a solid-colored AM band that turns onto inner margin to ST area. **HOSTS** Black almond and willow.

HIEROGLYPHIC MOTH

actual size

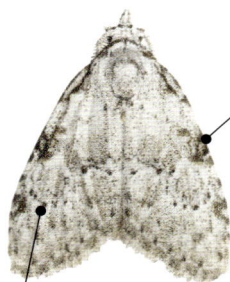

CONFUSED
MEGANOLA

actual size

CEANOTHUS NOLA

SMALL GARELLA

465

FRIGID OWLET

Rare

RANGE-WIDE

Nycteola frigidana — 93-1142 (8975)

TL 12–14 mm Variable. Resembles Gray Midget but rusty reniform spot is smaller, roundish, and has black scales at center. Basal and ST areas are usually more uniformly light to medium gray, other than dusky marks along ST line. Often has whitish scaling within double AM and/or PM lines, and/or a whitish patch in central basal area. Some individuals have a blackish band adjacent to thorax from costa to inner margin. Thorax rarely has brown and never extensively. **HOSTS** Poplar and willow.

COLUMBIA OWLET

Uncommon

RANGE-WIDE

Nycteola columbiana — 93-1143 (8976)

TL 13–16 mm Resembles Gray Midget, but markings are darker and more bold, and brown reniform spot is round with black scales at center. Median area is more extensively shaded dusky. Central AM area has a roundish black spot encircled by light gray. Thorax has two solid black spots dorsally, or black-outlined spots shaded dark gray inside, and is rarely marked with brown. Some individuals have a slight brownish wash in median area, but can be separated from Rusty Owlet by bold markings, round reniform spot, and thorax pattern. **HOSTS** Unknown; likely poplar or willow.

RUSTY OWLET

Uncommon

RANGE-WIDE

Nycteola n. sp. — 93-114x (8976.x)

TL 13–16 mm Light gray FW has wavy AM and PM lines that are double (outer line sometimes faint) and pale-edged brown reniform spot. Basal area has dusky shading along top edge to inner margin, sometimes hidden beneath thorax scales when wings are folded; thin, short basal dash is usually still visible. Median area has dusky shading at costa and a streak near inner margin. Thorax has two hollow circles dorsally and often a brown patch at posterior. Some individuals have a gray checkered pattern in basal and/or ST area. **HOSTS** Unknown; likely poplar or willow.

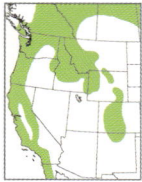

GRAY MIDGET

Uncommon

RANGE-WIDE

Nycteola cinereana — 93-1144 (8977)

TL 13–16 mm Light gray FW has wavy AM and PM lines that are double (outer line sometimes faint) and pale-edged, kidney-shaped, brown reniform spot. Basal area has dusky shading along top edge to inner margin, sometimes hidden beneath thorax scales when wings are folded; thin, short basal dash is usually still visible. Median area has dusky shading at costa and a streak near inner margin. Rare individuals have a blackish band adjacent to thorax from costa to inner margin. Thorax has two hollow circles dorsally and often a brown patch at posterior. Some individuals have a gray checkered pattern in basal and/or ST area. Individuals with a brownish wash and brown thorax can be separated from Rusty Owlet by thorax pattern. **HOSTS** Poplar.

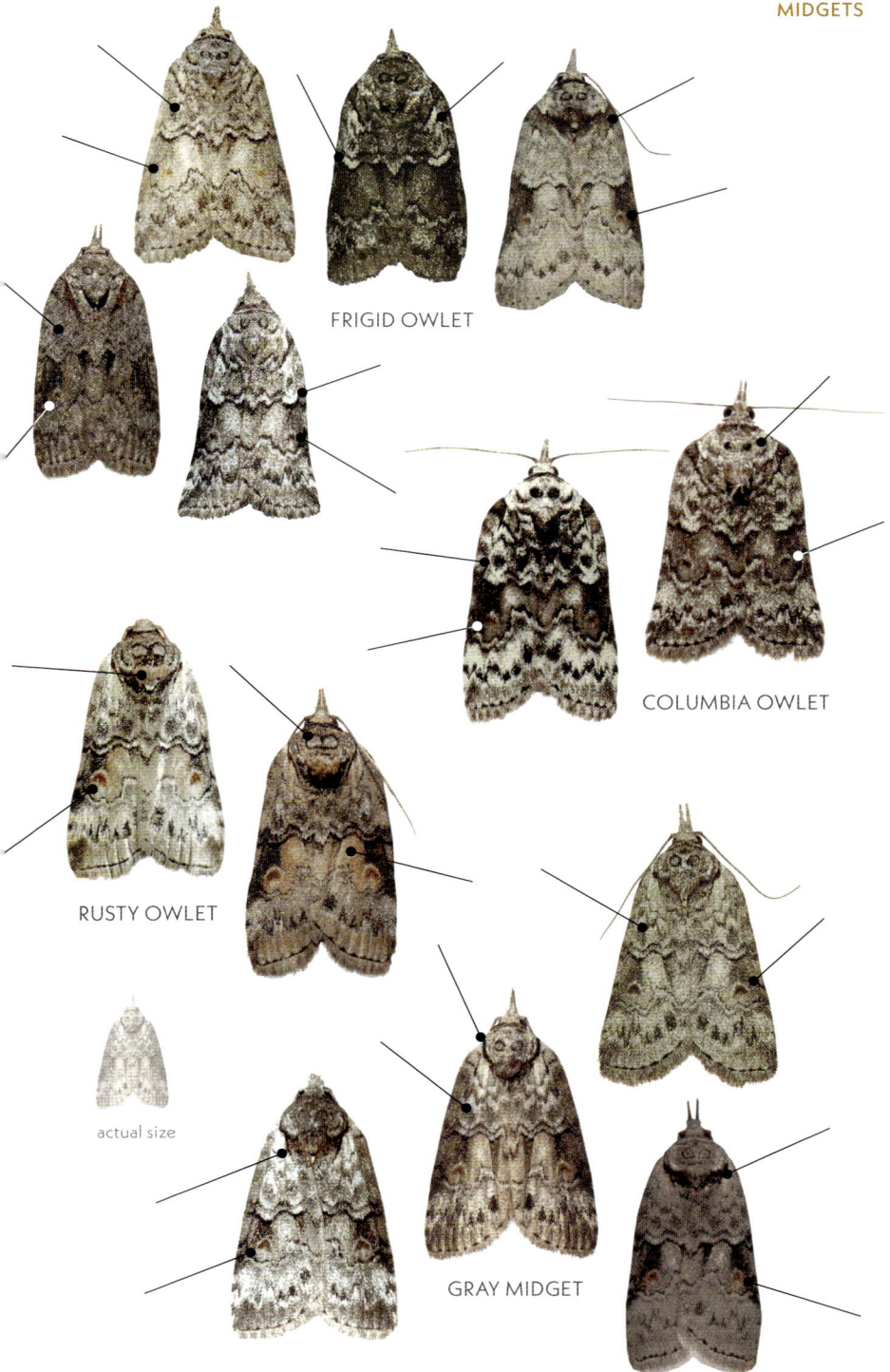

FRIGID OWLET

COLUMBIA OWLET

RUSTY OWLET

actual size

GRAY MIDGET

PALM BORERS
SUPERFAMILY Noctuoidea (93)
FAMILY Noctuidae SUBFAMILY Dyopsinae

A monogeneric subfamily of brown to pale tan moths with a small eyespot on the hindwing; this is the only species in the West. Caterpillars feed on palms and may be a minor household nuisance, as they sometimes come indoors and use carpet or other natural fibers to construct their cocoon. Adults are nocturnal and will come to lights.

PALM FLOWER MOTH
Litoprosopus coachella

Common

93-1160.8 (8558)

RANGE-WIDE

TL 18–30 mm Pearly, ivory to pale tan FW has thin brown AM and PM lines most visible at costa. At anal angle of HW are three white dashes, innermost two set in large black spots. **HOSTS** California fan palm.

LOOPERS
SUPERFAMILY Noctuoidea (93)
FAMILY Noctuidae SUBFAMILY Plusiinae

Distinctive medium-sized moths that typically rest with wings tented over the abdomen. The thorax and inner margin have pronounced scale tufts, and most species have a white or satiny pale mark ("stigma") in the central median area. Cabbage, Soybean, and Bilobed Loopers migrate north every spring to establish summer populations, and a few species are habitually nomadic in the fall, often appearing well north of their usual range. Caterpillars of some species can be crop pests. Adults are nocturnal and will come to lights, and a few species may be encountered at flowers during daytime.

SPECTACLED NETTLE MOTH
Abrostola urentis

Rare

93-1162 (8881)

RANGE-WIDE

TL 17–19 mm Mottled brownish-gray FW has thin, curving, black AM and PM lines edged outwardly with grayish brown, as well as large, thinly outlined (sometimes incompletely) orbicular and reniform spots. Thorax has a large dorsal tuft. **HOSTS** Stinging nettle.

CABBAGE LOOPER
Trichoplusia ni

Very Common

93-1168 (8887)

NORTH

SOUTH

TL 17–19 mm Mottled grayish-brown FW has indistinct, tan-edged, black PM line and jagged, thin, black ST line edged with diffuse grayish shading. Silvery stigma is an inverted U-shaped claviform spot adjacent to a solid round subreniform spot; on some individuals, the two parts are joined. **HOSTS** Generalist on herbaceous plants, including asparagus, cabbage, corn, tobacco, and watermelon. **NOTE** Also known as Ni Moth.

SOYBEAN LOOPER
Chrysodeixis includens

Common

93-1170 (8890)

RANGE-WIDE

TL 16–20 mm Bronzy-brown FW has dark brown central median area; golden stigma has inverted U-shaped claviform spot adjacent to large round subreniform spot. Wavy golden AM and PM lines are edged with warm brown. **HOSTS** Generalist on herbaceous plants, including goldenrod, lettuce, soybean, and tobacco.

actual size

PALM FLOWER MOTH

SPECTACLED NETTLE MOTH

CABBAGE LOOPER

actual size

SOYBEAN LOOPER

LESSER BEAN LOOPER

Rare

RANGE-WIDE

Autoplusia egenoides   93-1174 (8893)

TL 18–21 mm Resembles Large Looper, but thin white stigma is thorn-shaped and terminal area has reddish-brown patch in inner half; ST area often shaded concolorous with median area. **HOSTS** Unknown.

GRAY LOOPER

Uncommon

RANGE-WIDE

Rachiplusia ou   93-1176 (8895)

TL 17–22 mm Resembles Cabbage Looper, but gray FW is less mottled and PM line is double and filled with tan. U-shaped claviform spot appears to extend diagonally to costa. ST area often shaded dusky, with black dash at midpoint. **HOSTS** Generalist on herbaceous plants, including clover, mint, corn, and cosmos.

ALFALFA LOOPER

Very Common

NORTH

Autographa californica   93-1193 (8914)

SOUTH

TL 15–22 mm Resembles Cabbage Looper, but U-shaped claviform spot connects to teardrop-shaped subreniform spot, and costal ST area is shaded pale gray. Separable from Gray Looper by stigma, and is usually overall more brown. **HOSTS** Broad generalist. **NOTE** Often seen at flowers during daytime.

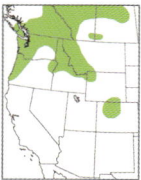

WAVY CHESTNUT Y

Rare

RANGE-WIDE

Autographa mappa   93-1194 (8912)

TL 20–22 mm Warm brown FW has golden AM and PM lines edged with brown and then outwardly with pink, as well as thin pink terminal line. U-shaped claviform and round subreniform spots form white stigma. Round orbicular and kidney-shaped reniform spots have thin black and yellowish outlines. **HOSTS** Blueberry and nettle, often in bogs and fens.

DELICATE SILVER Y

Rare

RANGE-WIDE

Autographa pseudogamma   93-1196 (8913)

TL 18–20 mm Resembles Alfalfa Looper, but ST area lacks black dash, and PM line has a single large tooth (rarely also a second tiny one) at warm brown spot of inner median area. Inner AM line is evenly curving, and orbicular spot is round or oval. **HOSTS** Unknown.

CARAMEL LOOPER

Uncommon

RANGE-WIDE

Autographa corusca   93-1201 (8918)

TL 16–18 mm Pinkish-gray FW has scalloped white AM, PM, and ST lines unevenly edged with black. White stigma is inverted Y-shaped claviform spot connected to elongate subreniform spot. Orbicular and reniform spots are thin, white outlines that are often fragmented and are thickly edged with black. Inner median area and portions of ST and terminal areas are shaded grayish brown. **HOSTS** Alder.

SHADED GOLD SPOT

Uncommon

RANGE-WIDE

Autographa metallica   93-1203 (8917)

TL 18–20 mm Light brown FW has golden-brown inner median area. Golden stigma is a single large blobby spot. Yellowish AM and PM lines are often indistinct. A dusky brown band runs from apex to inner margin of ST area; central portion is metallic gold. **HOSTS** Unknown.

LESSER BEAN LOOPER

GRAY LOOPER

ALFALFA LOOPER

WAVY CHESTNUT Y

DELICATE SILVER Y

actual size

CARAMEL LOOPER

SHADED GOLD SPOT

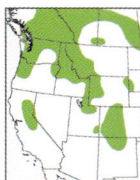

LARGE LOOPER
Autographa ampla

Uncommon
93-1204 (8923)

TL 21–23 mm Gray FW has dark brown inner median area, shaded blackish toward central wing. Thin white stigma is rounded, and reniform spot is a thin black outline. Pale zigzag ST line often indistinct except blackish section above black apical patch. Thoracic tufts are brown. HW is brownish gray. Separable from Epigaea Looper by stigma shape and apical patch, and from Saddled Looper by apical patch and HW. HOSTS Alder, birch, poplar, willow, and other deciduous trees.

BILOBED LOOPER
Megalographa biloba

Very Common
93-1209 (8907)

TL 19–20 mm Warm brown FW has reddish-brown median area with white sunglasses-shaped stigma spot and small white C-shaped reniform spot. Thin white AM and PM lines often indistinct. HOSTS Generalist on herbaceous plants, including alfalfa, cabbage, and tobacco.

SPRUCE FALSE LOOPER
Syngrapha viridisigma

Uncommon
93-1213 (8929)

TL 20–22 mm Mottled gray FW has variable mint-green stigma; claviform spot may be hollow or solid, and claviform and subreniform spots may be either discrete or connected in an M shape. Double AM and PM lines are often indistinct. Black ST line is irregularly pointed. HOSTS Coniferous trees, including spruce, fir, and pine.

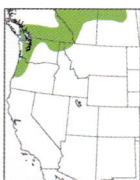

EPIGAEA LOOPER
Syngrapha epigaea

Rare
93-1215 (8927)

TL 21–23 mm Resembles Large Looper, but white stigma has small teardrop projection, and FW median area is grayish at inner margin, often with warm brown shading along PM line. Apex lacks black patch. HW is dusky brown. HOSTS Blueberry; also sheep laurel and some other shrubs later in season.

SADDLED LOOPER
Syngrapha orophila

Rare
93-1221 (8930)

TL 20–22 mm Resembles Large Looper, but FW apex lacks black patch and thoracic tuft is gray. Separable from Epigaea Looper by U shape of stigma and dark brown median area. HW is yellow orange, with black terminal band. HOSTS Blueberry.

MOUNTAIN BEAUTY
Syngrapha ignea

Rare
93-1223 (8949)

TL 16–19 mm Hoary brown FW has warm brown median patch with curving finger-like stigma that is notched at costal end, as well as small reddish-brown patch at inner PM line. AM and PM lines are gold, and terminal band is hoary gray. Front of thorax is warm brown; brown thoracic and AM tufts are tipped white. Can resemble Delicate Silver Y but easily separable by stigma shape, terminal line, and thorax. HOSTS Blueberry and willow; possibly others. NOTE Primarily diurnal and crepuscular; frequently encountered at flowers.

LARGE LOOPER

BILOBED LOOPER

actual size

SPRUCE FALSE LOOPER

EPIGAEA LOOPER

SADDLED LOOPER

MOUNTAIN BEAUTY

473

SALT-AND-PEPPER LOOPER
Syngrapha rectangula

Uncommon
93-1227 (8942)

TL 17–19 mm Striking FW has black median area with bold, white, Y-shaped stigma. Basal area is bold white. Black ST line is edged diffusely with black and cuts through marbled white and gray ST/terminal area. Thoracic tuft is brown. **HOSTS** Coniferous trees, including fir, hemlock, pine, and spruce.

ROCKY MOUNTAIN LOOPER
Syngrapha angulidens

Rare
93-1228 (8943)

TL 17–19 mm Resembles Pacific Mountain Looper, but double AM and PM lines are filled with gray (rarely whitish) in inner half, white to ivory stigma is separate (rarely touching slightly), upper half of ST area is light gray from costa to inner margin, and ST line is more jagged. **HOSTS** Spruce, fir, white pine, western hemlock, and other needled conifers.

PACIFIC MOUNTAIN LOOPER
Syngrapha celsa

Uncommon
93-1229 (8944)

TL 16–20 mm Gray FW has slightly darker median area and double AM and PM lines filled with tan to yellowish white in inner half. Golden to yellowish-white stigma has U-shaped claviform and small round subreniform spots that are touching or fused. Dark gray ST area has light gray shading in upper half from midpoint to costa. Thin black ST line is jagged. **HOSTS** Spruce, fir, white pine, western hemlock, and other needled conifers.

CELERY LOOPER
Anagrapha falcifera

Common
93-1234 (8924)

TL 18–22 mm Light brown to gray FW is brown across inner median area. White AM line curves evenly from inner margin to connect to narrow finger-like stigma spot. PM line indistinct; brown ST line usually darkest along apical dash. **HOSTS** Herbaceous plants, including beet, celery, clover, corn, and dandelion. **NOTE** May be encountered at flowers during daytime.

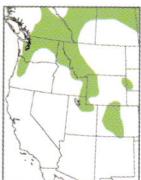

PUTNAM'S LOOPER
Plusia putnami

Rare
93-1236 (8950)

TL 18–20 mm Golden-orange FW has bold white stigma spots that are usually separate or barely touching, rarely fused; triangular claviform spot is larger than oval subreniform spot and is almost always cut through by central vein. Inner median and outer basal areas are washed paler. Brown cross-lines are crossed by thin brown-traced veins. Long brown apical dash borders pale subapical patch. In rare individuals, claviform spot does not clearly cross central vein, but size of spots and specimen location should be useful clues to identification. **HOSTS** Bur reed, grasses, and sedges in damp woodlands.

WEST COAST PLUSIA
Plusia nichollae

Uncommon
93-1237 (8951)

TL 15–20 mm Nearly identical to Putnam's Looper, but white stigma spots average smaller, and claviform spot does not cross central vein, or it crosses only as a tiny white dot. White subreniform spot is often more elongate than in Putnam's. Many individuals are a paler orangey tan. **HOSTS** Unknown.

SALT-AND-PEPPER
LOOPER

ROCKY MOUNTAIN
LOOPER

PACIFIC MOUNTAIN
LOOPER

CELERY LOOPER

actual size

PUTNAM'S
LOOPER

WEST COAST PLUSIA

475

TRIPUDIAS, COBUBATHAS, and GLYPHS

SUPERFAMILY Noctuoidea (93)
FAMILY Noctuidae SUBFAMILIES Bagisarinae, Cobubathinae, and Eustrotiinae

Small triangular moths that rest with wings curved or slightly tented over the body. Most of the tripudias and cobubathas are boldly marked with contrasting patches and bands. The Eight-spot is nomadic in autumn and will occasionally wander north of its regular range. Adults of all species will readily come to lights at night.

PALE YELLOW MALLOW MOTH
Bagisara buxea

Uncommon
93-1244 (9172)

RANGE-WIDE

TL 12–14 mm Tan FW has thin, slightly wavy, brown median, PM, and ST lines that cross from inner margin roughly straight, then angle sharply basally to meet costa. Reniform spot is a thin brown crescent. **HOSTS** Globemallow.

EIGHT-SPOT
Amyna stricta

Rare
93-1253 (9070)

RANGE-WIDE

TL 12–14 mm Grayish-brown, tan, or reddish FW is lightly speckled with white scales and has large, round, white or orangey-tan reniform spot. Zigzag dusky AM and PM lines are tipped with white specks and occasionally have sections of grayish edging near costa or inner margin. Costa has five or six small white spots along length, sometimes as short dashes at AM and PM lines. **HOSTS** Amaranth and croton.

BELTED TRIPUDIA
Tripudia balteata

Uncommon
93-1264 (9005)

RANGE-WIDE

TL 6–8 mm Brown to gray FW has broad orange to tan AM band edged by thin black median line and grayish basal area. PM area may be maroon or concolorous with ST area, and thin black PM line curves smoothly to frame round pale patch. Basal area is usually paler than lower wing; rarely, it is nearly concolorous with AM band. **HOSTS** Unknown.

PLAYFUL TRIPUDIA
Tripudia luda

Rare
93-1265 (9006)

RANGE-WIDE

TL 6–8 mm Resembles Belted Tripudia, but FW is usually darker and lacks round pale patch outlined by PM line. Most individuals have entire basal half of FW golden tan above thick black median line. Costa has reddish patches at PM area and apex. Terminal band is dark gray. **HOSTS** Unknown.

BICOLORED TRIPUDIA
Tripudia dimidata

Rare
93-1266 (9007)

RANGE-WIDE

TL 7–8 mm Resembles Playful Tripudia, but outer margin is edged with pale yellow and costa of basal area has large brown (sometimes gray) patches. Thin black PM line forms rounded loop with fill concolorous to ST area. Gray fringe has dusky sections at midpoint, apex, and anal angle. **HOSTS** Unknown.

PALE YELLOW
MALLOW MOTH

EIGHT-SPOT

BELTED TRIPUDIA

actual size

PLAYFUL TRIPUDIA

BICOLORED TRIPUDIA

LUXURIOUS TRIPUDIA
Tripudia luxuriosa

Uncommon
93-1267 (9008)

RANGE-WIDE

TL 7–8 mm Mottled brown to grayish-brown FW has narrow brownish- to grayish-tan AM band and large, round, pale yellow orbicular spot. Lower half of FW has mottled appearance. Outer margin is edged with pale yellow, and long gray fringe has dusky sections at midpoint, apex, and anal angle. **HOSTS** Unknown.

DARK-BANDED COBUBATHA
Cobubatha lixiva

Common
93-1275 (9014)

RANGE-WIDE

TL 7–8 mm FW has bold chocolate-brown AM band edged thinly in white, gray basal area, and tan to brownish lower half. Wavy ST line usually indistinct except for large chocolate-brown patch at costa. Dark brown terminal line is of uneven thickness. In form "basicinerea" band bulges at inner margin and/or is broken toward costa and terminal line is pale yellowish. **HOSTS** Unknown.

SHARP-BANDED COBUBATHA
Cobubatha orthozona

Rare
93-1276 (9017)

RANGE-WIDE

TL 9–11 mm Resembles Dark-banded Cobubatha, but AM band is narrower and has triangular point at midpoint. In some individuals, band is broken or incomplete near inner margin. **HOSTS** Unknown.

DIVIDED COBUBATHA
Cobubatha dividua

Uncommon
93-1277 (9018)

RANGE-WIDE

TL 7–9 mm FW has narrow, dark brown AM band, brown basal area, and straw-colored lower half. ST line indistinct except for brownish patch at costa. Reniform spot slightly brownish. **HOSTS** Shrimp plant.

PINK-BANDED GLYPH
Phoenicophanta bicolor

Rare
93-1281 (9028)

RANGE-WIDE

TL 8–10 mm Goldenrod-yellow FW has bright pink median and terminal bands and pink costa. Abdomen and thorax colors continue yellow and pink pattern from FW. **HOSTS** Unknown.

PALE GLYPH
Protodeltote albidula

Common
93-1291 (9048)

RANGE-WIDE

TL 12–14 mm Peppery, pale tan FW has wavy whitish AM, PM, and ST lines edged narrowly with tan. Indistinct orbicular and reniform spots are outlined in white. Markings may be difficult to see against pale wing, particularly on worn individuals, and read simply as a mottled tan FW. **HOSTS** Grasses.

SWIFT BROWN
Ozarba propera

Uncommon
93-1303 (9031)

RANGE-WIDE

TL 7–8 mm Dusky FW has brownish-tan ST/terminal area and whitish to tan claviform, orbicular, and reniform spots. Wavy dark AM line is edged in tan. **HOSTS** Unknown.

DARK-BANDED
COBUBATHA

LUXURIOUS TRIPUDIA

actual size

SHARP-BANDED COBUBATHA

DIVIDED COBUBATHA

PINK-BANDED GLYPH

PALE GLYPH

SWIFT BROWN

BIRD-DROPPING MOTHS
and **THIMBLES**

SUPERFAMILY **Noctuoidea (93)**
FAMILY **Noctuidae** SUBFAMILY **Acontiinae**

Small moths with long squared wings. Many species are white with mottled gray or olive patches, resembling at first glance small bird droppings, which the moths use as a form of camouflage when resting on foliage in the open. Though members are found throughout our area, this subfamily is far more abundant and diverse in the South. Adults will readily come to lights at night but can also be encountered resting in low vegetation or visiting flowers during daytime.

BROWN-HEADED HALF-YELLOW MOTH
Ponometia bicolorata   93-1307 (9084)   **Rare**   RANGE-WIDE

TL 9–13 mm Resembles Half-yellow Moth, but brownish-black median line is thicker and less crisply edged, often more vertical, and head and collar are olive. **HOSTS** Indian heliotrope and *Simsia foetida*.

HALF-YELLOW MOTH
Ponometia semiflava   93-1308 (9085)   **Common**   RANGE-WIDE

TL 9–13 mm Bicolored FW has a chocolate-brown lower half with a mustard-yellow basal half, divided by a diagonal, thin, black median line. Head and collar are yellow. Fringe is brownish gray. **HOSTS** Various composites, including goldenaster.

WHITE-LINED HALF-YELLOW MOTH
Ponometia clausula   93-1310 (9086)   **Uncommon**   RANGE-WIDE

TL 9–11 mm Resembles Half-yellow Moth, but median line is white and fringe is pale gray with large checkered dark patches. **HOSTS** Unknown.

PRETTY BIRD-DROPPING MOTH
Ponometia venustula   93-1311 (9087)   **Common**   RANGE-WIDE

TL 9–10 mm Sexually dimorphic. Male has pale orange to tan FW with a narrow, dusky brown median band that tapers to end at central FW. Female is more strongly orange, with a wide, white-edged, hoary, gray median band that bends downward at inner FW. **HOSTS** Unknown.

VIRGIN BIRD-DROPPING MOTH
Ponometia virginalis   93-1312 (9088)   **Rare**   RANGE-WIDE

TL 9–10 mm White FW has warm brown PM band that curves from inner margin, across silvery-gray reniform spot, to apex; inner half of PM band is filled with silvery scales. An indistinct brownish patch often crosses from reniform spot to costa. Discal spot and terminal line are absent. **HOSTS** Unknown.

PRAIRIE BIRD-DROPPING MOTH
Ponometia binocula   93-1313 (9089)   **Uncommon**   RANGE-WIDE

TL 9–11 mm Resembles Virgin Bird-dropping Moth, but dark gray of PM band crosses behind white-outlined, dark gray reniform spot into ST area. Brownish-orange median band adjoins brown PM band in inner half but crosses straight to costa; costal section is often a disjointed squarish patch. Terminal area is usually faintly grayish with white adterminal line. **HOSTS** Unknown.

BROWN-HEADED
HALF-YELLOW MOTH

HALF-YELLOW MOTH

WHITE-LINED
HALF-YELLOW MOTH

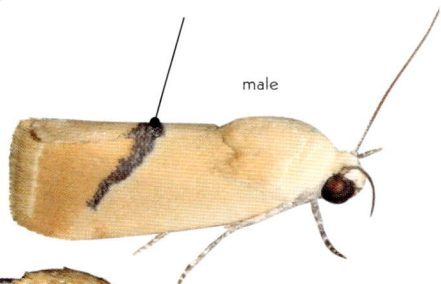

male

PRETTY
BIRD-DROPPING MOTH

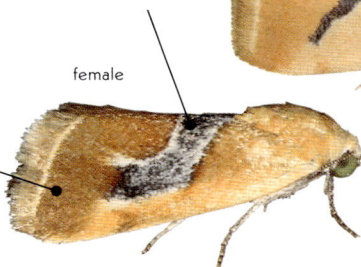

female

actual size

VIRGIN
BIRD-DROPPING MOTH

PRAIRIE
BIRD-DROPPING MOTH

481

OLIVE-SHADED BIRD-DROPPING MOTH   Common

Ponometia candefacta   93-1314 (9090)

TL 10–12 mm White FW has warm brown and gray markings of variable extent but generally running from inner median area to apex, across large, round, gray reniform spot. Inner median band is olive brown, while outer median band is warm brown and more diffuse. PM band is either warm brown with gray center or primarily gray with brown edges. Tiny black discal spot and black dashed terminal line are present. Inner basal area and posterior thorax have a variable amount of gray shading. **HOSTS** Ragweed.

HALF-CIRCLE BIRD-DROPPING MOTH   Uncommon

Ponometia cuta   93-1318 (9094)

TL 9–10 mm White FW has mottled, dark gray and brown lower half with large white semicircle at costa below V-shaped median band. Gray fringe has white patch at midpoint. **HOSTS** Unknown.

GRAY-RINGED BIRD-DROPPING MOTH   Uncommon

Ponometia libedis   93-1320 (9096)

TL 10–11 mm Ivory FW has dark patch at inner margin composed of gray PM and brown median areas, edged basally with black. Reniform spot is a gray outline, and tiny discal dot is black. AM, median, and PM lines are olive, often indistinct, and angle sharply basally near costa. Wavy ST line divides olive ST and gray terminal areas. Terminal line is black dots. Thorax is ivory with small olive patch at posterior. **HOSTS** Ragged marsh-elder.

BROWN-RINGED BIRD-DROPPING MOTH   Uncommon

Ponometia nannodes   93-1321 (9097)

TL 10–11 mm Resembles Gray-ringed Bird-dropping Moth, but FW is ivory to pale brown, reniform spot has a dark brown ring and pale outline, and ST and terminal areas are brown, often without pale ST line. Thorax washed brown dorsally. **HOSTS** Unknown.

ONE-SPOTTED BIRD-DROPPING MOTH   Uncommon

Ponometia phecolisca   93-1322 (9098)

TL 10–11 mm Resembles Brown-ringed Bird-dropping Moth, but FW is white, blackish-ringed reniform spot does not have white outer edging, and AM, median, and PM lines are absent except for at costa. Brown FW markings along inner margin may terminate either below or beside orbicular dot. Thorax is white. **HOSTS** Unknown.

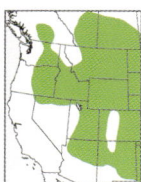

SUNNY BIRD-DROPPING MOTH   Uncommon

Ponometia tortricina   93-1326 (9101)

TL 10–11 mm Pale yellowish to brownish-gray FW has mottled gray to blackish patch at inner median area and small black discal dot. Fringe is gray, usually with dotted terminal line. **HOSTS** Unknown; possibly sunflower.

OLIVE-SHADED
BIRD-DROPPING MOTH

HALF-CIRCLE
BIRD-DROPPING MOTH

GRAY-RINGED
BIRD-DROPPING MOTH

BROWN-RINGED
BIRD-DROPPING MOTH

ONE-SPOTTED
BIRD-DROPPING MOTH

actual size

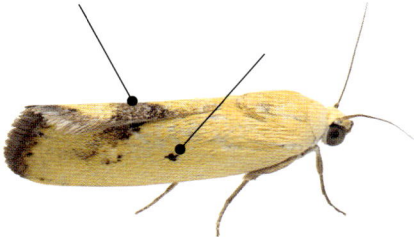

SUNNY
BIRD-DROPPING MOTH

SPECKLED BIRD-DROPPING MOTH
Ponometia fasciatella

Uncommon
93-1329 (9102)

TL 9–10 mm Sexually dimorphic. Male has pale tan to orange FW with broad, slightly darker (sometimes indistinct) AM and PM lines that curve downward, then angle basally near costa. Terminal line and fringe are brown. Female is hoary grayish brown with paler costa. PM band is most obvious at inner margin and costa. Dark reniform spot may be hollow or solid, and orbicular dot is dusky. Fringe is broadly checkered. HOSTS Unknown. NOTE A disjunct population exists in eastern TX and OK in which the females are heavily speckled gray along inner half of FW.

BROWN BIRD-DROPPING MOTH
Ponometia hutsoni

Uncommon
93-1330 (9103)

TL 9–10 mm Resembles female Speckled Bird-dropping Moth, but costa has strong brown patches at costa and median bands. Terminal area has a white line edged with black. Orbicular dot is small and blackish, and inner FW is washed faintly brown. Fringe is uniformly pale with faded brown at base. Some individuals can be very pale, with dark reniform spot and adterminal line. HOSTS Unknown.

INTENSE BIRD-DROPPING MOTH
Ponometia acutus

Uncommon
93-1332 (9105)

TL 10–13 mm White FW has thick gray PM and dark brown median bands from inner margin to costa. White-ringed reniform spot is connected to white basal half by pale stem, bisecting median band. ST area is light brown, and terminal area is gray. HOSTS Ragweed.

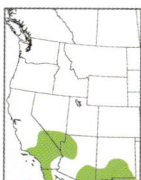

SCALLOPED BIRD-DROPPING MOTH
Ponometia altera

Rare
93-1333 (9107)

TL 10–13 mm White FW has a black-edged, scalloped, brown patch at inner median line and gray reniform spot that is ringed with black, thicker on apical side. PM band and ST area are grayish tan. Posterior of thorax has grayish band. HOSTS Goldenbush, rabbitbrush, and goldenweed.

ARIZONA BIRD-DROPPING MOTH
Ponometia elegantula

Common
93-1334 (9109)

TL 9–11 mm White FW has straight, thick, dark brown median band and brown PM band and ST area. Brown reniform spot is ringed with white and connected to white basal half by brown stem, bisecting median band. Usually appears to have white costal patch beside reniform spot. HOSTS Unknown.

EXPOSED BIRD-DROPPING MOTH
Tarache aprica

Uncommon
93-1343 (9136)

TL 9–15 mm Sexually dimorphic. Male has white FW with mottled brown and dark gray lower half, two large brownish-gray squares at costa, solid brownish-gray reniform spot, and small black discal dot. Median line at inner margin is black, edged basally with brown. Black ST line is fragmented. Female has predominantly gray FW with two large white patches at costa; anterior patch contains black discal dot. White thorax is shaded with gray. Black median, PM, and ST lines as in male. Male resembles One-spotted Bird-dropping Moth, but costal patches are larger and reniform spot is solid. HOSTS Hollyhock.

SPECKLED
BIRD-DROPPING MOTH

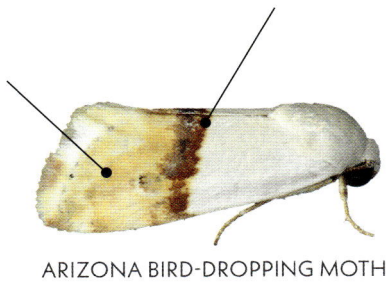

male

female

BROWN
BIRD-DROPPING MOTH

actual size

INTENSE
BIRD-DROPPING MOTH

SCALLOPED
BIRD-DROPPING MOTH

ARIZONA BIRD-DROPPING MOTH

male

female

EXPOSED
BIRD-DROPPING MOTH

FOUR-PATCHED BIRD-DROPPING MOTH
Common
Tarache quadriplaga   93-1353 (9142)

RANGE-WIDE

TL 10–13 mm Sexually dimorphic. Male and female resemble Exposed Bird-dropping Moth but lack black discal dot, inner PM and terminal areas are reddish brown, ST line is brown, edged with white only on outer side, and head is white. Male usually lacks gray basal shading, and costal patches are narrower; female has brownish AM area and grayish basal area. **HOSTS** Unknown.

FOUR-SPOTTED BIRD-DROPPING MOTH
Uncommon
Tarache tetragona   93-1354 (9143)

RANGE-WIDE

TL 7–10 mm Sexually dimorphic. Resembles Exposed Bird-dropping Moth, but ST line is warm brown and head is white. Dark AM patch on costa is larger. Inner AM and basal areas are mottled gray in females, white or light gray in males. Separable from larger Four-patched Bird-dropping Moth by gray inner PM and ST areas and terminal area, and more distinct black median line. Black discal dot usually absent, but sometimes present in male. **HOSTS** Turk's cap and curly herissantia.

EYED BIRD-DROPPING MOTH
Uncommon
Tarache areli   93-1356 (9159)

RANGE-WIDE

TL 10–13 mm Pale tan FW has grayish-brown lower half containing large, black-ringed, metallic blue reniform and subreniform spots and bold white subapical patch. Reniform spot has black "pupil" at center. Orbicular spot is a roughly circular, fragmented, black ring that does not touch brown median line. Wavy AM and basal lines are brownish gray. Basal area averages darker tan in females. **HOSTS** Globemallow.

BROKEN-SPOTTED BIRD-DROPPING MOTH **Uncommon**
Tarache geminocula   93-1359 (9159.2)

RANGE-WIDE

TL 10–13 mm Resembles Eyed Bird-dropping Moth, but black "pupil" of reniform spot is at basal edge. Orbicular spot appears as black discal dot and black crescent touching edge of brown median line. Dark lower half is often gray along inner margin. **HOSTS** Unknown; possibly mesquite. **NOTE** Three other very similar species (not shown) occur in our area. *T. areloides* has brown area extending basally past orbicular spot, and larger squarish subapical patch; *T. toddi* is warmer brown, often with white mottling, and reniform spot lacks "pupil"; and *T. albifusa* has ivory to tan basal half and mottled brown lower half, with pale tan mottling between metallic blue spots, reniform without or with indistinct black "pupil," and gray terminal area.

SILVERY BIRD-DROPPING MOTH
Rare
Tarache arida   93-1360 (9150)

RANGE-WIDE

TL 9–11 mm White FW has olive-brown inner median and outer ST areas, with silvery-gray S-shaped band connecting them down their centers. Solid silvery reniform spot is joined to silvery band. Costa has large silvery patch at median area, and basal area is mottled silvery gray with an olive-brown patch at center. Terminal area is washed olive brown. Separable from Chalky Bird-dropping Moth by reniform spot and basal and terminal areas. **HOSTS** Unknown.

male

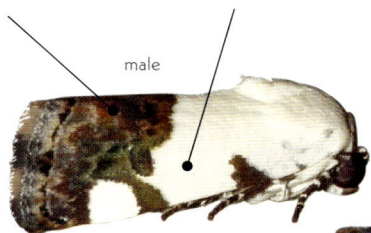

FOUR-PATCHED
BIRD-DROPPING MOTH

female

actual size

female

FOUR-SPOTTED
BIRD-DROPPING MOTH

EYED BIRD-DROPPING MOTH

BROKEN-SPOTTED
BIRD-DROPPING MOTH

SILVERY
BIRD-DROPPING MOTH

POLISHED BIRD-DROPPING MOTH

Uncommon

Tarache expolita   93-1363 (9149)

TL 9–11 mm   Dark grayish-brown FW has a broad white band along costa that continues across shoulders and white collar of thorax, and a large, pale tan patch at anal angle. Brownish PM line sometimes visible across costal band. White patch lacks dotted black terminal line. Separable from Gray-marked Bird-dropping Moth by AM and PM bands and terminal line. **HOSTS** Unknown.

ZEBRA BIRD-DROPPING MOTH

Rare

Tarache idella   93-1364 (9120)

TL 11–14 mm   White FW has bold black dashes and spots in rows perpendicular to inner margin. Apex shaded yellow orange. HW is yellow orange. Thorax is spotted, and legs are striped black and white. HW and abdomen bright yellow orange. **HOSTS** Sonoran indian mallow. **NOTE** Strikingly unique among bird-dropping moths; superficially resembles eastern Ladder-backed Ethmia (*Ethmia delliella*, not shown).

NARROW-WINGED BIRD-DROPPING MOTH

Uncommon

Tarache augustipennis   93-1365 (9111)

TL 11–14 mm   Sexually dimorphic. Dark grayish-brown FW has broad white stripe along costa that is crossed by thick gray PM and ST bands, as well as faint yellow-brown (male) to solid gray (female) median band. Gray reniform spot has a crescent of white edging around inner side. Terminal area is mottled whitish. Thin black PM line crosses inner half of FW. **HOSTS** Globemallow and other mallows.

WHITE-BANDED BIRD-DROPPING MOTH

Rare

Tarache huachuca   93-1366 (9113)

TL 10–13 mm   Sexually dimorphic. Resembles Narrow-winged Bird-dropping Moth, but males have a broad white costa crossed by tan median and PM bands, and reduced reniform spot; and females have a solid blackish-brown orbicular spot disconnected from inner brown area. Both sexes (more commonly in males) may show variable amounts of white in inner median area, from light mottling to a broad white band. **HOSTS** Unknown.

LARGE BIRD-DROPPING MOTH

Rare

Tarache major   93-1371 (9152)

TL 13–16 mm   Resembles the smaller Polished Bird-dropping Moth, but white costal band has gray patches at median and PM areas. Pale patch at anal angle is mottled with gray and has dotted black terminal line, and costal half of ST/terminal areas is olive brown. Separable from Gray-marked Bird-dropping Moth by PM line that is black at inner margin and gray reniform spot that is thinly outlined in black. **HOSTS** California flannelbush and apricot mallow.

RUSTY-PATCHED BIRD-DROPPING MOTH

Rare

Tarache lanceolata   93-1372 (9153)

TL 10–13 mm   Resembles larger Large Bird-dropping Moth, but white costal band has olive patches at median and PM areas, and rusty red apical patch marked with gray. White at anal angle is mottled with gray, and often appears to have a solid dark patch in middle where the fringe of the opposite FW folds over at rest. Separable from Polished Bird-dropping Moth by rusty apex and costal median patch. **HOSTS** Unknown.

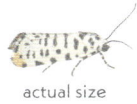

POLISHED
BIRD-DROPPING MOTH

ZEBRA BIRD-DROPPING MOTH

male

actual size

NARROW-WINGED
BIRD-DROPPING MOTH

male

female

male

female

female

WHITE-BANDED
BIRD-DROPPING MOTH

LARGE
BIRD-DROPPING MOTH

RUSTY-PATCHED BIRD-DROPPING MOTH

489

BROWN-BACKED BIRD-DROPPING MOTH Uncommon
Tarache lucasi 93-1373 (9148)

RANGE-WIDE

TL 10–13 mm Sexually dimorphic. Brown FW has broad white band along costa that is unmarked in males and crossed by thick brown PM and (usually) AM bands in females and extends across shoulder and collar of thorax (sometimes broadly). Black-edged brown reniform spot is inset into brown PM band. Thin white ST line is most visible at inner margin. Brown ST/terminal area sometimes mottled with white. Outer margin has large white patch near anal angle. **HOSTS** Arizona rosemallow, rock hibiscus, and likely other mallows.

CHALKY BIRD-DROPPING MOTH Uncommon
Acontia cretata 93-1375 (9161)

RANGE-WIDE

TL 12–14 mm Resembles Silvery Bird-dropping Moth, but white-outlined gray reniform spot is indistinct, terminal area is gray, and basal area lacks olive-brown shading. Thin white PM line usually visible at inner margin of brown patch as large looping scallops. **HOSTS** Unknown.

MAGNIFICENT SPRAGUEIA Uncommon
Spragueia magnifica 93-1381 (9121)

RANGE-WIDE

TL 8–9 mm Pale yellow FW has thick black stripes from base to inner median area, costal AM to inner median area (sometimes just as large black orbicular spot), and across ST area, and large, round, black reniform spot. Rusty, curving PM and straight ST lines pass behind black markings. Thorax and basal inner margin are black. **HOSTS** Unknown.

JAGUAR SPRAGUEIA Uncommon
Spragueia jaguaralis 93-1388 (9128)

RANGE-WIDE

TL 9–11 mm Sexually dimorphic. Male has wide rusty-brown AM, median, PM, and ST lines and rusty-brown orbicular and reniform spots, largely covering pale yellow FW. Central FW from base to anal angle has fragmented, wide, brown stripe. Female is similarly patterned, but lines are brown, spots are black, and central and inner FW are washed black. Both sexes have thorax either entirely orange or with wide black dorsal stripe and orange bands laterally. **HOSTS** Unknown.

SOMBER SPRAGUEIA Uncommon
Spragueia funeralis 93-1389 (9129)

RANGE-WIDE

TL 8–10 mm Pale yellow FW has wide black bands at basal, median, and PM lines connected to black orbicular and reniform spots. Hoary black ST/terminal area has pale yellow subapical patch. Narrow rusty lines edge median band. **HOSTS** Unknown.

IRON-BANDED SPRAGUEIA Uncommon
Spragueia obatra 93-1390 (9130)

RANGE-WIDE

TL 8–10 mm FW has wide, metallic gray AM and PM bands over rusty red in inner half and pale yellow in outer half of wing. Metallic gray reniform spot has rusty-red PM patch adjacent at costa. **HOSTS** Unknown.

male

BROWN-BACKED
BIRD-DROPPING MOTH

female

CHALKY BIRD-DROPPING
MOTH

actual size

MAGNIFICENT
SPRAGUEIA

male

JAGUAR SPRAGUEIA

female

SOMBER SPRAGUEIA

IRON-BANDED SPRAGUEIA

491

RASPBERRY THIMBLE
Uncommon

Chamaeclea pernana   93-1392.3 (9789)

RANGE-WIDE

TL 11–13 mm Rosy FW is golden yellow in ST/terminal area with darker shading toward inner margin. Inner AM line and midpoint of PM line are dark reddish, sometimes with a reddish wash in adjacent median area. Thorax is tan. **HOSTS** Unknown.

GREEN THIMBLE
Rare

Heminocloa mirabilis   93-1392.6 (9735)

RANGE-WIDE

TL 14–16 mm Olive-green FW is shaded in a gradient from pale costa to darker inner margin and has rounded dark green patches at outer basal and costal ST areas, anal angle, and posterior of thorax. Central terminal area is shaded dark green. Straight AM and V-shaped PM lines are pale yellowish, as are orbicular spot and reniform dash. **HOSTS** Possibly four o'clocks.

COTTON THIMBLE
Uncommon

Thurberiphaga diffusa   93-1392.7 (9817)

RANGE-WIDE

TL 14–16 mm Pale yellowish FW has diffuse, pink, V-shaped median band and pink shading along fringe and central terminal area, with dark pink apical dash. Legs are pink. Sometimes shows streaks of brownish shading in basal area and at reniform spot. **HOSTS** Desert cotton.

PANTHEAS and BROTHERS   SUPERFAMILY Noctuoidea (93)

FAMILY Noctuidae SUBFAMILIES Pantheinae and Raphiinae

Medium-sized gray moths with sturdy bodies and broad rounded wings crossed by black transverse lines. Melanistic individuals are rare but regular in the panthea species. Adults are nocturnal and will come to lights.

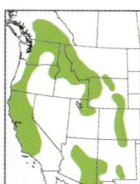

GIANT PANTHEA
Uncommon

Panthea gigantea   93-1395 (9181)

RANGE-WIDE

TL 24–30 mm Hoary gray FW has thick black AM, median, and PM lines, toothed white ST line shaded gray basally, and dark gray reniform crescent. AM line is straight. **HOSTS** Ponderosa pine.

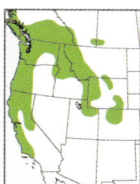

WESTERN PANTHEA
Very Common

Panthea virginarius   93-1399 (9178)

RANGE-WIDE

TL 18–22 mm Resembles the larger Giant Panthea, but FW is typically lighter and AM line is wavy. Median line often bolder than other lines. Basal area and inner margin usually shaded lightly grayish. **HOSTS** Douglas-fir and other conifers.

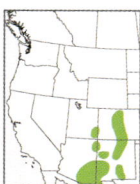

THE MONITOR
Rare

Meleneta antennata   93-1399.5 (9191)

RANGE-WIDE

TL 16–19 mm Resembles The Brother, but inner AM area and orbicular and reniform spots have diffuse, warm brown to tan shading. AM and PM lines are wavy and edged in paler gray basally and not connected by thin vertical line. Wavy white ST line has narrow dusky edge. **HOSTS** Unknown.

RASPBERRY THIMBLE

GREEN THIMBLE

COTTON THIMBLE

actual size

GIANT PANTHEA

WESTERN PANTHEA

THE MONITOR

actual size

493

SOUTHERN PANTHEA

Pseudopanthea palata   **Rare**

93-1405.5 (9186)

TL 17–20 mm Hoary gray FW has curving, dark gray AM, median, and ST lines and large, black-outlined, light gray orbicular and reniform spots. Gray PM line is often indistinct. Basal area has a light gray streak down center. Separable from The Brother by median and ST lines. **HOSTS** Unknown.

THE BROTHER

Raphia frater   **Very Common**

93-1412 (9193)

TL 17–19 mm Gray FW has curved, slightly wavy AM and toothed PM lines, usually connected by a thin vertical line parallel to inner margin. Hollow orbicular and dusky-centered reniform spots are narrowly outlined with black and sometimes tinged brownish. Pale dusky-edged ST line is sometimes indistinct. **HOSTS** Aspen; also alder, birch, cottonwood, and willow.

DAGGERS   SUPERFAMILY Noctuoidea (93)

FAMILY Noctuidae SUBFAMILY Acronictinae

Medium-sized moths with a broad, elongate, triangular appearance. Most species are shades of gray, and many have strong black basal and anal dashes; others have strong double AM lines and defined orbicular spots. Pale Green Dagger is our only green dagger (one other occurs in the East). Henry's Marsh Moth superficially resembles a wainscot but lacks dots or PM markings. Adults are nocturnal and will come to lights.

PALE GREEN DAGGER

Chloronycta tybo   **Uncommon**

93-1419.6 (9283)

TL 16–18 mm Pale mint-green FW has wavy, fragmented, black AM, median, and PM lines edged indistinctly with white, and large black-outlined orbicular and reniform spots with white at center. ST area is shaded diffusely dusky. **HOSTS** Velvet ash.

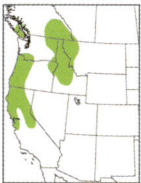

FUNERARY DAGGER

Acronicta funeralis   **Uncommon**

93-1419.71 (9221)

TL 17–21 mm Resembles Gentle Dagger but FW averages paler gray to whitish, stripes along inner margin are broader and dark black, and AM and PM lines are less distinct and most prominent at costa and behind dashes at inner margin. Orbicular spot is usually completely pale. **HOSTS** Deciduous trees, including apple, birch, cottonwood, elm, maple, oak, and others; also blueberry.

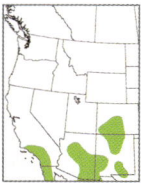

BLACK-STRIPED DAGGER

Acronicta atristrigatus   **Rare**

93-1419.74 (9232)

TL 20–22 mm Medium to light gray FW has broad black basal and anal dashes and thinner inner-median-area dash, which in well-marked individuals can appear as a solid black streak from base to anal angle. Double AM and PM lines are strongly toothed. Large orbicular and reniform spots are thinly outlined in black, with dusky centers. Costa is strongly barred with black. **HOSTS** Oak.

SOUTHERN PANTHEA

THE BROTHER

actual size

PALE GREEN DAGGER

FUNERARY DAGGER

actual size

BLACK-STRIPED DAGGER

MARBLED OAK DAGGER

Common

Acronicta marmorata   93-1419.78 (9256)

RANGE-WIDE

TL 18–20 mm Gray FW has strongly scalloped double AM, median, and PM lines with shaded patches, creating a marbled appearance. Base of inner margin has a pale brown spot. Conspicuous pale orbicular spot is thinly outlined in black and has a dusky "pupil." Reniform spot is filled with dusky shading. Narrow black basal dash reaches AM line. White ST line is zigzagged. **HOSTS** Garry oak and other oaks.

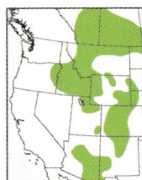

AMERICAN DAGGER

Common

Acronicta americana   93-1421 (9200)

RANGE-WIDE

TL 27–38 mm The largest of our daggers. Resembles Large Gray Dagger, but PM line has longer, sharp teeth, orbicular spot is larger, reniform spot is outlined completely, and thin anal dash is well defined. AM line is often indistinct. HW is brownish to dusky. **HOSTS** Trees and woody plants, including alder, ash, basswood, elm, chestnut, and hickory.

LARGE GRAY DAGGER

Very Common

Acronicta insita   93-1423 (9202)

RANGE-WIDE

TL 24–28 mm Hoary, pale gray FW has hollow orbicular spot strongly outlined with black and black-outlined reniform spot with dusky center. Black PM line is row of scallops edged in white, intersected by indistinct anal dash. Dusky AM line has thick pointed tooth inside from orbicular spot. HW is pale gray. **HOSTS** Birch, alder, aspen, poplar, and willow.

COTTONWOOD DAGGER

Uncommon

Acronicta lepusculina   93-1425 (9205)

RANGE-WIDE

TL 24–27 mm Pale gray FW has dusky black marks along costa at AM, median, and PM lines; dusky mark at inner margin of PM line is bisected by black anal dash. AM and PM lines are indistinct. FW has thin, black basal dash. In some individuals, orbicular and reniform spots are indistinct or nearly absent. **HOSTS** Poplar and willow.

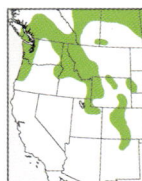

MILLER DAGGER

Rare

Acronicta vulpina   93-1427 (9206)

RANGE-WIDE

TL 20–24 mm Resembles Cottonwood Dagger, but FW averages paler and markings more crisply defined. Indistinct AM and PM lines have strong black chevron near midpoint. Basal dash is very short. **HOSTS** Quaking aspen, balsam poplar, paper birch, speckled alder, and willow.

FRAGILE DAGGER

Uncommon

Acronicta fragilis   93-1427.1 (9241)

RANGE-WIDE

TL 15–17 mm White to gray FW has strongly scalloped, double, black AM and PM lines filled with white. Conspicuous white orbicular spot is outlined with black; black-outlined reniform spot has dusky center. Basal, median, and ST areas often shaded darker. Basal, inner median, anal, and apical dashes are short and sometimes hidden within dusky shading. Veins in median area streaked white in boldly marked individuals. **HOSTS** Apple, birch, plum, willow, and white spruce.

MARBLED OAK DAGGER

AMERICAN DAGGER

LARGE GRAY DAGGER

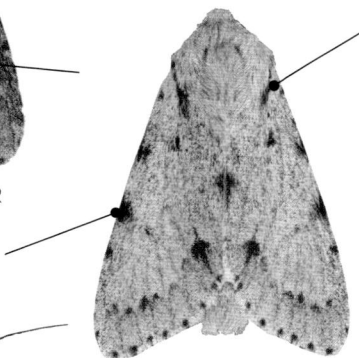

COTTONWOOD DAGGER

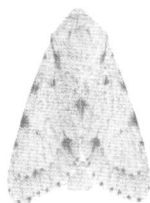

actual size

MILLER DAGGER

FRAGILE DAGGER

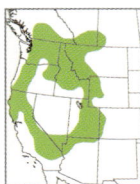

GENTLE DAGGER

Acronicta mansueta

Uncommon

93-1427.4 (9218)

TL 14–20 mm Dark gray FW has thick black basal, inner median, and anal dashes. Double AM and PM lines are slightly scalloped; PM line has whitish fill. Large round orbicular and reniform spots are shaded dusky at center; both are thinly outlined in fragmented black with pale edging and may be indistinct or nearly absent in some individuals. Separable from Funerary Dagger by narrower dashes along inner margin, and visible AM and PM lines. **HOSTS** Unknown.

GRAY DAGGER

Acronicta grisea

Common (uncommon in South)

93-1433 (9212)

TL 18–20 mm Resembles Gentle Dagger, but hoary gray FW averages lighter, black basal and anal dashes are thinner, and costal and inner median area lacks blackish shading. PM line lacks pale fill at inner margin. **HOSTS** Deciduous trees, including alder, apple, birch, and cherry.

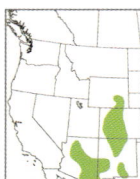

TAWNY DAGGER

Acronicta thoracica

Rare

93-1434.3 (9230)

TL 20–23 mm Resembles Cherry Dagger, but gray FW has faint, warm brown shading at central AM line, reniform spot, and ST area. Black dash connecting orbicular and reniform spots is thinner, and spots' outlines are not thickened. Diffuse dusky median-line dash connects reniform spot to costa. Thorax has warm brown dorsal patch behind collar. **HOSTS** Cherry and plum. **NOTE** Eastern Ochre Dagger (*A. morula*, not shown), which just barely enters our area, is also washed warm brown but lacks dash connecting central spots.

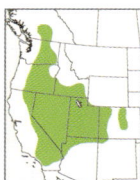

STRIGULOSE DAGGER

Acronicta strigulata

Rare

93-1434.4 (9231)

TL 19–21 mm Resembles Tawny Dagger and has similar brown thorax and black FW dashes but orbicular spot is indistinct, AM and PM lines are fainter, and brown shading is reduced or absent. Most individuals can be comfortably identified by range. **HOSTS** Unknown; possibly poplar and willow.

UNMARKED DAGGER

Acronicta innotata

Rare

93-1434.7 (9207)

TL 19–22 mm Ivory to pale gray FW has a thin, slightly scalloped, black PM line with light gray ST area. AM and median lines largely visible only as black dashes at costa. Reniform spot a thin black crescent. **HOSTS** Alder, birch, hickory, poplar, and willow.

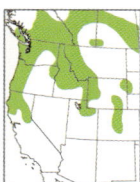

RADCLIFFE'S DAGGER

Acronicta radcliffei

Uncommon

93-1441.1 (9209)

TL 18–20 mm Pale to medium gray FW has thin, black, straight AM and curving PM lines edged medially with white, sometimes with a shadow of a doubled inner line. Large hollow orbicular and reniform spots are outlined with fragmented black and edged inwardly with white. Long, thin, black basal and anal dashes intersect AM and PM lines, respectively. **HOSTS** Apple, cherry, chokeberry, hawthorn, and others.

GENTLE DAGGER

GRAY DAGGER

TAWNY DAGGER

actual size

STRIGULOSE DAGGER

UNMARKED DAGGER

RADCLIFFE'S DAGGER

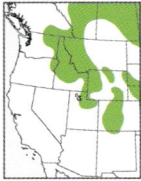

CHERRY DAGGER

Rare

Acronicta hasta   93-1441.2 (9229)

TL 22–23 mm Gray FW has bold black basal, apical, and anal dashes. Dusky AM line is double, and black PM line is edged with white. Round orbicular spot is pale to whitish, and reniform spot is shaded dusky at center; both are thickly edged on medial side and connected by thick black bar. **HOSTS** Cherry, oak, and plum.

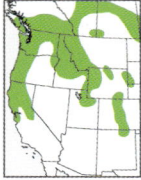

YELLOW-HAIRED DAGGER

Uncommon

Acronicta impleta   93-1474 (9257)

TL 21–27 mm Evenly gray FW has dusky, V-shaped median line, and scalloped, black PM line is edged broadly with pale gray medially. Orbicular spot is concolorous with reniform spot, and both are outlined in black and filled with dusky shading at center. Basal dash is absent, and anal angle is lightly shaded with dusky gray. **HOSTS** Hickory and walnut; also alder, ash, elm, and others.

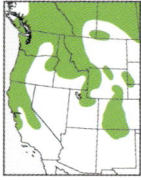

IMPRESSED DAGGER

Uncommon

Acronicta impressa   93-1477 (9261)

TL 18–24 mm Resembles Marbled Oak Dagger, but base of inner margin is white to gray, and dusky median band is flatter and less dark. Basal and anal dashes are diffusely shaded blackish. Orbicular spot is concolorous with FW with a dusky dot at center, and not connected to dusky-centered reniform spot. **HOSTS** Deciduous trees and woody plants.

MARSH DAGGER

Uncommon

Acronicta insularis   93-1477.2 (9280)

TL 20–22 mm Pale tan FW has contrasting whitish veins and streaks of dark shading (sometimes indistinct) parallel to inner margin and central vein, and at apex. FW apex is pointed. Resembles the wainscots but lacks markings in PM area. **HOSTS** Cattail, grass, sedge, smartweed, poplar, willow, and others. **NOTE** Also commonly known as Henry's Marsh Moth.

PACIFIC DAGGER

Uncommon

Acronicta perdita   93-1482 (9268)

TL 22–24 mm Light gray FW has broad black shading along inner half, obscuring FW markings except at costa. Median line is broadly black, and AM line is double. Orbicular spot is sometimes visible as a thick black ring with no shading. In some individuals, the black shading of inner margin and median lines is so broadly diffuse that it creates the appearance of a blackish wash to FW. **HOSTS** Ceanothus, willow, cliffrose, and bitterbrush.

OTHELLO DAGGER

Rare

Acronicta othello   93-1484 (9270)

TL 21–24 mm Smooth gray FW is pale gray in costal half of basal and median areas and diffusely pale gray in terminal area. Thin black AM and PM lines are strongly scalloped; PM line has whitish edge medially. Orbicular spot is absent; black-outlined reniform spot is obscured by diffuse dusky median line. Anal dash is dusky. Thorax has black-outlined, pale gray stripes laterally. Dark-form individuals have indistinct lines; the most distinct markings are medium gray costal patches and terminal area. **HOSTS** Laurel sumac.

CHERRY DAGGER

YELLOW-HAIRED DAGGER

IMPRESSED DAGGER

actual size

MARSH DAGGER

PACIFIC DAGGER

OTHELLO DAGGER

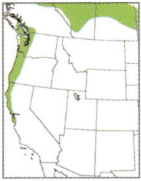

SMEARED DAGGER
Acronicta oblinita   **Uncommon**   RANGE-WIDE

93-1485 (9272)

TL 20–28 mm Pale gray FW is unevenly shaded with hoary shading, usually with darker patches along inner margin. Curving PM line is a row of black dots; other lines absent. Round orbicular spot is slightly paler, and reniform spot is dusky; both spots sometimes indistinctly outlined with fragmented black. **HOSTS** Trees, shrubs, and herbaceous plants, including apple, corn, elm, pine, and willow.

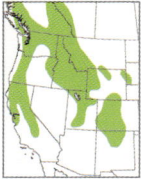

LUPINE DAGGER
Acronicta lupini   **Uncommon**   RANGE-WIDE

93-1489 (9275)

TL 18–20 mm Hoary gray FW has scalloped, black AM and PM lines broadly edged basally with pale gray, sometimes with a black doubled line. Small orbicular spot is pale gray and outlined black. Reniform spot usually has thick black outline along top and bottom. Basal and anal dashes are absent. **HOSTS** Lupines and other herbaceous plants.

HOODED OWLETS   SUPERFAMILY Noctuoidea (93)
FAMILY Noctuidae SUBFAMILY Cuculliinae

Medium-sized gray or brownish moths with long, pointed wings and a thick thoracic crest that projects forward over the head like a hood. The eye-catching caterpillars are boldly patterned, generally in black, red, and yellow; all of our species feed on species of Asteraceae. Adults are nocturnal and will come to lights.

GOLDENROD HOODED OWLET
Cucullia asteroides   **Uncommon**   RANGE-WIDE

93-1504 (10200)

TL 22–24 mm Light gray (sometimes brownish) FW gradients through warm brown into chocolate brown at costa. Veins are thinly traced with black, with thin whitish interveinal streaks in ST/terminal area. Thicker black anal dash is edged inwardly with warm brown, bounded above by white-edged black curve of inner PM line. Orbicular and reniform spots frequently visible as large rounded tan patches with mottled brown centers. Thorax has a dark brown dorsal stripe that connects to brown stripe along inner margin of FW. **HOSTS** Goldenrod and aster.

MOUNTAIN HOODED OWLET
Cucullia montanae   **Rare**   RANGE-WIDE

93-1505 (10201)

TL 22–24 mm Resembles Goldenrod Hooded Owlet and replaces this species from the Rocky Mountains west. Where they co-occur, separable by narrow blackish outline to pale orbicular and reniform spots and zigzag double AM line visible at costa. **HOSTS** Gumweed and rabbitbrush.

GRAY HOODED OWLET
Cucullia florea   **Uncommon**   RANGE-WIDE

93-1508 (10197)

TL 22–24 mm Resembles Goldenrod Hooded Owlet, but FW is iron gray with little to no brown. Inner margin sometimes has a defined, narrow, blackish edge. **HOSTS** Aster and goldenrod.

SMEARED DAGGER

LUPINE DAGGER

actual size

GOLDENROD HOODED OWLET

MOUNTAIN HOODED OWLET

GRAY HOODED OWLET

actual size

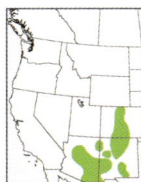

BROWN-PATCHED HOODED OWLET

Cucullia lilacina   **Rare**   93-1510 (10196)

TL 20–23 mm Gray FW has warm brown shading from reniform spot to apex, within orbicular spot, and across base and basal inner margin. Anal dash, inner PM line, and costal AM line are shaded diffuse blackish. **HOSTS** Fleabane.

INTERMEDIATE HOODED OWLET

Cucullia intermedia   **Uncommon**   93-1514 (10194)

TL 23–27 mm Resembles Tansyaster Hooded Owlet but lacks anal dash and black U-shaped curve of inner PM line. Most individuals show indistinct dusky AM band at costa. **HOSTS** Wild lettuce. **NOTE** Also known as Dusky Hooded Owlet.

SPEYER'S HOODED OWLET

Cucullia speyeri   **Rare**   93-1516 (10190)

TL 23–26 mm Resembles Tansyaster Hooded Owlet, but anal dash is thin or nearly absent and lines of wing are overall finer. Thorax usually has dusky to medium gray stripe down middle. **HOSTS** Horseweed; also sunflower. **NOTE** Most commonly encountered in its brightly colored larval form.

TANSYASTER HOODED OWLET

Cucullia dorsalis   **Common**   93-1518 (10190.2)

TL 23–26 mm Silvery-gray FW has veins thinly traced with black and a thick black anal dash. PM line visible as darker U-shaped curve over basal end of anal dash. Thorax is uniformly gray or with medium gray stripe down middle. **HOSTS** Hoary tansyaster. **NOTE** Two southwestern species are very similar: Charon Hooded Owlet (*C. charon*, not shown) has long, thin, black basal dash, and Truncate Hooded Owlet (*C. eccissica*, also not shown) has shorter, slightly thicker, black basal dash and pale patch at inner margin medial to PM line.

RABBITBRUSH HOODED OWLET

Cucullia laetifica   **Uncommon**   93-1519 (10191)

TL 20–23 mm Resembles Tansyaster Hooded Owlet, but FW has tawny shading at orbicular and reniform spots, at inner basal area, and along fringe. **HOSTS** Rabbitbrush and spiderling.

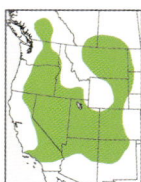

SPOTTED HOODED OWLET

Cucullia antipoda   **Rare**   93-1524 (10206)

TL 18–21 mm Hoary, light gray FW has prominent black-ringed orbicular and reniform spots that are shaded dusky at center, and thick black anal dash. V-shaped, black inner PM line frames top of anal dash. Veins are thinly traced with black, and basal half of inner margin is edged in black. **HOSTS** Unknown.

SMALL HOODED OWLET

Cucullia eulepis   **Uncommon**   93-1529 (10209)

TL 17–19 mm Resembles Spotted Hooded Owlet, but orbicular and reniform spots, and curving inner PM line, are absent. Black streak along inner margin reaches anal angle. **HOSTS** Wirelettuce.

BROWN-PATCHED
HOODED OWLET

INTERMEDIATE HOODED OWLET

SPEYER'S HOODED OWLET

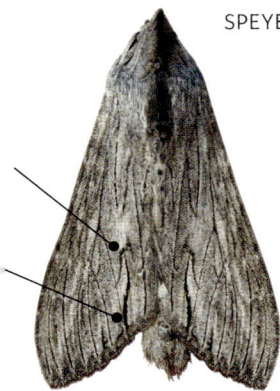

RABBITBRUSH HOODED OWLET

actual size

TANSYASTER HOODED
OWLET

SPOTTED HOODED OWLET

SMALL HOODED OWLET

DIMORPHIC HOODED OWLET

Cucullia serraticornis

Uncommon

93-1534 (10184)

TL 22–26 mm Sexually dimorphic. Male has medium gray FW with veins thinly traced with black and thicker black interveinal dashes in terminal area. Costa and inner margin have diffuse black dashes at cross-lines, and inner PM line is pale and zigzagged. FW is lightly washed warm brown, sometimes indistinctly. Female FW is similar but dark gray through median area with elongate pale spot at inner margin. ST/terminal area is lighter gray with blackish interveinal dashes. Orbicular and reniform spot sometimes visible as slightly lighter, diffuse patches. **HOSTS** Unknown; possibly goldenrod.

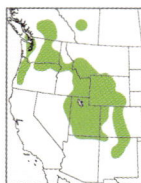

STREAKED HOODED OWLET

Cucullia strigata

Rare

93-1536 (10183)

RANGE-WIDE

TL 22–26 mm Gray FW has thick black interveinal dashes in terminal area and deeply zigzagged, black AM line (sometimes also PM line). Orbicular and reniform spot usually visible as indistinct tan or pale gray patches. Inner margin is sometimes shaded dusky. **HOSTS** Unknown; possibly gumweed.

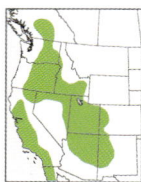

DARK HOODED OWLET

Cucullia pulla

Rare

93-1538 (10180)

RANGE-WIDE

TL 17–23 mm Dusky gray FW has paler costa and ST/terminal area. Terminal line is bold, dark dashes. Strongly jagged, black AM and PM lines usually visible; PM line has diffuse whitish edge below. Inner median area has a diffuse whitish spot. Some individuals have warm brown shading in reniform spot and basal and terminal areas. **HOSTS** Rabbitbrush, snakeweed, and turpentine bush.

AMPHIPYRINE SALLOWS

SUPERFAMILY Noctuoidea (93)

FAMILY Noctuidae SUBFAMILY Amphipyrinae

Most of the species in this group are medium sized, with a chunky, hairy thorax and long, slightly flared forewings that they fold tented over their body at rest. Copper Underwing and Mouse Moth rest with wings flat. Most moths are gray, brown, or tan, but several species are notably green. Adults are nocturnal and will come to lights. The *Amphipyra* will also regularly come to sugar bait.

COPPER UNDERWING

Amphipyra pyramidoides

Common

93-1544 (9638)

RANGE-WIDE

TL 23–28 mm Broad grayish-brown FW has a paler ST/terminal area with short black dashes. Double black AM and PM lines are scalloped and filled with pale tan; AM line is fragmented. Pale orbicular spot has dark spot in center. **HOSTS** Trees and vines, including birch, elm, oak, Virginia creeper, and willow.

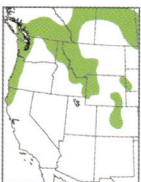

MOUSE MOTH

Amphipyra tragopoginis

[IN] Uncommon

93-1545 (9639)

RANGE-WIDE

TL 18–22 mm Hoary brown FW has three large dusky dots at orbicular and reniform spots. Veins are faintly marked with dusky brown. **HOSTS** Hawthorn and a variety of herbaceous plants, including columbine, geranium, plantain, and stinging nettle.

male

male

DIMORPHIC HOODED OWLET

female

actual size

STREAKED HOODED OWLET

DARK HOODED OWLET

actual size

COPPER UNDERWING

MOUSE MOTH

THREAD-LINED SALLOW
Pleromella opter

Uncommon 93-1559 (10026)

TL 15–19 mm Hoary gray FW has veins thinly traced with black. An indistinct dusky band runs diagonally from apex across ST/terminal area, slightly thickening black vein where it crosses. Thorax has a narrow dusky dorsal band. Superficially resembles Fine-lined Sallow but separable by apical band and thorax and by absence of orbicular and claviform spots. **HOSTS** Manzanita; possibly others.

THE JOKER
Feralia jocosa

Rare 93-1561 (10005)

TL 18–22 mm Pale green FW (rarely warm brown) has wavy, double, black AM and PM lines that are filled with white. Green orbicular and reniform spots are outlined with black and edged interiorly with white. Median area is speckled with black scales. **HOSTS** Coniferous trees, including balsam fir, hemlock, spruce, and tamarack.

DECEPTIVE SALLOW
Feralia deceptiva

Common 93-1562 (10006)

TL 18–20 mm Pale green FW has irregularly toothed black AM and PM lines that are edged boldly with white medially. Sometimes-indistinct black median line is fractured. Green orbicular and reniform spots are incompletely ringed with black and edged with white interiorly. **HOSTS** Douglas-fir.

COMSTOCK'S SALLOW
Feralia comstocki

Uncommon 93-1564 (10008)

TL 18–21 mm Resembles Deceptive Sallow, but median and PM areas have blackish shading, which varies from indistinct speckling to nearly fully black. **HOSTS** Hemlock, black spruce, and white pine.

FEBRUARY SALLOW
Feralia februalis

Common 93-1565 (10009)

TL 17–19 mm Resembles Deceptive Sallow, but orbicular and reniform spots are larger, and thin black terminal line is scalloped. PM line is double, filled with white. Median area is sometimes sprinkled with black scales. Most individuals are mint green; rare individuals may be olive or yellow-orange. **HOSTS** Mountain mahogany, blue oak, and greenbark ceanothus; possibly also elder and *Prunus*.

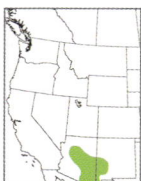

BRILLIANT MIRACAVIRA
Miracavira brillians

Rare 93-1567 (9620)

TL 19–22 mm Superficially similar to *Feralia* species, but toothed PM line touches large mostly-white reniform spot to form a black-edged, dark green patch. Anal angle has a rusty brown patch. Dark green, round orbicular spot is bordered with white and edged black. **HOSTS** Unknown.

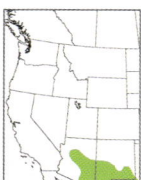

GREEN OSLARIA
Oslaria viridifera

Common 93-1575 (9791)

TL 14–16 mm Light yellow-green FW has darker olive-green patches at inner median area and along costa in median and ST areas. Dark green AM and PM lines are thin and slightly wavy. Thorax is tinged olive. **HOSTS** Unknown.

THREAD-LINED SALLOW

THE JOKER

DECEPTIVE
SALLOW

COMSTOCK'S SALLOW

actual size

FEBRUARY SALLOW

BRILLIANT MIRACAVIRA

GREEN OSLARIA

509

RIVULET NOCLOA
Nocloa rivulosa **Rare** 93-1578 (9794)

TL 14–16 mm Tan FW has wavy, double, warm to olive-brown AM, PM, and ST lines. Diffuse median band is broad and angles basally to meet costa. Inner basal and outer ST areas are shaded brown. Brown thorax is densely hairy. **HOSTS** Unknown; adults have been found sleeping on leaves of four o'clock.

PALE NOCLOA
Nocloa pallens **Uncommon** 93-1579 (9795)

TL 12–15 mm Pale tan or mint-green FW has a broad, diffuse, dark tan or green, V-shaped median band. White AM and PM lines are indistinct against pale FW. Terminal area sometimes shaded slightly darker. Pale thorax is densely hairy. **HOSTS** Unknown.

MASKED NOCLOA
Nocloa cordova **Rare** 93-1580 (9796)

TL 14–16 mm Yellow-orange to tan FW has purplish-brown shading in a C shape around reniform spot, and a small patch above orbicular spot. Thin, scalloped AM, median, and PM lines are indistinct. ST line is a row of dark dots. Thorax is densely hairy. **HOSTS** Unknown.

CHIHUAHUAN NOCLOA
Nocloa nanata **Rare** 93-1581 (9797)

TL 16–17 mm Orangey-yellow FW has irregular hoary brown patch in median area and smaller patch in outer AM area. ST line a row of widely spaced brown dots; AM and PM lines largely absent except for a few brown dots. Thorax is densely hairy. **HOSTS** Unknown.

GOLDEN NOCLOA
Nocloa alcandra **Uncommon** 93-1583 (9799)

TL 16–19 mm Heavily speckled, golden-brown FW has large roundish orbicular and reniform spots and darker brown shading through central median and outer basal areas, sometimes filling much of median area. Brown AM line is strongly scalloped, and straight PM line is crossed by dark veins. Thorax is densely hairy. **HOSTS** Unknown.

LEMON NOCLOA
Nocloa aliaga **Uncommon** 93-1584 (9800)

TL 13–15 mm Light yellow FW has indistinct, wavy, brown AM and PM lines and usually small, warm brown patches near costa at AM, median, and sometimes PM areas. Thorax and fringe are brown. **HOSTS** Unknown.

PERIDOT SALLOW
Paramiana smaragdina **Uncommon** 93-1587 (9803)

TL 13–15 mm Dusky FW has broad streaks (sometimes indistinct or incomplete) of yellow-green shading along inner margin, from central wing to anal angle, and across orbicular and reniform spots. Claviform spot is boldly outlined with black. Black PM line has a thick patch at inner margin. Lower costa is regularly marked with white dots. **HOSTS** Unknown.

RIVULET NOCLOA

PALE NOCLOA

MASKED NOCLOA

CHIHUAHUAN NOCLOA

GOLDEN NOCLOA

actual size

LEMON NOCLOA

PERIDOT SALLOW

TURQUOISE-BANDED SALLOW
Paramiana perissa

Rare

93-1589 (9805)

RANGE-WIDE

TL 15–17 mm Dusky FW has mint-green basal and ST areas bounded by scalloped black AM and PM lines. Mint-green orbicular and reniform spots have dusky centers. **HOSTS** Unknown.

WEBBED RUACODES
Ruacodes tela

Uncommon

93-1596 (9810)

RANGE-WIDE

TL 15–17 mm Brown FW has black-ringed, light brown orbicular and reniform spots shaded slightly dusky inside, and a bold black line connecting inner AM and PM lines and attached to black-outlined claviform spot. Black AM and PM lines are often indistinct. ST area has blackish interveinal dashes, and ST line is a row of small white dots. Brown thorax has two narrow black stripes laterally. **HOSTS** Unknown.

SANDY CRESCENT
Prothrinax luteomedia

Rare

93-1603 (9829)

RANGE-WIDE

TL 11–13 mm FW has a bold tan crescent curving from outer basal area to middle of outer margin. Costal half of wing has dark gray median area with a black-centered white orbicular spot, and light gray PM/ST area. Inner half of ST line is boldly white and edged in dark brown basally. **HOSTS** Unknown.

BELOVED EMARGINEA
Emarginea percara

Uncommon

93-1606 (9718)

RANGE-WIDE

TL 12–14 mm Light green FW has blackish shading to outer median area, surrounding a large whitish-green patch at costa. Wavy black AM and median lines are fractured and indistinct, not connecting to inner margin. Inner ST area has large black patch. Collar is whitish. **HOSTS** Mistletoe.

DARK-COLLARED EMARGINEA
Emarginea dulcinea

Rare

93-1607 (9719)

RANGE-WIDE

TL 12–14 mm Resembles Beloved Emarginea but lacks black patch at inner ST area, and collar is black. **HOSTS** Unknown.

SPLENDID SALLOW
Triocnemis saporis

Uncommon

93-1609 (10174)

RANGE-WIDE

TL 12–14 mm Ivory FW has iron-gray median area mottled with brown. V-shaped AM line points to black-ringed claviform spot. Orbicular and reniform spots are edged interiorly with white and shaded dusky; reniform spot sometimes a diffuse white patch. ST area has blackish patches at midpoint and costa. **HOSTS** California buckwheat, yellow turbans, and desert trumpet.

WHITE-MARGINED SALLOW
Oxycnemis advena

Rare

93-1611 (10039)

RANGE-WIDE

TL 12–14 mm Hoary gray FW has large black-outlined claviform, orbicular, and reniform spots filled with paler gray, and thin, black, straight AM and slightly toothed PM lines. Dusky shading runs from AM to PM lines, through claviform spot, and from reniform spot to apex. Costa from orbicular spot to PM line has a whitish wash. **HOSTS** Trailing rhatany; possibly also other rhatany species.

TURQUOISE-BANDED
SALLOW

WEBBED RUACODES

SANDY CRESCENT

actual size

BELOVED EMARGINEA

DARK-COLLARED EMARGINEA

SPLENDID SALLOW

WHITE-MARGINED SALLOW

FUSED SALLOW

Oxycnemis fusimacula

Uncommon
93-1615 (10046)

TL 11–15 mm Hoary gray FW has sharply zigzag AM line; tooth at midpoint projects into paler gray claviform spot. Teardrop-shaped to round orbicular and triangular reniform spots may be separate or joined and paler gray; reniform spot bleeds along costa to PM line. Thin black PM line has black shading along outer half. Basal and terminal areas are paler gray; terminal area has black dashes at veins. Some individuals have indistinct spots. **HOSTS** Rhatany.

FERVENT SALLOW

Unciella primula

Rare
93-1623 (10111)

TL 16–20 mm Ivory FW has hoary gray median area containing small, black-centered, white orbicular spot and large, white, kidney-shaped reniform spot. Basal and ST/terminal areas are ivory, irregularly mottled with gray. AM and PM lines are indistinct. Superficially resembles Splendid Sallow, but dark areas are more mottled. **HOSTS** Unknown. **NOTE** Very similar to sympatric Ardent Sallow (*U. flagrantis*, not shown), which has a round (not kidney-shaped) reniform spot.

ANNAPHILAS, GROTELLAS, and GOLDEN FLOWER MOTHS

SUPERFAMILY Noctuoidea (93)
FAMILY Noctuidae SUBFAMILY Stiriinae

Small to medium-sized moths. The annaphilas have brownish-gray forewings with scattered metallic scales and pale markings in the PM area, and orange, white, or pink hindwings that are often held slightly exposed while the moth nectars at flowers. The grotellas are white with crisp black markings. Most of the other moths in this group have tan, yellow, or golden forewings that are held tented over the body at rest. Most of these species are nocturnal and will come to lights, but the annaphilas and many others can be found at flowers during daytime.

ASTRAL ANNAPHILA

Annaphila astrologa

Uncommon
93-1638.9 (9857)

TL 10–13 mm Resembles White Annaphila, but HW is orange with black border; sometimes has a black basal area and small black discal dot. FW pattern overlaps with White Annaphila. In general, individuals where white PM line is thin and well defined, teeth of inner PM line do not touch reniform spot, and ST area lacks a white wash, are usually Astral Annaphila. Intermediate individuals may not be identifiable by FW alone, though range can be useful in some instances. **HOSTS** Whispering bells. **NOTE** Diurnal; often seen at blackbrush and other flowers during daytime.

ILLUSTRATED ANNAPHILA

Annaphila depicta

Uncommon
93-1639.9 (9866)

TL 10–13 mm Brownish-gray FW has thin black median line and light brown PM area containing solid gray reniform spot. ST area is brownish gray. AM line is gray and indistinct, sometimes absent. HW is orange with narrow black terminal band, large black discal spot, and faded blackish median line. **HOSTS** Unknown. **NOTE** Diurnal; often seen at flowers during daytime.

FUSED SALLOW

FERVENT SALLOW

actual size

ASTRAL ANNAPHILA

actual size

ILLUSTRATED ANNAPHILA

TENTH ANNAPHILA

Uncommon

Annaphila decia   93-1640.2 (9868)

TL 10–13 mm Resembles Astral Annaphila, but orange HW has black median band and large black discal spot. FW has bold black median band. Reniform spot is thinly and evenly outlined in grayish white. Grayish PM line is thin and often indistinct in costal half. **HOSTS** Unknown. **NOTE** Diurnal; often seen at flowers during daytime.

WHITE ANNAPHILA

Common

Annaphila diva   93-1640.3 (9869)

TL 10–13 mm Dark brownish to gray FW is sprinkled with metallic blue scales. Scalloped white PM line connects to large, white, beige-centered reniform spot. ST area is washed with grayish white. FW pattern overlaps with some Astral Annaphila. HW is white with thick black outer margin. **HOSTS** Spring beauty. **NOTE** Diurnal; often seen at ceanothus flowers during daytime.

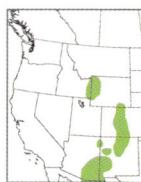

WHITE-SHOULDERED BROWN

Rare

Plagiomimicus dimidiata   93-1648 (9745)

TL 14–16 mm White FW has brown lower half fading to grayish brown at outer margin, with white V-shaped PM line. Median edge of brown area has two downward-pointing teeth. Dark brown subapical patch is framed by white apical dash that connects to PM line. Orbicular and reniform spots (sometimes indistinct) are outlined in darker brown. **HOSTS** Possibly sunflower.

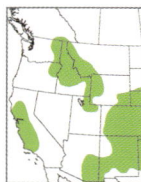

FROTHY MOTH

Uncommon

Plagiomimicus spumosum   93-1651 (9748)

TL 14–18 mm Hoary brown FW has a darker brown median area with whitish, often-indistinct, straight AM and V-shaped PM lines. Apex is pointed with whitish apical dash. **HOSTS** Seeds of sunflower.

OLIVE BROWN

Uncommon

Plagiomimicus tepperi   93-1659 (9755)

TL 14–16 mm Olive-green to olive-brown FW has bold, white, V-shaped PM and straight AM and ST lines. Inner median area and subapical patch shaded darker. Darker shading basally along inner ST line and outer terminal line. Central vein slightly whitish. **HOSTS** Tasselflower brickellia and false boneset.

SUNNY WHITEBAND

Uncommon

Lineostriastiria hachita   93-1665 (9758)

TL 12–14 mm Yellow-orange FW has crisp black AM and PM lines bordering bold white median area. ST line is a row of black spots. **HOSTS** Hairyseed bahia. **NOTE** Regularly seen on flowers during daytime.

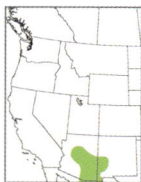

SNOWY WHITEBAND

Rare

Lineostriastiria sexseriata   93-1666 (9759)

TL 12–14 mm Resembles Sunny Whiteband, but FW is entirely white, with a yellowish fringe. Thorax is slightly yellowish. **HOSTS** Unknown. **NOTE** Regularly seen on flowers during daytime.

TENTH ANNAPHILA

WHITE ANNAPHILA

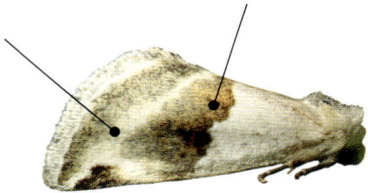

WHITE-SHOULDERED
BROWN

FROTHY MOTH

OLIVE BROWN

SUNNY
WHITEBAND

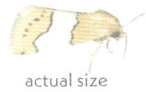

actual size

SNOWY WHITEBAND

517

CAREFREE FLOWER MOTH
Xanthothrix neumoegeni

Rare 93-1670 (9771)

TL 9–12 mm Orange FW has soft-edged, dark orange AM, PM, and ST lines, broad median band, and veins thinly traced with dark orange. Base color of FW is patchy, with paler and darker areas. **HOSTS** Asteraceae; likely coreopsis. **NOTE** Regularly seen on flowers during daytime.

HOARY FLOWER MOTH
Xanthothrix ranunculi

Rare 93-1671 (9772)

RANGE-WIDE

TL 9–12 mm FW has yellowish (sometimes grayish) scales over a black base, creating a hoary look. Form "albipunctata" of northern CA often has a large white reniform spot. **HOSTS** Coreopsis. **NOTE** Regularly seen on flowers during daytime.

CARAMEL FLOWER MOTH
Chrysoecia scira

Uncommon 93-1672 (9761)

RANGE-WIDE

TL 15–17 mm Rounded caramel-brown FW has white ST/terminal band and dark brown at outer median area, with hoary bluish-gray patches along costa and below reniform spot. Thin white AM and PM lines are scalloped. Orbicular spot is a small white dot; reniform spot is a thin white crescent. **HOSTS** Unknown. **NOTE** White-spotted Flower Moth (*C. gladiola*, not shown) is very similar but has a white patch at outer median area.

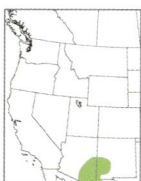

BLACK-ETCHED FLOWER MOTH
Chrysoecia atrolinea

Uncommon 93-1674 (9764)

RANGE-WIDE

TL 13–15 mm Yellow-orange FW has crisp, black, double AM and PM and single median lines, and outlined orbicular and reniform spots. Lines are incomplete or fragmented, particularly PM line. **HOSTS** Unknown. **NOTE** Regularly seen on flowers during daytime.

GILDED SEEDCROPPER
Basilodes chrysopis

Common 93-1678 (9780)

RANGE-WIDE

TL 16–18 mm Golden FW has a dark brownish median area. Rounded golden orbicular and reniform spots are outlined in brown; reniform spot usually has a small black dot at middle. Terminal area is shaded grayish. **HOSTS** Cowpen daisy.

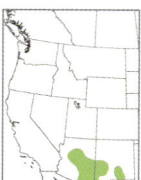

TOOTHED STOWAWAY
Cirrhophanus dyari

Uncommon 93-1680 (9765)

RANGE-WIDE

TL 16–18 mm Golden-orange FW has a crisp PM line with a single basal-pointing tooth bordering a paler ST/terminal area. Veins are darker gold. Diffuse AM, median, and ST lines are darker orange. **HOSTS** Unknown. **NOTE** Sometimes encountered on flowers during daytime.

CAREFREE FLOWER MOTH

HOARY
FLOWER MOTH

actual size

CARAMEL FLOWER MOTH

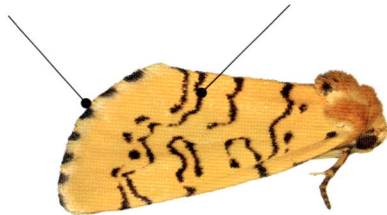

BLACK-ETCHED FLOWER MOTH

GILDED SEEDCROPPER

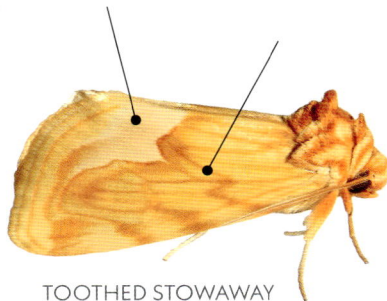

TOOTHED STOWAWAY

GRAY-EDGED YELLOW
Eulithosia composita   **Rare**   RANGE-WIDE

93-1683 (9773)

TL 12–13 mm Mustard-yellow FW is unmarked. Fringe is iron gray. **HOSTS** Unknown.

STRIPED EULITHOSIA
Eulithosia discistriga   **Rare**   RANGE-WIDE

93-1684 (9769)

TL 12–15 mm Ivory FW has brown veins that widen toward outer margin and extend into fringe. Basal and PM/ST areas have diffuse brown shading. Thorax is stripy brown. **HOSTS** Unknown.

BEAUTIFUL EULITHOSIA
Eulithosia plesioglauca   **Uncommon**   RANGE-WIDE

93-1685 (9767)

TL 13–15 mm Iridescent rosy-violet FW has diffuse white median band and peach ST/terminal band. Veins are traced with dark brown in basal and ST/terminal areas. Thorax is orange. **HOSTS** Yerba porosa.

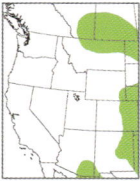

YELLOW SUNFLOWER MOTH
Stiria rugifrons   **Uncommon**   RANGE-WIDE

93-1688 (9785)

TL 18–20 mm Yellow FW has chocolate-brown saddle along inner margin. Brown fringe spills into middle of terminal area about one-third along outer margin from apex. Scalloped AM and PM lines are indistinct. Reniform spot most obvious as a tiny brown dot. Thorax is brown. **HOSTS** Sunflower. **NOTE** It has been traditionally believed that the range of Yellow Sunflower Moth does not overlap with that of Western. DNA barcoding data suggests the AZ population of Yellow may actually be a separate, undescribed species; further study is needed.

WESTERN SUNFLOWER MOTH
Stiria intermixta   **Rare**   RANGE-WIDE

93-1689 (9785.1)

TL 18–21 mm Resembles Yellow Sunflower Moth, but brown shading in terminal area tapers evenly to apex. FW pattern overlaps with Yellow Sunflower Moth and some intermediate individuals may not be visually identifiable to species where ranges meet. **HOSTS** Sunflower.

SPLENDID GOLDEN MOTH
Chalcopasta territans   **Rare**   RANGE-WIDE

93-1695 (9775)

TL 15–16 mm Metallic gold FW has broad brown patches at AM and PM areas that connect to narrow brown costal edge. Fringe and thorax are brown. **HOSTS** Unknown.

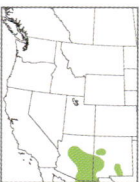

BEAUTIFUL GOLDEN MOTH
Chalcopasta howardi   **Rare**   RANGE-WIDE

93-1696 (9776)

TL 16–19 mm Resembles Splendid Golden Moth, but PM patch is broader with a watermarked reniform spot, and basal area is brown with a gold patch in outer half; AM patch is absent. **HOSTS** Unknown.

GRAY-EDGED YELLOW

STRIPED EULITHOSIA

BEAUTIFUL
EULITHOSIA

actual size

YELLOW SUNFLOWER
MOTH

WESTERN SUNFLOWER
MOTH

SPLENDID GOLDEN MOTH

BEAUTIFUL GOLDEN MOTH

SHINING GOLDEN MOTH

Rare   RANGE-WIDE

Chalcopasta fulgens   93-1697 (9777)

TL 18–20 mm Resembles Splendid Golden Moth, but PM patch is broader and inner basal and AM areas are brown. Costa is narrowly brown, but costal AM patch is reduced or absent. **HOSTS** Unknown.

POETRY MOTH

Common   RANGE-WIDE

Neumoegenia poetica   93-1699 (9737)

TL 11–13 mm Metallic gold FW has broad white costal stripe connected to narrow white terminal band and white thoracic collar. White reniform spot has two long sharp teeth along veins. **HOSTS** Fewflower beggarticks.

POWDERED OWLET

Uncommon   RANGE-WIDE

Narthecophora pulverea   93-1699.3 (9731)

TL 10–12 mm Hoary brown FW has jagged, dark brown AM and PM lines and pale kidney-shaped reniform spot indistinctly outlined with dark brown. **HOSTS** Unknown.

GOLD-LINED OWLET

Uncommon   RANGE-WIDE

Argentostiria koebelei   93-1699.4 (9779)

TL 11–13 mm White FW has golden-brown, V-shaped, double PM line with a small black dot at trough of V. Golden-brown basal line is straight, and tan terminal area is bordered by golden-brown ST line that spikes at midpoint to touch PM line. Thorax is golden brown with white patches laterally. **HOSTS** Unknown.

SPOTTED GROTELLA

Uncommon   RANGE-WIDE

Grotella sampita   93-1702 (11215)

TL 10–12 mm Snow-white FW has crisp black spots along basal (two spots), AM (three), PM (four), and terminal (seven) lines. Legs are striped black and white. **HOSTS** Unknown.

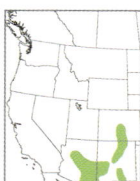

WHITE GROTELLA

Uncommon   RANGE-WIDE

Grotella blanca   93-1703 (11216)

TL 11–13 mm Resembles Spotted Grotella but lacks terminal line of spots and central spot of PM line is displaced toward outer margin. **HOSTS** Unknown.

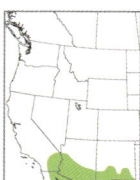

THICK-LINED GROTELLA

Uncommon   RANGE-WIDE

Grotella binda   93-1708 (11221)

TL 11–14 mm White FW has thick, fragmented, black AM and PM lines and thick black costal dashes at median and ST lines. Reniform spot is a small black dot. Thick black terminal line is usually crossed by thin white veins. **HOSTS** Unknown.

SHINING GOLDEN MOTH

actual size

POETRY MOTH

POWDERED OWLET

GOLD-LINED OWLET

SPOTTED GROTELLA

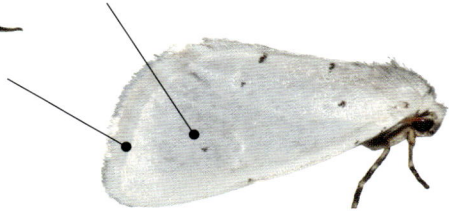

WHITE GROTELLA

THICK-LINED GROTELLA

TRICOLORED GROTELLA
Grotella tricolor

Uncommon
93-1709 (11222)

RANGE-WIDE

TL 11–13 mm Resembles Thick-lined Grotella, but ST area is brown and bordered by fragmented PM and ST lines. Terminal area has brown patches at apex and toward anal angle. **HOSTS** Unknown.

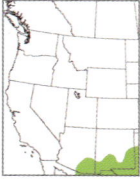

OBTUSE YELLOW
Azenia obtusa

Uncommon
93-1724 (9725)

RANGE-WIDE

TL 10–13 mm Lemon-yellow FW has square brown patches at costal ST and median lines and at inner margin of median line. Median line has an angled, narrow rectangular bar between patches, sometimes connected. ST line is a row of brown spots. AM and PM lines are present as three or four brown spots each. Thorax is brown. **HOSTS** Dodder; possibly also ragweed.

MODEST YELLOW
Azenia edentata

Uncommon
93-1725 (9726)

RANGE-WIDE

TL 10–13 mm Yellow FW has small brown patch at costal median line and sometimes also at inner margin. AM and PM lines are represented as one or two dots each in central wing. Rare, well-marked individuals may have a full ST line and resemble Obtuse Yellow but lack rectangular bar at central median line. Thorax is yellow. **HOSTS** Unknown.

CHECKERED YELLOW
Azenia implora

Uncommon
93-1728 (9729)

RANGE-WIDE

TL 11–13 mm Pale yellow FW has irregular, soft, brown median band and distinct dark brown spots at inner AM and PM lines. AM, PM, and ST lines are present as variable numbers of sometimes-indistinct brown dots; ST line may sometimes appear as a fragmented brownish band. **HOSTS** Unknown.

ROGENHOFER'S SALLOW
Metaponpneumata rogenhoferi

Common
93-1732.6 (9074)

RANGE-WIDE

TL 11–13 mm Variable. Grayish to dark brown FW has a pale kidney-shaped reniform spot that is incompletely edged in black, with partial brownish fill; inner lobe slightly smaller than outer. Orbicular spot absent. Apex has small dusky patch. All other markings may or may not be present, including narrow, brown-edged, black AM and PM lines; blackish patches above and below reniform spot; pale basal area; warm brown shading at inner AM line; and pale wavy ST line. **HOSTS** Herbaceous plants, including corn and sorghum.

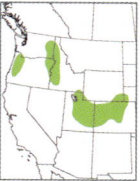

FOUR-SPOTTED MOTH
Tyta luctuosa

[IN] Very Common
93-1733.1 (9063.1)

RANGE-WIDE

TL 12–14 mm Brown FW has large white patch at outer PM area and fragmented black AM and PM lines edged slightly with paler brown. Inner ST line is mottled white; outer ST line is a row of black chevrons. **HOSTS** Bindweed. **NOTE** Intentionally introduced as a biological control for bindweed. Occasionally encountered during daytime.

TRICOLORED
GROTELLA

OBTUSE
YELLOW

MODEST YELLOW

actual size

CHECKERED
YELLOW

ROGENHOFER'S
SALLOW

FOUR-SPOTTED MOTH

ONCOCNEMIDINE SALLOWS

SUPERFAMILY Noctuoidea (93)
FAMILY Noctuidae SUBFAMILY Oncocnemidinae

Medium-sized, predominantly gray to grayish-brown moths with long rounded wings that are often held partially or fully tented at rest. Many species have crisp black markings. Adults are nocturnal and will come to lights; a few species, such as some *Sympistis* and *Neogalea*, may be found nectaring at flowers in daytime.

FINE-LINED SALLOW
Catabena lineolata

Uncommon
93-1765 (10033)

RANGE-WIDE

TL 13–14 mm Pale gray FW is tinged brownish and has black veins finely edged in white. Orbicular and claviform spots are usually visible as long pale teardrops thinly outlined in black. Zigzag PM line is usually indistinct except for blackish spots at veins. Superficially resembles Thread-lined Sallow but separable by lack of dusky apical band or thoracic stripe and presence of orbicular and claviform spots. Thorax has a forward-pointing crest like those of hooded owlets. **HOSTS** Goldenrod and vervain.

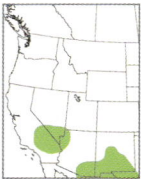

DIMORPHIC SALLOW
Catabenoides terminellus

Uncommon
93-1768 (10037)

RANGE-WIDE

TL 12–14 mm Sexually dimorphic. Male has pale gray FW with veins thinly traced with black and thicker black interveinal dashes in terminal area. Inner PM and ST areas shaded dusky. AM and PM lines are black dots, primarily visible at inner margin of PM line. Female the same, but with blackish stripe from base to center of outer margin. Thorax has a forward-pointing crest like those of hooded owlets. **HOSTS** Unknown.

LANTANA STICK MOTH
Neogalea sunia

Uncommon
93-1770 (10032)

RANGE-WIDE

TL 16–18 mm Hoary grayish-brown FW has indistinct blackish veins. Thin white orbicular spot is usually set in a long blackish streak. Reniform spot sometimes present as an indistinct grayish dot. Black interveinal dashes in terminal area are tipped by indistinct white dashes at ST line. Thorax has a forward-pointing crest like those of hooded owlets. **HOSTS** Lantana.

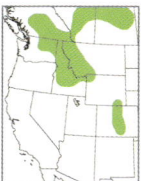

TOADFLAX BROCADE
Calophasia lunula

[IN] Uncommon
93-1771 (10177)

RANGE-WIDE

TL 14–17 mm Yellow-brown FW has blackish median line pinched along inner half between pale black-edged AM and PM lines. Bold white reniform spot is curving, and small white orbicular spot is outlined in black. Claviform spot is a long white teardrop pierced by end of thin black basal dash. ST area has thick black interveinal dashes. **HOSTS** Butter-and-eggs (toadflax). **NOTE** Intentionally introduced as biocontrol for toadflax.

FULL MOON SALLOW
Behrensia conchiformis

Common
93-1775 (10178)

RANGE-WIDE

TL 12–14 mm Gray to brown FW has blackish median area containing large, round, whitish orbicular spot filled with dusky shading at center, and gray reniform spot thinly outlined in black with blackish bar at center and inner half partially obscured by dark shading. Basal and ST areas usually have areas of yellow-green scales, and terminal line is yellow-green dashes. Wavy AM and PM lines are double. Some individuals have pale greenish edging to AM and PM lines. **HOSTS** Southern honeysuckle and common snowberry.

FINE-LINED SALLOW

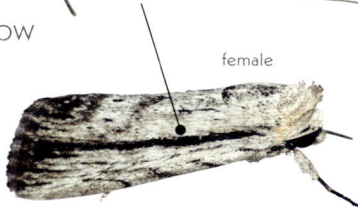

female

male

DIMORPHIC
SALLOW

male

actual size

LANTANA STICK MOTH

TOADFLAX BROCADE

FULL MOON SALLOW

527

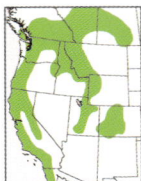

SHROUDED SALLOW
Pleromelloida conserta

Uncommon
93-1777 (10027)

TL 15–17 mm Variable. Hoary gray FW has paler gray costa (often indistinct) and blackish-traced veins. Black AM and PM lines are indistinctly edged with pale gray; AM line is wavy where it meets costa. Median line is dusky. Pale gray ST line is a row of chevrons, often indistinct. Some individuals have very dark gray FW with contrasting costa and reniform spot and indistinctly black veins. Hoary gray thorax is densely hairy. **HOSTS** Common snowberry and honeysuckle.

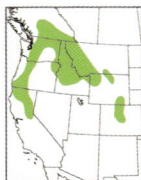

CLOAKED SALLOW
Pleromelloida bonuscula

Rare
93-1778 (10029)

TL 15–17 mm Resembles Shrouded Sallow but often has a brownish wash and a thin brown line parallel to inner margin from base to dusky gray ST area, sometimes visible primarily as an elongated white dash at ST line. Pale gray ST line is a solid line in inner half. AM line is straight and strongly slanting at costa. **HOSTS** Unknown; likely snowberry and honeysuckle.

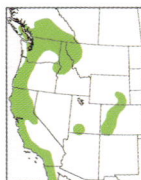

ASHY SALLOW
Pleromelloida cinerea

Uncommon
93-1780 (10031)

TL 14–16 mm Gray FW has ST area with dusky shading slanting from apex, and black interveinal dashes tipped by white ST line spots. Inner PM line indistinctly visible where it crosses bold, black, pale-edged dash. AM and median lines primarily visible as diffuse dusky dashes along costa. **HOSTS** Southern honeysuckle and snowberry.

GOLD-WINGED COPANARTA
Copanarta aurea

Uncommon
93-1781 (10169)

TL 10–12 mm Dark gray FW has large white reniform spot with dusky shading at center; black-outlined orbicular spot usually indistinct. Thin black PM line is backed by whitish shading. ST area has thin black interveinal dashes. Jagged white ST line is often indistinct except near anal angle. Bright orange HW has a thick black border. **HOSTS** Unknown.

LONG-SPOTTED SALLOW
Sympistis gracillinea

Rare
93-1785.5 (10043)

TL 10–11 mm Hoary gray FW has very elongated, slightly paler gray claviform and orbicular spots that stretch across median area; orbicular spot usually touches or joins with indistinct reniform spot. Costa and ST/terminal area shaded dark gray. AM and PM lines are absent. **HOSTS** Littleleaf rhatany.

SCRIBBLED SALLOW
Sympistis perscripta

Uncommon
93-1797 (10154)

TL 16–17 mm Light gray FW has thick, wavy, black AM line and thinner, toothed PM line edged outwardly with warm brown. Large round orbicular, claviform, and reniform spots are thinly outlined in black with pale edging inside. Black interveinal dashes in ST area are tipped by white dots of ST line. Veins are thinly dusky, particularly in ST/terminal area. **HOSTS** Snapdragon and toadflax.

SHROUDED SALLOW

actual size

CLOAKED SALLOW

ASHY SALLOW

GOLD-WINGED COPANARTA

LONG-SPOTTED SALLOW

SCRIBBLED SALLOW

CALIFORNIA SALLOW
Sympistis behrensi

Rare

93-1798 (10155)

TL 15–17 mm Resembles Scribbled Sallow, but FW is tan to brownish, AM and PM lines are thinner and less even, and orbicular and reniform spots are indistinctly outlined. Grey veins cross PM line at scallop points. ST area has short black interveinal dashes tipped with tan ST line. **HOSTS** Unknown.

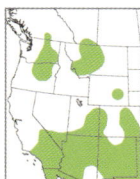

HARROW MOTH
Sympistis occata

Uncommon

93-1801 (10101)

NORTH

SOUTH

TL 12–15 mm Light brown FW has blackish patch at inner median area and thin, scalloped, black AM and PM lines. Large round orbicular and reniform spots and smaller claviform spot are black outlines. Black interveinal dashes in ST area are tipped by white dots of ST line. **HOSTS** Unknown.

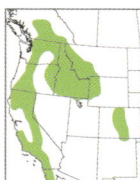

SADDLED SALLOW
Sympistis umbrifascia

Rare

93-1802 (10122)

RANGE-WIDE

TL 14–16 mm Peppery, light gray FW has blackish median area and small dusky orbicular and reniform spots outlined with white that connect to thin white AM and PM lines. Costa of median area is grayish. ST/terminal area is diffusely shaded dusky. Head is dark. **HOSTS** Unknown; possibly penstemon.

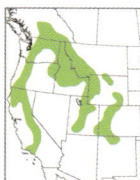

TWO-LINED SALLOW
Sympistis fifia

Rare

93-1822 (10066)

RANGE-WIDE

TL 16–19 mm Brownish-gray FW has curving black AM and PM lines and thick black streaks from base to inner PM line and from outer AM line to central outer margin. ST area has thick black interveinal dashes. **HOSTS** Snowberry.

HAPPY SALLOW
Sympistis poliochroa

Rare

93-1853 (10094.4)

RANGE-WIDE

TL 17–20 mm Mottled gray FW has elongated orbicular spot that connects to middle of reniform spot; entire shape is outlined in white and shaded dusky at center. Elongate claviform spot is whitish with black streak down middle. Indistinct median line is a diffuse blackish band. ST area is paler gray basally. Lower ST area has thick black interveinal dashes tipped with white dots of ST line. **HOSTS** Unknown; possibly penstemon.

AUGUST SALLOW
Sympistis augustus

Uncommon

93-1865 (10096)

RANGE-WIDE

TL 15–19 mm Warm brown FW has darker brown median area containing large, warm brown orbicular and reniform spots outlined with black and edged inwardly with whitish color, and oblong claviform spot outlined in black. Whitish, toothed ST line bordered broadly by dark brown shading in ST area. Veins in central FW have whitish tracing. **HOSTS** Unknown; possibly penstemon.

GARLAND SALLOW
Sympistis greyi

Uncommon

93-1950 (10131.1)

RANGE-WIDE

TL 15–18 mm Light gray FW has thin, black, looping PM line that is connected by a short bar near inner margin to thick, black, straight median line. Terminal area has black interveinal dashes; the most central stretches to cross PM line. Head is blackish. **HOSTS** Snowberry.

CALIFORNIA SALLOW

HARROW MOTH

SADDLED SALLOW

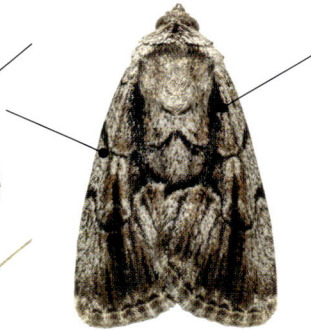

TWO-LINED SALLOW

actual size

HAPPY SALLOW

AUGUST SALLOW

GARLAND SALLOW

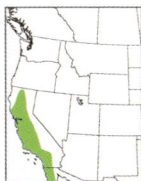

DELTA-LINED SALLOW
Sympistis ragani

Uncommon
93-1951 (10132)

RANGE-WIDE

TL 14–16 mm Resembles Garland Sallow, but median line is thinner, widening into triangular patches at costa and inner margin, and bar connecting median and PM lines is usually a long thin dash extending beyond lines. Most individuals reliably separated by range. **HOSTS** Southern honeysuckle.

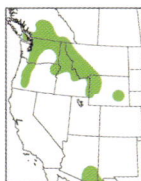

SEMICOLLARED SALLOW
Sympistis semicollaris

Rare
93-1952 (10133)

RANGE-WIDE

TL 15–18 mm Resembles Garland Sallow, but median line is thin, and median and PM lines form points where they attach to vertical bar at inner median area. Head is gray. **HOSTS** Snowberry.

WOOD-NYMPHS and FORESTERS

SUPERFAMILY Noctuoidea (93)

FAMILY Noctuidae SUBFAMILY Agaristinae

The small to medium-sized moths in this group have very different phenotypes and habits. The wood-nymphs are white moths with colorful markings along the margin of the forewing; they stretch their fuzzy legs in front of their head at rest. The closely related *Gerra* and *Gerrodes* have brown forewings that are partly orange on the underside, and orange hindwings. The *Euscirrhopterus* are grayish brown with a bold pale arc across the forewing. All tent their wings over their body. They are nocturnal and will come to lights. Foresters are black moths with bold white to yellowish spots and orange tufts at the base of the legs; they rest with wings spread or folded flat. They are diurnal and usually found in meadows or forest edges among low vegetation or at flowers.

WILSON'S WOOD-NYMPH
Xerociris wilsonii

Uncommon
93-1963 (9298)

RANGE-WIDE

TL 18–20 mm White FW has wine-red ST/terminal area overlaid with wispy white ST line. Wavy olive-green PM line runs up inner margin to blackish-blue AM band. Orbicular and reniform spots are olive green, set in grayish costal patch. Thorax has a raised dorsal stripe of blackish-blue scales. **HOSTS** Sorrelvine.

WHITE-PATCHED REVEALER MOTH
Gerra sevorsa

Uncommon
93-1968 (9303)

RANGE-WIDE

TL 16–20 mm Hoary brown FW has whitish-gray patch in outer PM area. Brown-outlined orbicular and reniform spots have dark brown dash at center (sometimes indistinct). Grayish wavy ST line is broadly shaded warm brown basally. HW is orange with broad black terminal band. When walking, usually holds wings folded above body, revealing bold yellow-orange patches on underside of FW and HW. **HOSTS** Canyon wild grape and Virginia creeper.

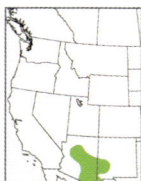

WHITE-STRIPED REVEALER MOTH
Gerrodes minatea

Uncommon
93-1969 (9304)

RANGE-WIDE

TL 23–28 mm Grayish FW has black double ST line that curves around inner margin to AM area, bordered by narrow white stripe along lower inner margin. Central FW has broad brown streaks on either side of thinly white-traced central vein. HW is orange with broad black terminal band. Underside of wings is orange with brown terminal band on both FW and HW. **HOSTS** Wild grape.

DELTA-LINED SALLOW

SEMICOLLARED SALLOW

actual size

WILSON'S
WOOD-NYMPH

actual size

WHITE-PATCHED
REVEALER MOTH

WHITE-STRIPED REVEALER MOTH

PURSLANE MOTH
Euscirrhopterus gloveri

Very Common
93-1972 (9307)

TL 16–21 mm Hoary grayish-brown FW has broad white arc that passes behind orbicular and reniform spots. Spots have black outer ring, yellow inner ring, and gray center. HW is orange with broad black terminal band and white fringe. **HOSTS** Moss rose, shrubby and common purslane, beet, and fameflower.

STAGHORN CHOLLA MOTH
Euscirrhopterus cosyra

Very Common
93-1974 (9308)

TL 18–21 mm Resembles Purslane Moth but lacks orbicular and reniform spots. **HOSTS** Cholla.

TWO-SPOTTED FORESTER
Alypiodes bimaculata

Uncommon
93-1977 (9312)

TL 19–23 mm Black FW has bold white patches in outer AM, median, and PM areas, with metallic blue orbicular and reniform spots in between. Black thorax and collar have a total of six white spots. Fringe at apex is white. HW is black with white spot (sometimes indistinct) at center and white apical fringe. **HOSTS** Four o'clock. **NOTE** Diurnal.

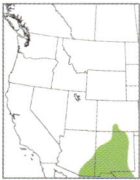

BLUE-SPOTTED FORESTER
Alypiodes flavilinguis

Uncommon
93-1977.1 (9312.1)

TL 19–22 mm Resembles Two-spotted Forester, but with only four spots on black thorax and collar (rare individuals may have an indistinct third pair), and inner basal, and sometimes inner median, areas usually have metallic blue patches. White patch in AM area of FW is smaller or sometimes nearly absent. HW is entirely black or with just a small, indistinct, white spot. **HOSTS** Four o'clock. **NOTE** Diurnal.

GERONIMO FORESTER
Alypiodes geronimo

Uncommon
93-1978 (9313)

TL 17–20 mm Sexually dimorphic. Black FW of female has bold white patches at inner and outer PM area; male has a third white patch at outer median area. Long thin reniform spot is metallic blue; FW sometimes shows additional short metallic blue streaks in inner median area. Sides of thorax underneath base of FW have tuft of bright orange scales. HW is completely black in female but has bold white median area in male. **HOSTS** Spiderling. **NOTE** Diurnal.

EIGHT-SPOTTED FORESTER
Alypia octomaculata

Uncommon
93-1979 (9314)

TL 16–21 mm Sexually dimorphic. Resembles Langton's Forester, but antennae are entirely black and FW spots are larger on average. White HW basal patch is present in both sexes and lacks black streaks; patch is larger in male, often continuing along inner margin (may be hidden). Abdomen of male has white dorsal stripe; that of female is entirely black or has a few white dots at base and tip. **HOSTS** Grape, peppervine, and Virginia creeper. **NOTE** Diurnal.

PURSLANE MOTH

STAGHORN
CHOLLA MOTH

TWO-SPOTTED FORESTER

actual size

BLUE-SPOTTED FORESTER

male

GERONIMO FORESTER

male

EIGHT-SPOTTED FORESTER

LANGTON'S FORESTER
Alypia langtoni
Common 93-1981 (9318)

TL 14–17 mm Sexually dimorphic. Black FW has two large, rounded, pale yellowish spots and metallic blue scales in narrow patches around yellow spots. Black thorax has two fluffy yellow stripes laterally, and legs have large orange puffs at base. HW of female has a single, rounded, pale yellowish spot; HW of male also has white basal area usually with one or more black streaks along veins. Abdomen of male usually has white dorsal stripe; that of female is entirely black. Antennae have white rings around basal half. **HOSTS** Willowherb and fireweed. **NOTE** Diurnal.

RIDINGS' FORESTER
Alypia ridingsii
Common 93-1982 (9319)

TL 11–14 mm Black FW has triangular white AM patch, small round median patch, and curving PM patch crossed by thin black veins. HW has small white AM spot and roundish PM spot crossed by black veins. Black thorax and abdomen are unmarked, and collar has two white spots. Middle segment of forelegs has orange tufts. Antennae are black. **HOSTS** Evening primrose. **NOTE** Diurnal.

MARIPOSA FORESTER
Alypia mariposa
Uncommon 93-1983 (9320)

TL 11–14 mm Resembles Ridings' Forester, but PM patch is crescent-shaped and unmarked by veins. **HOSTS** Clarkia. **NOTE** Diurnal.

MACCULLOCH'S FORESTER
Androloma maccullochii
Common 93-1984 (9321)

TL 11–14 mm Black FW resembles that of Ridings' Forester, but triangular AM patch is more pointed at medial end and is usually bisected by a black vein, and PM patch is straight along medial edge. Black thorax has yellowish stripes laterally, and abdomen has a dorsal line of small white dots at base. Middle segment of legs is orange. Basal half of antennae is ringed with white. **HOSTS** Willowherb and fireweed. **NOTE** Diurnal.

GROUNDLINGS
SUPERFAMILY Noctuoidea (93)
FAMILY Noctuidae SUBFAMILY Condicinae

Small to medium-sized moths with relatively short wings and a triangular shape. Most rest with wings slightly tented. They are typically somewhat mottled or streaky in appearance, sometimes speckled with pale scales, but Whaleback Moth is boldly patterned. Adults are nocturnal and will come to lights.

BLACK-DASHED GROUNDLING
Condica temecula
Rare 93-1990 (9691)

TL 15–17 mm Warm brown FW has rectangular white reniform bar (sometimes with a rounder spot at one end) with long black interveinal dashes below in ST/terminal area. Veins are traced with white-speckled dusky shading, with a dark brown wash behind central vein. Veins become pale yellowish in terminal area, extending into fringe. **HOSTS** Unknown.

female

LANGTON'S FORESTER

actual size

RIDINGS' FORESTER

MARIPOSA FORESTER

MACCULLOCH'S FORESTER

actual size

BLACK-DASHED GROUNDLING

GRAY-STRIPED GROUNDLING

Condica discistriga

Rare

93-1991 (9692)

RANGE-WIDE

TL 17–20 mm Tan FW has broad dusky shading down center. Dusky round orbicular and bar-shaped white or dusky reniform spots are outlined with tan. PM line is a row of black dots. Dusky veins turn white at terminal line. **HOSTS** Rabbitbrush.

BLACK GROUNDLING

Condica albolabes

Uncommon

93-1994 (9695)

RANGE-WIDE

TL 18–21 mm Dusky blackish FW has a light speckling of white scales and a small, round, white reniform spot rarely surrounded by tan marks. Orbicular spot sometimes indistinctly visible as tan ring. Often-indistinct AM and PM lines are rows of tiny white spots along nearly invisible jagged black line, rarely edged by fragmented tan. ST line sometimes visible as black-tipped tan dots. Costa has white dots at AM line and from median line to apex. Terminal line is white dots. **HOSTS** Unknown.

WHALEBACK MOTH

Stibaera thyatiroides

Uncommon

93-2017 (9716)

RANGE-WIDE

TL 19–22 mm FW has distinctive pattern of striated, warm brown median area, dark brown triangles along costa, and thin, lightly scalloped AM and PM lines. Tan thorax has dark brown central stripe. **HOSTS** Morning glory.

VERBENA MOTH

Crambodes talidiformis

Rare

93-2030 (9661)

RANGE-WIDE

TL 15–17 mm Tan FW has a dark brown inner median area and costa, and a white or tan black-edged reniform bar. Strongly jagged AM and scalloped PM lines are thin black (sometimes indistinct), and edged with tan and warm brown to create a streaked appearance. ST/terminal area below reniform bar has two long blackish interveinal dashes. Fringe is checkered. **HOSTS** Vervain.

TRIPLEX CUTWORM

Micrathetis triplex

Uncommon

93-2031 (9644)

INLAND

CALIFORNIA

TL 9–13 mm FW variably straw-colored to light brown or reddish, sometimes with light speckling. AM and PM lines are double rows of blackish dots. Reniform spot usually blackish and joined to costa by dark median-line bar. ST area shaded blackish. **HOSTS** Unknown.

EXPRESSIVE SALLOW

Micrathetis costiplaga

Uncommon

93-2032 (9645)

RANGE-WIDE

TL 10–12 mm Tan to reddish-brown FW has a dark brown triangular mark at costa, sometimes broken into a bar and a spot. Zigzag AM and PM lines are often indistinct but for black dots at tips. Fringe is silvery gray. **HOSTS** Unknown.

INCA SALLOW

Aleptina inca

Uncommon

93-2034.71 (9071)

RANGE-WIDE

TL 11–13 mm Medium gray FW has an orangish patch in inner basal area. Pale orbicular spot has dark center and black outline. Pale gray shading inside reniform spot bleeds to costa. Claviform spot incompletely outlined with black. **HOSTS** Unknown.

GRAY-STRIPED GROUNDLING

BLACK GROUNDLING

WHALEBACK MOTH

VERBENA MOTH

TRIPLEX
CUTWORM

EXPRESSIVE
SALLOW

actual size

INCA
SALLOW

FLOWER MOTHS SUPERFAMILY Noctuoidea (93)
FAMILY Noctuidae SUBFAMILY Heliothinae

Small to medium-sized, often brightly colored or boldly patterned moths with triangular wings that are typically partially or entirely tented when at rest. Most species are shades of brown, tan, or white, but a number are pink or pale green. A few species, particularly Corn Earworm, cannot survive northern winters but will migrate and recolonize northern areas each summer. Adults of many species are diurnal and can be found visiting flowers, but most species will also come to lights at night.

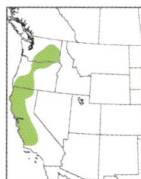

LITTLE FLOWER MOTH
Microhelia angelica
Rare
93-2035 (11056)
RANGE-WIDE

TL 6–8 mm Tiny. Black wings are lightly sprinkled with white scales and have white PM band of two large, often-connected, white spots. **HOSTS** Common hareleaf. **NOTE** Diurnal visitor to flowers.

SMALL HELIOTHODES
Heliothodes diminutiva
Very Common
93-2036 (11058)
RANGE-WIDE

TL 9–11 mm When fresh, FW is pink with gold AM and PM bands; midpoint of AM and PM and inner end of PM bands have paler whitish spots. ST band is gold. Pink and gold scales frequently wear off to reveal black ground color beneath. HW is black with whitish PM band. **HOSTS** Unknown; possibly tidytips, goldfields, and tarweed. **NOTE** Diurnal visitor to flowers.

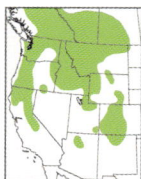

WHITE-SPOTTED MIDGET
Eutricopis nexilis
Uncommon
93-2038 (11062)
RANGE-WIDE

TL 9–11 mm Resembles Small Heliothodes, but spot at inner PM line extends into central median area. FW is often more heavily suffused with gold across central parts of pink areas. HW is black with two-spotted white PM band. **HOSTS** Pussytoes. **NOTE** Diurnal visitor to flowers.

PURPLE-LINED SALLOW
Pyrrhia exprimens
Uncommon
93-2041 (11064)
RANGE-WIDE

TL 17–19 mm Warm brown FW has blackish V-shaped median and curving PM lines that are diffusely shaded with dark brown below; PM line does not reach costa. Veins are traced with dark brown, and thin, dark brown AM and ST lines scallop between them. Hollow orbicular and shaded reniform spots are outlined with brown. **HOSTS** Deciduous trees and herbaceous plants, including cherry, knotweed, rose, sweet fern, and willow.

PINK PRAIRIE MOTH
Psectrotarsia suavis
Uncommon
93-2043 (11066)
RANGE-WIDE

TL 15–18 mm Yellow FW has bright pink AM and PM bands that appear to be made of densely arranged small spots. Costa and fringe are also pink. Orbicular spot is a pink dot, and kidney-shaped reniform spot is outlined in pink. Terminal line is a row of brown dots. **HOSTS** Unknown. **NOTE** Sometimes visits flowers during daytime.

FLOWER MOTHS

LITTLE FLOWER MOTH

SMALL
HELIOTHODES

WHITE-SPOTTED MIDGET

PURPLE-LINED SALLOW

actual size

PINK PRAIRIE MOTH

541

CORN EARWORM
Helicoverpa zea

Very Common
93-2045 (11068)

NORTH

SOUTH

TL 18–22 mm Tan to light brown FW has scalloped PM line tipped with white dots, and dark brown shading in ST area. Round orbicular spot has a dark dot in center, and reniform spot typically has a dusky dot in inner half. Brown AM and median lines are narrow and often indistinct. White HW has broad blackish terminal band with a thin yellow terminal line and white fringe when fresh. **HOSTS** Herbaceous plants and crops, including corn, cotton, tomato, tobacco, and many others. Can be a serious agricultural pest, particularly of corn.

DARKER-SPOTTED STRAW MOTH
Heliothis phloxiphaga

Very Common
93-2046 (11072)

RANGE-WIDE

TL 17–20 mm Pale to warm brown FW has strong, warm brown shading below V-shaped median line, shifting to dark brown near costa. Reniform spot is outlined with black dots and shaded dark brown. ST area is dark brown at costa. **HOSTS** Flowers and seedheads of various herbaceous plants.

OREGON GEM
Heliothis oregonica

Uncommon
93-2050 (11078)

RANGE-WIDE

TL 14–16 mm FW is warm brown in basal half, with large, solid brown orbicular and reniform spots narrowly separated by a tan line. PM area and ST line are tan, and ST/terminal area is olive, with a warm brown patch at costa of ST area. HW has a wide black terminal band containing a white patch at midpoint, and large black reniform spot that blends into blackish basal shading. Underside of FW is distinctively white with bold black orbicular and reniform spots and black ST band. **HOSTS** Paintbrush; also other herbaceous plants.

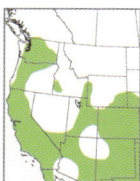

TOBACCO BUDWORM
Chloridea virescens

Very Common
93-2054 (11071)

NORTH

SOUTH

TL 15–19 mm Olive FW has parallel, dark AM, median, and PM lines; AM and PM are broadly edged with white basally. Round orbicular and reniform spots often have dark dot at center. Scalloped dark ST line often indistinct. White HW has a dark terminal band, and central veins often are dusky gray. **HOSTS** *Abutilon*, cotton, geranium; also ground cherry, tobacco, and other members of nightshade family.

PARADOXICAL GRASS MOTH
Heliocheilus paradoxus

Uncommon
93-2058 (11074)

RANGE-WIDE

TL 13–15 mm Sexually dimorphic. Straw-colored FW has a zigzag PM line often visible only as a row of black dots or sometimes indistinct. Often has a dusky smudge at reniform spot. Male has a semitranslucent bulge in outer median area. Female is often darker, with a dark ST area and jagged PM and median lines. HW has a dark terminal band with a pale patch at midpoint and large dusky reniform spot. **HOSTS** Grass.

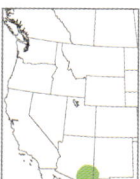

JULIA GRASS MOTH
Heliocheilus julia

Uncommon
93-2059 (11075)

RANGE-WIDE

TL 12–15 mm Bright pink FW has a single, pale yellowish streak down center and pale yellowish patch at inner basal area. Hairy thorax is pale yellow with pink wash across head and shoulders. **HOSTS** Likely grass.

CORN EARWORM

OREGON
GEM

DARKER-SPOTTED
STRAW MOTH

female

female

actual size

female

TOBACCO BUDWORM

male

female

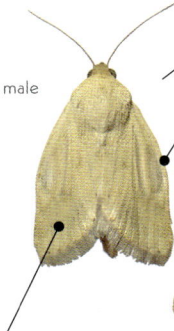

male

PARADOXICAL GRASS MOTH

JULIA GRASS MOTH

PINK-BANDED GRASS MOTH

Uncommon · RANGE-WIDE

Heliocheilus toralis   93-2060 (11076)

TL 12–14 mm   Light yellow FW has pink ST band connecting to apex, pink fringe, and small pink patch at base. Some individuals have entirely pink ST/terminal area. HOSTS Grass.

SPOTTED CLOVER MOTH

Rare · RANGE-WIDE

Protoschinia nuchalis   93-2061 (11082)

TL 15–18 mm   Tan FW has bold brown claviform, orbicular, and reniform spots outlined in black and with pale center. Pale veins are narrowly edged with dusky brown. Curving blackish AM and PM lines are crossed by veins. ST/terminal area is brown with pale ST line. HOSTS Tarragon.

GREEN FLOWER MOTH

Rare · RANGE-WIDE

Schinia simplex   93-2064 (11162)

TL 13–17 mm   Pale green FW is unmarked. HW is apricot orange with cream fringe. Hairy thorax is pale green. HOSTS Bush morning glory.

SHINING FLOWER MOTH

Uncommon · RANGE-WIDE

Schinia luxa   93-2066 (11203)

TL 15–17 mm   Ivory FW has very faint, brown AM and PM lines, usually visible as one or two small brown dots. Similar ST line is sometimes visible. Brown reniform spot may be visible as a crescent of brown shading. Tiny dusky terminal dots often present. HOSTS Blazingstar.

CITRUS GEM

Rare · RANGE-WIDE

Schinia citrinellus   93-2067 (11204)

TL 11–14 mm   Ivory FW has tiny black reniform dot and very faint grayish ST band and terminal line. Collar is pale yellowish orange. HOSTS Croton. NOTE Despite common name, not associated with citrus.

JAGUAR FLOWER MOTH

Uncommon · RANGE-WIDE

Schinia jaguarina   93-2073 (11132)

TL 14–16 mm   Light brown FW has warm brown basal area that fades toward base, and brown ST area. White AM line is strongly angled at midpoint. PM line is gently wavy. Golden-yellow HW has a dark terminal line with a pale patch at midpoint. HOSTS *Psoralea* species.

FAMILIAR FLOWER MOTH

Uncommon · NORTH / SOUTH

Schinia suetus   93-2075 (11088)

TL 13–15 mm   Tan or sometimes reddish-pink FW has white AM and PM bands that connect at inner margin. Terminal area is shaded slightly lighter. As scales wear off, blackish ground color may show through. HW is black with two white patches (sometimes connected) as median band. Hairy thorax is tan. Sometimes resembles bright individuals of Small Heliothodes but is larger and AM line is lower and better defined. HOSTS Lupine.

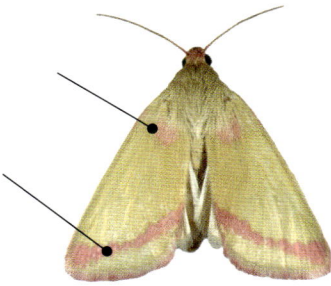

PINK-BANDED
GRASS MOTH

actual size

SPOTTED CLOVER MOTH

GREEN FLOWER MOTH

SHINING FLOWER MOTH

CITRUS GEM

JAGUAR FLOWER MOTH

FAMILIAR FLOWER MOTH

MEAD'S FLOWER MOTH

Uncommon

Schinia meadi   93-2078 (11175)

TL 13–15 mm   Pale yellowish FW has mottled pattern made of golden-tan basal and ST areas, median band, orbicular and reniform spots, and patch at central terminal area. White basal, AM, and PM lines are thinly and incompletely edged with black. Fringe is checkered. **HOSTS** Unknown.

PRIMROSE MOTH

Rare

Schinia florida   93-2082 (11164)

TL 15–19 mm   Bubble-gum-pink FW has pale yellow terminal band. Indistinct, soft AM and PM lines are pale. Thorax is pale yellow. **HOSTS** Seed capsules of evening primrose. **NOTE** Often found roosting on evening primrose flowers during daytime.

CLOUDED CRIMSON

Uncommon

Schinia gaurae   93-2083 (11168)

TL 14–16 mm   Ivory FW has pink wash in basal and terminal areas and strong pink band from costal ST area to central PM area. Median area sometimes has a tan wash. Thorax is tan in posterior half. **HOSTS** Gaura.

MEXICAN FLOWER MOTH

Uncommon

Schinia mexicana   93-2087 (11134.3)

TL 14–16 mm   Brown FW has slightly lighter median area bounded by double, black, tan-filled, straightish AM and PM lines, and with pale tan veins. Grayish claviform, orbicular, and reniform spots are outlined with tan and black. Scalloped tan ST line is edged basally with black chevrons. **HOSTS** Unknown.

BEAUTIFUL FLOWER MOTH

Uncommon

Schinia pulchripennis   93-2088 (11097)

TL 9–12 mm   Bright pink FW has curving gray AM and PM lines and golden-tan central median area. Hairy thorax is gray tipped with pink. As scales are lost with wear, black ground color shows through, and thorax becomes gray. **HOSTS** Purple owl's-clover.

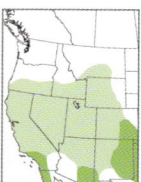

BRICKELLBUSH FLOWER MOTH

Rare

Schinia oleagina   93-2094 (11148)

TL 12–15 mm   Warm brown FW has slightly grayish-brown median and terminal areas. Smooth, pale brown AM and PM lines curve basally at costa; pale ST line is straight. Separable from Jaguar Flower Moth by relatively straight PM line. **HOSTS** Brickellbush.

BONESET FLOWER MOTH

Rare

Schinia grandimedia   93-2095 (11148.1)

TL 12–15 mm   Resembles Brickellbush Flower Moth but is slightly lighter with an olive tinge, and brown basal area fades toward base. Separable from Jaguar Flower Moth by relatively straight PM line. **HOSTS** False boneset.

MEAD'S FLOWER MOTH

roosting

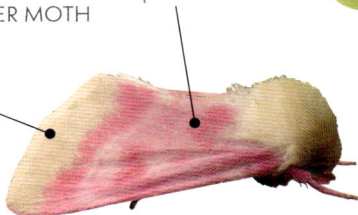

PRIMROSE MOTH

actual size

CLOUDED CRIMSON

MEXICAN
FLOWER MOTH

BRICKELLBUSH
FLOWER MOTH

BEAUTIFUL FLOWER
MOTH

BONESET
FLOWER MOTH

WHITE-BANDED SCHINIA

Schinia vacciniae

Rare

93-2100 (11087)

TL 11–12 mm Dark brown FW has ivory median area bounded by double, white-filled, black PM line and containing soft, dusky reniform-spot outline (sometimes warm brown). Irregular, pale yellow ST line is edged basally with black spots. **HOSTS** Unknown; likely fleabane and aster.

LITTLE DARK GEM

Schinia villosa

Rare

93-2104 (11083)

TL 10–12 mm Resembles larger Familiar Flower Moth, but white orbicular spot creates a tooth on AM line, PM line is narrower and white patch at inner median area is broader. ST line has a tooth near midpoint. White median spots on black HW are reduced or sometimes indistinctly defined. **HOSTS** Fleabane.

WANDERING FLOWER MOTH

Schinia errans

Uncommon

93-2110 (11124)

TL 12–14 mm Resembles Little Dark Gem, but median area is broader, white basal line is present, and olive to orange-brown reniform spot often has some black scales. **HOSTS** Tahoka daisy.

LYNX FLOWER MOTH

Schinia lynx

Rare

93-2120 (11117)

TL 10–12 mm Resembles Wandering Flower Moth, but inner AM line curves basally to nearly touch thorax, and curving PM line thins near midpoint. AM and PM lines are usually shaded outwardly with diffuse blackish. Reniform spot is often diffusely shaded blackish. **HOSTS** Fleabane and camphorweed.

ALLURING SCHINIA

Schinia siren

Uncommon

93-2123 (11115)

TL 12–13 mm Golden-tan FW has brownish-orange basal and ST areas and diffuse brownish median band. Thin white AM and PM lines are often indistinct. Veins are faintly traced with whitish color, especially in median area. Thorax is orange. **HOSTS** Camphorweed.

GUMWEED FLOWER MOTH

Schinia mortua

Rare

93-2126 (11131)

TL 12–14 mm Brownish-tan FW has brown basal and ST areas and tan basal, AM, and PM lines. Entire AM line is wavy, and PM line usually curves basally and slightly flares to meet costa. Darker reniform spot lacks outline. Some individuals are light tan with golden-tan basal and ST areas. Separable from Alluring Schinia by costal AM and PM lines and by lack of vein markings, and Lynx Flower Moth by inner AM line and less-contrasting terminal area. **HOSTS** Gumweed.

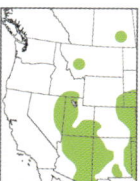

ARCIGERA FLOWER MOTH

Schinia arcigera

Uncommon

93-2134 (11128)

TL 12–13 mm Sexually dimorphic. Maroon to warm brown FW has a paler median area and terminal line. White AM line is evenly curved. Nearly straight PM line is broadly bordered by blackish shading above. Male has yellow HW with a black terminal line and white fringe; female's HW is all black with white fringe. **HOSTS** Aster, camphorweed, horseweed, and sea lavender.

WHITE-BANDED SCHINIA

LITTLE DARK GEM

WANDERING FLOWER MOTH

LYNX FLOWER MOTH

ALLURING SCHINIA

actual size

GUMWEED FLOWER MOTH

ARCIGERA FLOWER MOTH

DESERT MARIGOLD MOTH
Schinia miniana

Uncommon
93-2138 (11155)

TL 11–13 mm Golden-tan FW has thick, curving, white AM and PM lines and reddish dash at outer ST line. Some individuals have a complete reddish ST line. Thorax is yellow orange. **HOSTS** Desert marigold.

PAINTED SCHINIA
Schinia volupia

Uncommon
93-2139 (11106)

TL 11–12 mm Bright pink FW has thin, white, zigzag AM and PM lines that often show up just as white dashes at veins and patch at costa. White ST line is diffuse. Thorax is yellow orange. **HOSTS** Indian blanket.

SPANISH NEEDLES FLOWER MOTH
Schinia niveicosta

Rare
93-2146 (11167)

TL 14–16 mm Ivory FW has long pink basal dash to central median area and pink ST band. Central wing is washed pink with a touch of golden tan. Inner margin is pinkish. Terminal area and thorax are tan. **HOSTS** Spanish needle.

REGINIA PRIMROSE MOTH
Schinia regina

Uncommon
93-2148 (11166.1)

TL 13–15 mm Pink FW has golden-tan median area bordered by diffuse whitish AM and PM lines, and golden-tan terminal area. Outer median area usually has a diffuse pink patch. Basal area often has a whitish wash at center. Thorax is pale yellowish. Some individuals may be quite pale overall. **HOSTS** Palafox.

ACUTE-LINED FLOWER MOTH
Schinia acutilinea

Uncommon
93-2153 (11150.1)

TL 11–14 mm Warm brown to brownish-gray FW has darker basal and ST areas. Thin, dark, V-shaped AM and bulging PM lines are edged medially with white. Irregular ST line is white. Reniform spot is dusky. **HOSTS** Sagebrush.

TOOTHED FLOWER MOTH
Schinia walsinghami

Rare
93-2154 (11184)

TL 11–15 mm Resembles Acute-lined Flower Moth, but PM line is toothed and median area has an indistinct, diffuse, darker median band. **HOSTS** Rubber rabbitbrush; possibly also other rabbitbrush.

THIRD FLOWER MOTH
Schinia tertia

Uncommon
93-2158 (11179)

TL 12–14 mm Brown FW has white AM band and fractured black AM and PM lines. Reniform spot is metallic gray with incomplete black outline. Basal area has metallic gray patches, and ST area has metallic gray along veins. White thorax has sparse, short, black striations. **HOSTS** Probably dotted blazing star.

actual size

DESERT MARIGOLD MOTH

PAINTED SCHINIA

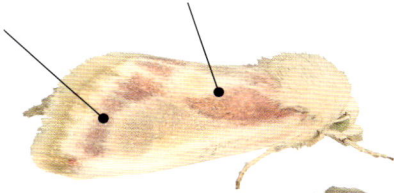

SPANISH NEEDLES
FLOWER MOTH

REGINIA
PRIMROSE MOTH

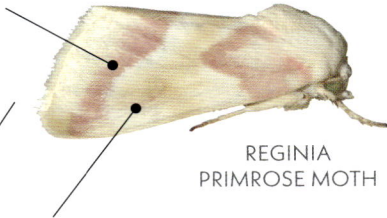

ACUTE-LINED FLOWER MOTH

TOOTHED FLOWER MOTH

THIRD FLOWER MOTH

SILVER-SPOTTED FLOWER MOTH
Schinia albafascia   **Uncommon**
93-2159 (11181)

TL 13–15 mm Dark brown FW has broad white median area bounded by fractured black AM and PM lines. Orbicular and reniform spots are incompletely outlined with black and lightly shaded with gray. Faint grayish-brown median band passes behind reniform spot. Basal and ST areas have metallic gray patches. **HOSTS** Rubber rabbitbrush.

WHITE-BANDED FLOWER MOTH
Schinia argentifascia   **Rare**
93-2163 (11180)

TL 13–15 mm Reddish-brown FW has white AM band, warm brown PM and terminal bands, and curving, dark brown AM and PM lines. Irregular ST line is white. Veins in basal and ST areas are faintly marked with pale tan. **HOSTS** Goldenbush.

APPROACHABLE FLOWER MOTH
Schinia accessa   **Uncommon**
93-2165 (11150)

TL 13–15 mm Olive-gray FW has bold, white, V-shaped AM and curving PM lines and thin white ST line. Veins are faintly marked with white, strongest in median area. PM line is edged below with black, and incomplete reniform spot and short basal dash are black. **HOSTS** Sagebrush.

RABBITBRUSH FLOWER MOTH
Schinia unimacula   **Uncommon**
93-2168 (11188)

TL 12–13 mm Ivory FW has warm brown basal area washed whitish at base, and warm brown ST area with strong, brown, squarish patch at costa. Golden-brown PM band has dusky reniform spot, sometimes reduced to one or a few blackish dots. Pale individuals may have fairly light-colored markings. **HOSTS** Rubber and black-banded rabbitbrush.

EYED FLOWER MOTH
Schinia oculata   **Rare**
93-2170 (11197)

TL 12–14 mm Satiny white FW has warm brown to brown basal area and reniform spot. Inner ST area and apical patch of terminal area are brown, shading along a gradient to white at opposite end of respective bands. Thorax is concolorous with basal area. **HOSTS** Desert broom.

CHRYSELLUS FLOWER MOTH
Schinia chrysellus   **Uncommon**
93-2171 (11199)

TL 11–13 mm Satiny white FW has golden-brown basal area, V-shaped median band, and terminal area. Thorax is warm brown. **HOSTS** Prairie broomweed.

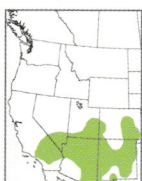

SNAKEWEED FLOWER MOTH
Schinia ciliata   **Rare**
93-2174 (11200)

TL 11–13 mm Resembles Chrysellus Flower Moth, but point of V-shaped median band bulges into reniform spot in a teardrop shape, and terminal area is ivory with warm brown shading near apex. PM line is sometimes indistinctly present as a row of darkish spots. **HOSTS** Broom snakeweed.

SILVER-SPOTTED FLOWER MOTH

WHITE-BANDED FLOWER MOTH

APPROACHABLE FLOWER MOTH

RABBITBRUSH FLOWER MOTH

EYED FLOWER MOTH

actual size

CHRYSELLUS FLOWER MOTH

SNAKEWEED FLOWER MOTH

GRAY-LINED FLOWER MOTH

Schinia reniformis   **Rare**   93-2176 (11195)

RANGE-WIDE

TL 10–12 mm Satiny white FW has thick, gray, curving AM and PM lines and straightish ST line; PM line flares at costa. Gray reniform spot is pinched at middle, has a rounded tooth at inner end, and is partially outlined in black. **HOSTS** Unknown.

SILVER-BANDED GEM

Schinia cumatilis   **Rare**   93-2177 (11192)

RANGE-WIDE

TL 13–16 mm White FW has olive-gray basal and ST areas and gray terminal area. Thin blackish AM and PM lines are toothed. ST line is white. Faint gray reniform spot sometimes visible only as tiny black dot. **HOSTS** Possibly fringed sagebrush.

HULST'S FLOWER MOTH

Schinia hulstia   **Uncommon**   93-2178 (11193)

RANGE-WIDE

TL 12–14 mm Resembles Silver-banded Gem, but terminal area is olive tan and median area lacks reniform spot or dot. Bumpy AM and PM lines are edged with black. Broad, soft-edged ST line is white. Outer margin has black terminal line toward apex. **HOSTS** Unknown.

PINK-WASHED FLOWER MOTH

Schinia scarletina   **Rare**   93-2181 (11096)

RANGE-WIDE

TL 11–14 mm Golden-brown FW has thin white AM and PM lines that angle to form an X near inner margin; center of X may sometimes form a white patch. Terminal area is lighter to pale brown. Many individuals have a pink wash, often most apparent in fringe and thorax. **HOSTS** Wirelettuce.

AGOSERIS FLOWER MOTH

Heliolonche modicella   **Rare**   93-2184 (11207)

RANGE-WIDE

TL 10–11 mm Resembles Chicory Flower Moth, but yellow stripe is straighter and has a dusky border, and spot in outer median area is reduced or absent. Fringe is silvery gray. HW is completely black. **HOSTS** Annual agoseris.

CHICORY FLOWER MOTH

Heliolonche carolus   **Rare**   93-2185 (11208)

RANGE-WIDE

TL 12–15 mm Pink FW has light yellow boomerang-shaped stripe parallel to inner margin and bending into ST area, and light yellow teardrop shape in outer median area. Fringe is pink. As wing becomes worn, black ground color shows through. HW is black with an orange center. Thorax is grayish yellow. **HOSTS** Desert chicory.

RED-LINED MOTH

Heliolonche pictipennis   **Uncommon**   93-2187 (11210)

RANGE-WIDE

TL 9–10 mm Pale yellow FW has brown to reddish-pink margins and AM and PM lines; in some individuals, colored areas may be broad, giving the appearance of a brown or pink moth with square, pale yellow patches in median area and pale terminal area. As wing becomes worn, black ground color shows through. Thorax is grayish yellow. **HOSTS** Desert dandelion and desert chicory.

GRAY-LINED FLOWER MOTH

SILVER-BANDED GEM

HULST'S FLOWER MOTH

PINK-WASHED
FLOWER MOTH

AGOSERIS
FLOWER MOTH

CHICORY FLOWER MOTH

RED-LINED MOTH

actual size

ASSORTED NOCTUIDS

SUPERFAMILY Noctuoidea (93)

FAMILY Noctuidae SUBFAMILY Noctuinae TRIBES Bryophilini, Pseudeustrotiini, Dypterygiini, Prodeniini, and Elaphriini

This catchall category contains several small tribes that do not fit into other noctuid groups. The moths represent an assortment of shapes and patterns. Most are small, but that is about the extent of their similarity. Some of these species can be quite common, particularly the armyworms and The Wedgling. A few species migrate north each summer and so may not be present at a location in all years. Adults are nocturnal and will come to lights; a few will also visit sugar bait.

EYED CRYPHIA
"Cryphia" olivacea

Rare RANGE-WIDE

93-2195 (9287)

TL 15–17 mm Variable. FW is either pale gray or yellow orange; some individuals have orange only in basal and ST/terminal areas. Wavy AM and scalloped PM lines are thick and black, bordering a dusky median area; PM line appears evenly scalloped in inner half. Large, pale orbicular spot is round, with a dusky center. Pale basal and ST areas are mottled with dusky shading. **HOSTS** Lichen. **NOTE** Provisionally placed in genus *Cryphia* pending further taxonomic study.

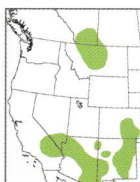

CAMOUFLAGED CRYPHIA
"Cryphia" cuerva

Rare RANGE-WIDE

93-2200 (9292)

TL 13–15 mm Resembles gray individuals of Eyed Cryphia, but orbicular spot is relatively indistinct, with just a narrow pale ring. AM and Pm lines are narrower, with pale gray edging; Pm line appears to have a flat section below indistinct reniform spot. Basal and ST areas are filled with dusky shading. Some individuals are lightly washed with olive or ochre. **HOSTS** Unknown; possibly lichen. **NOTE** Provisionally placed in genus *Cryphia* pending further taxonomic study.

GOLD-SPECKED ACOPA
Acopa perpallida

Rare RANGE-WIDE

93-2204.52 (9826)

TL 16–20 mm White FW has brown, strongly zigzag AM and slightly toothed PM lines and white-ringed brown reniform spot. Entire wing peppered with brown scales, sometimes quite heavily. Dark brown subapical patch sometimes reduced to indistinct diagonal band below reniform spot. Terminal line a row of brown dashes. **HOSTS** Unknown.

PINK-BARRED PSEUDEUSTROTIA
Pseudeustrotia carneola

Uncommon RANGE-WIDE

93-2205 (9053)

TL 11–13 mm Brown FW has strongly slanting, broad, pale pink band from costal AM to inner median area, where it touches pale brown ST area. Terminal area is gray below uneven white ST line. Reniform spot is bluish gray, thinly outlined in white and black. **HOSTS** Dock and smartweed.

BLACK BIRD'S-WING
Dypterygia patina

Uncommon RANGE-WIDE

93-2213.13 (9561)

TL 14–16 mm Dusky FW has broad tan (sometimes brown to dusky brown) inner margin with paired black lines at PM line, and thin white bar at bottom of reniform spot. Two veins at inner end of reniform spot are edged with tan to brown, down to outer margin. Basal area has thick black basal dash separating dusky and tan areas. Scalloped double AM and PM lines are usually indistinct except at margins. **HOSTS** Unknown.

EYED CRYPHIA

CAMOUFLAGED
CRYPHIA

GOLD-SPECKED
ACOPA

actual size

PINK-BARRED
PSEUDEUSTROTIA

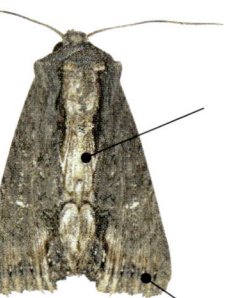

BLACK
BIRD'S-WING

VARIABLE NARROW-WING
Magusa divaricata

Uncommon
93-2213.62 (9637.1)

TL 15–19 mm Highly variable. Long narrow FW always has large apical patch, usually pale tan but sometimes brown, with white ST line at basal edge. Bicolored form has a brown, dark brown, or dusky costal half and a tan or grayish inner half, with a thin white line in between from base to AM area. Lines and spots are dark and indistinct. Unicolorous form is entirely brown or gray, with brown/gray apical patch defined by diffuse white edging along ST line. Lighter orbicular spot has a dark outline and dusky shading at center; brown to dusky reniform spot has dark outline and a pale basal edge. Thick median line and scalloped double AM and PM lines are most visible in inner half. Veins in ST/terminal area are traced with black. Intermediate form has the line/spot markings of unicolorous form and the dual wing colors and pale apical patch of bicolored form. **HOSTS** Leadwood and Humboldt coyotillo. **NOTE** Adults may wander north as far as Canada in late summer and fall. Once considered conspecific with Orbed Narrow-wing (*M. orbifera*, not shown), which is restricted to southern FL and the Caribbean.

BEET ARMYWORM
Spodoptera exigua

[IN] Very Common
93-2215 (9665)

TL 15–16 mm Mottled, gray to grayish-brown FW has small, pale, brown-centered orbicular and dusky-centered reniform spots outlined with blackish; reniform spot is bisected by a thin white line. Wavy black AM and PM lines are double, often filled with pale gray at inner end. Terminal line is a row of small black triangles, often connected to thin black to brown dashes crossed by pale ST line. Some individuals are relatively unmarked except for orbicular and reniform spots and terminal line. **HOSTS** Grass and crops, such as bean, beet, corn, and potato. **NOTE** Also known as Small Mottled Willow Moth. Cannot survive winter freezes, so migrates north late each summer to repopulate colder regions.

FALL ARMYWORM
Spodoptera frugiperda

Uncommon
93-2216 (9666)

TL 16–19 mm Sexually dimorphic. Male resembles a lightly marked Yellow-striped Armyworm, but inner edge of orange-brown orbicular spot is defined and does not extend past central vein, white in ST area is restricted to apex, and white spot and vein at inner reniform spot does not have a second branch. Basal area lacks white veins, and AM line meets inner margin perpendicularly. Female is mousy gray, with small, elongate, slanting, dusky-centered, tan to gray orbicular spot and indistinct dusky reniform spot with tan to gray section at lower edge. Dusky AM and PM lines are indistinct. **HOSTS** Generalist on grass, crops, and herbaceous plants. **NOTE** Year-round resident of southern TX, but migrates north and west late each summer to repopulate cooler regions.

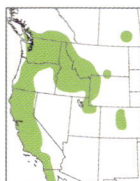

WESTERN YELLOW-STRIPED ARMYWORM
Spodoptera praefica

Common
93-2217 (9667)

TL 18–20 mm Very similar to plain form of Yellow-striped Armyworm, but AM and ST bands are grayer, and white of ST line never bleeds to outer margin and fringe at apex. **HOSTS** Generalist on fruit trees, woody vines, and herbaceous plants.

VARIABLE NARROW-WING

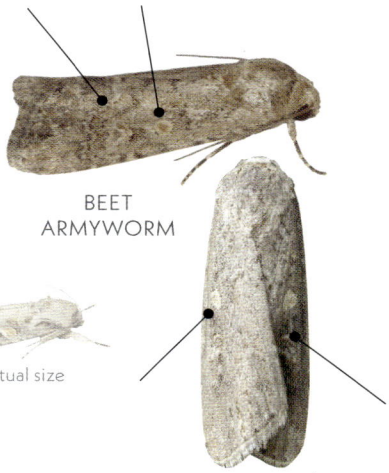

female

male

BEET
ARMYWORM

actual size

female

male

male

FALL ARMYWORM

WESTERN YELLOW-STRIPED
ARMYWORM

YELLOW-STRIPED ARMYWORM

Spodoptera ornithogalli

Very Common
93-2219 (9669)

TL 18–24 mm Dimorphic. Bold form has dark brown FW with elongate, slanting, brown-centered, tan orbicular spot that connects to tan inner median area, appearing as a long tan band. Dark brown reniform spot has pale-outlined brown spot at costal end. Central vein is traced with yellow across median area to where it forks beside reniform spot. Double AM line is edged with violet-gray wash. Whitish-gray band runs from inner ST area to apex. Straightish ST line is pale yellow. Plain form is similar but less contrasting, lacking pale median area and with brownish AM and ST areas. Double, black AM and PM lines are more distinct and filled with light brown. White ST line on both forms bleeds to outer margin and fringe at apex. **HOSTS** Wide variety of grasses, crops, and low and woody plants. **NOTE** Migrates north from Mexico each spring and summer to repopulate cooler regions.

THE WEDGLING

Galgula partita

Common
93-2224.5 (9688)

TL 11–13 mm Sexually dimorphic and variable. Male has warm brown to light grayish FW; female has maroon to dusky FW. Both sexes have blackish spot at costa between indistinct pale-outlined orbicular and reniform spots. Thin, pale yellow PM line is edged in darker brown, is wider at inner margin, and does not reach costa; in some individuals, the line is primarily darker brown, with minimal pale yellow. Pale AM line is usually indistinct. Basal and ST areas have white-edged black dots at veins, more evident on paler individuals. **HOSTS** Wood sorrel.

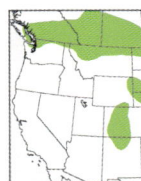

PALE-WINGED MIDGET

Elaphria alapallida

Rare
93-2234 (9681.1)

TL 13–15 mm Light gray to grayish-brown FW has round orbicular, small claviform, and kidney-shaped reniform spots with incomplete dark outline and dusky shading at center, with a blackish patch behind reniform. Wavy double AM and PM lines usually just touch edge of spots. ST/terminal area is shaded darker, with pale apical patch. Gray thorax has brown to dark brown collar and tan head. **HOSTS** Has been reared on Manitoba maple.

HALF-BANDED BRYOLYMNIA

Bryolymnia semifascia

Uncommon
93-2246 (9686)

TL 13–15 mm Brown FW has white patch at inner margin of median area, occasionally extending to costa, or rarely reduced to a white wash. Central FW is often shaded warm brown. Thin black basal and anal dashes are sometimes indistinct. Thorax has a bold white dorsal patch at anterior and mottled whitish wash at posterior. **HOSTS** Unknown.

GREEN BRYOLYMNIA

Bryolymnia viridata

Common
93-2247 (9296)

TL 13–16 mm Grayish-green FW has large, round, green orbicular, claviform, and fused reniform-subreniform spots outlined in black and edged with white inside. AM, PM, and ST lines are strong black scallops edged with white. Basal, outer median, and ST areas are shaded dusky. Fringe is checkered. Similar to *Feralia* species, but lines are often fragmented, and ST line and claviform spot are present. **HOSTS** Unknown.

YELLOW-STRIPED ARMYWORM

male

female

THE WEDGLING

PALE-WINGED MIDGET

actual size

HALF-BANDED
BRYOLYMNIA

GREEN BRYOLYMNIA

RUSTICS

SUPERFAMILY Noctuoidea (93)

FAMILY **Noctuidae** SUBFAMILY **Noctuinae** TRIBES Caradrinini and "Nacopa Clade" (unplaced)

Small brown or brownish-gray moths with relatively short, broad, slightly rounded forewings that are held flat or slightly tented at rest. Except for the *Proxenus*, the forewings usually have a mottled or speckled appearance. Most species have a pale reniform spot; it is dark in the *Caradrina*. Adults are nocturnal and will come to lights.

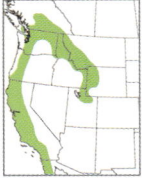

WHITE-SPOTTED PROPERIGEA
Properigea albimacula

Uncommon
93-2249.41 (9588)

NORTH

SOUTH

TL 13–18 mm Brown FW has bold white reniform spot with brown center at costal end, tiny pale orbicular spot outlined in brown, and tightly scalloped, dark brown AM and PM lines edged outwardly with pale brown. Light brown terminal area is bordered by indistinct, pale brown ST line. **HOSTS** Unknown.

TRICOLORED PROPERIGEA
Properigea continens

Uncommon
93-2249.43 (9590)

RANGE-WIDE

TL 13–16 mm FW has light brown basal area, dark chestnut-brown median area, and reddish-brown ST/terminal area; some individuals are tan with dark chocolate and reddish-brown washes. White reniform spot has a dark brown central bar; small white orbicular spot is outlined with dark brown. Dark brown AM and PM lines are shallowly scalloped and edged outwardly with tan. **HOSTS** Unknown.

MOTTLED PROPERIGEA
Properigea suffusa

Rare
93-2249.47 (9594)

RANGE-WIDE

TL 14–16 mm Mottled, tan, reddish-brown, dark brown, or grayish-brown FW has deeply scalloped, tan-edged, dark brown AM and PM lines with white dots at points on veins, and curving tan to white reniform spot with brown bar at center. Orbicular spot is absent or rarely indistinct. Inner and usually central median areas have diffuse dusky patches. ST area is not mottled and has dusky gray veins. **HOSTS** Unknown.

GRAY PROPERIGEA
Properigea niveirena

Rare
93-2249.49 (9596)

RANGE-WIDE

TL 16–18 mm Gray to grayish-brown FW has scalloped black AM and PM lines that are edged outwardly with light gray, and bold, white, curving reniform spot with dusky bar at center. Small gray orbicular spot has indistinctly dusky outline. Light gray ST line is often indistinct. **HOSTS** Unknown.

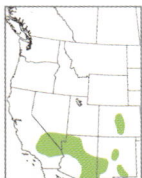

MASKED HEMIBRYOMIMA
Hemibryomima chryselectra

Uncommon
93-2249.5 (9597)

RANGE-WIDE

TL 12–15 mm Warm brown FW has light yellow-green median area mottled with brown, bounded by irregularly wavy, double, white-filled, black AM and PM lines. White round orbicular and curved reniform spots are outlined with black; reniform spot has a black crescent at center. Basal area is light green above double basal line and below line along inner margin. Terminal area is hoary. Uncommon individuals have the same pattern but are peppery and light gray with darker gray basal and ST areas. **HOSTS** Unknown.

RUSTICS

WHITE-SPOTTED PROPERIGEA

TRICOLORED PROPERIGEA

MOTTLED PROPERIGEA

GRAY PROPERIGEA

MASKED HEMIBRYOMIMA

actual size

563

DECEPTIVE PSEUDOBRYOMIMA

Pseudobryomima fallax

Uncommon

93-2249.63 (9600)

TL 13–16 mm Brown to tan FW has heavy dusky mottling. Black scalloped AM and PM lines are edged outwardly with tan; PM line has white dots at points on veins. Orbicular and reniform spots are tan, outlined in dark brown; reniform spot usually has a brown bar at center and, sometimes, small white spots at inner end. Separable from Mottled Properigea by presence of orbicular spot. **HOSTS** Coffee fern. **NOTE** The similar but less common Mossy Pseudobryomima (*P. muscosa*, not shown) is darker, dusky, with a less prominent orbicular spot, and streaked vertically with ochre or olive green.

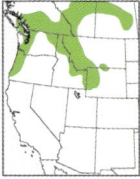

MOTTLED RUSTIC

Caradrina morpheus

[IN] Uncommon

93-2256 (9653)

TL 16–19 mm Light brown to brownish-tan FW has soft-edged, dark brown orbicular and reniform spots indistinctly washed with rusty brown. Dark brown ST area has smoothly curving basal edge and irregular lower edge. Thin, dark brown AM and PM lines are scalloped. Thick, dark brown median line touches reniform spot. **HOSTS** Generalist on herbaceous plants, including common dandelion, prostrate knotweed, and stinging nettle.

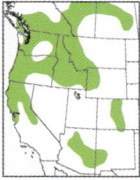

RARE SAND QUAKER

Caradrina meralis

Uncommon

93-2257 (9654)

TL 14–16 mm Resembles Civil Rustic, but FW is grayish to grayish brown, and orbicular and reniform spots are dark brown to blackish, outlined with black; reniform lacks dusky shading in inner half and white spots are reduced. Dark AM and PM lines, and pale ST line, are sometimes indistinct. ST/terminal area is often darker brown. Many individuals have a hoary appearance. **HOSTS** Unknown.

CIVIL RUSTIC

Caradrina montana

Very Common

93-2264.1 (9656)

TL 16–19 mm Peppery tan FW has tiny brown orbicular spot and brown reniform spot edged with tiny white dots above and below at dusky inner end. Pale ST line borders dusky terminal area and is edged basally with warm brown shading. Costa has blackish spots at positions of indistinct or absent cross-lines. **HOSTS** Unknown.

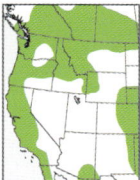

MIRANDA MOTH

Proxenus miranda

Uncommon

93-2266 (9647)

TL 13–15 mm Dark gray to bronzy-brown FW has small white reniform dot. Some individuals have curving black AM and PM lines that are indistinct on darker individuals. **HOSTS** Herbaceous plants, including alfalfa, dandelion, and strawberry.

DECEPTIVE
PSEUDOBRYOMIMA

MOTTLED RUSTIC

RARE SAND
QUAKER

CIVIL
RUSTIC

MIRANDA MOTH

actual size

HALF-SPOTS
SUPERFAMILY Noctuoidea (93)

FAMILY Noctuidae SUBFAMILY Noctuinae TRIBE Actinotiini

Medium-sized moths with streaky brown to brownish-gray wings; both of our species are bicolored, with the inner half of the forewing darker. Except for Bicolored Alastria, all Actinotiini have a pale, outlined reniform spot that is open on the costal side. Caterpillars of all species feed on St. John's wort. Adults are nocturnal and will come to lights.

BROWN HALF-SPOT
Nedra stewarti

Rare
93-2284 (9583)

RANGE-WIDE

TL 14–16 mm Hoary gray FW has brown shading from inner side of wing spots to anal angle, along terminal area, and in inner half of basal area from thin black basal dash. Grayish-brown orbicular spot is outlined in black, and pale tan reniform spot has a black outline on basal half and grayish-brown shading at center. Veins are thinly black, terminating at pale dashes in brown fringe. **HOSTS** St. John's wort.

BICOLORED ALASTRIA
Alastria chico

Rare
93-2288 (9522.1)

RANGE-WIDE

TL 17–20 mm Bicolored FW is tan in costal half and brown in inner half, separated by a very thin, sinuous, dark brown line. Veins are very thinly traced with dark brown. Terminal area has thick, dark brown interveinal dashes. Thorax is dark brown at center and pale tan at sides. **HOSTS** St. John's wort.

ANGLE SHADES
SUPERFAMILY Noctuoidea (93)

FAMILY Noctuidae SUBFAMILY Noctuinae TRIBE Phlogophorini

Medium-sized brown moths that characteristically roll the costal edge of their forewing when at rest. The AM and PM lines converge at the inner margin, creating a dark chevron in the median area. Adults are nocturnal and will come to lights.

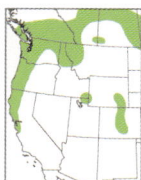

AMERICAN ANGLE SHADES
Euplexia benesimilis

Uncommon
93-2290 (9545)

RANGE-WIDE

TL 16–18 mm Brown FW has a blackish median area that narrows toward inner margin and contains slanted dusky orbicular spot and tan reniform spot with dusky shading at costal end. **HOSTS** Ferns, deciduous trees, and herbaceous plants, including aster, huckleberry, and willow.

BROWN ANGLE SHADES
Phlogophora periculosa

Uncommon
93-2292 (9547)

RANGE-WIDE

TL 26–28 mm Dimorphic. Warm brown FW has median area shaded dark brown in inner half, bounded by thin, pale brown AM and PM lines that converge toward (sometimes curving to connect at) inner margin. Pale outlines of orbicular and reniform spots slant to connect at inner end; reniform spot is filled with darker brown in bottom half. Some individuals have a medium brown median area and dusky double AM and PM lines. Typically rests with costal margin of wing rolled. **HOSTS** Trees and plants, including alder, balsam fir, cranberry, and plum. **NOTE** The two forms were once thought to be male and female but are actually not sex linked.

BROWN
HALF-SPOT

actual size

BICOLORED ALASTRIA

AMERICAN ANGLE SHADES

actual size

BROWN ANGLE SHADES

APAMEAS, BROCADES, and ALLIES

SUPERFAMILY Noctuoidea (93)
FAMILY Noctuidae SUBFAMILY Noctuinae
TRIBE Apameini and Arzamini

Medium-sized, brown or tan, often-mottled moths with a squared outer margin, sometimes with a pointed apex. The apameas generally have large orbicular and reniform spots and scalloped double AM and PM lines, and many have a pale W shape at the outer margin where the central veins cross the terminal area. The brocades are more boldly colored; the eye-catching Raspberry Brocade is particularly unusual among this group. The distinctive stem borers are mostly golden with fractured white orbicular and reniform spots. Adults of all species are nocturnal and will come to lights.

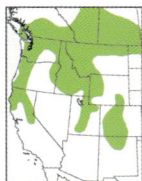

FOX APAMEA
Apamea alia

Rare RANGE-WIDE
93-2307 (9351)

TL 18–20 mm Pale tan FW has tan to rusty shading broadly along costal half. Orbicular and reniform spots are indistinctly defined, with tan, rusty, or sometimes dusky shading at center; reniform is always dusky at inner end. AM and ST areas have short black dashes at veins. HOSTS Grasses.

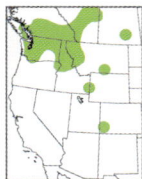

SMALL CLOUDED BRINDLE
Apamea unanimis

[IN] Uncommon RANGE-WIDE
93-2308 (9362.2)

TL 16–18 mm Mottled brown to brownish-gray FW has slightly angled orbicular and curving reniform spots outlined thinly in black with white inner edging; reniform is paler in lower half. Indistinct claviform spot is outlined with black. Median area is indistinctly darker brown, bounded by indistinct lighter brown AM and PM lines. Indistinct, irregular, pale ST line bounds dusky terminal area. HOSTS Beach grasses, including phragmites, canary grass, and mannagrass.

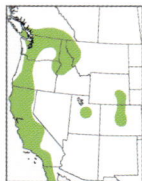

WHITE-SPURRED APAMEA
Apamea cuculliformis

Rare RANGE-WIDE
93-2313 (9325)

TL 22–25 mm Warm brown FW has darker brown along inner margin and terminal area, crossed by thin white crescent of PM line. Black-traced veins are thinly edged in tan, with thicker, paired, black-and-white dots in ST area. Reniform spot is visible as a row of small black spots. HOSTS Grasses.

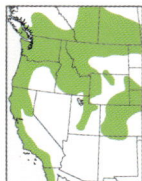

BORDERED APAMEA
Apamea sordens

Uncommon RANGE-WIDE
93-2314 (9364)

TL 20–22 mm Light brown to brownish-gray FW has brown median area bounded by thin scalloped black lines edged outwardly with pale brown; AM line sometimes appears double. Rounded orbicular and reniform spots are filled with FW ground color, thinly outlined with black, and edged inwardly with white scales; reniform has dusky patch at inner end. HOSTS Grasses. NOTE Very similar Traced Apamea (*A. digitula*, not shown) is found from OR to CA and often best identified by range.

SPALDING'S QUAKER
Apamea spaldingi

Rare RANGE-WIDE
93-2317 (9356)

TL 18–22 mm Brownish gray to dark gray FW has thick black line parallel to inner margin that connects doubled AM and PM lines; PM line is lightly toothed. Slanting orbicular spot connects to pale inner median area. PM area beneath indistinct reniform spot is pale gray. Inner terminal area is blackish, with pale ST line and W shape at veins. Pale HW has dusky border. Separable from Ashy Apamea by orbicular spot, inner median area, and HW. HOSTS Unknown; likely grasses.

actual size

FOX APAMEA

SMALL CLOUDED
BRINDLE

WHITE-SPURRED APAMEA

BORDERED
APAMEA

SPALDING'S QUAKER

ASHY APAMEA
Apamea cinefacta

Uncommon
93-2318 (9357)

NORTH

SOUTH

TL 17–19 mm Gray to brownish-gray FW has black, double AM and scalloped PM lines with whitish fill in inner half; PM line is strongly toothed. Elongate orbicular spot and kidney-shaped reniform spot are brownish and thinly outlined black. Median area is slightly darker, usually with a diffuse, dark brown, V-shaped median band. Inner terminal area is blackish, with a white to tan W shape and indistinct, fragmented, tan ST line. Many individuals have stripes of brownish wash parallel to inner margin and through reniform spot. HW is entirely ashy gray. Separable from Spalding's Quaker by orbicular spot, inner median area, inner ST line, and HW. **HOSTS** Unknown; likely grasses.

WOOD-COLORED APAMEA
Apamea lignicolora

Rare
93-2319 (9333)

RANGE-WIDE

TL 24–26 mm Warm brown FW has a dark blackish terminal area interrupted by a pale tan W shape, and darker brown shading along costa. Elongate orbicular and kidney-shaped reniform spots are warm brown with pale tan outline. Pale tan AM and PM lines have black dashes at veins. **HOSTS** Quack grass and other grasses.

FROSTED APAMEA
Apamea antennata

Uncommon
93-2323 (9334)

NORTH

SOUTH

TL 21–24 mm Variable. Warm brown to grayish brown FW has a grayish wash along inner margin and inner half of ST area. Elongate orbicular spot is tan, outlined in black. Tan reniform spot is outlined in lighter tan with black edging along basal half; lower edge bleeds into a diffuse tan patch in PM area. Scalloped black AM and PM lines converge near inner margin, where they are connected by a thin black line. **HOSTS** Unknown; likely grasses.

WESTERN APAMEA
Apamea occidens

Uncommon
93-2332 (9346)

RANGE-WIDE

TL 21–25 mm Gray FW has broad, pale gray ST area shaded with warm brown along pale ST line and crossed by dark gray veins. Light gray orbicular and reniform spots are shaded with dusky gray inside and outlined with pale gray and black. Double, black AM and PM lines are filled with pale gray. Median area is often partly or largely brick red. Central thorax is tan to brick red. **HOSTS** Grasses.

YELLOW-HEADED CUTWORM
Apamea amputatrix

Common
93-2333 (9348)

INLAND

COASTAL

TL 26–27 mm Variable. Wine-red to warm brown FW has hoary gray (rarely brownish or tan) ST and lower basal areas. Scalloped double AM and PM lines are filled with grayish. Reniform spot has tan patch at lower center and paired small white spots at each side. Uneven yellow ST line borders dusky gray terminal area. **HOSTS** Fruit-bearing trees, grasses, and crops, including cabbage, corn, and lettuce.

BURGESS' APAMEA
Apamea burgessi

Rare
93-2339 (9378)

RANGE-WIDE

TL 17–23 mm Grayish-brown to dark gray FW has scalloped, black, double AM and PM lines connected by a black bar parallel to inner margin. Black-outlined reniform spot has reddish-tan scales along lower edge. ST line is tiny tan dots edged with strong black chevrons. Some lighter individuals have darker median and terminal areas. **HOSTS** Unknown; likely grasses.

ASHY APAMEA

WOOD-COLORED
APAMEA

actual size

FROSTED APAMEA

WESTERN
APAMEA

YELLOW-HEADED
CUTWORM

BURGESS' APAMEA

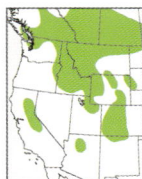

FAINT-SPOTTED APAMEA

Apamea scoparia

Uncommon
93-2343 (9365)

TL 22–24 mm Reddish-brown FW has dusky veins, often with paired black-and-white dots in ST area. Reniform spot is incompletely outlined in ivory, with indistinct dusky spot at inner end. Indistinct orbicular spot has a faint ivory outline and is sometimes shaded dusky. **HOSTS** Unknown; probably grasses.

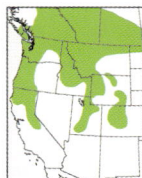

THOUGHTFUL APAMEA

Apamea cogitata

Uncommon
93-2345 (9367.1)

TL 19–21 mm Resembles Faint-spotted Apamea, but FW is darker maroon brown, reniform spot is almost entirely filled with ivory, and irregular ST line is ivory, widening slightly at costa. **HOSTS** Grasses.

LINED QUAKER

Apamea inficita

Rare
93-2347 (9369)

TL 17–24 mm Yellow-orange to reddish-brown FW has distinct dusky spot at inner end of indistinct reniform spot that appears to bleed along dusky central vein. Scalloped AM and PM lines are sometimes indistinct. Slightly lighter ST line borders blackish dashes along veins in terminal area. **HOSTS** Unknown; probably grasses.

GLASSY CUTWORM

Apamea devastator

Common
93-2350 (9382)

TL 22–26 mm Grayish-brown FW has faint, warm brown stripes down central and inner wing from basal to ST area. Scalloped double AM and PM lines are filled with light brown. Small, pale-outlined, slanting orbicular and level reniform spots are dusky at center. Black-outlined claviform spot sometimes has blackish fill. Pale ST line is shaded dusky in ST area, edged with indistinct dusky chevrons. Costa has small whitish dots at ST area. **HOSTS** Grasses and crops. **NOTE** Sometimes a serious pest.

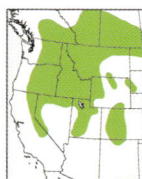

SNOWY-VEINED APAMEA

Apamea niveivenosa

Rare
93-2355 (9374)

TL 18–20 mm Boldly patterned FW is dusky gray to dusky brown with tan costa, inner margin, and central stripe, tan orbicular and reniform spots, and pale ivory veins. ST area is pale gray to tan with black interveinal chevrons, and terminal area is violet gray to dusky brown. **HOSTS** Bluegrass.

DOUBLE LOBED MOTH

Lateroligia ophiogramma

[IN] Uncommon
93-2363 (9385.1)

TL 15–17 mm Tan FW has dusky brown outer median area that stretches along costa to basal and ST lines; inner side of dusky brown area is marked by elongate black claviform spot. Indistinctly outlined reniform spot has large tan spot in lower half. Orbicular spot is brown. **HOSTS** Reed canary grass and other coarse grasses.

THOUGHTFUL APAMEA

FAINT-SPOTTED APAMEA

LINED QUAKER

GLASSY CUTWORM

actual size

SNOWY-VEINED
APAMEA

DOUBLE LOBED
MOTH

COMMON RUSTIC

Mesapamea secalis

[IN] Uncommon
93-2371.1 (9407.1)

TL 13–15 mm Brown to dusky brown FW has black-outlined, dusky brown orbicular and reniform spots edged interiorly with tan scales: orbicular spot appears cut off at both ends by hoary gray veins; reniform spot has a white to tan spot on lower side cut through by a brown line. Median area is narrow and bounded by doubled AM and PM lines filled with slightly lighter brown; sometimes has a thin black bar parallel to inner margin connected to black-outlined claviform spot. Terminal area is often dusky gray. **HOSTS** Grasses, including wheat, rye, barley, oats, and corn.

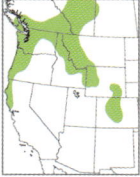

DARK-WINGED QUAKER

Eremobina claudens

Rare
93-2372 (9396)

TL 16–18 mm Gray to whitish FW has rounded, black-outlined orbicular and reniform spots shaded dusky at center, and elongate, black-outlined claviform spot that connects scalloped, white-edged AM and PM lines. Area between orbicular and reniform spots is shaded dusky. Basal area has thin black basal dash, and ST/terminal area has thin black along veins. **HOSTS** Unknown.

VIOLET BROCADE

"Oligia" violacea

Rare
93-2380 (9414)

TL 16–17 mm Strongly resembles Raspberry Brocade but colors are typically more muted, and ivory thorax has rusty scales down center. **HOSTS** Thread rush and other rushes. **NOTE** Provisionally placed in genus *Oligia* pending further taxonomic study, as is the case for Raspberry and Exposed Brocades below.

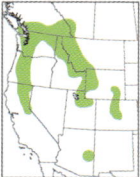

RASPBERRY BROCADE

"Oligia" rampartensis

Rare
93-2381 (9414.1)

TL 16–17 mm Beautiful tapestry-like FW has a hoary wine-red basal area, reddish median area with orange wash in inner half, whitewashed, rosy ST area, and pale yellow ST line connected to yellowish-white apical spot. Rounded, pale yellowish orbicular and reniform spots are lightly shaded violet inside. Reddish, scalloped AM and PM lines are narrowly edged with whitish. Thorax is ivory. **HOSTS** Thread rush and other rushes.

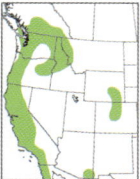

EXPOSED BROCADE

"Oligia" divesta

Uncommon
93-2386 (9559)

TL 17–19 mm Brownish-gray FW has black, scalloped, double AM and PM lines connected by a black bar parallel to inner margin, bordered by light brown shading. Gray, round orbicular and dusky-shaded reniform spots are thinly outlined with black. Inner basal area is washed light brown with thick black basal dash. Resembles Summer Hyppa, but FW is broader, and AM and PM lines are smoothly scalloped. **HOSTS** Coarse grasses.

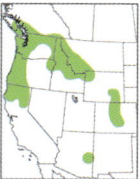

OBLIQUE BROCADE

Xylomoia indirecta

Rare
93-2417 (9401)

TL 15–17 mm Brown FW has dusky AM and PM lines framing dark inner median area containing a narrow, black, vertical bar. Slanting oval orbicular and kidney-shaped reniform spots are tan with brown centers, outlined thinly in black. Inner basal area is shaded dark brown with tan stripe along inner margin. Terminal area has blackish patches at midpoint and near anal angle, with a thick, black, triangular anal dash projecting into ST area. **HOSTS** Sedges and grasses.

COMMON RUSTIC

DARK-WINGED QUAKER

VIOLET BROCADE

RASPBERRY BROCADE

actual size

EXPOSED BROCADE

OBLIQUE BROCADE

OBLONG SEDGE BORER

Globia oblonga

Rare 93-2438 (9449)

TL 25–27 mm Lightly speckled tan FW has three faint gray stripes in central wing. AM, PM, and terminal lines are rows of black dots. Orbicular and reniform spots are tan circles with black dots at center. Tufted abdomen projects beyond outer margin when moth is at rest. **HOSTS** Cattail and bulrush.

KIDNEY-SPOTTED RUSTIC

Helotropha reniformis

Uncommon 93-2443 (9453)

TL 20–24 mm Variable. Well-marked individuals have dark brown FW with tan-outlined orbicular and reniform spots; reniform spot is split horizontally by a thin white line and often washed or filled with white in lower half, with veins marked white at inner end. ST area is light brown basally and dark brown distally. ST line is tan. Dark brown thorax has tan dorsal stripe. Dark individuals have a brown to dark brown FW, sometimes with brown wash along inner margin and ST area, and dusky veins. White lower half of reniform spot is present, but tan outlines of reniform and orbicular spots are indistinct or absent. **HOSTS** Sedges.

AMERICAN EAR MOTH

Amphipoea americana

Rare 93-2447 (9457)

TL 14–20 mm Golden- to orange-brown FW has rounded orange orbicular and reniform (rarely white) spots; reniform spot has a brown bar at center. Thin, brown, scalloped AM and PM lines are broadly double, filled with FW ground color. Thick, darker brown median line has a strong V shape inside from reniform spot. ST area is slightly paler. **HOSTS** Grasses and sedges, sometimes corn.

CRESCENT EAR MOTH

Amphipoea lunata

Rare 93-2452 (9455)

TL 13–17 mm Warm brown FW has curving white reniform bar and thin, curving, white AM and PM lines (sometimes indistinct); PM line curves strongly basally to meet costa. Terminal and inner ST areas are lighter brown. **HOSTS** Unknown.

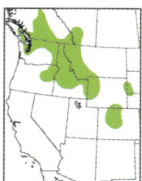

PALLID RUSTIC

Hydraecia medialis

Uncommon 93-2455 (9511)

TL 17–25 mm Light brown FW has steep white PM line that curves abruptly basally to meet costa, and wavy white AM line. Slightly darker median area has light brown orbicular and reniform spots. Costa is faded. Veins are lightly traced with gray. **HOSTS** Unknown; possibly lupines.

RAGWORT STEM BORER

Papaipema insulidens

Rare 93-2480 (9488)

TL 16–19 mm Golden- to reddish-brown FW has white reniform spot fragmented into multiple sections by veins with a thin bar at center, white orbicular spot with a brown dot at center, and white claviform spot split in two by a brown vein. White to pale tan basal patch is fragmented by double basal line and vein. Violet-brown ST area is scalloped along indistinct tan ST line. Apical patch is golden brown, and apex is pointed. Separable from Serrated Borer-Mimic by claviform spot and apex. **HOSTS** Tansy, water, and tall ragwort; also other herbaceous plants, including lupine, cow parsnip, and lilies. **NOTE** Also known as Umbellifer Borer.

OBLONG SEDGE BORER

KIDNEY-SPOTTED
RUSTIC

AMERICAN
EAR MOTH

CRESCENT EAR MOTH

PALLID RUSTIC

actual size

RAGWORT STEM BORER

CATTAIL BORER
Bellura obliqua

Uncommon
93-2517 (9525)

TL 22–28 mm Brown FW has crisp, slanting edge to pale tan basal area that runs from costal PM area to inner basal area and onto thoracic crest; posterior edge is diffusely shaded dark brown. Elongate, slanting reniform spot is pale tan with a brown central line. ST line is scalloped. Apex is sharply pointed. **HOSTS** Cattail.

SWORDGRASSES
SUPERFAMILY Noctuoidea (93)
FAMILY Noctuidae SUBFAMILY Noctuinae TRIBE Xylenini (part)

Medium-sized moths with long narrow wings that they characteristically wrap around their body at rest, overlapping at the outer end to form a tapered appearance. Adults emerge in the fall and overwinter as adults; they fly briefly in autumn, then again in late winter and spring. They are nocturnal and will come to lights but can sometimes also be found at sugar bait.

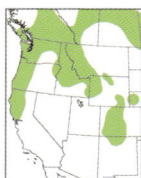

AMERICAN SWORDGRASS
Xylena nupera

Uncommon
93-2519 (9873)

RANGE-WIDE

TL 25–28 mm Long FW has warm brown inner half and tan outer half with blackish costa, and thick black lines in central basal and PM/ST areas. Reniform spot is shaded dusky inside and thinly outlined in black; thinly outlined orbicular spot is indistinct. Thorax is dusky gray with pale tan sides. **HOSTS** Generalist on herbaceous plants and trees, including cherry, poplar, and willow.

DOT-AND-DASH SWORDGRASS
Xylena curvimacula

Uncommon
93-2520 (9874)

RANGE-WIDE

TL 25–28 mm Long FW has violet-brown inner half and tan outer half with brown costa. Black-outlined orbicular spot is dusky, and tan reniform spot has brown spot. Deeply scalloped double AM line filled with brown creates loop at basal area. PM line is a row of tiny black dots. Brown thorax has tan sides.
HOSTS Generalist on dandelion and other herbaceous plants and trees, including alder, birch, and poplar.

GRAY SWORDGRASS
Xylena cineritia

Uncommon
93-2522 (9876)

RANGE-WIDE

TL 21–25 mm Long FW has gray inner half and tan outer half, with dark brown along costa. Dusky orbicular spot is outlined with black. Dusky reniform spot has warm tan spot at center. Thorax is dusky brown with light gray sides. **HOSTS** Generalist on trees and shrubs, including alder, birch, maple, meadowsweet, and poplar.

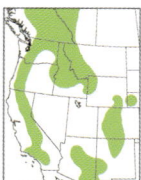

BRUCE'S SWORDGRASS
Xylena brucei

Rare
93-2523 (9877)

NORTH

SOUTH

TL 25–27 mm Gray FW is streaky, with thin black veins and indistinct, thin, blackish AM and PM lines; AM line sometimes forms a long double loop. Central FW has a faint brown wash. Orbicular spot has a blackish double outline. Large reniform spot is incompletely outlined with black, with a pale gray center. Central ST area has a long black dash. Thorax is gray with a dusky brown dorsal stripe. **HOSTS** Bitterbrush.

actual size

CATTAIL BORER

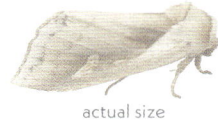

AMERICAN
SWORDGRASS

DOT-AND-DASH
SWORDGRASS

actual size

GRAY
SWORDGRASS

BRUCE'S
SWORDGRASS

PINIONS and XYLENINE SALLOWS

SUPERFAMILY Noctuoidea (93)
FAMILY Noctuidae SUBFAMILY Noctuinae TRIBE Xylenini (part)

Medium-sized, gray to brown noctuids with squared forewings that are held flat at rest. The pinions have a slightly convex costa and distinctive scale tufts on the sides of the thorax that give the appearance of shoulder pads. Many of the sallows are warm brown to orangish, with pale or weakly defined markings. The species in this group are often cold hardy and among the first noctuids seen in spring and the latest in autumn; in northern areas, many hibernate as adults. Adults are nocturnal and will come to lights; many species will also regularly visit sugar bait.

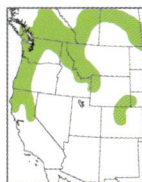

AMERICAN BRINDLE
Lithomoia germana

Rare   RANGE-WIDE

93-2524 (9878)

TL 26–28 mm Gray FW has pale gray reniform spot shaded dusky inside with a pale bar at center, and smaller, round, pale gray orbicular and claviform spots. Pale zigzagged AM line is present at costa. Veins are finely traced with black. Diffuse dusky median patch is often indistinct. Indistinct whitish ST line has black interveinal dashes at tips of central portion. Usually rests with wings rolled into a tube like some *Schizura* prominents. **HOSTS** Deciduous trees and woody shrubs, including alder, birch, blueberry, poplar, and willow.

BLACK-BARRED HOMOGLAEA
Homoglaea dives

Rare   NORTH

93-2528 (9882)   SOUTH

TL 20–22 mm Resembles Twin-spotted Homoglaea, but costal ST line is a single black bar. FW color averages darker, often with dark purplish-brown tones, and double AM and PM lines are dark brown to blackish. Darker individuals may be unicolorous with just black elements of pattern, sometimes with colored reniform spot. **HOSTS** Willow, poplar, and aspen.

TWIN-SPOTTED HOMOGLAEA
Homoglaea carbonaria

Uncommon   NORTH

93-2529 (9883)   SOUTH

TL 20–22 mm Reddish- to maroon-brown FW has slightly paler terminal area and reddish-brown wash along inner margin. Wavy, brown, double AM and PM lines are filled with light brown. Costal end of ST line has a pair of black spots. Outlines of orbicular and reniform spots are whitish to tan; rarely, reniform spot is filled with whitish peppering. **HOSTS** Willow, poplar, and aspen.

FALSE PINION
Litholomia napaea

Uncommon   RANGE-WIDE

93-2530 (9884)

TL 16–17 mm Light gray FW has pale gray orbicular spot and pale-ringed, dark gray reniform spot that is incompletely outlined with black. Median line is bold black. Wavy black AM and PM lines are double, usually with medial line darker. **HOSTS** Quaking aspen.

NAMELESS PINION
Lithophane innominata

Uncommon   RANGE-WIDE

93-2534 (9888)

TL 19–20 mm Pale golden- to warm brown FW has darker shading below brown-outlined orbicular spot and surrounding reniform spot, and a squarish black spot at inner margin of median area. AM and PM areas have veins marked with paired black dots. ST line is a curving row of brown spots. Anal angle and midpoint of terminal area are smudged with blackish color. **HOSTS** Deciduous trees and shrubs, including alder, apple, hemlock, oak, spruce, and willow.

AMERICAN
BRINDLE

actual size

BLACK-BARRED HOMOGLAEA

TWIN-SPOTTED
HOMOGLAEA

FALSE PINION

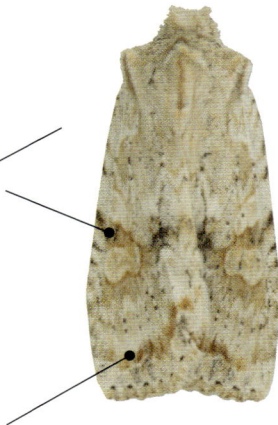

NAMELESS
PINION

WANTON PINION
Lithophane petulca

Uncommon
93-2536 (9889)

RANGE-WIDE

TL 18–20 mm Dimorphic. Light form has light brown FW with slanting golden-brown orbicular and reniform spots that connect at inner end and are shaded slightly darker in between. Central median area, and anal angle and midpoint of terminal area, have indistinct dusky patches. AM and ST areas often have indistinct paired black dots. Dark morph is the same but has dark brown shading across inner wing from base to ST line, creating pale costal stripe. **HOSTS** Alder and birch; also ash, cherry, maple, oak, and willow.

LARGE GRAY PINION
Lithophane georgii

Uncommon
93-2573 (9913)

NORTH

SOUTH

TL 18–22 mm Long, medium gray FW has slightly paler, round orbicular, reniform, and small claviform spots outlined with black and shaded dusky inside, and thin black basal dash. ST line is a row of dusky gray chevrons. **HOSTS** Generalist on woody trees and shrubs, especially birch and *Prunus*.

TORRID PINION
Lithophane pertorrida

Uncommon
93-2579 (9912)

NORTH

SOUTH

TL 22–25 mm Resembles Large Gray Pinion but has wavy double AM and PM lines, orbicular spot is often mostly filled with pale gray, and reniform spot is shaded brown at inner end. Outer basal area is paler gray. **HOSTS** Plum and willow.

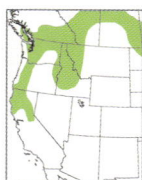

THREE-SPOTTED SALLOW
Eupsilia tristigmata

Uncommon
93-2590 (9935)

RANGE-WIDE

TL 17–19 mm Orange-brown FW has orange orbicular, small claviform, and reniform spots outlined in brown; reniform spot has a black spot at inner end and tiny tan to white dots at lower corners. Straight AM and scalloped PM lines are brown. Median area often has an orange wash, with soft brown median line passing behind reniform spot. **HOSTS** Deciduous trees, including birch, butternut, cherry, oak, and willow.

SMOOTH-LINED SALLOW
Mesogona olivata

Uncommon
93-2610 (9953)

RANGE-WIDE

TL 17–21 mm Brown FW has darker, round orbicular and reniform spots outlined in pale tan and thin, pale tan AM and PM lines spottily edged medially with black; costal end of AM line angles basally, while inner end slants toward PM line. Thin, pale tan, sometimes-indistinct ST line is edged basally with blackish spots. Sometimes has a dusky patch at costal ST area. HW is tan to dusky. **HOSTS** Generalist on woody plants, including white oak, ceanothus, currant, bitterbrush, and others.

COPPER-WINGED SALLOW
Mesogona subcuprea

Rare
93-2612 (9953.2)

RANGE-WIDE

TL 20–23 mm Resembles Smooth-lined Sallow, but ST line is unmarked or with even dusky shading, and HW is peach orange to warm brown. Costal margin and legs may sometimes be tinged peach orange. Rare individuals are entirely washed rosy pink. **HOSTS** Oak.

WANTON PINION

LARGE GRAY
PINION

TORRID PINION

THREE-SPOTTED
SALLOW

SMOOTH-LINED
SALLOW

COPPER-WINGED SALLOW

actual size

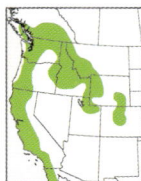

LUPINE SALLOW
Agrochola purpurea

Uncommon
93-2613 (9954)

TL 17–20 mm Grayish- to orange-brown FW has brown-outlined reniform spot with dusky shading in inner half and at apex of V-shaped, slightly wavy, brown median line. ST line of black dots has blackish shading at costa of ST area; entire ST area is sometimes slightly shaded dusky. Indistinct AM and PM lines are double, with outermost of pair often darker. **HOSTS** Lupine.

BICOLORED SALLOW
Sunira bicolorago

Uncommon
93-2616 (9957)

TL 18–20 mm Golden-brown FW has reniform spot with blackish spot in inner half, adjacent to slightly wavy, straightish, brown median line. Light golden-brown ST line is bordered basally by darker brown shading. Veins in ST/terminal area are grayish, with black dots at basal end in ST area. Eastern individuals are sometimes completely washed with dark reddish brown below median line, but this form is rare in the West. **HOSTS** Deciduous trees, including cherry, elm, maple, oak, and others.

SHIELD-BACKED CUTWORM
Sunira decipiens

Uncommon
93-2617 (9958)

TL 16–20 mm Resembles Bicolored Sallow, but FW is reddish brown with darker reddish-brown markings. Black vein dots in ST area are tiny, and veins are not noticeably grayish. Orbicular spot is roughly circular. **HOSTS** Unknown.

PUTA SALLOW
Anathix puta

Common (uncommon in South)
93-2620 (9962)

TL 14–15 mm Variable. FW color may be brownish tan, reddish brown, or brownish gray, with or without dusky AM, ST, and terminal bands. Lighter ST line has dark brown to black interveinal dots in central portion. Brown to dusky median line is wavy, touching dusky end of reniform spot. Double, scalloped AM and smoothly curving PM lines are filled with gray or tan. Veins are traced with gray or tan. Often resembles Lupine Sallow but lacks black bar and shading at costa of dotted ST line. **HOSTS** Quaking aspen.

DUSKY SHOULDER-KNOT
Aseptis perfumosa

Uncommon
93-2628 (9529)

TL 16–19 mm Dark gray FW has scalloped double AM line with connected, black-outlined, U-shaped claviform spot. Black-outlined orbicular and reniform spots are often sparsely edged inside with light gray scales. Brownish patches are usually present between spots, below reniform spot, and in inner median area. **HOSTS** Big berry manzanita.

RUSTY SHOULDER-KNOT
Aseptis binotata

Uncommon
93-2634 (9532)

TL 15–17 mm Variable. FW may be warm brown, dark brown, or hoary gray. Wavy AM and PM lines are double and filled with tan. Area beneath black-outlined reniform spot is tan, crossed by thin black upper PM line. Inner AM line curves almost parallel to inner margin to terminate near base. Basal area is darker, without basal dash. ST/terminal area has black interveinal dashes near midpoint. **HOSTS** Generalist on both deciduous and coniferous shrubs and trees.

LUPINE SALLOW

BICOLORED
SALLOW

SHIELD-BACKED CUTWORM

PUTA SALLOW

actual size

DUSKY SHOULDER-KNOT

RUSTY SHOULDER-KNOT

ASHY SHOULDER-KNOT

Aseptis characta

Rare

93-2648 (9543)

TL 15–16 mm Mottled gray FW has black-outlined orbicular and reniform spots edged with white scales inside, and indistinct double AM and PM lines. Inner basal area and inner and central ST area are washed warm brown (sometimes faded). Black basal dash does not reach AM line. HOSTS Sagebrush and mugwort.

BROWN SHOULDER-KNOT

Paraseptis adnixa

Uncommon

93-2648.5 (9533)

TL 20–22 mm Resembles Rusty Shoulder-Knot, but double AM line does not curve basally at inner margin and basal area is often less dark and has thin black basal dash that reaches AM line. HOSTS Osoberry.

POPLAR SHOULDER-KNOT

Brachylomia populi

Rare

93-2653 (9993)

TL 15–17 mm Variable. Light gray FW is washed tan, with blackish basal area and median band (sometimes entire median area). Wavy double AM line is relatively straight and filled with tan. Curving double PM is often indistinct. Pale reniform spot is shaded with brown inside along long sides. Some individuals have a rusty-brown median area. Black terminal line is dashed, with dusky patch at anal angle. HOSTS Balsam poplar and quaking aspen; also hazel, oak, and chokecherry.

SUMMER HYPPA

Hyppa contrasta

Uncommon

93-2665 (9579)

TL 20–23 mm Light gray FW has thick black bar parallel to inner margin that connects jagged AM and curving PM lines, bordered by warm brown shading. Orbicular and reniform spots are indistinctly outlined with black. Terminal area has black dashes. Basal area has thick black basal dash connected to brownish lateral thoracic stripes. Separable from Exposed Brocade by narrower FW and jagged AM line. HOSTS Has been reared from velvetleaf blueberry but is probably a generalist of hardwood trees and herbaceous plants.

VARIABLE DUN-BAR

Cosmia praeacuta

Uncommon

93-2670 (9814)

TL 16–18 mm Peppery gray to light brown FW has thick, curving, black AM line and thin, wavy, black PM line. Pale, round orbicular and reniform spots are lightly shaded brown. Inner ST area usually has a dusky patch. Terminal area is usually slightly paler. Some individuals have dark gray FW with light gray terminal area and pale spots. HOSTS Douglas-fir and white fir.

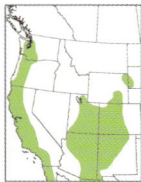

AMERICAN DUN-BAR

Cosmia calami

Common

93-2672 (9815)

TL 15–17 mm Pale to golden-tan FW has whitish, smooth, slanting, straight AM and curving PM lines narrowly edged medially with dark tan or brown. Orbicular and reniform spots are outlined in white and are sometimes shaded dark inside. Some individuals have a diffuse, dark tan, V-shaped median band. HOSTS Oak; also caterpillars of other lepidopteran species.

ASHY
SHOULDER-KNOT

BROWN
SHOULDER-KNOT

SUMMER HYPPA

POPLAR
SHOULDER-KNOT

actual size

AMERICAN DUN-BAR

VARIABLE DUN-BAR

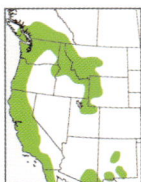

ELDER MOTH
Zotheca tranquilla

Common

93-2673 (9816)

TL 18–20 mm Ivory FW is usually liberally washed with spring green, with a strong green median band. Thin zigzag AM and V-shaped PM lines are dark green. Some individuals are warm brown instead of green. **HOSTS** Elder.

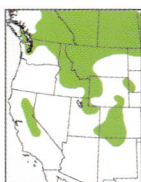

PALE ENARGIA
Enargia decolor

Common (uncommon in South)

93-2674 (9549)

TL 17–23 mm Pale reddish-tan to darker reddish-orange FW has large, rounded, pale tan to orange orbicular and reniform spots outlined with reddish brown; reniform spot is unmarked or has a diffuse smudge at inner end. Reddish-brown V-shaped AM and curving PM lines are thin; both lines usually bend at costa to meet it perpendicularly. Diffuse reddish-brown V-shaped median band and pale, wavy, broad ST line are often indistinct. **HOSTS** Quaking aspen, bigtooth aspen, alder, white birch, and balsam poplar. **NOTE** May experience outbreaks, making it become very abundant in some regions.

SMOKED SALLOW
Enargia infumata

Common

93-2675 (9550)

TL 18–20 mm Resembles Pale Enargia, but inner part of reniform spot has a well-defined, rounded, dusky (sometimes brown) spot. AM and PM lines usually meet costa at an angle. Some individuals are peppered with dusky scales; others may have dusky shading through PM/ST area. **HOSTS** Quaking aspen, white birch, willow, and balsam poplar.

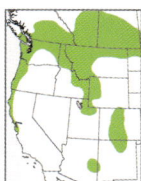

EVEN-LINED SALLOW
Ipimorpha pleonectusa

Uncommon

93-2679 (9555)

TL 18–20 mm Brownish-tan to grayish-brown FW has straightish, pale tan basal, AM and PM lines, and darker, rounded orbicular, claviform (connected to AM line), and reniform spots outlined in pale tan. Thin, irregular ST line is often indistinct. **HOSTS** Aspen, balsam poplar, and willow.

OBSERVANT SALLOW
Andropolia diversilineata

Uncommon

93-2680 (9563)

TL 19–22 mm Gray FW has slightly scalloped, straightish AM and curving PM lines (sometimes indistinct) and jagged black ST line with prominent spikes into inner and central ST area. Light brown orbicular, reniform, and thin orbicular spots are narrowly outlined in black; reniform spot often has a dusky center. Dusky V-shaped median band is often indistinct. Terminal line is flat dusky triangles. Veins are thinly traced with black. **HOSTS** Antelope bitterbrush; likely also other bitterbrush and cliffrose.

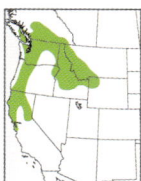

BLIND SALLOW
Andropolia aedon

Uncommon

93-2684 (9570)

TL 21–23 mm Resembles Observant Sallow, but AM and PM lines are strongly scalloped and reniform spot lacks dark crescent at center. ST line is usually fragmented into individual chevrons. Terminal line is black dots. **HOSTS** Alder, maple, ocean spray, and Pacific ninebark.

ELDER MOTH

PALE ENARGIA

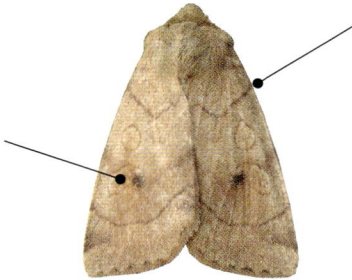

SMOKED SALLOW

actual size

EVEN-LINED SALLOW

OBSERVANT
SALLOW

BLIND SALLOW

589

SLEEPING SALLOW
Andropolia theodori

Uncommon
93-2685 (9571)

RANGE-WIDE

TL 21–23 mm Resembles Observant Sallow, but orbicular and reniform spots are only indistinctly outlined and PM line is zigzagged, though often indistinct. FW has a rusty-brown wash (sometimes faint) across reniform spot and outer median area, inner basal area, and terminal area. Black chevrons of ST line are sometimes boldly shaded. **HOSTS** Willow and ceanothus; also ocean spray and plum.

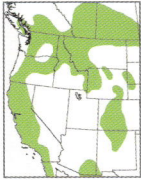

DISCORDANT BROCADE
Fishia discors

Rare
93-2691 (9970)

RANGE-WIDE

TL 20–23 mm Resembles Wandering Brocade, but inner basal area has strong black basal dash, reniform spot is light brown, and terminal area has black interveinal dashes. Inner PM line is edged thinly in white and bordered with light brown shading. **HOSTS** Generalist on woody shrubs, including spirea, bitterbrush, manzanita, and ceanothus.

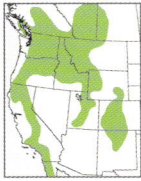

DARK GRAY FISHIA
Fishia yosemitae

Rare
93-2693 (9972)

RANGE-WIDE

TL 19–23 mm Light gray FW has strongly zigzag, thin, black AM and PM lines that connect at point tips near inner margin. Orbicular and reniform spots are thinly outlined in black and usually touch nearest point tips of zigzags. Central ST area has black dashes with white dots at base of each. **HOSTS** Douglas-fir, wild buckwheat, goldenrod, mullein, and Welsh poppy.

WANDERING BROCADE
Fishia illocata

Uncommon
93-2695 (9420)

RANGE-WIDE

TL 22–24 mm Dark brown FW has bold black bar connecting black, wavy, double AM and PM lines parallel to inner margin, bordered by warm brown shading; PM line has whitish fill adjacent to inner margin. Brown orbicular and white reniform spots are thinly outlined in brown. Thin, pale tan ST line is sometimes indistinct. **HOSTS** Alder, birch, and willow.

GROTE'S SATYR
Ufeus satyricus

Uncommon
93-2709 (11051)

NORTH

SOUTH

TL 23–25 mm Reddish-brown FW has thick, curving, soft black PM (and sometimes indistinct AM) line and short black dashes in central terminal area. Central median area usually has an elongate dusky patch. Dusky AM line is usually indistinct. Subspecies *U. s. sagittarius* lacks PM line and has a short pale streak in central median area edged costally with black. **HOSTS** Quaking aspen. **NOTE** *U. s. satyricus* is found east of the Rocky Mountains; *U. s. sagittarius* occurs in the Rocky Mountains and west.

MOSSY BROCADE
Viridiseptis marina

Uncommon
93-2749.5 (9403)

RANGE-WIDE

TL 16–18 mm Olive-green to olive-brown FW has thick gray veins and white-outlined orbicular and reniform spots. Indistinct black AM and PM lines are scalloped and sometimes edged with pale scales. Inner median and basal areas have dusky shading. **HOSTS** Unknown.

SLEEPING SALLOW

DISCORDANT BROCADE

DARK GRAY FISHIA

WANDERING BROCADE

actual size

subspecies
U. s. sagittarius

GROTE'S SATYR

subspecies
U. s. satyricus

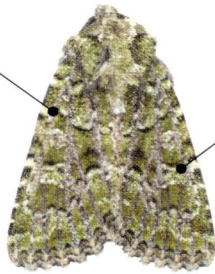

MOSSY BROCADE

SPRING QUAKERS and WOODLINGS

SUPERFAMILY Noctuoidea (93)
FAMILY Noctuidae SUBFAMILY Noctuinae
TRIBE Orthosiini

Small to medium-sized moths with long squared wings and a sturdy, hairy thorax; the anterior of the thorax has a squared, shoulder-pad appearance similar to the pinions, but less pronounced. Most species have a rounded apex, but a few, notably the *Perigonica*, are strongly pointed. Most species fly in the winter (in warmer regions) and early spring. Adults are nocturnal and will come to lights; some species may also visit sugar bait.

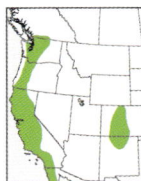

THIRD QUAKER
Perigonica tertia   **Uncommon**   93-2750 (10464)   RANGE-WIDE

TL 18–20 mm Outer margin of warm brown FW has pointed apex and point at midpoint. AM and ST areas have paired black dots at veins. Wavy pale AM, PM, and ST lines are edged with brown. Pale-outlined orbicular and reniform spots are often shaded darker brown. Brown median line is usually darkest near costa, often indistinct in inner half. Fringe is brown, sometimes checkered. Thorax brown, sometimes with a sprinkling of gray scales at posterior. **HOSTS** Garry oak and tanoak.

ANGLED QUAKER
Perigonica angulata   **Uncommon**   93-2753 (10468)   RANGE-WIDE

TL 15–18 mm Resembles Third Quaker, but thorax has a patch of dark gray scales at posterior and darker median line usually extends to inner margin. Pale subapical ST patch is well defined. Fringe is dusky brown or checkered, with a dusky gray section at apex. Pale individuals lack dark brown markings but can usually be identified by thorax and fringe markings. **HOSTS** Golden chinquapin and oak.

ORANGE-RIMMED QUAKER
Perigonica pectinata   **Rare**   93-2754 (10469)   RANGE-WIDE

TL 17–18 mm Peppery reddish-tan to lavender-gray FW has pale tan lower edge to reniform spot, sometimes bordered with warm brown; reniform spot sometimes filled with dark brown. AM and PM lines have black dots at veins. Thin, pale tan ST line often indistinct. Some individuals have dark ST/terminal area crossed by pale veins. **HOSTS** Golden chinquapin and coast live oak; possibly also other oak.

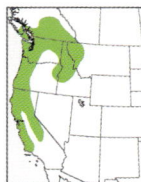

NORMAL QUAKER
Acerra normalis   **Uncommon**   93-2755 (10470)   RANGE-WIDE

TL 18–23 mm Warm grayish-brown FW has fused orbicular and reniform spots with thin, whitish outline surrounded by dark brown in central median area. **HOSTS** Serviceberry, mountain mahogany, ocean spray, cliffrose, bitterbrush, and *Prunus*.

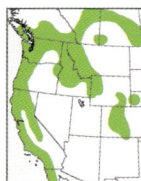

ARCHED QUAKER
Stretchia muricina   **Uncommon**   93-2758 (10473)   RANGE-WIDE

TL 14–16 mm Gray FW has fused gray orbicular and brown-shaded reniform spots set in warm brown central median area. Light gray AM and PM lines are scalloped. Costal ST area has dark brown patch. **HOSTS** Currant and gooseberry.

actual size

THIRD QUAKER

ANGLED QUAKER

ORANGE-RIMMED
QUAKER

NORMAL QUAKER

ARCHED
QUAKER

593

FANCY QUAKER

Orthosia erythrolita

Common

93-2762 (10477)

TL 18–20 mm Variable. Tan to gray, rarely dusky, FW has narrow reniform spot lightly shaded dusky and blackish at inner end, and gently wavy ST line that is offset at costal end; ST line is edged by dark brown shading, sometimes only at inner margin and inside from offset portion. ST area has a row of paired dots at veins. Some individuals have dark double AM, median, and PM lines; rare individuals have dusky shading across PM/ST area. **HOSTS** Unknown; likely a generalist.

BEAUTIFUL QUAKER

Orthosia pulchella

Uncommon

93-2763 (10478)

TL 18–20 mm Highly variable. In all phenotypes, pale-outlined orbicular and reniform spots are shaded dusky (sometimes just faintly), and median area strongly narrows to inner margin. AM line is straightish with a small dark projection (black dot in pale forms) near inner margin and small bumps near costa. Gently S-shaped PM line has squarish black projections (paired black dots in pale forms) at veins. FW color may be pale gray, light tan, golden brown, reddish brown, or chocolate. Dark forms have median area filled with dark brown except along costa, and dark brown AM and PM lines; in some individuals, basal and ST areas are also filled with brown except at costa. Rare individuals are entirely darker except for pale terminal area. Pale forms lack median-area shading, and AM and PM lines are indistinct; terminal area is lightly shaded dark brown, as is also the median line sometimes. **HOSTS** Manzanita, particularly greenleaf and hairy manzanita.

TRANSPARENT QUAKER

Orthosia transparens

Common

93-2764 (10479)

TL 17–20 mm Maroon, warm brown, or dusky brown FW has indistinctly outlined orbicular and reniform spots shaded faintly dusky; reniform spot is slightly darker at inner end. ST area has paired black dots at veins. Median line sometimes faintly visible. Apex is pointed. **HOSTS** Arbutus and rhododendron.

PROTECTOR QUAKER

Orthosia praeses

Common

93-2765 (10480)

TL 16–18 mm Variable. In all phenotypes, large orbicular spot is slanting, and reniform is level, shaded slightly at center; spots are usually connected or almost so at inner end. Thin dark AM and PM lines are scalloped and closer together at inner margin than costa. Terminal area is paler than rest of FW, often contrastingly so, and terminal line is a row of black dots. FW may be tan, warm brown, dusky, or light gray. Median area may be concolorous with rest of FW, contrastingly darker, or darker only in a patch behind wing spots. Spots and terminal area may be strongly contrasting or nearly concolorous with rest of wing. **HOSTS** Generalist on many deciduous trees and shrubs.

PINK-LEGGED QUAKER

Orthosia mys

Uncommon

93-2766 (10481)

TL 18–21 mm Variable. Resembles Transparent Quaker, but ST area lacks black dots on veins and legs are pinkish brown to pinkish orange. Orbicular and reniform spots lack outlines and are evenly shaded darker or dusky; spots may be extremely faint on some individuals. Edge of costa and often inner margin are usually pinkish orange. **HOSTS** Arbutus and manzanita.

FANCY
QUAKER

BEAUTIFUL
QUAKER

TRANSPARENT QUAKER

PROTECTOR
QUAKER

actual size

PINK-LEGGED
QUAKER

FERRUGINOUS QUAKER
Orthosia ferrigera

Uncommon
93-2767 (10482)

TL 17–19 mm Peppery orange-brown FW has brown, lightly scalloped AM and PM and straight median lines, and dusky veins. Brown reniform spot is surrounded by ochre shading. Costa has long white marks at AM and median areas, and smaller white marks at basal and ST areas. Indistinct pale ST line is edged basally by darker shading. Some individuals have a reddish-brown FW lacking dark cross-lines or vein markings, identifiable by white costal markings and brown reniform spot with yellowish surrounding shading; AM line is usually pale. **HOSTS** Oak.

SANDY QUAKER
Orthosia behrensiana

Uncommon
93-2769 (10485)

TL 17–22 mm Variable. Common form is sandy brown peppered with black scales, with dusky V-shaped median line that is blackish at costa and where it crosses through inner end of reniform spot, and blackish patch at costa of ST line. Large, rounded, tan outlines of orbicular and reniform spots usually (or nearly) touch at inner end. Some individuals are brownish gray with pale brown AM and PM lines, orbicular and reniform spot outlines, and lower veins. **HOSTS** Uncertain, but probably oak, golden chinquapin, and tanoak.

HUMBLE QUAKER
Orthosia arthrolita

Rare
93-2772 (10489)

TL 18–20 mm Light brown FW has dark brown patch between indistinct, pale-outlined orbicular and reniform spots that fades to brown at costa, and faint brown patch above orbicular spot. ST and terminal lines are a row of dark brown dots; ST line has a dark brown patch at costal end that fades to brown at costa. Scalloped brown AM and PM lines are sometimes indistinct. **HOSTS** Unknown.

SUBDUED QUAKER
Orthosia revicta

Uncommon
93-2773 (10490)

TL 19–22 mm Violet-gray to reddish-brown FW has rounded orbicular and reniform spots outlined in rusty red that are edged inwardly with pale yellow, and rusty-red ST line edged basally with pale yellow. Indistinct wavy AM and scalloped PM lines have black dots at veins. Some individuals have reddish FW with contrasting gray terminal and inner median areas. **HOSTS** Deciduous trees, including ash, birch, beech, elm, maple oak, poplar, and willow.

PACIFIC QUAKER
Orthosia pacifica

Common
93-2777 (10494)

TL 18–21 mm Peppery, warm brown FW has slightly dusky reniform spot with darker inner end, indistinct orbicular spot, and darker brown to dusky shading along indistinct ST line. Veins are marked with hoary gray. Diffuse brown median band is usually at least faintly visible, but curving AM and PM lines are frequently indistinct or absent. Fringe is checkered when fresh. Separable from Speckled Green Fruitworm Moth by median band and ST line. **HOSTS** Generalist on many deciduous trees and shrubs, particularly oak, ceanothus, manzanita, and arbutus.

FERRUGINOUS QUAKER

SANDY QUAKER

HUMBLE QUAKER

actual size

SUBDUED QUAKER

PACIFIC QUAKER

SPECKLED GREEN FRUITWORM MOTH

Orthosia hibisci

Very Common
93-2778 (10495)

TL 20–23 mm Variable. Peppery FW is reddish brown to brownish gray with thin yellowish ST line that steps basally at costa and is bordered basally by reddish-brown shading at costa and midpoint (sometimes entire line). Rounded, slightly slanting orbicular and level reniform spots have yellowish outline and may touch at inner side in some individuals; reniform spot has dusky shading at inner end. Curving brownish to dusky AM and PM lines are sometimes indistinct or nearly absent. Veins are traced with gray. Fringe is brown. Separable from Pacific Quaker by ST line and lack of median band. **HOSTS** Deciduous and coniferous trees, including apple, chokecherry, elm, hickory, poplar, spruce, tamarack, and willow.

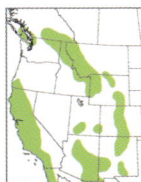

VARIABLE WOODLING

Egira variabilis

Uncommon (rare in North)
93-2787 (10504)

TL 16–18 mm Variable. Pale to brownish-gray FW has dusky median area containing very large, rounded, pale orbicular and reniform spots that are often fused and backed by a narrow blackish streak through central median area. Claviform and reniform spots are often washed with rusty brown. ST line is pale. AM and PM lines are usually indistinct. **HOSTS** Pine and other conifers.

WINTER WOODLING

Egira hiemalis

Common
93-2788 (10505)

TL 16–20 mm Variable. Gray FW has double, black, AM line and rounded, black-outlined, pale orbicular and reniform spots. Dusky shading along AM line and around reniform spot creates the appearance of a broad, angled, pale bar from orbicular spot to inner median area. Reniform spot and (often) inner median area are washed with warm brown. Some individuals are heavily suffused with brown below reniform spot, in inner median and basal areas, and along ST line. **HOSTS** Douglas-fir, beaked hazelnut, antelope bitterbrush, and ash.

SIMPLE WOODLING

Egira simplex

Common
93-2789 (10506)

TL 18–20 mm Resembles Crucial Woodling but also has a black patch at inner median line; sometimes entire median line is faintly visible. Large rounded claviform and orbicular spots are outlined with black. Black interveinal dashes form ST line, not extending to outer margin. **HOSTS** Alder, willow, *Prunus*, Sitka spruce, Douglas-fir, grand fir, and other deciduous and coniferous trees.

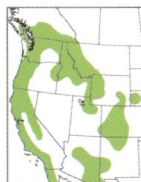

CRUCIAL WOODLING

Egira crucialis

Very Common
93-2791 (10508)

TL 17–21 mm Gray FW has an angled black median bar at costa that crosses inner reniform spot. Pale reniform spot is faintly washed with brown; small claviform and orbicular spots are incompletely and indistinctly outlined with black. Terminal area has thick black interveinal dashes at midpoint and anal angle. Thin black basal dash extends onto thorax as lateral black stripe. **HOSTS** Alder, snowbrush ceanothus, antelope bitterbrush, ocean spray, salmonberry, salal, oak, rose, and spirea.

SPECKLED GREEN FRUITWORM MOTH

melanistic

VARIABLE
WOODLING

WINTER WOODLING

actual size

SIMPLE WOODLING

CRUCIAL WOODLING

FAMILIAR WOODLING

Uncommon

Egira cognata

93-2792 (10509)

RANGE-WIDE

TL 17–20 mm Variable. Pale to brownish-gray FW has large, rounded, black-outlined orbicular, claviform, and reniform spots. Paler band angles from costa across orbicular spot to inner median area, bordered below by diffuse dusky median line. Pale ST line is edged outwardly with black shading. Double PM line is often indistinct except at inner margin. Lacks basal dash. **HOSTS** Garry oak.

MOTTLED OAK WOODLING

Uncommon

Egira februalis

93-2793 (10510)

RANGE-WIDE

TL 15–18 mm Resembles Familiar Woodling, but pale gray, rounded claviform and reniform spots are shaded dusky inside. Basal area has a blackish, smudged basal dash. Dusky to blackish shading extends from costal median line across inner reniform spot to inner end of dusky blackish terminal area. **HOSTS** Oak.

CITRUS CUTWORM

Common

Egira curialis

93-2794 (10511)

RANGE-WIDE

TL 17–20 mm Variable. Light to dark gray FW has dusky veins often speckled with white, and a row of blackish interveinal spots along ST line (sometimes indistinct). Reniform spot usually has a blackish partial outline and smudge at inner end and, often, brown wash (sometimes faint) from middle into PM area. Irregular orbicular and U-shaped claviform spots are outlined with black and have dusky centers but often are indistinct; claviform spot occasionally has a brown wash. **HOSTS** Oak, cherry, hackberry, and other hardwood trees. **NOTE** Can be a pest on citrus.

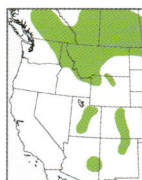

GRIEVING WOODLING

Uncommon

Egira dolosa

93-2795 (10513)

RANGE-WIDE

TL 19–20 mm Dark gray FW has light gray patch in inner median area that connects to pale-outlined orbicular spot. Pale-outlined reniform spot is often indistinct. Sometimes-indistinct white ST line is edged with black spots. Double black AM line is often visible only along pale median patch. Thorax has a thin white line across crest. **HOSTS** Birch, poplar, and willow.

PEREGRINE WOODLING

Common

Egira rubrica

93-2796 (10514)

NORTH

SOUTH

TL 16–19 mm Variable. Grayish-brown FW has flattened, slanting orbicular spot backed with dark brown shading (sometimes faint) that extends across inner reniform spot to outer margin. Reniform spot is often washed with warm brown. Wavy AM and PM lines are double; black-outlined claviform spot connects to AM line. Basal dash extends onto thorax as lateral stripes. **HOSTS** Ocean spray, and deerbrush and snowbrush ceanothus.

BROWN WOODLING

Common

Egira perlubens

93-2797 (10515)

NORTH

SOUTH

TL 17–20 mm Resembles more weakly marked individuals of Peregrine Woodling, but orbicular spot is large and round, ST line has uneven blackish edging basally, and basal dash is absent. Often has a narrow blackish streak in ST area below inner reniform spot. **HOSTS** Generalist on needled conifers and some deciduous trees.

FAMILIAR
WOODLING

CITRUS
CUTWORM

MOTTLED OAK
WOODLING

GRIEVING
WOODLING

actual size

PEREGRINE
WOODLING

BROWN
WOODLING

LARGE ARCHES

SUPERFAMILY Noctuoidea (93)

FAMILY Noctuidae SUBFAMILY Noctuinae TRIBES Tholerini and Hadenini

Medium-sized gray or brown moths, typically with pronounced orbicular and reniform spots and scalloped AM and PM lines. Many species have a paler ST area. Olive Green Cutworm resembles some darts, while the other two *Dargida* resemble wainscots. Adults are nocturnal and will come to lights.

AMERICAN GOTHIC
Tholera americana

Rare
93-2809 (10523)

RANGE-WIDE

TL 16–20 mm Brownish-gray FW has white-traced veins that turn black in ST area. Scalloped ST line is tan and bordered by dusky chevrons in ST area. Elongate claviform spot is shaded slightly darker, round orbicular spot is light brown with a dusky spot at center, and reniform spot is concolorous with FW. HOSTS Unknown; likely grass.

BRONZED CUTWORM
Nephelodes minians

Common
93-2810 (10524)

RANGE-WIDE

TL 17–23 mm Light brown FW has medium brown median area, lighter along costa and inner margin, containing light brown orbicular and reniform spots. Scalloped AM and PM lines are double, with inner of pair darker. Scalloped ST line is edged with dark brown shading in ST area. Straight, dusky median line is usually visible. HOSTS Grass, including corn and cereal crops; possibly also other herbaceous plants.

SIMPLE ARCHES
Trichocosmia inornata

Rare
93-2819 (10219)

RANGE-WIDE

TL 12–14 mm Light tan to pale brown FW has very slightly darker reniform spot and ST area, and narrow, curving, whitish AM, PM, and ST lines. Some individuals have a sprinkling of black scales and blackish scales in the reniform spot. HOSTS Unknown.

THE NUTMEG
Anarta trifolii

Common
93-2826 (10223)

RANGE-WIDE

TL 17–19 mm Tan to light brown FW is mottled with hoary dusky patches. Round orbicular spot has a dusky center, and reniform spot is shaded dusky at inner end. V-shaped claviform spot touches AM line and is often shaded dusky. Double, dark brown basal, AM, PM, and ST lines are filled with tan; inner ST line forms a W shape that touches outer margin. HOSTS Various woody and herbaceous plants. NOTE A minor crop pest.

THE MUTANT
Anarta mutata

Rare
93-2827 (10224)

RANGE-WIDE

TL 19–22 mm Resembles The Nutmeg, but light brown FW has smaller orbicular and claviform spots, and reniform spot is dark gray at inner end. Mottling, particularly in median and ST areas, that is not hoary, and. HOSTS Unknown.

AMERICAN GOTHIC

BRONZED CUTWORM

SIMPLE ARCHES

actual size

THE NUTMEG

THE MUTANT

ROYAL ARCHES
Anarta farnhami

Rare

93-2837 (10232)

TL 16–18 mm Dusky brown FW has violet-brown shading edged with rusty brown in basal, inner median, and ST areas. Orbicular and reniform spots are elongate and curving, ringed with black, and with a narrow, pale tan outline and dusky fill inside. Inner median area has a tan patch along dark brown veins that touches inner orbicular spot. Crisp, thin, tan ST line is edged with blackish chevrons basally and has a strong W shape in inner half that touches outer margin. Thin, black, double AM and PM lines are filled with tan. Central veins in upper FW are usually traced with pale tan. **HOSTS** Unknown.

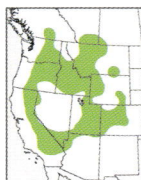

WOODEN ARCHES
Anarta crotchii

Rare

93-2838 (10233)

TL 16–19 mm Light brown to light tan FW has elongate, triangular, dusky-shaded claviform spot, tall and narrow orbicular spot edged with pale tan inside, and reniform spot with bulbous inner end shaded dusky. Thin, pale tan ST line is bordered by blackish chevrons in inner half, framing pale W shape containing dark brown veins; a second, smaller, pale W shape touches outer margin near apex. **HOSTS** Unknown.

GRANITE ARCHES
Tridepia nova

Rare

93-2862 (10253)

TL 15–19 mm Hoary gray FW has a pale round orbicular spot with a small dusky center, and reniform spot filled nearly entirely with dusky shading. Scalloped, double, dusky AM and PM lines and pale gray ST line are often indistinct. Median line and ST areas are sometimes shaded dusky. Costa is checkered. **HOSTS** Unknown.

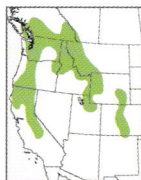

WINDY ARCHES
Polia piniae

Uncommon

93-2866 (10274)

TL 26–28 mm Gray FW has large rounded orbicular and slightly pointed reniform spots that are usually incompletely outlined in dark gray and shaded in between with dusky gray. Indistinct pale ST line has dusky shading basally at costa, at midpoint, and near inner margin. Black-outlined triangular claviform spot is visible on some individuals. **HOSTS** Alder and willow.

STORMY ARCHES
Polia nimbosa

Uncommon

93-2867 (10275)

TL 28–30 mm Resembles Windy Arches but has scalloped, double, blackish AM and PM lines filled with light gray, and larger rounded claviform spot. **HOSTS** Deciduous trees and herbaceous plants, including alder, gooseberry, huckleberry, and maple.

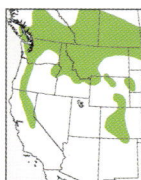

PURPLE ARCHES
Polia purpurissata

Rare

93-2872 (10280)

TL 26–28 mm Gray FW often has a violet tinge. Rounded orbicular and reniform spots are outlined in black, shaded dusky at center, and backed by a warm brown wash. Dusky double AM and PM lines are filled with light gray; small claviform spot is attached to AM line and has brownish fill. Pale ST line is edged basally with patchy, dark brown shading. Veins in ST/terminal area are dusky gray. **HOSTS** Trees and herbaceous plants, including alder, birch, blueberry, sweet fern, and willow.

ROYAL ARCHES

WOODEN ARCHES

GRANITE ARCHES

WINDY ARCHES

actual size

STORMY ARCHES

PURPLE ARCHES

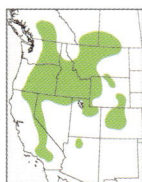

CLEVER ARCHES
Polia nugatis

Uncommon
93-2873 (10281)

RANGE-WIDE

TL 21–26 mm Hoary gray FW has round gray orbicular spot and curving reniform spot filled with dusky shading and outlined in black; reniform spot has a spike at inner end that often connects to orbicular spot, a blackish patch behind and below inner end, and warm brown below curve. Black outline of claviform spot is often shaded brown inside. Pale ST line is edged with brown basally; inner end of ST line has a W shape that touches outer margin. Some individuals have a pale gray orbicular spot. **HOSTS** Big sagebrush.

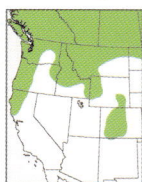

HITCHED ARCHES
Melanchra adjuncta

Uncommon
93-2874 (10292)

RANGE-WIDE

TL 19–21 mm Dusky blackish FW has bold white reniform spot with dusky bar at center, and dusky orbicular spot outlined with white. Upper basal and terminal areas are white mottled with dusky blackish. Inner basal area is often shaded tan. **HOSTS** Generalist on deciduous trees and herbaceous plants, including alder, clover, elm, and plantain.

ZEBRA CATERPILLAR MOTH
Melanchra picta

Uncommon
93-2875 (10293)

RANGE-WIDE

TL 18–22 mm Reddish-brown to maroon FW is shaded reddish orange in inner half and below reniform spot. Small orbicular and large reniform spots are outlined with speckled white; reniform spot has a speckled white bar at center and a long basal projection on inner end. Speckled whitish ST line is often fragmented. Veins are usually indistinctly traced with maroon. **HOSTS** Various trees, crops, and herbaceous plants.

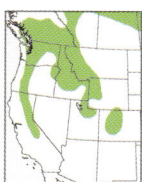

NEVADA ARCHES
Lacanobia nevadae

Rare
93-2878 (10296)

RANGE-WIDE

TL 18–20 mm Dusky gray FW has rusty-brownish patch on costal side of black basal dash, from black-outlined claviform spot to white ST line, and from outer half of reniform spot to ST line. Rounded orbicular and reniform spots are outlined with black. ST line has white W shape in inner half that touches outer margin. **HOSTS** Generalist on trees, including birch, alder, and some conifers.

SPECKLED CUTWORM
Lacanobia subjuncta

Uncommon
93-2881 (10299)

RANGE-WIDE

TL 18–22 mm Brownish-gray to grayish-brown FW has large rounded orbicular, claviform, and reniform spots. Claviform spot is connected by vertical black bar to scalloped, gray-edged, black PM line. Bold black basal dash borders rusty-brown basal patch. PM area below reniform spot is washed rusty brown. **HOSTS** Deciduous trees and woody and herbaceous plants; sometimes a pest.

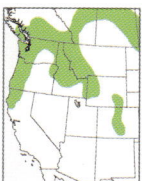

OTTER ARCHES
Spiramater lutra

Uncommon
93-2883 (10301)

RANGE-WIDE

TL 20–22 mm Dusky blackish FW has orange-tinged reniform spot, black-outlined gray orbicular spot, and a large rusty patch in inner basal area. ST area is gray with rusty shading along tan ST line. Scalloped AM and PM lines are usually inconspicuous. **HOSTS** Deciduous trees, including aspen, birch, maple, and willow.

CLEVER ARCHES

HITCHED ARCHES

ZEBRA
CATERPILLAR
MOTH

actual size

NEVADA ARCHES

SPECKLED CUTWORM

OTTER ARCHES

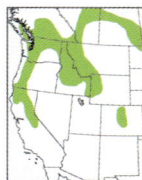

TACOMA ARCHES
Trichordestra tacoma

Uncommon
93-2885 (10303)

TL 18–20 mm Reddish-brown FW has violet-gray ST area, inner basal area, and orbicular spot, and warm brown reniform spot. Claviform spot is outlined with black. Pale yellow ST line is edged basally with dark brown. **HOSTS** Various trees and herbaceous plants, including black chokeberry, elderberry, and willow.

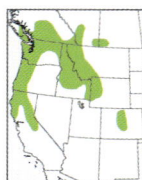

FLOWING ARCHES
Trichordestra liquida

Uncommon
93-2890 (10308)

TL 17–20 mm Gray FW has pale gray orbicular, black claviform, and rusty-washed reniform spots; PM area is shaded rusty and black behind and below reniform spot. Lightly scalloped AM and PM lines are double, filled with light gray. Pale yellowish ST line has W shape in inner half. Basal and ST areas are often washed pale gray. **HOSTS** Generalist on herbaceous plants, including aster and grass; also alder.

INTRICATE ARCHES
Trichordestra prodeniformis

Uncommon
93-2891 (10309)

TL 17–20 mm Complexly patterned, light brown FW has elongate black claviform, brown-centered ivory orbicular, and brown-centered tan reniform spots; median area between claviform and orbicular spots has an angled, pale wash along white vein. Inner margin is tan. Wavy black AM and PM lines are double. White ST line has W shape in inner half and is edged basally with dark brown shading and black interveinal chevrons. Veins in upper wing are traced with white; those in lower half are dusky. **HOSTS** Unknown; probably a generalist.

CONFIGURED ARCHES
Mamestra configurata

Uncommon
93-2898 (10271)

TL 17–20 mm Dusky gray FW has mint- to spring-green patches in inner basal and PM areas, and basally along white ST line. Large white reniform spot has dusky double bar at center. Dusky orbicular spot is outlined in black and edged inside with white scales. Some individuals have warm brown shading adjacent to green PM patch and below reniform spot. **HOSTS** Generalist on herbaceous plants, particularly those in amaranth and legume families. **NOTE** Larvae are known as Bertha Armyworm and can be a significant crop pest on canola.

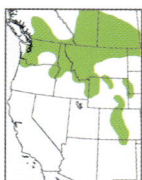

ROSEWING
Sideridis rosea

Uncommon
93-2906 (10265)

TL 23–28 mm Warm brown FW has thin brown outlines of large rounded orbicular, claviform, and reniform spots and thin, brown, scalloped AM and PM lines. Diffuse V-shaped median band touches dusky patch at inner reniform spot. Scalloped ST line is edged basally by maroon-brown shading. Terminal area is rosy brown. Hairy thorax is contrastingly reddish brown. **HOSTS** Russian olive, gooseberry, soapberry, and willow.

MAROONWING
Sideridis maryx

Rare
93-2908 (10268)

TL 20–23 mm Velvety maroon FW has an orange wash in central PM and inner median and PM areas. Indistinct orbicular and reniform spots are hoary gray. Hoary gray ST line is sometimes visible. Costa has small tan spots along lower half. Veins are traced with gray, often sprinkled with tan scales. Maroon thorax has an orange dorsal stripe. **HOSTS** Unknown; possibly bearberry.

TACOMA ARCHES

FLOWING ARCHES

INTRICATE ARCHES

actual size

CONFIGURED ARCHES

ROSEWING

MAROONWING

SMALL RANUNCULUS

Hecatera dysodea

[IN] Common
93-2909 (10270.1)

RANGE-WIDE

TL 15–17 mm Hoary gray FW has hoary blackish median area bounded by scalloped black AM and PM lines, and ochre outlines of gray orbicular and reniform spots, ochre bars in inner basal and median areas, and ochre spots below indistinct dusky ST line. Hoary gray thorax has a pair of ochre stripes. **HOSTS** Wild lettuce, hawksbeard, and sow thistle.

VARIABLE CAPSULE MOTH

Hadena variolata

Rare
93-2910 (10326)

RANGE-WIDE

TL 16–20 mm Gray to olive-gray FW has white mottling in basal area, at inner margin of median and ST areas, behind orbicular spot, and at apex. Whitish orbicular spot is outlined in black with dusky center; reniform spot is unevenly mottled white and gray. Scalloped, black, double AM, PM, and ST lines are filled with white at inner margin. Thorax is mottled white and gray. **HOSTS** Campion.

SIMILAR BROKEN-TWIG MOTH

Admetovis similaris

Uncommon
93-2924.52 (10270)

NORTH

SOUTH

TL 18–20 mm Ash-gray FW has bold tan patches in central ST/terminal area, below tan-bottomed reniform spot, that shade to darkish brown at apex and inner margin. Costa, and rounded sections of inner and central terminal area, are gray. Inner basal area is tan bordered by dark brown, adjoining brown-edged tan dorsal stripe on gray thorax. Scalloped black AM and PM lines and black outline of orbicular spot are thin. HW is white. **HOSTS** Unknown; possibly elder. **NOTE** Different Broken-twig Moth (*A. oxymorus*, not shown) is similar but has a paler gray FW with rosy-brown ST area and tan HW.

OLIVE GREEN CUTWORM

Dargida procinctus

Very Common
93-2925 (10428)

INLAND

COASTAL

TL 20–22 mm Dark brown FW has bold, pale, tan stripe from inner margin of median area to apex; a short, brown-centered, tan stripe from inner median area to base; and a vertical, brown-edged, tan stripe from base to outer margin. Elongate, brown-centered, tan orbicular and reniform spots are wedged between central stripe and brown costal shading. Black veins are edged with brown shading. Thorax is boldly striped with brown, black, and tan. **HOSTS** Grass, particularly reed canary grass.

WHEAT HEAD ARMYWORM

Dargida diffusa

Common
93-2928 (10431)

RANGE-WIDE

TL 15–19 mm Tan FW has long white central vein broadly bordered by brown shading on inner side, and long brown triangle from outer margin to costal side of white vein. A small blackish spot is adjacent to terminus of white vein, costal to tip of brown triangle. Costa is lightly shaded grayish brown. Separable from Stola Wainscot by stripes on thorax and brown ST triangle that flares along outer margin. **HOSTS** Grass seedheads, including many cereal grains.

BLUSHING GRAY-STREAK

Dargida tetera

Uncommon
93-2930 (10433)

RANGE-WIDE

TL 15–18 mm Resembles Wheat Head Armyworm, but shading and triangle adjacent to white central vein are gray and edged with brown. Terminal area is gray, and costa is shaded brown. **HOSTS** Orchard grass and crabgrass.

SMALL RANUNCULUS

VARIABLE
CAPSULE MOTH

actual size

SIMILAR BROKEN-
TWIG MOTH

OLIVE GREEN
CUTWORM

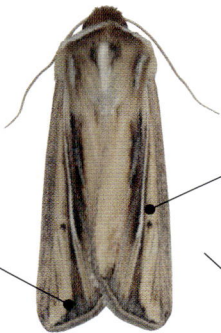

WHEAT HEAD
ARMYWORM

BLUSHING
GRAY-STREAK

WAINSCOTS

SUPERFAMILY Noctuoidea (93)

FAMILY Noctuidae SUBFAMILY Noctuinae TRIBE Leucaniini

Medium-sized tan to light brown moths with streaky veins and minimal markings; most species have dusky dots along the PM line and a pale discal spot, with few other distinct markings. Nearly all species feed on grasses and sedges as caterpillars. Adults are nocturnal and will come to lights.

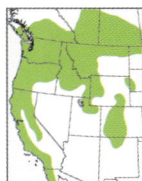

LESSER WAINSCOT
Mythimna oxygala

Uncommon
93-2933 (10436)

RANGE-WIDE

TL 17–19 mm Tan to light brown FW has paler veins edged with slightly darker tan, creating a streaked appearance. Central vein is shaded dark brown, sometimes with black speckles, along inner side of basal half and costal side of lower half of central vein. Three dark brown spots form a triangle at junction of central vein and PM line. Thorax is unmarked. Pale HW has a dusky wash. **HOSTS** Grasses.

THE WHITE-SPECK
Mythimna unipuncta

Very Common
93-2935 (10438)

NORTH

SOUTH

TL 20–25 mm Peppery tan FW has indistinct white central vein terminating in small white dot at PM area. PM line is a curving row of black dots. Warm tan orbicular and reniform spots are often indistinct. Apex has a long dusky apical dash. **HOSTS** Generalist on grass; also woody plants and cereal grains.

MEADOW WAINSCOT
Leucania farcta

Uncommon
93-2939 (10441)

NORTH

SOUTH

TL 19–23 mm Resembles Lesser Wainscot, but FW often has a reddish tinge, central vein has light brown or no shading, and thorax usually has a pair of thin stripes (sometimes faint). HW is ivory to tan, without dusky wash. **HOSTS** Cocks-foot, orchard grass, wild rye, wheatgrass, and other broad-leaved grass.

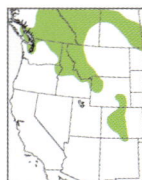

MANY-LINED WAINSCOT
Leucania multilinea

Rare
93-2945 (10446)

RANGE-WIDE

TL 19–20 mm Resembles Lesser Wainscot, but FW has dark basal dash in inner basal area with diffuse shading parallel to and along full length of inner margin. Shading along central vein is dark brown and extends to outer margin. **HOSTS** Grass and sedge.

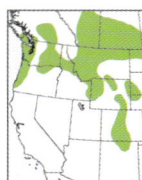

TWO-LINED WAINSCOT
Leucania commoides

Uncommon
93-2947 (10447)

RANGE-WIDE

TL 20–21 mm Resembles Many-lined Wainscot, but dots at PM line are usually indistinct, ST/terminal area is shaded dusky with dark brown interveinal dashes at center, and dark brown basal dash adjacent to inner margin is distinct. **HOSTS** Grasses.

IMPERFECT WAINSCOT
Leucania imperfecta

Rare
93-2953 (10452)

RANGE-WIDE

TL 17–18 mm Brownish- to grayish-tan, often hoary or peppery FW has a fishhook-shaped white mark in central wing, sometimes with connected veins also white. Central FW has dark, brown to grayish-brown shading along central vein to large triangle at outer margin. Curving row of black dots at PM line is often indistinct. **HOSTS** Unknown; likely grasses.

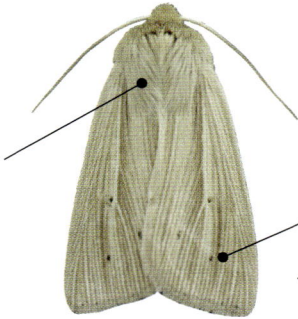

LESSER WAINSCOT

THE WHITE-SPECK

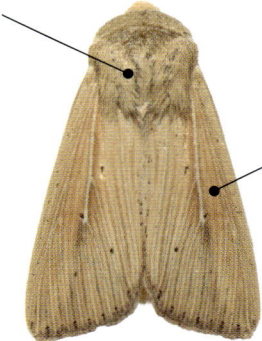

MEADOW WAINSCOT

MANY-LINED
WAINSCOT

actual size

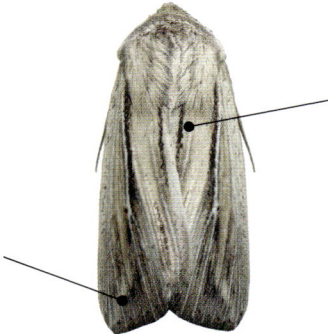

TWO-LINED
WAINSCOT

IMPERFECT
WAINSCOT

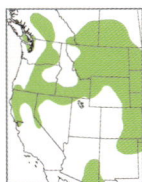

HETERODOX WAINSCOT

Leucania insueta

Rare   93-2948 (10449)

TL 18–19 mm Lightly peppered tan FW has a contrastingly grayish costa peppered with black scales, and light gray veins with black dots along curving PM line. Pale central vein is contrastingly shaded on both sides with dark brown. Basal area has a brown dash adjacent to inner margin, and a longer dash beside central vein. Dark brown terminal area extends as triangle into ST area. HW is light brownish gray with brown veins, grading darker along outer margin. **HOSTS** Grasses.

DIA WAINSCOT

Leucania dia

Uncommon   93-2949 (10449.1)

TL 18–19 mm Resembles Heterodox Wainscot, but costa is tan to grayish tan and shading along central vein and in terminal area is brown, both contrasting only weakly with tan to warm brown FW. Central basal dash is usually brown but sometimes bold blackish, and basal dash at inner margin is usually weak to absent. HW is brownish gray with brown veins, sometimes grading darker along outer margin. **HOSTS** Grasses. **NOTE** Dia Wainscot was once considered a subspecies of Heterodox Wainscot and catalogued as such in collections. Due to this, the exact range boundaries of these two species remain unclear; presented is an approximation based on available DNA barcoding and known specimen data. In general, Heterodox is found at lower elevations and latitudes, while Dia occurs at higher elevations and boreal habitats, though they may be sympatric in areas of habitat and latitude overlap such as mountain foothills.

SPECKLED WAINSCOT

Leucania oaxacana

Rare   93-2952 (10451)

TL 18–21 mm Peppery tan to light brown FW has whitish veins eged with gray, and curving row of black dots along PM line. Central vein is narrowly shaded dusky blackish, with a small black dot at midpoint on inner side. Large white spot at terminal fork of central vein cups small black dot. Eastern and Mexican individuals are browner, often with a thin dusky line connecting black PM dots. **HOSTS** Unknown; likely grasses.

OREGON WAINSCOT

Leucania oregona

Rare   93-2953.1 (10441.1)

TL 17–20 mm Resembles Speckled Wainscot, but FW is warmer brown, shading along central vein is brown to dusky brown and not noticeably darker at base, and vein lacks black dot at midpoint. Dusky lateral stripes on thorax are lighter gray. **HOSTS** Unknown; likely grasses.

STOLA WAINSCOT

Leucania stolata

Rare   93-2954 (10453)

TL 18–20 mm Resembles Wheat Head Armyworm, but thorax is more uniformly tan with just a pale dorsal stripe, FW has diffuse brown shading parallel to inner margin, and brown shading along central vein is narrower and less dark. Brown triangle in ST area is directly below black discal dot and does not flare along outer margin. Separable from other wainscots by dark shading along central vein that runs from base to outer margin **HOSTS** Unknown; likely grasses.

actual size

HETERODOX
WAINSCOT

DIA WAINSCOT

SPECKLED WAINSCOT

OREGON WAINSCOT

STOLA WAINSCOT

SMALL ARCHES and SUMMER QUAKERS

SUPERFAMILY Noctuoidea (93)
FAMILY Noctuidae SUBFAMILY Noctuinae TRIBE Eriopygini

Small noctuids with relatively short wings compared with larger arches and quakers. Most species are brown or grayish brown, but many *Lacinipolia* are pale green. These species fly on average later in the summer and into the fall, though some may also be encountered in spring. Adults are nocturnal and will come to lights; a few species may also be encountered at flowers.

BROWN-PATCHED GOLDEN ARCHES
Pseudanarta caeca
Uncommon
93-2967.54 (9604)
RANGE-WIDE

TL 10–13 mm Hoary gray FW has warm brown median area with pale gray to brown orbicular and reniform spots and a thin, black bar parallel to inner margin. Thin, black PM line is edged white in inner half. Lower ST area is warm brown. Orange HW has a broad black terminal band. **HOSTS** Unknown; possibly bunchgrasses. **NOTE** Occasionally encountered nectaring at yellow flowers.

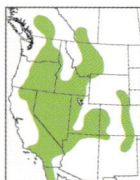

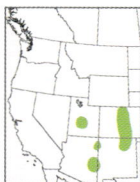

CROSSED GOLDEN ARCHES
Pseudanarta crocea
Uncommon
93-2967.55 (9605)
RANGE-WIDE

TL 10–13 mm Hoary gray FW has whitish outlines of elongate, slanting orbicular spot and reniform spot; inner ends of spots touch a diffuse pale tan to light brown patch. Thin, black PM line has a broad white wash below at inner margin and costa. Orange HW has a broad black terminal band. **HOSTS** Unknown; possibly bunchgrasses. **NOTE** Occasionally encountered nectaring at yellow flowers.

LINED GOLDEN ARCHES
Pseudanarta flava
Uncommon
93-2967.56 (9606)
RANGE-WIDE

TL 11–13 mm Resembles Crossed Golden Arches, but orbicular and reniform spots are darker gray to brown, often hidden within brown median area. White in ST area is reduced, primarily present as edging along black PM line. Black AM line may or may not have white edging. Black border on HW has a wavy basal edge. **HOSTS** Grasses. **NOTE** Occasionally encountered nectaring at yellow flowers.

SINGULAR GOLDEN ARCHES
Pseudanarta singula
Uncommon
93-2967.57 (9607)
RANGE-WIDE

TL 11–13 mm Resembles Lined Golden Arches, but white edging below PM line is narrow or absent. Orbicular and reniform spots are lighter brown, often contrasting against brownish gray median area. Basal edge of black border on HW is smooth. **HOSTS** Unknown; possibly bunchgrasses. **NOTE** Occasionally encountered nectaring at flowers.

SCHINIESQUE ARCHES
Pseudanthoecia tumida
Rare
93-2967.8 (9617)
RANGE-WIDE

TL 12–14 mm Rusty-orange FW has a mottled appearance created by paler orange terminal area and thick, broadly zigzag AM and PM lines that connect to each other at points. PM line has two small white patches near center. Fringe is checkered. HW is black with orange base and white fringe. Usually rests with wings tented over body. **HOSTS** Possibly sunflower. **NOTE** Often encountered at flowers during daytime. Superficially similar to *Schinia* species.

BROWN-PATCHED
GOLDEN ARCHES

CROSSED GOLDEN ARCHES

actual size

LINED GOLDEN
ARCHES

SINGULAR GOLDEN
ARCHES

SCHINIESQUE ARCHES

COASTAL ARCHES
Psammopolia wyatti

Rare
93-3012 (10365)

TL 20–26 mm Light to pale gray FW has rounded orbicular and reniform spots incompletely outlined with black and shaded gray at center; inner end of reniform spot often has a basal-pointing tooth and touches PM line. Black-outlined claviform spot touches pale gray AM line and often also PM line. **HOSTS** Generalist on herbaceous plants, including sand verbena, beach knotweed, dune tansy, and beach grasses.

CUNEATE ARCHES
Lacinipolia cuneata

Uncommon
93-3018 (10371)

TL 15–18 mm Gray to dusky FW has a darker median area that narrows toward inner margin. ST line is a row of mustard-yellow dots (sometimes indistinct), with a large yellow spot at inner margin. Inner basal area is washed brown. Elongate, light gray orbicular and reniform spots are outlined in black and shaded dusky at center; inner end of reniform touches scalloped double PM line. Separable from Pendant and Friendly Arches by inner basal area and darker yellow ST line spot. **HOSTS** Ocean spray; also reared from dandelion, chickweed, and lettuce.

PENDANT ARCHES
Lacinipolia pensilis

Uncommon
93-3042 (10395)

TL 15–18 mm Variable. Gray to dark gray FW has light gray, round orbicular and flat-topped reniform spots outlined in black and shaded dusky at center. Central median area is washed warm brown, adjacent to black-outlined claviform spot and beneath reniform spot. Indistinct ST line is usually pale yellowish, broadening to distinct pale yellowish spot at inner margin. Blackish basal line borders pale basal area. **HOSTS** Generalist on many woody shrubs and herbaceous plants. **NOTE** Generally found at higher elevations than Friendly Arches. Similar individuals from western WA to CA are likely *L. dimocki* (not shown).

FRIENDLY ARCHES
Lacinipolia acutipennis

Uncommon
93-3042.1 (10394.1)

TL 12–15 mm Resembles Pendant Arches and some intermediates may not be visually identifiable to species. Averages smaller, with less brown in median area. Basal line and dash are usually faint. Basal edge of reniform spot is curving, and thin, black claviform spot doesn't usually touch PM line. Pale yellow ST line and inner spot are edged with blackish brown spots basally. **HOSTS** Alfalfa and dandelion. **NOTE** Generally found at lower elevations than Pendant Arches.

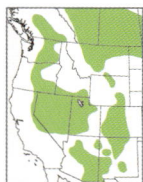

SERRATED ARCHES
Lacinipolia sareta

Uncommon
93-3042.2 (10395.1)

TL 14–16 mm Resembles Friendly Arches but typically flies earlier in season. FW usually grayer, often with a faint brown wash to median area and brown edging to ST line. Black claviform spot usually connects AM and PM lines. Long, black basal dash and PM line are usually distinct. **HOSTS** Alfalfa and dandelion.

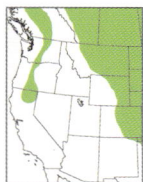

BRISTLY CUTWORM
Lacinipolia renigera

Common (uncommon in South)
93-3044 (10397)

TL 14–15 mm Dusky brown to gray FW has rectangular, white-outlined reniform spot filled with spring green, and green patches at central basal and inner ST areas. Claviform spot and thick basal dash are black. Straightish AM and shallowly convex PM line are double. **HOSTS** Wide variety of crops and herbaceous plants.

COASTAL ARCHES

CUNEATE ARCHES

PENDANT ARCHES

FRIENDLY ARCHES

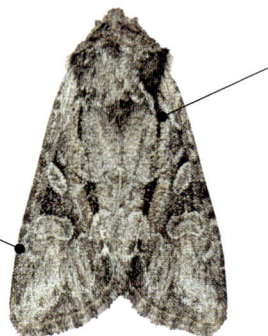

actual size

SERRATED ARCHES

BRISTLY CUTWORM

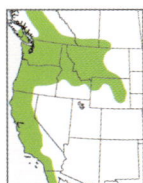

BROWN ARCHES
Lacinipolia stricta

Uncommon

93-3045 (10398)

TL 13–16 mm Warm to dark brown FW has bold white outline to narrow reniform spot. Orbicular spot has a thin, incomplete ring of whitish scales but is often indistinct. Median area may be concolorous or slightly darker, bounded by often-indistinct double AM and PM lines. Terminal area is dark brown. **HOSTS** Generalist on herbaceous plants. **NOTE** Can be a pest on some crops.

GIRDLED ARCHES
Lacinipolia circumcincta

Rare

93-3046 (10399)

TL 13–15 mm Resembles Brown Arches, but FW is dark chestnut brown with dark chocolate-brown median area. Both orbicular and reniform spots are dark brown outlined with white (sometimes thinly). Inner ST area is usually lighter brown to tan. Pale brown ST line contrasts against dark brown terminal area. **HOSTS** Unknown; likely a generalist.

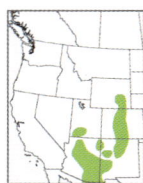

BRONZE ARCHES
Lacinipolia lepidula

Rare

93-3048 (10401)

TL 13–15 mm Resembles Girdled Arches, but median area is unevenly rusty and dark brown, and both orbicular and reniform spots are filled with rusty brown. Light brown AM line is widely zigzagged. Tan ST line widens into tan apical patch. **HOSTS** Unknown; likely a generalist.

OLIVE ARCHES
Lacinipolia olivacea

Uncommon

93-3053 (10406)

TL 13–15 mm FW has a dusky olive to gray base, darker median area, and whitish to light gray ST area; some or all areas are usually washed with olive green. Reniform spot is mostly white; dusky gray orbicular spot is outlined black. Double, black AM and PM lines are filled light gray to whitish; inner PM line is smoothly convex. Some individuals have a streak of rusty-brown wash parallel to inner margin from base to anal angle. **HOSTS** Herbaceous plants, including dandelion, phlox, and plantain.

RED-SPOT ARCHES
Lacinipolia davena

Uncommon

93-3061 (10407)

TL 14–17 mm FW has dark gray base, darker gray median area, and light gray ST/terminal; ST area has a rusty (rarely olive or whitish) patch at inner margin. Orbicular and reniform spots are outlined black; reniform is usually washed white. Curving, double AM and PM lines have light gray to whitish fill. Thorax usually has a rusty patch at posterior. Southern individuals are usually more extensively washed rusty brown, particularly parallel to inner margin. **HOSTS** Unknown.

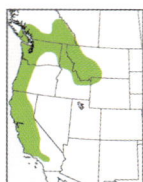

EMERALD ARCHES
Lacinipolia comis

Rare

93-3062 (10408)

TL 14–17 mm Variable. Resembles Olive and Red-spot Arches, but ST/terminal area has a pale crescent above a diffuse blackish patch beside inner margin and an emerald-green wash along pale ST line, usually most visible in the crescent. Emerald green is often also present as a streak in central basal area. FW often has golden-orange scales, sometimes extensively. Thorax often has small brownish spots at posterior. Some females may be entirely charcoal gray but still show green patch at inner ST area. **HOSTS** Generalist on herbaceous plants.

GIRDLED ARCHES

BRONZE ARCHES

BROWN ARCHES

OLIVE ARCHES

actual size

RED-SPOT ARCHES

EMERALD ARCHES

COLLARED ARCHES COMPLEX

Lacinipolia strigicollis/buscki

Common

93-3069 (10415)

TL 15–17 mm Two species overlapping in phenotype. Brownish- to grayish-green (rarely mint-green or light brown) FW has blackish median area bounded by scalloped black lines edged outwardly with white; AM line is often double. Median area narrows and is often washed paler toward inner margin. Black-outlined, small orbicular and rounded reniform spots have olive to dusky fill; reniform spot has a whitish inner outline. Base is blackish above double, white-filled basal line. Some individuals have warm brown streaks in inner and central median area. Individuals with little to no white along AM/PM lines and a uniformly colored thorax can generally be considered *L. buscki* (Weathered Arches), while individuals with crisp white borders to AM/PM lines and whitish epaulets on thorax can be considered *L. strigicollis* (Collared Arches). Intermediate individuals may be identifiable only by DNA. **HOSTS** Generalist on herbaceous plants, particularly those in the legume and aster families. **NOTE** There is some uncertainty as to the true identity of *L. strigicollis* as the original specimen used to describe the species was presumed lost for nearly a century before being rediscovered in 2022. Recent treatments of the genus have relied on published descriptions that have probably resulted in incorrect synonymies. The phenotype illustrated here as *L. strigicollis* almost certainly represents *L. illaudabilis*, a synonym that will likely eventually be raised to species. True *L. strigicollis* is possibly not that common. Further study using the rediscovered holotype is needed.

FOUR-LINED ARCHES

Lacinipolia quadrilineata

Common

93-3075 (10422)

TL 13–15 mm Sexually dimorphic. Mottled brownish to gray FW has dusky median area bounded by straightish, slightly angled, black AM line and curving, slightly scalloped PM line edged with white to tan in inner half; median area narrows strongly toward inner margin. Light brown to gray orbicular and reniform spots have dusky centers. Anal dash is a dusky smudge. Basal area is light gray in males, dark gray in females. An uncommon brown form occurs in southern CA that is visually inseparable from Ashen Arches; see Note for Ashen. **HOSTS** Generalist on herbaceous plants, particularly those in the legume and aster families.

ASHEN ARCHES

Lacinipolia martini

Rare

93-3076 (10422.1)

TL 13–15 mm Resembles male Four-lined Arches, but FW is browner with less contrast between median and basal areas. Some individuals may have streaks of rusty wash through central and inner FW. **HOSTS** Unknown. **NOTE** The extent of this species' CA range is unclear as there are no consistent phenotypic traits that can be used to separate it from the brown form of Four-lined Arches. DNA barcoding for brown individuals in CA has thus far only been completed for specimens from the San Diego area, where both species are shown to be present. However, brown individuals occur at least as far north as San Francisco area.

FUSED ARCHES

Lacinipolia patalis

Uncommon

93-3077 (10423)

TL 15–17 mm Gray FW has light gray, elongate, and slanting orbicular and kidney-shaped reniform spots that fuse at inner end, are outlined in black, and have dusky centers. Black-outlined U-shaped claviform spot is sometimes filled with light gray. ST line is a row of black spots, often with a diffuse dusky spot near inner margin. **HOSTS** *Rubus*, roses, and other woody Rosaceae.

Collared Arches
type

COLLARED ARCHES
COMPLEX

intermediate
type

intermediate
type

Weathered Arches
type

brown form

unidentified
Central CA

male

female

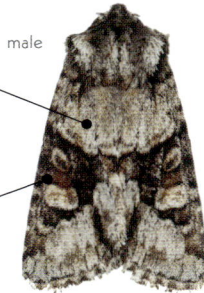

FOUR-LINED
ARCHES

actual size

ASHEN
ARCHES

FUSED
ARCHES

623

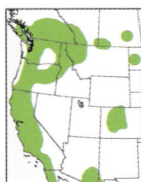

NORTHERN SCURFY QUAKER

Rare | RANGE-WIDE

Homorthodes furfurata 93-3088 (10532)

TL 13–14 mm Warm brown to grayish-brown FW often has a mottled appearance. Grayish veins are sprinkled with white scales, most noticeable at tips of scalloped brown AM and PM lines. Orbicular and reniform spots are light brown; reniform spot has a dusky smudge at inner end with white dots around edge. **HOSTS** Deciduous trees, including cherry and maple.

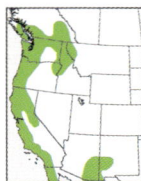

ALDER QUAKER

Uncommon | RANGE-WIDE

Homorthodes communis 93-3090 (10533)

TL 12–15 mm Resembles light form Fractured Quaker, but FW is more evenly colored with fewer black speckles. Orbicular and reniform spots usually have incomplete, thin, black outlines, and slightly darker shading at center. Veins usually have white scales at tips of PM line scallops. **HOSTS** Alder; also other hardwoods.

FRACTURED QUAKER

Uncommon | RANGE-WIDE

Homorthodes fractura 93-3091 (10534)

TL 11–14 mm Dark form has dark brown FW with lighter brown median area bounded by double, brown AM and PM lines filled light brown; PM line has pale dashes below at veins. Light form resembles Northern Scurfy Quaker, but lacks gray and white scales along veins. Often has a warm reddish brown wash around reniform spot. **HOSTS** Unknown; likely alders and other hardwoods.

HANHAM'S QUAKER

Uncommon | RANGE-WIDE

Homorthodes hanhami 93-3096 (10539)

TL 13–16 mm Brown FW has a thick, dark brown median line connected to a large dark patch at inner reniform spot, often with a patch of whitish scales below. Orbicular spot appears absent. Veins are traced with dark brown, sprinkled with white scales. Cross-lines are usually indistinct, but darker individuals show pale AM, PM and ST lines. Separable from Northern Scurfy Quaker by larger black spot at reniform, and indistinct AM line and orbicular spot. **HOSTS** Unknown.

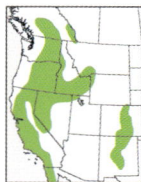

NUTMEG QUAKER

Rare | RANGE-WIDE

Trichopolia curtica 93-3103 (10546)

TL 14–16 mm Warm to dusky brown FW has median area washed rusty brown. Orbicular and reniform spots are usually indistinct except for blackish spot at inner reniform. Gray, deeply scalloped AM and curving PM lines are edged in dark brown medially. Indistinct yellowish ST line is shaded dark brown basally. Veins are indistinctly traced with gray. **HOSTS** Generalist on herbaceous plants.

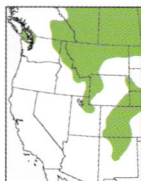

RUDDY QUAKER

Rare | RANGE-WIDE

Trichopolia oviduca 93-3113 (10563)

TL 14–16 mm Reddish-brown, dark brown, or brownish-gray FW has rounded dusky-washed orbicular and reniform spots outlined in pale yellowish; reniform usually has a dark patch at inner end. AM and PM lines are gray, edged medially with dark brown; AM line crosses straight from inner margin to edge of orbicular spot, then bends over the top. ST line is brownish yellow bordering lighter-colored terminal area. Veins are marked with dusky gray, with paired black and gray dots below PM line. **HOSTS** Dandelion, grass, plantain, and other herbaceous plants.

NORTHERN
SCURFY QUAKER

ALDER QUAKER

FRACTURED
QUAKER

HANHAM'S
QUAKER

actual size

NUTMEG QUAKER

RUDDY
QUAKER

DARK-SPOTTED QUAKER
Trichopolia melanopis

Common
93-3115 (10566)

RANGE-WIDE

TL 14–18 mm Grayish-brown FW has rounded dusky orbicular and reniform spots that have pale yellowish outline. Straightish ST line is contrastingly pale yellow. Faded, light brown AM and PM lines are edged medially with sparse black scales; AM line is smoothly curving. Veins in ST/terminal area are traced with gray, with dark dots below PM line. Rare individuals have dusky basal and ST/terminal areas. HOSTS Unknown.

RUFOUS QUAKER
Trichopolia rufula

Uncommon
93-3115.3 (10557)

NORTH

SOUTH

TL 15–18 mm Resembles Dark-spotted Quaker, but orbicular and reniform spots are concolorous with FW except for blackish spot at inner reniform, and AM line is deeply scalloped with black dots at scallop tips. Pale yellowish ST line is often faded or indistinct. FW color highly variable, including brown, rusty brown, tan, gray, and dusky brown. Median area is often very slightly darker. HOSTS Apple, plum, and pear, and other trees and shrubs in Rosaceae.

LIGHT-SPOTTED QUAKER
Trichopolia alfkenii

Common
93-3115.5 (10559)

RANGE-WIDE

TL 14–16 mm Brown to dusky brown FW has light-colored, rounded orbicular spot with brown to dusky center and black outline, and dusky, rounded reniform spot with thin, light brown outline around costal half and blackish patch at inner end. Wavy, pale yellowish ST line has thicker sections at midpoint and near inner margin. AM and PM lines are usually indistinct. Veins are indistinctly traced with gray. HOSTS Generalist on herbaceous plants.

BICOLORED QUAKER
Ulolonche disticha

Uncommon
93-3124 (10573)

RANGE-WIDE

TL 13–15 mm Lightly peppered gray to brownish-gray FW has thick black median line that touches inner corner of brown reniform spot. Lower half of FW is washed dusky, sometimes only lightly. Indistinct dotted ST line has thick black bar at costa. HOSTS Unknown.

ORBICULATE QUAKER
Ulolonche orbiculata

Uncommon
93-3125 (10574)

RANGE-WIDE

TL 13–15 mm Brown to gray FW has black-outlined, slanting orbicular spot; indistinctly defined reniform spot has white mark at inner end where central vein forks. ST area is dusky with black chevrons along irregular ST line. Basal area has a black basal dash that connects to black-outlined claviform spot. Wavy black AM and PM lines are double. HOSTS Unknown.

VARIED QUAKER
Pseudorthodes irrorata

Rare
93-3132 (10582)

RANGE-WIDE

TL 13–17 mm Brown to reddish-brown FW has soft-lined, scalloped, double AM and PM lines filled with light brown; PM line is edged below with black dots at veins. Light brown reniform spot has white scales along outline; orbicular spot is usually indistinct. Pale ST line is edged basally by small black chevrons. Terminal line has tiny tan dots between black dashes. HOSTS Generalist on deciduous and coniferous trees and shrubs.

DARK-SPOTTED QUAKER

RUFOUS QUAKER

LIGHT-SPOTTED
QUAKER

BICOLORED QUAKER

ORBICULATE
QUAKER

actual size

VARIED QUAKER

ACCURATE QUAKER
Hexorthodes accurata

Uncommon
93-3156.1 (10601)

TL 15–17 mm Chestnut- to golden-brown FW has pale-outlined, dusky, slanting orbicular and level reniform spots that conjoin at inner end, and smooth, pale tan PM line. Inner median and terminal areas are shaded darker brown, and ST area is dusky brown. **HOSTS** Brickellbush.

WHITE-SPECKED QUAKER
Hexorthodes nipana

Uncommon
93-3160 (10286)

RANGE-WIDE

TL 13–16 mm Mottled brown to grayish-brown FW has dusky to dark brown shading in central median area, behind black-outlined (often incompletely), rounded orbicular and reniform spots. Inner end of reniform spot has two small white marks where vein splits. Terminal area is dusky to dark brown. Gray veins are most obvious in ST/terminal area. **HOSTS** Unknown; possibly brickellbush.

V-LINED QUAKER
Zosteropoda hirtipes

Common
93-3163 (10607)

NORTH

SOUTH

TL 13–14 mm Brownish-tan FW has crisp, brown, V-shaped AM and PM lines. Central median area is indistinctly washed brown to dusky, sometimes with a dusky discal spot. Apex is slightly falcate. Tan thorax is hairy. **HOSTS** Deciduous trees and shrubs, including alder and willow, as well as grass and possibly St. John's wort.

BLACK MOON QUAKER
Hypotrix lunata

Rare
93-3182 (10606)

RANGE-WIDE

TL 17–19 mm Brownish-gray FW has large, pale-outlined, dark brown to blackish crescent in central median area narrowly surrounded by rusty shading. Thin pale ST line is smoothly curving. Thin dusky AM and PM lines are often indistinct. Terminal area is shaded darker, with pale subapical patch. **HOSTS** Unknown.

GRAY-VEINED QUAKER
Miodera stigmata

Uncommon
93-3188 (10623)

RANGE-WIDE

TL 13–16 mm Tan-brown FW has speckled gray veins. Curving tan ST line is boldly edged with black on both sides. Orbicular, claviform, and reniform spots are outlined with black; reniform is tan and sometimes also orbicular. Scalloped, dusky, double AM and PM lines are sometimes visible. **HOSTS** California sagebrush.

SHARP-SPOTTED QUAKER
Tricholita chipeta

Uncommon
93-3199 (10631)

RANGE-WIDE

TL 15–16 mm Hoary gray FW has distinctive, fragmented, white reniform spot with long basal-pointing tooth at inner end. Round orbicular and U-shaped claviform spots are gray outlined with black. Terminal line is a row of tiny white dots that continue up lower costa. **HOSTS** Wright's snakeweed.

SERRATED BORER-MIMIC
Hydroeciodes serrata

Rare
93-3207 (10637)

RANGE-WIDE

TL 14–17 mm Golden-brown FW has fragmented white reniform spot with narrow bar at center, round white orbicular spot with brown dot at center, and bisected white claviform spot against AM line. White to tan basal patch is fragmented. Apical patch is tan. Separable from similar Ragwort Stem Borer by claviform spot and rounded apex. **HOSTS** Unknown.

ACCURATE QUAKER

WHITE-SPECKED QUAKER

V-LINED QUAKER

BLACK MOON
QUAKER

actual size

GRAY-VEINED QUAKER

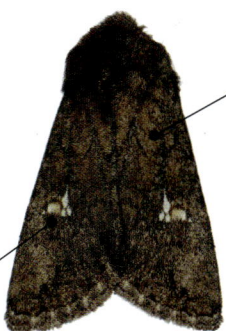

SHARP-SPOTTED
QUAKER

SERRATED
BORER-MIMIC

DARTS SUPERFAMILY Noctuoidea (93)
FAMILY Noctuidae SUBFAMILY Noctuinae TRIBE Noctuini

Medium-sized brown to brownish-gray moths with long squared wings. Species often have strong orbicular and reniform spots, sometimes with wide pale streaks along the central veins or dark shading behind the reniform spot. As a group, they are often associated with late summer and autumn, though some species may fly year-round. Adults are nocturnal and will come to lights; a number of species will also come to sugar bait. Some species, such as Army Cutworm, may occasionally be found at flowers in daytime.

VARIEGATED CUTWORM
Peridroma saucia

Very Common
93-3211 (10915)

NORTH
SOUTH

TL 21–25 mm Variable. Mottled FW is pale tan to chestnut brown. Scalloped double AM and PM lines are often indistinct except at costa. Large rounded orbicular and reniform spots are outlined with black and edged inside with pale scales; reniform spot is usually filled with dusky shading. Thorax has raised white-tipped dorsal stripe. HOSTS A generalist on a wide variety of trees, crops, and herbaceous plants. NOTE Also called Pearly Underwing. Can be common at sugar bait.

GREEN CUTWORM
Anicla infecta

Common
93-3212 (10911)

RANGE-WIDE

TL 17–20 mm Variable. Tan to reddish- or violet-gray FW is lightly to heavily peppered with dark brown. Reniform spot has blackish fill, crossed by a thin gray bar, and is outlined in reddish brown with pale yellowish inner edging; often-indistinct orbicular spot is not filled, outlined in reddish brown, and pale yellowish. Terminal area is dark brown with pale apical patch. Pale yellowish ST line is edged basally with reddish brown. Thoracic collar is dark brown. HOSTS Grass; also herbaceous plants, including beet, clover, and tobacco.

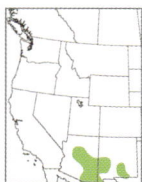

YELLOW-EYED DART
Anicla biformata

Uncommon
93-3220 (10908.1)

RANGE-WIDE

TL 17–19 mm Reddish-brown to brownish-gray FW has a grayish costa with dark brown marks at AM and PM lines. Dark brown reniform spot (sometimes fragmented or indistinct) has thin to broad pale yellowish outline. ST line is usually present as indistinct, pale yellow dots. Thoracic collar is dark brown. HOSTS Unknown; possibly grasses.

EXUBERANT DART
Anicla exuberans

Rare
93-3221 (10905)

RANGE-WIDE

TL 17–19 mm Lightly peppery, tan to light warm brown FW has a light gray costa. Gray reniform spot is thinly edged above and below with black. Cross-lines are usually indistinct except for dark brown marks at costa; sometimes faint scalloped AM/PM lines and thick median line may be present. Dark brown thoracic collar is even in width across entire span from one side of head to the other; collar is narrowish and often not visible when moth viewed dorsally. HOSTS Unknown; possibly grasses. NOTE Similar Poetic Anicla (*A. espoetia*, not shown) has more visible cross-lines, a black reniform spot, and thoracic collar that bows wider at midpoint.

VARIEGATED
CUTWORM

GREEN
CUTWORM

actual size

YELLOW-EYED DART

EXUBERANT DART

CREAKY DART
Hemieuxoa rudens

Very Common
93-3227 (10914)

TL 14–16 mm Dimorphic. Typical form has a light brownish-gray FW with black-outlined, round, gray orbicular and dusky-centered reniform spots backed by a black and brown, bicolored, triangular median patch. Central vein from base to reniform spot is whitish, bordered inside by wide black basal dash. Costal basal area is often pale, and costal PM/ST area washed brownish. In less common "pellucida" form, peppery or lightly striated, light tan to light brownish-gray FW has brown-outlined orbicular and reniform spots with dusky centers. Costa is often washed with brown and has dusky dashes along length. Inner basal and costal ST areas have diffuse dusky patches. Dusky double AM and black-dotted PM lines are usually visible. **HOSTS** Snakeweed.

WESTERN BEAN CUTWORM
Striacosta albicosta

Uncommon
93-3228 (10878)

TL 19–20 mm Dark brown FW has bold, pale tan costal stripe from base to PM area. Small brown-centered orbicular and reniform spots are outlined in pale tan and backed by narrow black stripe. Wide black basal dash crosses wavy, double, brown AM line to join black claviform spot. Scalloped, dusky, double PM line is faint or indistinct. Thorax is brown. Separable from Olive Dart and Army Cutworm by inner margin and ST/terminal area. **HOSTS** Herbaceous plants and crops, including bean, corn, and tomato.

FINNISH DART
Actebia fennica

Uncommon
93-3229 (10924)

TL 20–22 mm Dark grayish-brown FW has bold, pale tan orbicular and reniform spots shaded warm brown at center, and elongate black claviform spot. Inner margin is broadly shaded tan to reddish from base to PM area. Double AM and PM lines are black. ST area has black dashes at midpoint and costa of indistinct pale ST line. **HOSTS** Deciduous trees and herbaceous plants, including blueberry, elm, and clover.

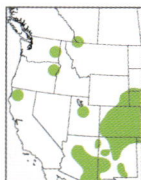

GROTE'S BLACK-TIPPED QUAKER
Dichagyris grotei

Uncommon
93-3231 (10869)

TL 16–18 mm Brown FW has black double AM and PM lines that are darker in costal half; medial of paired lines is thicker black. Central ST area has a thin, vertical, black dash, and ST/terminal area costal to this is shaded dusky. Thinly outlined brown orbicular and reniform spots are joined by a short thin line; spots are often indistinct. **HOSTS** False boneset; possibly also other aster family members.

PERPENDICULAR DART
Dichagyris cataclivis

Uncommon
93-3233 (10870.1)

TL 14–17 mm Brown FW has tan, elongate orbicular and reniform spots that are stretched perpendicular to each other; both have brown center and are outlined with black. Inner median and outer ST/terminal areas are shaded dusky. Black, double, scalloped AM and curving PM lines are filled with brown; inner line of pair is darker. Claviform spot is blackish, sometimes dusky brown. ST line is tiny tan dots at veins. **HOSTS** Unknown.

CREAKY DART

form
"pellucida"

actual size

WESTERN BEAN
CUTWORM

FINNISH DART

GROTE'S BLACK-TIPPED
QUAKER

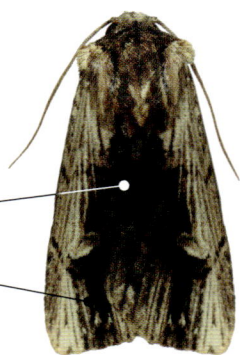

PERPENDICULAR
DART

SKIRTED DART
Dichagyris capota

Rare   RANGE-WIDE

93-3236 (10876)

TL 14–17 mm Resembles Western Bean Cutworm with broad tan costal stripe, but ST/terminal area is light brown, black double PM line is well defined, claviform spot is brown, and orbicular spot opens on costal side to join costal stripe. Thorax is tan with dark brown collar. HOSTS Unknown.

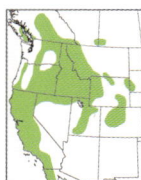

YELLOW DART
Dichagyris variabilis

Uncommon   RANGE-WIDE

93-3251 (10889)

TL 18–21 mm Tan to light brown FW has small, brown (sometimes dark brown), slanting rectangular orbicular and level reniform spots outlined in pale tan; reniform spot has a blackish spot at inner end. Wavy brown AM and curving double PM lines are usually faint. Indistinct pale ST line is shaded basally with brown. HOSTS Generalist on herbaceous plants.

LAUGHING DART
Richia chortalis

Rare   RANGE-WIDE

93-3263 (10881)

TL 20–25 mm Tan to light brown FW is often peppery and has large, rounded, pale yellowish outlines of orbicular and reniform spots. Double wavy AM and curving PM lines are usually faint; PM line sometimes has black dots at veins. In many individuals, double AM is filled with black in outer half, and basal and outer PM lines are also edged with black. Dusky, diffuse median line is sometimes visible. Inner margin is usually edged in peach pink. HOSTS Unknown.

PARENTAL DART
Richia parentalis

Rare   RANGE-WIDE

93-3264 (10882)

TL 18–21 mm Resembles larger Laughing Dart but has a dusky brown patch between orbicular and reniform spots, black of AM line is broader, ST line is a chain of pale yellowish spots, and reniform spot has a dusky smudge at inner end. Rarely, black AM line may be absent. Some individuals (not shown) have a dusky median area that is often darker behind brown spots, and dusky shading in ST area along ST line. HOSTS Unknown.

SERANO DART
Richia serano

Rare   RANGE-WIDE

93-3265 (10879)

TL 20–22 mm Brown to brownish-gray FW has broad, pale brown AM band. Rectangular black median patch passes behind dusky orbicular spot and terminates at basal edge of indistinct reniform spot. Orbicular spot is triangular, open to the costa. Claviform spot is a small black patch against indistinct double AM line. AM and PM lines are most visible at costa as paired black dashes. Black basal line is double. Thorax is reddish brown. HOSTS Unknown.

MILLER'S DART
Protogygia milleri

Rare   RANGE-WIDE

93-3306 (10898)

TL 18–20 mm Hoary gray FW has slightly darker median area bounded by thick, slightly scalloped, black AM and PM lines. Orbicular, claviform, and reniform spots are light gray outlined with black. Orbicular spot is usually connected to AM line by thick black stem; reniform spot has smudge of warm brown shading from lower edge to PM line. Basal area has warm brown basal dash. HOSTS Unknown.

SKIRTED DART

YELLOW DART

LAUGHING DART

PARENTAL DART

SERANO DART

actual size

MILLER'S DART

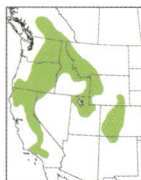

BLURRY DART
Euxoa bochus

Uncommon
93-3307 (10913)

TL 16–19 mm Tan FW has darker brown shading along costal edge and short brown striations. Soft-edged reniform spot is blackish brown. PM line is a row of blackish dots at veins; AM line is absent or present only as darkish bar within dark costal shading. ST area is tan; terminal area is washed dark brown. Light gray thorax has dark brown collar, often not visible when viewed dorsally. **HOSTS** Has been reared on varied herbaceous plants but suspected to feed naturally on grass.

ARMY CUTWORM
Euxoa auxiliaris

Very Common
93-3309 (10731)

TL 19–24 mm Dimorphic. Dark form has dark brown FW with bold tan to gray costa and inner margin, and thick tan streak in inner median area. Slanting orbicular and kidney-shaped reniform spots are outlined in pale yellow and shaded dusky at center. Pale gray to tan AM line is V-shaped adjacent to inner margin. Curving PM line is well defined. Tan ST area has black chevrons along pale yellowish ST line. Light form has a similar overall pattern but without the pale costa, inner margin, and tan streak of inner median area. Well-defined PM line does not touch reniform spot. ST area is light brown, connecting with a slanting light brown apical patch. **HOSTS** Generalist on herbaceous plants and grasses. **NOTE** Occasionally found nectaring at flowers in daytime. Can be a pest on cereal crops.

ADAPTABLE DART
Euxoa septentrionalis

Uncommon
93-3317 (10739)

TL 18–21 mm Resembles Reaper Dart, but FW is browner, central and costal veins are gray and ST line has dark spots between veins. Orbicular spot usually has dusky center. Median area behind spots usually shaded slightly darker brown. **HOSTS** Unknown.

OLIVE DART
Euxoa olivia

Rare
93-3318 (10741)

TL 15–19 mm Sexually dimorphic and variable. Bold-form males resemble dark form Army Cutworm, but PM line is indistinct to absent. AM line is not lighter than basal area, and claviform spot is usually blackish. ST line is indistinct. Plain-form males are similar in pattern but lighter brown, without pale costal streaks. Double, black AM and basal lines are distinct, and claviform spot is a black outline. Terminal area is often contrastingly blackish. Females resemble light-form Army Cutworm, but PM line touches inner end of reniform spot and ST line is indistinct. **HOSTS** Generalist on herbaceous plants.

REAPER DART
Euxoa messoria

Uncommon
93-3319 (10705)

TL 15–20 mm Grayish-brown to brown FW has double, black AM and PM lines; medial side is strongly scalloped, creating a beaded appearance. Rounded orbicular and reniform spots are outlined in black; orbicular spot is pale gray, and reniform spot is shaded dusky at center. Dusky median line often creates appearance of dark spot between spots. Pale ST line borders dusky-washed terminal area and dusky shading at costal ST area. Separable from light-form Army Cutworm and female Olive Dart by large, black-outlined orbicular spot that is concolorous with median area, and PM line does not touch reniform spot. **HOSTS** Many forbs and woody plants; sometimes a pest.

BLURRY DART

ARMY
CUTWORM

ADAPTABLE
DART

male

male

actual size

female

OLIVE DART

REAPER
DART

637

DIVERGENT DART
Euxoa divergens

Common (uncommon in South)

93-3320 (10702)

RANGE-WIDE

TL 17–21 mm Dark brown FW has pale tan, V-shaped central and costal veins framing pale-outlined, dusky orbicular and reniform spots. Wavy double basal and AM lines are filled with light brown. Scalloped black PM line is usually indistinct. **HOSTS** Unknown.

MEDIAN-BANDED DART
Euxoa medialis

Rare

93-3354 (10813)

RANGE-WIDE

TL 17–22 mm Variable. Light gray, tan, brown, or reddish-brown FW has double, tightly scalloped basal, AM, and PM lines filled with pale tan; PM line rounds basally to run parallel with costa to median line. Median line may be bold or indistinct but is strongly zigzagged and shaded below with darker ground color, sometimes filling entire PM area. Round orbicular and kidney-shaped reniform spots are outlined pale tan; reniform has a blackish spot at inner end; spots are usually filled with FW ground color but are sometimes washed dusky. Pale ST line is usually edged basally with bold, dark shading. **HOSTS** Unknown.

AUROUS DART
Euxoa pluralis

Rare

93-3372 (10795)

RANGE-WIDE

TL 17–20 mm FW has broad, warm brown streaks parallel to inner margin and below reniform spot, and hoary gray along inner margin, costa, and veins. Orbicular and reniform spots have whitish outline, are filled with hoary gray and blackish at inner end of reniform. Orbicular spot is usually irregularly shaped, and reniform is pinched at middle, often forming two spots. **HOSTS** Unknown; possibly mustard and other herbaceous plants.

TESSELLATE DART
Euxoa tessellata

Common

93-3395 (10805)

RANGE-WIDE

TL 15–19 mm Grayish-brown to violet-gray FW has large, rounded, pale orbicular and reniform spots backed by a black triangle from AM line to top of reniform. Dusky basal, AM, and PM lines are double, sometimes indistinct. ST line is pale. **HOSTS** Herbaceous plants and crops, including corn, squash, and tobacco.

WHITE-WINGED DART
Euxoa albipennis

Uncommon

93-3397 (10807)

RANGE-WIDE

TL 18–19 mm Light gray to warm brown FW has large, gray orbicular and reniform spots backed by a black rectangle from AM line to reniform and bordered by gray costa. Spots are edged inside with yellowish scales and reniform is filled with dusky shading at inner end. Black, scalloped AM line is straightish. Thorax is dusky gray. Some individuals are entirely gray through inner FW. **HOSTS** Unknown. **NOTE** Scientific and common names apparently refer to whitish HW of males.

ZIGZAG DART
Euxoa hollemani

Uncommon

93-3401 (10820)

RANGE-WIDE

TL 12–18 mm Light brown FW has grayish-brown costal stripe and dark brown shading in central wing. Grayish-brown orbicular and reniform spots are fused at costal end, forming a distinctive L shape. Veins are marked with dusky gray. Terminal area is dusky brown. **HOSTS** Unknown.

DIVERGENT DART

MEDIAN-BANDED
DART

AUROUS DART

TESSELLATE DART

WHITE-WINGED
DART

ZIGZAG DART

actual size

639

HAIRY DART
Euxoa comosa

TL 14–20 mm Variable. Light gray to brown FW has scalloped, double, black AM and PM lines and often dusky median line; cross-lines may be well defined or faint and indistinct. Orbicular and reniform spots are usually indistinct to absent but are sometimes visible as faint pale marks and very rarely as defined dark tan outlines. Well-marked brown individuals resemble Reaper Dart but lack defined orbicular and reniform spots. **HOSTS** Generalist on herbaceous plants and grasses.

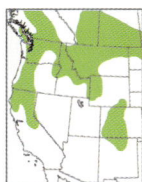

RED-BACKED CUTWORM
Euxoa ochrogaster

TL 18–21 mm Variable. Lighter forms resemble Olive Dart, but reniform spot has dusky spot at inner end, and central vein is distinctly pale tan above orbicular spot. ST area is uniformly tan. PM line does not touch reniform spot. Dark forms have reddish brown FW with dusky gray veins. Orbicular and reniform spots are outlined whitish; reniform has a dark gray spot at inner end. AM and PM lines are indistinct. **HOSTS** Generalist on herbaceous plants and grasses.

OBELISK DART
Euxoa obeliscoides

TL 15–21 mm Dark grayish-brown FW has a pale tan costal streak from base to PM line and a rusty wash to ST area. Pale-edged orbicular and reniform spots are backed by a long blackish triangle from AM line to reniform; orbicular is filled with warm brown, and reniform is pale yellowish with warm brown in upper half. Scalloped, black, double AM and PM lines are often indistinct against ground color. **HOSTS** Unknown.

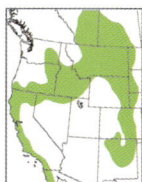

YELLOW-BASED DART
Euxoa basalis

TL 20–22 mm Brown FW has pale yellowish basal area bounded by scalloped double AM line. Pale yellowish orbicular and reniform spots are shaded brown at center, with a dusky spot at inner reniform, and backed by darker brown from AM line to reniform. Black-outlined U-shaped claviform spot sometimes has pale yellowish fill. **HOSTS** Generalist on herbaceous plants.

MURDOCK'S DART
Euxoa murdocki

TL 14–17 mm Brown to rusty FW has light gray median area bounded by bold black AM and PM lines; lines are doubled by a brown line outwardly. Light gray orbicular and reniform spots are shaded faintly dusky inside, and separated by a rusty brown patch. **HOSTS** Generalist on herbaceous plants.

FOUR-TOOTHED DART
Euxoa quadridentata

TL 14–18 mm Dusky brown FW has broad, pale tan to grayish-tan stripes along costa and inner margin. Tan-washed ST area has four pale-edged brown veins that project as teeth into dark brown terminal area. Pale tan orbicular and reniform spots are shaded dusky at center. Central vein is whitish, and vein branches in inner median area are broadly edged with tan. Separable from Dingy Cutworm by shape of orbicular spot and number of teeth in terminal area. **HOSTS** Largely unknown but has been found feeding on wheat crops.

HAIRY DART

OBELISK
DART

RED-BACKED
CUTWORM

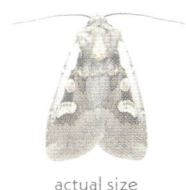

actual size

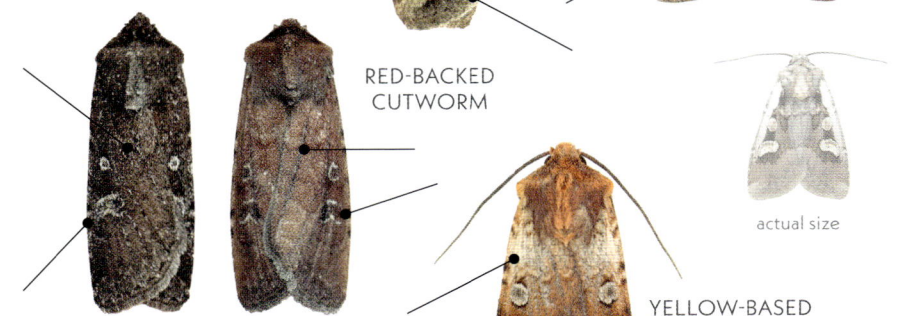

YELLOW-BASED
DART

MURDOCK'S
DART

FOUR-TOOTHED
DART

DINGY CUTWORM
Feltia jaculifera

Very Common
93-3498 (10670)

TL 19–21 mm Dark brown FW has dusky tan along costa and inner margin and along bordering grayish central vein and its branches. ST area is washed tan, with paired teeth at veins projecting into dark brown terminal area in inner half. Reniform spot is washed warm brown. Pale dusky tan orbicular spot is triangular, opening onto costa. Thin, pale tan AM and basal lines cross dark brown inner basal area. **HOSTS** Generalist on various shrubs, crops, and grasses. **NOTE** Occasionally found nectaring at flowers late in the day.

MASTER'S DART
Feltia herilis

Uncommon
93-3503 (10676)

TL 22–24 mm Dusky brown FW has blackish brown in central wing from base to PM line, crossed by tan-edged central vein. Reniform spot is tan. Dusky brown, pale-edged orbicular spot is triangular, opening toward costa. Thin black PM line is often indistinct. Light brown basal and AM lines cross dark brown inner basal area. **HOSTS** Generalist on various herbaceous plants and crops.

SUBTERRANEAN DART
Feltia subterranea

Very Common
93-3504 (10664)

TL 21–24 mm Dusky brown FW has broad tan costal stripe from base to reniform spot, and tan terminal area. Small orbicular and reniform spots are shaded dusky at center and backed by narrow blackish stripe from AM line to reniform. Strongly scalloped, double, black AM and PM lines are filled with tan when visible but are often indistinct. **HOSTS** Generalist on grass, crops and herbaceous plants.

OLD MAN DART
Agrotis vetusta

Uncommon
93-3506 (10641)

TL 18–24 mm Lightly peppered, pale brown to light gray FW has indistinct reniform spot primarily visible as dusky patch at inner end, and curving PM line of white-edged black dots. Terminal line is small black triangles. Some individuals also have dusky patch at orbicular spot. **HOSTS** Bean, corn, lettuce, tobacco, tomato, and other crops and herbaceous plants.

PALE WESTERN CUTWORM
Agrotis orthogonia

Rare
93-3512 (10645)

TL 15–19 mm Light brown to tan FW has paler basal and ST areas and darker terminal area. Large, rounded, dusky orbicular and reniform spots are outlined with whitish and black; elongate claviform spot is dusky, outlined with black. Scalloped AM and PM lines are black, edged outwardly with white. Veins are whitish. **HOSTS** Generalist on herbaceous plants and grass.

VENERABLE DART
Agrotis venerabilis

Uncommon
93-3516 (10651)

TL 18–22 mm Tan to light brown FW has broad dusky costa and veins traced with dark gray. Large, dusky, rounded reniform and elongate orbicular spots are thinly outlined in pale tan and black and often blend into costal band. Wide basal dash connects to claviform spot; both are dusky with black outline, and claviform has a thin dusky tail into median area. Terminal area has a triangle of dusky shading at midpoint. **HOSTS** Generalist on herbaceous plants and crops, including alfalfa, chickweed, corn, tobacco, and tomato.

DINGY CUTWORM

MASTER'S DART

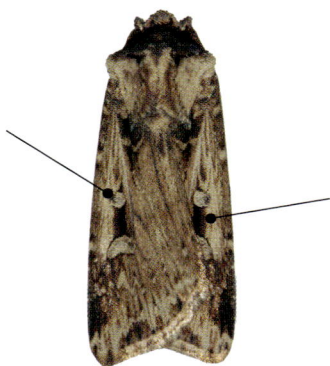

SUBTERRANEAN
DART

OLD MAN DART

actual size

PALE WESTERN CUTWORM

VENERABLE
DART

VANCOUVER DART
Common

Agrotis vancouverensis   93-3517 (10652)

TL 16–20 mm Variable. Tan to warm brown FW has thick, dark brown basal dash that connects to dark brown claviform spot and rounded orbicular and reniform spots shaded dusky at center with a pale yellow outer ring. Spots are backed by rectangular, dark brown median patch from AM to PM lines. AM and PM lines are edged with pale brown, particularly visible at costa. Uncommonly, some individuals are brown with slightly darker brown basal dash and median patch. **HOSTS** Generalist on herbaceous plants, including clover, strawberry, and grasses.

VOLUBLE DART
Uncommon

Agrotis volubilis   93-3521 (10659)

TL 20–22 mm Resembles Vancouver Dart, but orbicular and reniform spots typically lack a pale yellow inner ring, and AM and PM lines are usually faint or indistinct. Separable from Venerable Dart by round, gray orbicular spot, and basal dash connected to claviform spot. **HOSTS** Many herbaceous plants and crops. **NOTE** Historical misidentifications with look-alike species, and overlapping DNA barcoding data in this genus, has resulted in many false records outside the known western range of this species. Specimens resembling Voluble Dart in the West are most likely Venerable Dart, Oblique Dart or Antique Dart (*A. antica*, not shown).

OBLIQUE DART
Uncommon

Agrotis obliqua   93-3522 (10660)

TL 20–22 mm Resembles Voluble Dart, but inner FW is usually more reddish brown, AM and PM lines are typically visible across inner FW, and claviform spot and basal dash are generally thin and hollow. Dark shading behind orbicular and reniform spots, and in terminal area, is often reduced. **HOSTS** Unknown.

RASCAL DART
Uncommon

Agrotis malefida   93-3526 (10661)

TL 22–25 mm Resembles lighter individuals of Ipsilon Dart, but dusky reniform spot lacks black dash beneath, dusky triangle in terminal area is smaller and lacks black dashes in ST area, and thorax is tan with a dusky gray collar. **HOSTS** Generalist on herbaceous plants and crops.

IPSILON DART
Common

Agrotis ipsilon   93-3528 (10663)

TL 21–24 mm Long tan to light brown FW has dusky brown costal band and dusky reniform, orbicular, and claviform spots; reniform spot has a short black dash beneath. Wavy double AM and PM lines are dusky. Terminal area has a dusky triangle at midpoint, with black dashes in adjacent ST area. Some individuals have median area completely dusky gray. **HOSTS** Generalist on herbaceous plants and crops. **NOTE** A southern resident that migrates northward each spring.

FLAME-SHOULDERED DART
Uncommon

Ochropleura implecta   93-3529 (10891)

TL 14–15 mm Reddish-brown FW has pale tan costal stripe from base to PM area, and small, dusky-centered, gray orbicular and reniform spots backed by a black stripe adjacent to costa. PM line is a curved row of black dots. Thorax is brown with tan collar. Separable from Western Bean Cutworm by AM line and thorax. **HOSTS** Aster, clover, dock, and other herbaceous plants.

VANCOUVER
DART

actual size

VOLUBLE DART

OBLIQUE DART

RASCAL DART

IPSILON DART

FLAME-SHOULDERED
DART

HUNGRY DART
Diarsia esurialis

Uncommon
93-3530 (10920)

TL 15–18 mm Light brown FW has rounded, light brown orbicular and pale-outlined, brown reniform spots backed by rectangular, dark brown median patch. Brown AM and PM lines are strongly scalloped. ST area has brown dashes on veins and dark brown shading at costa. Diffuse, dark brown median band is sometimes indistinct. **HOSTS** Alder, hazelnut, and sword fern.

UMBER DART
Diarsia rosaria

Uncommon
93-3535 (10921)

TL 15–17 mm Warm brown to reddish, sometimes grayish-brown, FW has orbicular and reniform spots with pale yellowish outline, separated by a squarish, darker brown patch. Outer ST area is shaded broadly darker brown, with pale tan spots along costa. Terminal area is darker brownish or dusky gray. Dusky, double, scalloped AM and PM lines are filled with lighter brown but are often indistinct. Veins are traced with gray, most obvious in lighter inner ST area. **HOSTS** Grasses; also strawberry and avens.

ENIGMATIC DART
Cerastis enigmatica

Uncommon
93-3540 (10995.2)

TL 13–15 mm Warm brown FW has elongate, pale-outlined, light brown orbicular and reniform spots that connect at inner end, backed by dark brown shading in median area. Straightish double AM line has deep scoop at inner margin; double PM line is scalloped. ST area has dark brown dashes at midpoint and costa. Terminal area is light brown. **HOSTS** Salmonberry.

LABRADOR DART
Paradiarsia littoralis

Uncommon
93-3544 (10992)

TL 16–19 mm Warm brown FW has curving brown AM and PM lines and large round orbicular and reniform spots; reniform spot has a dusky smudge at inner end. ST area has dark brown shading at costa. Veins are thinly traced with brown. **HOSTS** Uncertain; probably a generalist on deciduous trees and herbaceous vegetation. **NOTE** Occasionally encountered during daytime.

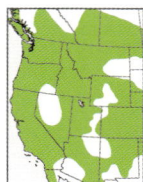

LARGE YELLOW UNDERWING
Noctua pronuba

[IN] Very Common
93-3551 (11003.1)

TL 30–35 mm Highly variable. Long FW has dusky reniform and lighter orbicular spots outlined in paler color, and short black bar at costal ST line. Ground color may be tan, warm brown, orange brown, dark brown, or brownish gray, mottled or not, and with or without pale costal band from base to median area. Double AM and PM lines are scalloped, sometimes filled with a paler hue but often indistinct. Orange HW has thick black ST band. **HOSTS** Grass and herbaceous plants.

LESSER YELLOW UNDERWING
Noctua comes

[IN] Common
93-3552 (11003.2)

TL 19–22 mm Light to reddish-brown FW has rounded, darker, slanting orbicular and level reniform spots outlined in pale color; reniform is shaded dusky at inner end. Wide, double, dark brown AM and PM lines are sometimes filled with lighter brown but often visible only as dark dots at veins. Terminal area is lighter brown, bordered by dusky shading along ST area. Orange HW has black ST band and narrow spot in median area. **HOSTS** Various grasses and herbaceous plants.

HUNGRY DART

UMBER DART

ENIGMATIC
DART

LABRADOR
DART

actual size

LESSER YELLOW
UNDERWING

LARGE YELLOW
UNDERWING

actual size

647

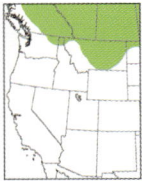

CATOCALINE DART
Cryptocala acadiensis

Uncommon
93-3553 (11012)

TL 15–18 mm Rusty to reddish brown FW has slanting orbicular and pinched reniform spots outlined pale yellow, frequently fused at inner end; orbicular is sometimes filled pale yellow. Costa has pale yellow patches at indistinct, dark brown AM and PM lines. Dark brown ST line has a large dark patch at costa. Separable from Enigmatic Dart by pale patches on costa and lack of dark shading along AM line. **HOSTS** Generalist of woody shrubs and herbaceous plants, including elderberry, cherry, meadowsweet, St. John's-wort, and yarrow.

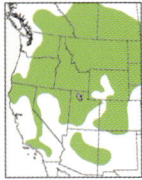

CLANDESTINE DART
Spaelotis clandestina

Common
93-3554 (10926)

TL 19–22 mm Long brownish-gray FW has light gray, oval orbicular spot thinly outlined in black and connected to incompletely outlined dusky reniform spot by a short black bar. Scalloped, double, dusky AM and PM lines are often indistinct. Basal dash, if present, is needle thin. **HOSTS** Trees and herbaceous plants, including apple, blueberry, maple, and pine.

WESTERN W-MARKED CUTWORM
Spaelotis bicava

Rare
93-3555 (10926.1)

TL 18–20 mm Resembles Clandestine Dart, but FW has distinct black basal dash and black bar connecting orbicular and reniform spots is usually poorly defined. **HOSTS** Probably buckwheat.

GREAT BROCADE
Eurois occulta

Uncommon
93-3560 (10929)

TL 29–35 mm Lightly peppered, pale gray FW appears mottled with dusky blackish color. Elongate gray claviform, slanting pale orbicular, and dusky-shaded, kidney-shaped reniform spots are outlined with black. Scalloped double AM and PM lines are filled with light gray. ST line is edged basally with black chevrons. **HOSTS** Tamarack, quaking aspen, alder, and willow.

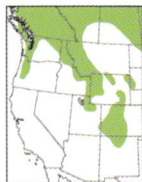

GREAT BROWN DART
Eurois astricta

Uncommon (rare in South)
93-3561 (10930)

TL 23–28 mm Brown to brownish gray FW has slightly lighter brown median area with a darker brown patch behind claviform, orbicular and reniform spots. Orbicular spot is usually lighter brown and shaded dusky at center. Lightly scalloped basal, AM and PM lines are broadly edged with pale brown. Pale ST line is edged with black chevrons, and terminal area is light brown. **HOSTS** Alder, birch, maple, quaking aspen, blueberry and huckleberry, and viburnum.

DOUBLE DART
Graphiphora augur

Uncommon
93-3563 (10928)

TL 27–29 mm Brown FW has elongate claviform and rounded orbicular and reniform spots outlined with black; kidney-shaped reniform spot has black spot in bend at lower side. Blackish, deeply scalloped AM and PM lines are sometimes indistinct. ST/terminal area shaded dusky. **HOSTS** Birch and willow. **NOTE** Also known as Soothsayer Dart.

CATOCALINE
DART

CLANDESTINE
DART

WESTERN W-MARKED
CUTWORM

GREAT BROCADE

actual size

GREAT BROWN
DART

DOUBLE DART

649

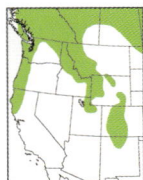

GREEN ARCHES
Uncommon
Anaplectoides prasina   93-3564 (11000)

TL 27–28 mm Grayish-brown FW has streaks of moss green that wash down center of wing and along wing margins. Scalloped double AM and PM lines are filled with white, and reniform spot has a bold white patch beneath. Large round orbicular and reniform spots are grayish brown and outlined with black. ST line is edged basally with black chevrons. **HOSTS** Trees and herbaceous plants, including apple, cranberry, currant, honeysuckle, and knotweed.

DAPPLED DART
Rare
Anaplectoides pressus   93-3565 (11001)

TL 18–20 mm Light gray to pale grayish-brown FW has large, rounded, pale orbicular and reniform spots thinly outlined in black and with a dusky inner ring, backed by a dusky to blackish patch from AM line to top of reniform. Claviform spot is incompletely outlined. Double wavy AM and scalloped PM lines are often indistinct in outer half. ST line is edged with black chevrons at midpoint and costa. Most individuals show very light streaks of yellow-green wash in central wing. **HOSTS** Cornsalad.

QUIVERING DART
Uncommon
Aplectoides condita   93-3567 (10999)

TL 18–22 mm Peppery, light gray FW has warm brown shading in basal, outer median, and ST areas. Large, rounded, pale orbicular and reniform spots are shaded dusky at center and incompletely outlined with black; thick black line runs behind spots from basal area to top of reniform spot. Black, wavy AM and scalloped PM lines are edged outwardly with white. **HOSTS** Fir and larch.

SQUARE-SPOT RUSTIC
[IN] Common
Xestia xanthographa   93-3571 (10945)

TL 17–19 mm Warm brown FW has pale yellowish, slanting orbicular and level reniform spots that are shaded brown at each end; orbicular spot sometimes completely washed brown inside. Thin, indistinct, black AM and PM lines are deeply scalloped; sometimes only the thick black points are visible as a row of dots. Pale yellowish ST line is usually a row of dots. **HOSTS** Rye and orchard grass; also other grasses and herbaceous plants.

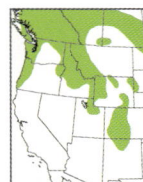

SMITH'S DART
Uncommon
Xestia smithii   93-3572 (10944)

TL 20–23 mm FW may be reddish brown, grayish, tan, or pale brown. Round brown-outlined orbicular and reniform spots are backed by darker shading; reniform spot has dusky shading at inner end. Curving double AM and PM lines are edged by black dots at veins. Indistinct pale ST line has paired black spots (sometimes joined into thick bar) at costa; FW sometimes also has small black dots along central line. **HOSTS** Trees and herbaceous plants, including alder, birch, and violet.

TRICOLORED DART
Rare
Xestia conchis   93-3574 (10946)

TL 18–22 mm Brown to brownish gray FW has a striking pattern of warm brown reniform and tan orbicular spots outlined in pale tan and backed by a blackish rectangle. Curving AM line is double; PM line is usually indistinct. Brown thorax has a broad gray dorsal stripe. **HOSTS** Unknown; likely a generalist.

GREEN ARCHES

DAPPLED DART

actual size

QUIVERING DART

SQUARE-SPOT
RUSTIC

SMITH'S DART

TRICOLORED
DART

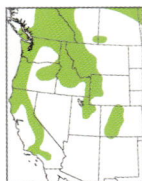

ROSY DART
Xestia oblata

Uncommon
93-3575 (10947)

TL 15–18 mm Brightly patterned FW has rosy basal area, brown median area with reddish brown in outer half, reddish ST area washed white along PM line, and orange-tan terminal area. Slanting rosy orbicular and level rusty-brown reniform spots are outlined in light yellowish and black and have pale shading at center. **HOSTS** Willow, spirea, and blueberry and huckleberry.

KNOTWOOD DART
Xestia infimatis

Uncommon
93-3581 (10972)

TL 17–22 mm Variable. Dark brown to light tan FW has pale-outlined dusky reniform spot with rounded basal-pointing finger on inner half. Elongate claviform and reniform spots are outlined in pale color; claviform spot connects to black basal dash. Curving PM line is a row of black dots. Central wing has dusky smudge from AM area to outer margin. Dusky median band often indistinctly visible. **HOSTS** Serviceberry and alder; may also feed on herbaceous plants.

SETACEOUS HEBREW CHARACTER
Xestia c-nigrum

Common
93-3588 (10942)

TL 15–19 mm Sexually dimorphic. Dark violet-gray (female) or warm brown (male) FW has triangular tan patch from orbicular spot to costa set in blackish outer median area. Reniform spot is thinly outlined in tan and filled with brownish gray. Double AM line is straightish and filled with lighter brownish gray; curving double PM line is less distinct. Indistinct pale ST line has black bar at costa. **HOSTS** Various herbaceous plants, crops, and grasses. **NOTE** Also known as Lesser Black-letter Dart.

COLLARED DART
Agnorisma bugrai

Uncommon
93-3627 (10954)

TL 17–19 mm Brownish gray FW has black squares above and between large, rounded, gray orbicular and reniform spots. Double, dusky AM and PM lines are filled pale gray; AM line is relatively straight, and PM line is smoothly curving beneath reniform spot. Head and collar are black. Somewhat resembles Tessellate Dart, but separable by AM and PM lines, and collar. **HOSTS** Unknown.

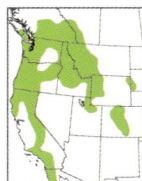

RED-WASHED DART
Setagrotis pallidicollis

Uncommon
93-3633 (10975)

TL 17–19 mm Hoary brownish-gray FW frequently has a rusty wash in inner half. Blackish AM and PM lines are toothed, and dusky median band is diffuse. Reniform and elongate orbicular spots are outlined in black and sometimes indistinct. Basal area usually has a thickish dusky basal dash. **HOSTS** Alder and serviceberry.

STELLAR DART
Adelphagrotis stellaris

Rare
93-3640 (10989)

TL 17–19 mm Brownish-gray to grayish-brown FW has reddish-brown to warm tan reniform spot and ground-colored orbicular spot, thinly outlined in whitish and black. Thick, double, scalloped AM and PM lines are filled with light gray. Indistinct, pale tan ST line has a dusky smudge at costa. **HOSTS** Generalist on woody shrubs.

ROSY DART

KNOTWOOD
DART

male

SETACEOUS
HEBREW
CHARACTER

female

COLLARED
DART

actual size

RED-WASHED DART

STELLAR DART

INDEFINITE DART
Adelphagrotis indeterminata

Uncommon
93-3641 (10991)

TL 16–17 mm Dark to grayish-brown FW has pale tan reniform spot shaded brown at outside and dusky at inside end, and ground-color orbicular spot, backed by long black rectangle in central median area. Basal area has sharp basal dash. ST line is a row of pale tan dots. Double wavy AM and PM lines are faint. Collar is blackish brown. **HOSTS** Willow, ocean spray, *Spiraea*, and raspberry and blackberry.

FORMAL DART
Parabagrotis formalis

Uncommon
93-3643 (11047.1)

TL 17–19 mm Variable. Warm to grayish-brown FW has blackish-brown triangle in central median area that passes from AM line behind triangular orbicular spot to top of reniform spot. Orbicular and reniform spots are dusky and outlined in pale tan; orbicular opens at a right angle or wider. Costa is contrastingly pale from base to PM line. Double, black, wavy AM and scalloped PM lines are filled with light brown. Short black basal dash is sometimes indistinct. Collar is bicolored, with pale tan anteriorly and dark brown posteriorly. **HOSTS** Unknown.

INSULAR DART
Parabagrotis insularis

Uncommon
93-3644 (11048)

TL 15–18 mm Highly variable. FW may be gray, brown, dark brown, or reddish, sometimes with colored inner median area or blackish median bar accentuating triangular orbicular spot that opens to costa. ST line is pale yellow, edged basally with faint shading. Orbicular and reniform spots faintly outlined in pale tan in darker morphs. Double basal line is present in most individuals, while double AM and PM lines may be indistinct in many. Thorax usually contrasts with FW, typically with a pale dorsal stripe. **HOSTS** Generalist on herbaceous plants and grasses. Outbreaks have occasionally occurred on cultivated peppermint fields.

BROWN-EDGED DART
Parabagrotis exsertistigma

Uncommon
93-3646 (11047)

TL 16–20 mm Variable. Resembles Formal Dart, but ST line is pale tan, black collar is narrower and often interrupted by dorsal stripe, and triangular orbicular spot opens less widely on costal side (usually narrower than right angle). Orbicular spot usually contrasts with costa. **HOSTS** Grasses.

VIBRANT DART
Parabagrotis sulinaris

Uncommon
93-3647 (11047.2)

TL 20–23 mm Variable. Resembles the smaller Formal Dart but is warmer brown and lacks black collar. Also may resemble Brown-edged Dart, but spots are washed light brown to pale yellowish, and ST area is shaded dusky along ST line. **HOSTS** Grasses.

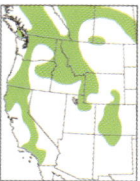

RED-BREASTED DART
Protolampra rufipectus

Uncommon (rare in South)
93-3648 (11004)

TL 16–21 mm Lightly peppered grayish-brown FW has thin, dusky, irregular AM and scalloped PM lines and often-indistinct orbicular and reniform spots thinly outlined in brown. Some individuals have diffuse brown median band that shades into reniform spot and brown shading in ST/terminal area. Thorax is reddish brown with dark brown collar anterior, separated by a thin whitish line; visibility in photos often depends on camera angle. **HOSTS** *Spiraea*.

**INDEFINITE
DART**

actual size

**FORMAL
DART**

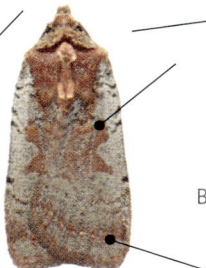

**BROWN-EDGED
DART**

**INSULAR
DART**

VIBRANT DART

RED-BREASTED DART

655

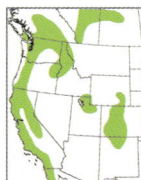

LUTEOUS DART
Abagrotis trigona

Uncommon
93-3653 (11018)

RANGE-WIDE

TL 13–17 mm Warm brown to brownish-gray FW has orbicular and reniform spots outlined in pale color; reniform always has a dark spot at inner end, often also costal end, and may sometimes be filled entirely blackish. Faint, dusky, double AM and PM lines are often most visible at costa and as dark dots at veins. Pale yellowish ST line is edged with dusky shading in ST area. Uncommonly, orbicular spot may be near FW ground color. Lighter individuals separable from Well-marked Cutworm by black dots along stronger PM line and terminal line, and dark obicular spot when present. **HOSTS** Willow.

CHESTNUT DART
Abagrotis apposita

Uncommon
93-3654 (11037)

RANGE-WIDE

TL 16–19 mm Chestnut-brown FW has curving, scalloped, double AM and PM lines that are filled with light brown, and light brown orbicular spot outlined with dark brown and edged with a few white scales. Slightly lighter median area has darker, diffuse median band filling center and obscuring reniform spot. Pale ST line is often indistinct. **HOSTS** Serviceberry, ceanothus, and madrone.

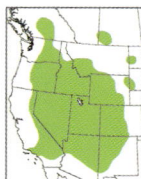

PALE-EDGED DART
Abagrotis vittifrons

Rare
93-3655 (11016)

RANGE-WIDE

TL 15–18 mm Dusky FW has thick white costal stripe from base to PM area. Pale reniform spot and ST line are usually faintly visible. Dusky thorax has orange-brown collar. **HOSTS** Unknown.

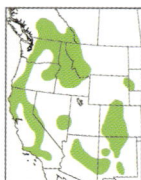

BLACK CEDAR DART
Abagrotis mirabilis

Rare
93-3657 (11019)

RANGE-WIDE

TL 18–20 mm Dusky FW has bold ivory reniform spot and small orbicular spot backed by a thick black streak. Terminal area is usually slightly lighter gray. Curving, scalloped, gray AM and PM lines sometimes visible. Dusky gray thorax has wide brown dorsal stripe edged with black. **HOSTS** Juniper and incense-cedar.

WELL-MARKED CUTWORM
Abagrotis orbis

Rare
93-3676 (11027)

NORTH

SOUTH

TL 18–20 mm Resembles the smaller Luteous Dart, but FW is yellowish tan to grayish tan, without a darker form. PM line is faint and lacks dots at veins; AM line is often indistinct. Orbicular and reniform spots may be dusky and reniform has a dark spot at inner or both ends, but neither spot are ever shaded dark brown. Pale ST line is usually less distinct. Terminal area is usually paler. Terminal line is wider black dashes that give outer margin a scalloped appearance. **HOSTS** Generalist on woody shrubs and deciduous trees.

VIOLACEOUS BROWN DART
Abagrotis scopeops

Rare
93-3679 (11033)

RANGE-WIDE

TL 17–20 mm Light to dark brown FW has large rounded orbicular and reniform spots that are shaded slightly darker with pale yellowish outline and pale brownish-gray terminal area. Dark brown, scalloped, double AM and PM lines are filled with light brownish gray. Basal area is lighter brown to pale brownish gray. **HOSTS** Unknown.

LUTEOUS DART

CHESTNUT
DART

PALE-EDGED DART

actual size

BLACK CEDAR
DART

WELL-MARKED
CUTWORM

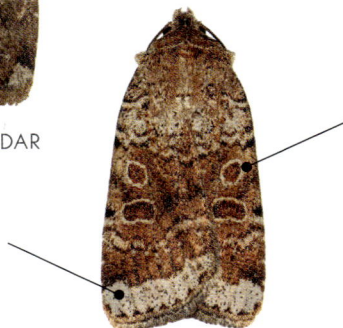

VIOLACEOUS
BROWN DART

GLOSSARY

Abdomen: The terminal body section of an insect, comprised of ten segments.

AM (antemedial) area: The portion of the wing between the AM and median lines. Second area from the base of wing.

AM (antemedial) line: The transverse line that separates the basal and median areas of the wing, usually used in reference to forewing markings. Second line from the base of wing.

Anal angle: The corner of the wing farthest from the thorax and closest to the abdomen.

Anal dash: A narrow line, usually short, near the anal angle and extending parallel to the inner margin.

Antennae: Long thread- or feather-like sensory appendages on the top of the head.

Apex: The corner of the wing farthest from both the thorax and abdomen.

Apical dash: A short line or band extending inward diagonally from the apex.

Band: A very broad line, usually transverse across the wing.

Basal area: The portion of the wing between the thorax and AM line.

Basal dash: A narrow line crossing the basal area parallel to the inner margin.

Basal line: A line crossing the basal area from inner margin to costa, dividing it in two. The line closest to the base of wing.

Base: The point where the wing attaches to the body.

Bipectinate antennae: Antennae with feather-like branches on both sides of the central stem.

Brindled: Patterned with thick, irregular, light and dark striping.

Brood: One generation, from eggs to adults.

Caterpillar: The larval form of moths and butterflies.

Claviform spot: A forewing marking in the inner AM area, often U-shaped and connected to the AM line. Present primarily in noctuids.

Collar: A band of scaling at the front of the thorax, behind the head; sometimes contrastingly colored to the thorax.

Costa: The edge of the wing from the thorax to the apex.

Costal: With reference to the costa.

Crepuscular: Active primarily at dawn and/or dusk.

Dash: A short, narrow, usually dark marking.

Dimorphic: Having two distinct forms. Usually refers to differences in color or markings, but may also refer to size or shape.

Discal spot or dot: A small, usually dark dot in the central median area, typically where the vein pattern forms an open, rounded lobe.

Diurnal: Active primarily during the day.

Dorsal: Referring to the upper side of the organism; the "back."

Falcate: Having a curved or hooked shape; usually in reference to the apex.

Filiform antennae: Thread-like antennae without bumps or branches. Also called simple antennae.

Flight period: The window of time within the calendar year in which a moth species is present in its adult stage.

Forelegs: The frontmost pair of legs.

Form: A distinct color or pattern variant commonly found within a regional population or the species as a whole.

Fringe: The long scales projecting from the outer margin.

FW: Forewing.

Hoary: Evenly and densely covered with pale scales interspersed among darker ones, creating a grizzled or frosty appearance.

Host plant or host: The plant species on which the moth feeds during its larval stage. Some moth species are very selective and use only a single or few host plant(s).

HW: Hindwing.

Inner margin: The edge of the wing from the thorax to the anal angle.

Introduced: Present in an area due to human activities; not historically native to an area.

Labial palps: Short, usually scale-covered sensory appendages on the front of the head. Frequently impart the appearance of a face or snout.

Larva: A moth in its juvenile stage. Also called the caterpillar.

Lateral: Referring to the sides of the organism.

Leaf mine: A feeding excavation by a caterpillar between the upper and lower leaf layers.

Macromoth: A broad term referring to adult moths of larger size, including all of the Noctuoidea and taxonomic neighbors, as well as some unrelated groups, such as the Saturniidae, Sphingidae, and Geometridae.

Median area: The area of the wing between the AM line and PM line. Composed of two halves: the AM area and PM area, divided by the median line.

Median line: The transverse line that crosses the center of the wing, bisecting the median area. Passes between the orbicular and reniform spots when those are present. Third line from the base of wing.

Micromoth: A broad term referring to adult moths of smaller size, including many that are tiny, that taxonomically precede the Geometridae, Noctuoidea, and neighbors. Includes a wide variety of groups, including the Tortricidae, Crambidae, Pyralidae, and some groups with larger species, such as carpenterworm moths.

Noctuid: A moth in the family Noctuidae; generally medium- to large-sized, large-bodied moths that rest with wings folded flat over the abdomen.

Nocturnal: Active primarily at night.

Orbicular spot: A forewing marking in the outer AM area. Usually a rounded outline with a different-colored fill. Present primarily in noctuids.

Outer margin: The edge of the wing from the anal margin to the apex, edged by the fringe.

Palps: See *Labial palps*.

Pectinate antennae: Antennae with feather-like branches on only one side of the central stem.

Peppered: Variably covered with scales of a darker color; may be sparse or dense.

Pheromone: A chemical substance released by organisms to communicate; often used to attract a mate. Some pheromones are commercially synthesized to attract target insects and sold in traps to catch household moth pests.

PM (postmedial) area: The portion of the wing between the median and PM lines. Third area from the base of wing.

PM (postmedial) line: The transverse line that separates the median and ST areas of the wing. Fourth line from the base of wing.

Reniform spot: A forewing marking in the outer PM area. Usually a rounded outline with a different-colored fill. Present primarily in noctuids.

Reticulated: Having a pattern of interconnected lines that superficially resembles a net.

Scales: Modified bristles that may be flattened or hair-like, generally covering both the wings and body of most moths and butterflies.

Scale tuft: A patch of scales that project from the wing or body surface, appearing raised or tufted.

Sexually dimorphic: Male and female each having a distinct form. Usually refers to differences in color or markings but may also refer to size or shape.

Shading: A broad area of color, usually soft-edged or diffuse and often bordering a line or wing edge.

ST (subterminal) area: The portion of the wing between the PM and ST lines. Fourth area from the base of wing.

ST (subterminal) line: The transverse line that separates the ST and terminal areas of the wing. Fifth line from the base of wing.

Stigma: A spot, usually white or silvery, formed from the orbicular and subreniform spots. Seen primarily in the Plusiinae.

Striated: Variably covered with very short dashes, usually of a darker color; may be sparse or dense.

Stripe: A very broad line, usually longitudinal along the wing and parallel to either inner margin or costa.

Subapical patch: A small colored area on the costa of the ST area; often bordered by an apical dash.

Subreniform spot: A forewing marking in the inner PM area. Usually a rounded outline with a different-colored fill. Uncommon; present primarily in noctuids.

Terminal area: The portion of the wing between the ST line and outer margin. The outermost area from the base of wing.

Terminal line: The transverse line that follows the outer margin from anal angle to apex.

Thorax: The middle section of the moth body, to which the wings and legs attach.

TL: Total length. The length of a moth from head to outer margin or abdomen tip when naturally at rest. Typically used for moths that rest with wings folded.

Tooth: A small point or projection of a line or marking.

Veins: The thin branching structures that provide strength and hemolymph (insect "blood") to the wings.

Ventral: Referring to the underside of the organism; the "belly."

WS: Wingspan. The width of a moth from apex to apex when naturally at rest. Typically used for moths that rest with wings outspread.

ADDITIONAL RESOURCES

If you've been bitten by the mothing bug, here are some additional resources that may help you expand your identification skills and learn about what can be found in your area. Some of these publications may be out of print, but used copies should be available from online retailers with a little digging. You might also be able to borrow a copy through your local library's interlibrary loan program.

Also try searching the internet for resources specific to your area. Many moth enthusiasts have started posting online photographs and lists of moths. A web search for "moths of <your state/city>" or "lepidoptera of <your state>" will often turn up local websites. Many of these moth-ers are happy to offer their help and may even be willing to get together for a night of mothing. There are several regional Facebook groups for moth enthusiasts; it is worth searching to see if there is one for your area. Some states and provinces have published official annotated checklists, and these too will usually come up in a web search.

Many counties and larger cities have entomologist or naturalist associations. These groups can be amazing fonts of knowledge, with friendly, enthusiastic members who are delighted to welcome new people to their passion. Many will hold public moth nights that you can attend.

PRINTED GUIDES AND CHECKLISTS

Beadle, David, and Seabrooke Leckie. 2012. *Peterson Field Guide to Moths of Northeastern North America*. Boston: Houghton Mifflin.

Eaton, Eric R., and Kenn Kaufman. 2007. *Kaufman Field Guide to Insects of North America*. Boston: Houghton Mifflin. [This excellent guide has a small but informative section on moths; the entire book is a rich resource for the general naturalist.]

Himmelman, John. 2002. *Discovering Moths: Nighttime Jewels in Your Own Backyard*. Camden, ME: Down East Books.

Holland, William Jacob. 1968. *The Moth Book: A Guide to the Moths of North America*. New York: Dover Publications.

Leckie, Seabrooke, and David Beadle. 2018. *Peterson Field Guide to Moths of Southeastern North America*. Boston: Houghton Mifflin.

Lees, David, and Alberto Zilli. 2019. *Moths: A Complete Guide to Biology and Behavior*. Washington, DC: Smithsonian Books.

Leverton, Roy. 2001. *Enjoying Moths*. London: Academic Press. [Although this book uses British moths as examples, the information it contains is applicable anywhere.]

Pohl, Gregory R., and Stephen R. Nanz, eds. 2023. *Annotated Taxonomic Checklist of the Lepidoptera of North America, North of Mexico*. Wedge Entomological Research Foundation.

Powell, Jerry A., and Paul A. Opler. 2009. *Moths of Western North America*. Berkeley: University of California Press.

Sourakov, Andrei, and Rachel Warren Chadd. 2022. *The Lives of Moths: A Natural History of Our Planet's Moth Life*. Princeton, NJ: Princeton University Press.

Wagner, David L. 2005. *Caterpillars of Eastern North America: A Guide to Identification and Natural History*. Princeton, NJ: Princeton University Press. [This guide focuses on eastern North America, but many of the species are found in the West, as well.]

INTERNET RESOURCES

BugGuide.net Hosted by Iowa State University Department of Entomology. An incredible resource for all insects, not just moths. Includes user-submitted photographs as well as information on identification, range, food plants, etc., where it is available. An "ID Request" feature allows you to upload your own photographs for expert identification. **bugguide.net**

Butterflies and Moths of North America A searchable database of lepidoptera records for the United States. Provides maps, written accounts, and photographs for many species. Butterflies are very well covered, but moths receive some treatment as well. **butterfliesandmoths.org**

iNaturalist This citizen science observation-reporting website covers all living organisms but is a great tool for documenting and sharing your moth records. The auto-identification feature does a good job with ID suggestions, and other users will help with confirmation. The website also offers the ability to filter records, including your own, by species, date, or location. A free app is available for submitting and managing records from mobile devices. **inaturalist.org**

The Lepidopterists' Society Some information is contained at the site itself, but it also provides an excellent list of additional resources, both in print and online. Although this is not strictly an identification or information website, the Lepidopterists' Society provides an excellent opportunity to learn more from like-minded people. This is the largest North American lepidopteran organization and is global in scope. You can also look for more regional societies or associations in your area. **lepsoc.org**

Leps by Fieldguide A citizen science website and app dedicated to moths and butterflies that uses an auto-identification tool to help you label your images. The free app allows you to submit and manage records from mobile devices; the platform also supports cross-sharing to iNaturalist. Images are organized taxonomically, and the app allows you to filter and search within your sightings, making it easier for list keepers to review their records than is currently possible with the iNaturalist app. **leps.fieldguide.ai**

Mothing and Moth-Watching One of the first Facebook groups dedicated to sharing the enjoyment of moths. Membership is open, and contributors are friendly and supportive, helping with identifications where they can. Although the group has a North American focus, images are often shared from elsewhere in the world, showcasing some outstanding diversity. **facebook.com/groups/137219092972521**

North American Moth Photographers Group (MPG) Hosted by the Mississippi Entomological Museum at Mississippi State University. The best moth-identification website out there; contains a compilation of images for the majority of North American species, accompanied by range maps displaying specimen records. **mothphotographersgroup.msstate.edu**

Pacific Northwest Moths This terrific reference website, hosted by Western Washington University and supported by many other regional institutions, covers about 1,200 macromoth species found in the Pacific Northwest, from British Columbia to northern California and Utah. The informative species profiles have information about identification, range, habitat, host plants, and flight seasons, as well as photos of representative specimens. **pnwmoths.biol.wwu.edu**

PUBLIC EVENTS

Moth-ers and naturalists of all skill levels are invited to take part in these public events. They can be a great way to learn from more experienced people and have fun discovering new-to-you species.

BugGuide Gatherings These informal events, held most years since 2007, invite BugGuide contributors to come together to hunt for bugs and have a good time with fellow naturalists. Nighttime blacklighting for moths and other nocturnal insects features prominently among activities, which also include group nature hikes through local state parks or wildlife reserves. The location is different each year, but past events have been held in Idaho, New Mexico, Louisiana, and Wisconsin, among other places. **bugguide.net/events**

Mothapalooza Begun in 2013, this now-annual weekend event is held in southern Ohio and attracts moth-ers from across the continent. Participants get to moth alongside numerous experts at night and look for day-flying moths (and many other species) on afternoon field trips through the Appalachian foothills. Afternoon presentations and evening keynote speeches are often by prominent figures in the field. The highlight for many attendees is the multiple "light stations" that attract a large variety of beautiful and diverse species. **arcofappalachia.org/mothapalooza**

National Moth Week (NMW) Despite its name, this is now a global event that invites moth enthusiasts to record the moths seen on any given night over the course of a week, usually held in July. Data can be submitted through the website or by joining NMW projects on iNaturalist and is used to build a citizen science database accessible to the scientific community. Participants can count on their own in their backyard or can join a public moth night in their area; a list of public events is available on the website. **nationalmothweek.org**

PHOTOGRAPHY CREDITS

This guide could not have been completed without the valuable contributions of many talented photographers. They are listed here in alphabetical order. The species for which they provided images are noted by P-number.

Daniel Antonaccio 91a-0400
Tara Armijo-Prewitt 57a-0008
R. & L. Avis 07-2003, 11-2007, 21a-0107, 21a-0110, 47a-0047, 51a-0629, 51a-0737.3, 55a-0061, 55a-0127, 55a-0223.5, 59a-0503, 59a-0809, 80a-0723, 80a-0837, 80a-1549, 89-0142, 89-0199, 91a-0284, 91a-0344, 91a-0397, 91a-0510, 91a-0524, 91a-0871, 91a-1038, 91a-1051, 91a-1132, 91a-1456, 91a-1457, 93-0215.1, 93-0283, 93-0365, 93-0585, 93-0896, 93-0923, 93-0924, 93-1196, 93-1775, 93-1777, 93-2256, 93-2371.1, 93-2386, 93-2665, 93-2670, 93-2673, 93-2685, 93-2796, 93-2949, 93-3424, 93-3535, 93-3540, 93-3571
Jorge Ayon 59a-0500
Parker Backstrom 57a-0093, 89-0086, 89-0090, 89-0092, 89-0204.3, 91a-0109, 91a-0401, 91a-0646, 91a-0989, 93-0762, 93-1343, 93-1427.1, 93-1441.1, 93-2054, 93-3504
James Bailey 21a-0098, 30-0033, 33a-0371, 36a-0021, 47a-0048.1, 51a-0485, 51a-0973.1, 51a-1065, 51a-1121.5, 51a-1122.1, 55a-0061, 55a-0083, 57a-0008, 59a-0063.6, 59a-0183, 59a-0399, 59a-0650, 59a-0944, 63a-0053, 63a-0157, 80a-0011, 80a-0042, 80a-0225, 80a-0514, 80a-0523, 80a-0745, 80a-0754, 80a-0757, 80a-0839, 80a-0844, 80a-0855, 80a-0871, 80a-0978, 80a-0996, 80a-1131, 80a-1132, 80a-1225, 80a-1488, 89-0044, 89-0053, 89-0184, 91a-0099, 91a-0453, 91a-0585, 91a-0642, 91a-0678, 91a-0697, 91a-0793, 91a-0832, 91a-0859, 91a-0867, 91a-0985, 91a-1007, 91a-1012, 91a-1209, 91a-1231, 91a-1256, 91a-1293, 91a-1304, 91a-1333, 91a-1398, 93-0025.1, 93-0182, 93-0183.1, 93-0193, 93-0200, 93-0253, 93-0336.52, 93-0372, 93-0380, 93-0423, 93-0427, 93-0429, 93-0438, 93-0468, 93-0701, 93-0881, 93-0885, 93-0982, 93-0996, 93-1011, 93-1031, 93-1244, 93-1267, 93-1312, 93-1313, 93-1318, 93-1321, 93-1330, 93-1354, 93-1373, 93-1388, 93-1392.3, 93-1419.6, 93-1587, 93-1589, 93-1666, 93-1683, 93-1689, 93-1770, 93-2034.71, 93-2043, 93-2061, 93-2148, 93-2170, 93-2249.43, 93-2967.55, 93-2967.56
David D. Beadle 16a-00xx, 21a-0057, 30-0046, 30-0168, 30-0181, 33a-0107, 33a-0207, 33a-0380, 36a-0034, 36a-0129, 36a-0179, 36a-0180, 47a-0046, 51a-0001, 51a-0010, 51a-0050, 51a-0057, 51a-0072, 51a-0221.1, 51a-0227, 51a-0300, 51a-0302, 51a-0303, 51a-0305.1, 51a-0323, 51a-0349, 51a-0357, 51a-0360, 51a-0364, 51a-0368, 51a-0390, 51a-0396, 51a-0419, 51a-0466, 51a-0478, 51a-0538, 51a-0636, 51a-0989.1, 51a-1133, 51a-1143, 51a-1291.7, 51a-1351, 51a-1380, 51a-1383, 53a-0025, 53a-0029, 55a-0057, 55a-0075, 55a-0094, 55a-0124, 57a-0006, 57a-0064, 59a-0002, 59a-0024, 59a-0054, 59a-0087, 59a-0091, 59a-0321, 59a-0401, 59a-0600, 59a-0670, 59a-1009, 59a-1599.1, 59a-1646.1, 59a-1766, 59a-1833, 61a-0001, 63a-0005, 63a-0012, 63a-0051, 63a-0104, 63a-0150, 65a-0010, 70-0003, 80a-0017, 80a-0041, 80a-0049, 80a-0076, 80a-0128, 80a-0134, 80a-0138, 80a-0145, 80a-0301, 80a-0332, 80a-0353, 80a-0409, 80a-0447, 80a-0667, 80a-0673, 80a-0674, 80a-0681, 80a-0689, 80a-0696, 80a-0715, 80a-0716, 80a-0717, 80a-0724, 80a-0735, 80a-0742, 80a-0772, 80a-0785, 80a-0802, 80a-0826, 80a-0877, 80a-0917, 80a-0944, 80a-0956, 80a-1019, 80a-1036, 80a-1037, 80a-1051, 80a-1065, 80a-1083, 80a-1097, 80a-1133, 80a-1156, 80a-1169, 80a-1221, 80a-1223, 80a-1284, 80a-1286, 80a-1308, 80a-1313, 80a-1369, 80a-1372, 80a-1456, 80a-1464, 80a-1465, 80a-1469, 80a-1472, 80a-1522, 80a-1529, 80a-1540, 80a-1544, 80a-1548, 80a-1553, 80a-1554, 80a-1565, 80a-1567, 85-0003, 85-0005, 85-0008, 85-0019, 85-0020, 87-0003, 87-0014, 87-0017, 87-0021, 89-0055, 89-0070, 89-0091, 89-0102, 89-0103, 89-0111, 89-0140, 89-0141, 89-0144, 89-0145, 89-0148, 89-0216, 89-0217, 91a-0002, 91a-0024, 91a-0026, 91a-0067, 91a-0079, 91a-0104, 91a-0115, 91a-0127, 91a-0143, 91a-0143, 91a-0152, 91a-0184, 91a-0278, 91a-0289, 91a-0308, 91a-0331, 91a-0332, 91a-0335, 91a-0336, 91a-0337, 91a-0341, 91a-0342, 91a-0344, 91a-0354, 91a-0360, 91a-0363, 91a-0388, 91a-0392, 91a-0393, 91a-0403, 91a-0411, 91a-0413, 91a-0431, 91a-0434, 91a-0446, 91a-0448, 91a-0454, 91a-0456, 91a-0458, 91a-0463, 91a-0465, 91a-0465, 91a-0661, 91a-0673, 91a-0691, 91a-0700, 91a-0704, 91a-0727, 91a-0751, 91a-0752, 91a-0767, 91a-0789, 91a-0991, 91a-0997, 91a-0998, 91a-1050, 91a-1060, 91a-1083, 91a-1087, 91a-1128, 91a-1130, 91a-1146, 91a-1147, 91a-1156, 91a-1183, 91a-1187, 91a-1188, 91a-1191, 91a-1211, 91a-1241, 91a-1242, 91a-1285, 91a-1306, 91a-1339, 91a-1344, 91a-1361, 91a-1366, 91a-1405, 91a-1410, 91a-1461, 93-0003, 93-0009, 93-0012, 93-0013, 93-0019, 93-0025, 93-0029, 93-0033.1, 93-0039, 93-0046, 93-0098, 93-0100, 93-0103, 93-0105, 93-0146, 93-0170, 93-0201, 93-0204, 93-0205, 93-0217, 93-0225, 93-0240, 93-0242, 93-0244, 93-0246, 93-0264, 93-0281, 93-0288, 93-0290, 93-0316, 93-0317, 93-0319, 93-0332, 93-0335, 93-0336.57, 93-0336.65, 93-0345, 93-0404, 93-0405, 93-0412, 93-0435, 93-0440, 93-0469, 93-0471, 93-0482, 93-0500, 93-0502, 93-0520, 93-0551, 93-0564, 93-0565, 93-0566, 93-0584, 93-0588, 93-0592, 93-0601, 93-0622, 93-0627, 93-0715, 93-0790, 93-0792, 93-0795, 93-0798, 93-0804, 93-0815, 93-0841, 93-0923, 93-0924, 93-0929, 93-1023, 93-1032, 93-1055, 93-1103, 93-1108, 93-1121, 93-1141, 93-1162, 93-1168, 93-1170, 93-1176, 93-1194, 93-1203, 93-1204, 93-1209, 93-1213, 93-1215, 93-1227, 93-1229, 93-1234, 93-1236, 93-1291, 93-1314, 93-1412, 93-1421, 93-1425, 93-1433, 93-1434.7, 93-1441.2, 93-1474, 93-1477, 93-1477.2, 93-1485, 93-1504, 93-1508, 93-1514, 93-1544, 93-1545, 93-1561, 93-1564, 93-1724, 93-1765, 93-1771, 93-1979, 93-1984, 93-2030, 93-2041, 93-2045, 93-2082, 93-2120, 93-2134, 93-2205, 93-2213.62, 93-2215, 93-2216, 93-2219, 93-2224.5, 93-2264.1, 93-2266, 93-2290, 93-2292, 93-2308, 93-2314, 93-2319, 93-2333, 93-2343, 93-2350, 93-2355, 93-2363, 93-2372, 93-2390, 93-2438, 93-2443, 93-2447, 93-2517, 93-2520, 93-2524, 93-2534, 93-2536, 93-2590, 93-2616, 93-2620, 93-2653,

93-2672, 93-2674, 93-2675, 93-2679, 93-2695, 93-2709, 93-2773, 93-2778, 93-2795, 93-2810, 93-2826, 93-2867, 93-2872, 93-2874, 93-2875, 93-2878, 93-2881, 93-2883, 93-2885, 93-2906, 93-2928, 93-2933, 93-2935, 93-2945, 93-2947, 93-2948, 93-3044, 93-3053, 93-3062, 93-3113, 93-3211, 93-3212, 93-3228, 93-3229, 93-3319, 93-3395, 93-3397, 93-3498, 93-3503, 93-3506, 93-3516, 93-3521, 93-3528, 93-3529, 93-3544, 93-3551, 93-3552, 93-3554, 93-3560, 93-3563, 93-3564, 93-3565, 93-3572, 93-3588, 93-3676

Joey Bom   80a-0031, 80a-0519, 80a-xxxx, 89-0048

Valerie Bugh   89-0078, 91a-0025, 91a-0050

Susan D. Carnahan   36a-0030, 51a-0313, 51a-0972.1, 53a-0010, 57a-0010, 57a-0047, 57a-0079, 57a-0088, 59a-0653, 59a-0837, 59a-0840, 59a-0841, 59a-1713, 70-0006, 80a-0001, 80a-0011, 80a-0171, 80a-0427, 80a-0510, 80a-0575, 80a-0768, 80a-0769, 80a-1045, 80a-1116, 80a-1396, 80a-1460, 80a-1489, 87-0016, 87-0019, 89-0021, 89-0026, 89-0028, 89-0059, 89-0069, 89-0071, 89-0115, 89-0141.1, 89-0154, 89-0210, 91a-0037, 91a-0566, 91a-0697, 91a-0742, 91a-0775, 91a-0788, 91a-0813, 91a-0816, 91a-0824, 91a-0835, 91a-0881, 91a-0992, 91a-1110, 91a-1252, 91a-1318, 91a-1322, 91a-1336, 91a-1356, 91a-1359, 91a-1419, 93-0136, 93-0174, 93-0195, 93-0198, 93-0209, 93-0214, 93-0709, 93-0869, 93-0871, 93-0880, 93-0936, 93-0999, 93-1001, 93-1051, 93-1079, 93-1092, 93-1264, 93-1265, 93-1276, 93-1277, 93-1307, 93-1322, 93-1329, 93-1434.3, 93-1584, 93-1596, 93-1732.6, 93-1768, 93-1801, 93-1963, 93-1978, 93-2058, 93-2059, 93-2094, 93-2165, 93-2246, 93-2930, 93-2967.57, 93-3220, 93-3354

Gary Chang   55a-0075.01

Yan Chun Su   36a-0146, 59a-0011, 80a-0270, 93-1365, 93-2104

Jack Cochran   91a-0795, 93-0707, 93-1519, 93-2126

Ann Cooper   51a-0969

Jason Cooper   91a-0452

Jillian Cowles   21a-0048, 21a-0105, 36a-0004, 36a-0031, 36a-0148, 51a-0440, 51a-0984.1, 51a-1327, 51a-1328, 53a-0008, 55a-0141, 55a-0149, 57a-0014, 57a-0087, 59a-0589, 59a-0594.5, 59a-0648, 59a-1001, 59a-1079, 59a-1148, 59a-1159, 80a-0004, 80a-0018, 80a-0027, 80a-0071, 80a-0183, 80a-0526, 80a-0620, 80a-0627, 80a-0676, 80a-0791, 80a-0795, 80a-0797, 80a-0912, 80a-0918, 80a-0943, 80a-0981, 80a-1030, 80a-1083, 80a-1083, 80a-1092, 80a-1109, 80a-1113, 80a-1133, 80a-1138, 80a-1158, 80a-1159, 80a-1172, 80a-1188/9, 80a-1232, 80a-1341, 87-0032, 89-0036, 89-0045, 89-0068, 89-0149, 91a-0060, 91a-0069, 91a-0070, 91a-0082, 91a-0657, 91a-0663, 91a-0685, 91a-0833, 91a-0835, 91a-0836, 91a-0885, 91a-0918, 91a-0977, 91a-1236, 91a-1398, 93-0062, 93-0083, 93-0093, 93-0187, 93-0199, 93-0227, 93-0325, 93-0406, 93-0416, 93-0707, 93-0737, 93-0884, 93-1004, 93-1266, 93-1281, 93-1303, 93-1381, 93-1389, 93-1412, 93-1412, 93-1575, 93-1580, 93-1581, 93-1583, 93-1584, 93-1603, 93-1606, 93-1609, 93-1611, 93-1615, 93-1648, 93-1659, 93-1665, 93-1672, 93-1674, 93-1684, 93-1697, 93-1709, 93-1725, 93-1728, 93-1732.6, 93-1781, 93-1785.5, 93-1972, 93-1974, 93-1977, 93-2032, 93-2060, 93-2066, 93-2110, 93-2138, 93-2159, 93-2163, 93-2178, 93-2185, 93-2794, 93-2827, 93-2891, 93-2953, 93-2954, 93-3076, 93-3115.5, 93-3125, 93-3156.1, 93-3227, 93-3233

John Davis   11-2007, 11-2015, 21a-0107, 21a-0111, 21a-0112, 30-0176, 36a-0139, 36a-0141, 36a-0182, 43a-0003, 47a-0039, 51a-0030, 51a-0073, 51a-0268, 51a-0277, 51a-0320, 51a-0371, 51a-0731, 51a-0735, 51a-0737.8, 51a-1144, 51a-1191, 51a-1202, 51a-1240, 51a-1250, 51a-1291.8, 51a-1316, 55a-0118, 59a-0040, 59a-0044, 59a-0051, 59a-0060, 59a-0108, 59a-0216, 59a-0691.9, 59a-0721, 63a-0051, 80a-0202, 80a-0335, 80a-0336, 80a-0355, 80a-0439, 80a-0627, 80a-0695, 80a-0833, 80a-0836, 80a-0837, 80a-0878, 80a-0879, 80a-1089, 80a-1218, 80a-1235, 80a-1236, 80a-1358, 80a-1376, 80a-1379, 80a-1388, 80a-1470, 80a-1537, 80a-1555, 85-0009, 87-0028, 89-0034, 89-0085, 91a-0020, 91a-0128, 91a-0142, 91a-0150, 91a-0181, 91a-0245, 91a-0290, 91a-0303, 91a-0304, 91a-0348, 91a-0354, 91a-0356, 91a-0368, 91a-0373, 91a-0379, 91a-0401, 91a-0429, 91a-0450, 91a-0455, 91a-0459, 91a-0475, 91a-0478, 91a-0479, 91a-0504, 91a-0523, 91a-0547, 91a-0562, 91a-0578, 91a-0583, 91a-0584, 91a-0601, 91a-0622, 91a-0681, 91a-0726, 91a-0734, 91a-0736, 91a-0760, 91a-0784, 91a-0786, 91a-0817, 91a-0860, 91a-0989, 91a-0990, 91a-1048, 91a-1051, 91a-1071, 91a-1079, 91a-1091, 91a-1092, 91a-1096, 91a-1099, 91a-1100, 91a-1101, 91a-1103, 91a-1133, 91a-1143, 91a-1144, 91a-1175, 91a-1244, 91a-1248, 91a-1279, 91a-1319, 91a-1341, 91a-1394, 91a-1397, 91a-1411, 91a-1414, 91a-1416, 91a-1439, 91a-1447, 91a-1457, 93-0047, 93-0107, 93-0164, 93-0312, 93-0314, 93-0336, 93-0347, 93-0356, 93-0368, 93-0381, 93-0585, 93-0654, 93-0676, 93-0753, 93-0791, 93-0803, 93-0838, 93-0866, 93-0868, 93-0896, 93-0898, 93-0899, 93-0902, 93-0910, 93-0913, 93-0922, 93-0930, 93-1133, 93-1142, 93-1193, 93-1237, 93-1399, 93-1419.71, 93-1419.78, 93-1427, 93-1441.1, 93-1536, 93-1777, 93-1780, 93-1802, 93-2035, 93-2036, 93-2038, 93-2075, 93-2200, 93-2217, 93-2249.41, 93-2249.49, 93-2313, 93-2345, 93-2417, 93-2530, 93-2573, 93-2579, 93-2613, 93-2617, 93-2634, 93-2648.5, 93-2665, 93-2670, 93-2673, 93-2684, 93-2750, 93-2764, 93-2765, 93-2767, 93-2777, 93-2788, 93-2789, 93-2793, 93-2797, 93-2866, 93-2890, 93-2909, 93-2924.52, 93-2925, 93-2953.1, 93-3018, 93-3042, 93-3045, 93-3061, 93-3069, 93-3077, 93-3090, 93-3096, 93-3103, 93-3132, 93-3163, 93-3251, 93-3307, 93-3317, 93-3320, 93-3405, 93-3530, 93-3555, 93-3561, 93-3567, 93-3633, 93-3641, 93-3643, 93-3644, 93-3646, 93-3647, 93-3648, 93-3653, 93-3679

Jason J. Dombroskie   11-2017, 91a-0347, 91a-1046, 91a-1070, 93-2530

Mark Dreiling   30-0044, 30-0048, 30-0075.5, 30-0079, 30-0087.8, 30-0140.82, 30-0141, 30-0155, 30-0164, 30-0195, 30-0198, 30-0214, 30-0222, 33a-0187, 36a-0002, 36a-0141, 36a-0147, 51a-0126.1, 51a-0232, 51a-0270, 51a-0273, 51a-0406, 51a-0449, 51a-0650/1, 51a-0845, 51a-0871, 51a-0904.1, 51a-0933.1, 51a-1121.1, 51a-1258, 57a-0008, 59a-0006, 59a-0064, 59a-0186, 59a-0187, 59a-0385, 59a-0568, 59a-0591, 59a-0608, 59a-0645, 59a-0804, 59a-0856, 59a-0906, 59a-1146, 59a-1147, 59a-1510.1, 59a-1569.1, 59a-1611.1, 63a-0014, 63a-0071, 63a-0111, 80a-0032, 80a-0067, 80a-0077, 80a-0146, 80a-0167, 80a-0281, 80a-0409, 80a-0462, 80a-0482, 80a-0571, 80a-0672, 80a-0746, 80a-0791, 80a-0820, 80a-0880, 80a-1138, 80a-1188/9, 80a-1197, 80a-1201, 80a-1212, 80a-1217, 80a-1218, 80a-1341, 80a-1364, 80a-1375, 80a-1393, 80a-1545, 80a-1580, 89-0114, 91a-0044, 91a-0288, 91a-0523, 91a-0550,

91a-0614, 91a-0628, 91a-0643, 91a-0655, 91a-0663, 91a-0681, 91a-0682, 91a-0702, 91a-0703, 91a-0740, 91a-0741, 91a-0775, 91a-0782, 91a-0849, 91a-0953, 91a-0985, 91a-1082, 91a-1095, 91a-1099, 91a-1101, 91a-1103, 91a-1233, 91a-1245, 91a-1262, 91a-1325, 91a-1420, 91a-1436, 93-0161, 93-0179, 93-0180, 93-0186, 93-0257, 93-0271, 93-0276, 93-0422, 93-0517, 93-0676, 93-0695, 93-0705, 93-0898, 93-0913, 93-0939, 93-1160.8, 93-1253, 93-1330, 93-1332, 93-1334, 93-1484, 93-1534, 93-1565, 93-1578, 93-1688, 93-1775, 93-2170, 93-2247, 93-2628, 93-2749.5, 93-2762, 93-2767, 93-2769, 93-2792, 93-2952, 93-3115, 93-3115.3, 93-3160, 93-3198, 93-3251, 93-3526, 93-3641, 93-3644

Eric R. Eaton   51a-0689, 51a-0898.1, 51a-0940.1, 59a-0920, 80a-1583, 93-0924, 93-1326, 93-2073, 93-2171, 93-2967.8

Heidi Eaton   91a-1024

J. Craig Emery   89-0087, 89-0156

Glenn Fine   36a-0146, 36a-0194, 51a-0251, 51a-0366, 51a-0673, 51a-0733, 51a-0897.1, 51a-0955.1, 51a-1235, 51a-1375, 53a-0026, 59a-0188, 59a-0810, 80a-0164, 80a-0221, 80a-0471, 80a-0571, 80a-0743, 80a-0801, 80a-0828, 80a-1230, 80a-1285, 80a-1472, 80a-1485, 80a-1495, 80a-1568, 80a-128x, 89-0125, 91a-0090, 91a-0116, 91a-0238, 91a-0256, 91a-0336, 91a-0365, 91a-0369, 91a-0624, 91a-0675, 91a-0781, 91a-0992, 91a-1217, 91a-1250, 91a-1254, 91a-1293, 91a-1318, 91a-1459, 93-0009.1, 93-0103.1, 93-0258, 93-0313, 93-0676, 93-0922, 93-1143, 93-1143.x, 93-1524, 93-1538, 93-1640.2, 93-1640.3, 93-1778, 93-1822, 93-1950, 93-2046, 93-2154, 93-2200, 93-2317, 93-2318, 93-2323, 93-2380, 93-2523, 93-2528, 93-2612, 93-2653, 93-2693, 93-2709, 93-2709, 93-2753, 93-2754, 93-2755, 93-2763, 93-2766, 93-2787, 93-2939, 93-2949, 93-3046, 93-3062, 93-3075, 93-3088, 93-3318, 93-3354, 93-3408, 93-3451, 93-3644

Graham Floyd   80a-1570, 89-0060, 93-2083

Catherine C. Galley   36a-0024

Laura Gaudette   51a-0089.1, 51a-1015.1, 53a-0018, 53a-0035, 57a-0008, 57a-0039, 57a-0066, 80a-0671, 80a-0763, 80a-0834, 80a-0838, 80a-0843, 80a-0851, 80a-0853, 80a-0896/8, 80a-0996, 80a-1200, 80a-1228, 80a-1285, 80a-1350, 87-0014, 87-0014, 87-0016, 87-0018, 87-0020, 87-0034, 89-0021, 89-0032, 89-0052, 89-0071, 89-0094, 89-0095, 89-0109, 89-0124, 89-0132, 89-0185, 89-0201, 89-0215, 91a-0025, 91a-0050, 91a-0050, 91a-0317, 91a-0365, 91a-0382, 91a-0568, 91a-0608, 91a-0623, 91a-0631, 91a-0644, 91a-0664, 91a-0678, 91a-0750, 91a-0776, 91a-0780, 91a-0796, 91a-0816, 91a-0818, 91a-0841, 91a-0882, 91a-0899, 91a-0973, 91a-1035, 91a-1106, 91a-1209, 91a-1210, 91a-1211, 91a-1231, 91a-1236, 91a-1252, 91a-1254, 91a-1256, 91a-1261, 91a-1322, 91a-1332, 91a-1333, 91a-1334, 91a-1337, 91a-1353, 91a-1356, 91a-1359, 91a-1392, 91a-1398, 91a-1452, 91a-1458, 93-0005, 93-0008, 93-0045.1, 93-0052, 93-0087.15, 93-0183.1, 93-0191, 93-0207, 93-0208, 93-0254, 93-0255, 93-0313, 93-0321, 93-0330, 93-0336.52, 93-0336.6, 93-0355, 93-0358, 93-0362, 93-0372, 93-0382, 93-0383, 93-0386, 93-0427, 93-0433, 93-0595, 93-0600, 93-0638, 93-0721, 93-0754, 93-0805, 93-0817, 93-0839, 93-0867, 93-0878, 93-0895, 93-0900, 93-0905, 93-0909, 93-0918, 93-0973, 93-0996, 93-1011, 93-1051, 93-1079, 93-1275, 93-1308, 93-1310, 93-1311, 93-1320, 93-1333, 93-1353, 93-1359, 93-1388, 93-1390,

93-1392.6, 93-1395, 93-1399.5, 93-1405.5, 93-1419.74, 93-1510, 93-1567, 93-1651, 93-1678, 93-1680, 93-1685, 93-1688, 93-1695, 93-1696, 93-1699.3, 93-1702, 93-1703, 93-1708, 93-1725, 93-1733.1, 93-1944, 93-1969, 93-2031, 93-2058, 93-2073, 93-2087, 93-2095, 93-2123, 93-2139, 93-2146, 93-2148, 93-2158, 93-2213.13, 93-2246, 93-2257, 93-2339, 93-2347, 93-2862, 93-2930, 93-3048, 93-3182, 93-3207, 93-3220, 93-3231, 93-3236, 93-3263, 93-3264, 93-3265, 93-3455, 93-3574, 93-3644, 93-3657, 93-3676

Ellyne Geurts   51a-0011, 93-1952

Scott Gilmore   33a-0158, 51a-0030, 51a-0048, 51a-0246, 51a-0359, 51a-0373, 51a-0635, 51a-1210, 51a-1220, 51a-1231, 51a-1259, 51a-1396, 59a-0061, 89-0204.6, 91a-0524, 93-0012.1, 93-0586, 93-2765, 93-2791, 93-2797

Joe Girgente   55a-0085

Leslie Goethals   91a-0564, 91a-0662, 91a-1258, 91a-1324, 91a-1332, 93-0871, 93-1143.x, 93-1434.4, 93-2046, 93-2181, 93-2332, 93-2862, 93-2873, 93-2967.54, 93-2967.55, 93-3227, 93-3405, 93-3512

Elliott Gordon   30-0015.51, 59a-0681.5, 80a-0002, 80a-0254, 93-1366

David Greenberger   93-1174

Jason Headley   51a-0002, 51a-1159, 80a-0447, 91a-0395, 91a-0606, 93-1144, 93-1365, 93-2050, 93-2967.56

David Heckard   53a-0046, 89-0050

Tony Iwane   93-0436

Kimi Jackson   89-0039

Jeanette M. Jaskula   30-0202

Jim T. Johnson   11-2007, 36a-0137, 51a-0050, 51a-0123.1, 59a-0971, 80a-0436, 80a-0843, 80a-1025, 85-0011, 87-0030, 89-0052, 89-0054, 91a-0116, 91a-0143, 91a-0143, 91a-0145, 91a-0166, 91a-0175, 91a-0186, 91a-0203, 91a-0247, 91a-0335, 91a-0337, 91a-0338, 91a-0505, 91a-0524, 91a-0653, 91a-0816, 91a-0869, 91a-0968, 91a-0975, 91a-1004, 91a-1030, 91a-1047, 91a-1326, 91a-1359, 93-0158, 93-0258, 93-0289, 93-0367, 93-0437, 93-1201, 93-1536, 93-1562, 93-1977.1, 93-1981, 93-1990, 93-2078, 93-2257, 93-2455, 93-2529, 93-2809, 93-3012, 93-3075, 93-3091, 93-3221, 93-3640, 93-3655

Paul G. Johnson   07-2002, 16a-00xx, 21a-0004, 21a-0009, 21a-0012, 21a-0014, 21a-0018, 21a-0104, 21a-0105, 21a-0107, 21a-0110, 21a-0111, 21a-0112, 21a-0113, 21a-0124, 30-0066.4, 30-0095, 30-0110, 30-0154, 30-0157, 33a-0070, 33a-0169, 33a-0196, 33a-0242, 33a-0335, 33a-0385, 36a-0006, 36a-0154, 51a-0073, 51a-0122.2, 51a-0155.1, 51a-0161.1, 51a-0206.1, 51a-0309, 51a-0373, 51a-0523, 51a-0650/1, 51a-0725, 51a-0737.4, 51a-0737.45, 51a-0737.6, 51a-0823, 51a-0886.1, 51a-0923.1, 51a-0951.1, 51a-1143, 51a-1163, 51a-1200, 51a-1212, 51a-1397, 53a-0007, 53a-0017, 55a-0122, 59a-0053, 59a-0394, 59a-0455, 59a-0466, 59a-0643, 59a-0802, 59a-0844, 59a-0938, 59a-0944, 59a-0948, 59a-1153, 59a-1163, 59a-1180, 59a-1231, 59a-1553.1, 59a-1731, 63a-0015, 63a-0142, 80a-0112, 80a-0137, 80a-0150, 80a-0181, 80a-0214, 80a-0440, 80a-0452, 80a-0464, 80a-0467, 80a-0481, 80a-0763, 80a-0787, 80a-0801, 80a-0822, 80a-0843, 80a-0852, 80a-0873, 80a-0897, 80a-0904, 80a-1116, 80a-1124, 80a-1149, 80a-1152, 80a-1238, 80a-1387, 80a-1526, 80a-1538, 87-0015, 87-0019, 87-0028, 87-0031, 89-0054, 89-0064, 89-0149, 89-0195, 89-0198, 89-0202, 91a-0047,

91a-0052, 91a-0068, 91a-0070, 91a-0109, 91a-0110, 91a-0141, 91a-0142, 91a-0236, 91a-0284, 91a-0290, 91a-0303, 91a-0354, 91a-0368, 91a-0369, 91a-0405, 91a-0406, 91a-0453, 91a-0508, 91a-0548, 91a-0554, 91a-0573, 91a-0602, 91a-0604, 91a-0606, 91a-0637, 91a-0653, 91a-0685, 91a-0774, 91a-0794, 91a-0800, 91a-0809, 91a-0818, 91a-0885, 91a-0889, 91a-0893, 91a-0992, 91a-1047, 91a-1069, 91a-1082, 91a-1093, 91a-1094, 91a-1258, 91a-1284, 91a-1288, 91a-1320, 91a-1323, 91a-1327, 91a-1330, 91a-1386, 91a-1437, 91a-1455, 91a-1462, 93-0012.1, 93-0025.1, 93-0028, 93-0095, 93-0103.1, 93-0289, 93-0327, 93-0353, 93-0483, 93-0693, 93-0744, 93-0753, 93-0803, 93-0809, 93-0825, 93-0878, 93-0885, 93-0888, 93-0894, 93-0901, 93-0911, 93-0982, 93-0998, 93-1001, 93-1047, 93-1121, 93-1125, 93-1142, 93-1330, 93-1360, 93-1371, 93-1373, 93-1375, 93-1423, 93-1529, 93-1534, 93-1559, 93-1565, 93-1579, 93-1623, 93-1638.9, 93-1639.9, 93-1640.2, 93-1640.3, 93-1670, 93-1671, 93-1699.4, 93-1797, 93-1798, 93-1853, 93-1951, 93-1982, 93-1983, 93-1991, 93-1994, 93-2031, 93-2034.71, 93-2050, 93-2067, 93-2088, 93-2100, 93-2153, 93-2168, 93-2187, 93-2195, 93-2247, 93-2249.63, 93-2264.1, 93-2452, 93-2610, 93-2648, 93-2680, 93-2691, 93-2772, 93-2787, 93-2819, 93-2910, 93-3042.1, 93-3061, 93-3069, 93-3075, 93-3124, 93-3188, 93-3199, 93-3306, 93-3309, 93-3318, 93-3372, 93-3401, 93-3528, 93-3581

Thaddeus Charles Jones   55a-0143, 93-0247

Ben Keen   36a-0159, 51a-0068, 53a-0026, 55a-0125, 80a-0828, 80a-1025, 89-0084, 89-0218, 91a-0335, 91a-0385, 91a-0439, 91a-0708, 91a-0781, 91a-1209, 91a-1210, 91a-1217, 93-0158, 93-0258, 93-0313, 93-0797, 93-1142, 93-1144, 93-1427.4, 93-1482, 93-2347, 93-2480, 93-2653, 93-2838, 93-2898, 93-2908, 93-3042.2, 93-3091, 93-3096, 93-3424, 93-3517, 93-3535, 93-3553, 93-3561, 93-3571, 93-3575, 93-3588, 93-3627, 93-3653, 93-3654

Michael H. King   30-0112, 51a-1143, 51a-1159, 51a-1218, 59a-1856, 80a-0334, 80a-0503, 80a-0897, 93-1516, 93-3428

Rich Kostecke   53a-0024

Karl Kroeker   93-2483

H. McIlvaine Lewis   51a-0885.1

William Lisowsky   59a-0030, 89-0156, 93-1518

Don Loarie   55a-0123, 55a-0167

Mason Maron   51a-0089.1, 51a-0359, 55a-0120, 59a-0339, 80a-0571, 80a-0759, 91a-0145, 93-0420

Robert A. Martin   91a-0738

William Mason   93-0799

Ian Maton   11-2008, 59a-0194, 80a-0800, 80a-1337, 89-0112, 89-0204.5, 91a-0107, 91a-0720, 93-1223, 93-1505, 93-2381, 93-2483, 93-2758, 93-2796, 93-2837, 93-3431

Graham Montgomery   36a-0015, 80a-0805, 93-2184

Christian Nunes   89-0197

Darrin O'Brien   89-0042

Mike Ostrowski   30-0075.1

George Pollock   55a-0072, 93-0759

M. Quinn   89-0006

Jon Rapp   51a-0629, 51a-0905.1, 57a-0042, 57a-0043, 57a-0087, 80a-0620, 80a-0761, 80a-0912, 87-0008, 87-0011, 87-0012, 89-0002, 89-0011, 89-0013, 89-0028, 91a-0033, 91a-0560, 91a-0844, 91a-1014, 93-0052, 93-0055, 93-0120, 93-0131, 93-0318, 93-0359, 93-0371, 93-0390, 93-0919, 93-1364, 93-1607, 93-1699, 93-1968, 93-2017, 93-3156.1

John D. Reynolds   51a-0001, 51a-0359, 55a-0122, 59a-0063, 59a-1733, 80a-1383, 91a-0145, 91a-0385, 91a-0464, 91a-0832, 93-0023, 93-0160, 93-1201, 93-1981, 93-3654

JoAnne Russo   91a-0336, 93-1366

Harvey Schmidt   47a-0047, 55a-0069, 55a-0112, 80a-1356, 91a-0354, 91a-1394, 91a-1416, 93-0276, 93-0283, 93-0906, 93-2234

Mark Silverstein   80a-0514

Julie Spencer   89-0049

Brian Starzomski   51a-0002

Van Truan   51a-0267, 51a-0823, 51a-0964.1, 63a-0017, 80a-0905, 93-0206, 93-0253, 93-0269, 93-1228, 93-2064, 93-2195

Ken-ichi Ueda   36a-0032, 47a-0044, 51a-0155.1, 51a-0169.1, 51a-0938.1, 51a-1186.2, 51a-1235, 59a-1841, 65a-0001, 80a-0725, 80a-0743, 80a-0811, 80a-1113, 80a-1135, 80a-1238, 80a-1346, 89-0052, 89-0154, 91a-0284, 91a-0471, 91a-0574, 91a-0587, 91a-1110, 91a-1210, 91a-1331, 93-0138, 93-0181, 93-0336.54, 93-0678, 93-0881, 93-1120.5, 93-1356, 93-1363, 93-1392.7, 93-2031, 93-2249.47

Trevor Van Loon   51a-0477, 80a-0896/8, 91a-0335, 91a-0395, 93-1221, 93-2288

Carrie Voss   93-1142, 93-1489, 93-2177

Jim & Lynne Weber   89-0069, 89-0186, 91a-1007, 93-0358, 93-0434, 93-1009, 93-1977.1

Michael Woodruff   93-0799, 93-2284, 93-2455

Lena Zappia   51a-1121.5, 80a-0548, 89-0132, 93-0759, 93-1326, 93-1366, 93-1372, 93-1865, 93-2034.71, 93-2174, 93-2176, 93-2204.52, 93-2213.13, 93-3309

INDEX

A

Abagrotis apposita 656
  mirabilis 656
  orbis 656
  scopeops 656
  trigona 656
  vittifrons 656
Abegesta reluctalis 198
  remellalis 198
Abrostola urentis 468
Acallis, Orange-lined 144
Acallis griphalis 144
Acerra normalis 592
Achroia grisella 146
Achyra occidentalis 170
  rantalis 170
Acleris, Great 48
  Red-edged 46
  Snowy-shouldered 46
Acleris albicomana 46
  forsskaleana 46
  gloveranus 48
  maccana 48
  maximana 48
  nivisellana 46
  rhombana 46
  variegana 46
Acontia cretata 490
Acopa, Gold-specked 556
Acopa perpallida 556
Acorn Moth 134
Acossus centerensis 94
  populi 94
Acrobasis, Large Oak 152
  Small Oak 150
  Tricolored 150
Acrobasis caliginella 152
  comptella 150
  tricolorella 150
Acrolophus cockerelli 24
  griseus 24
  kearfotti 24
  laticapitana 24
  pyramellus 26
  variabilis 26
Acronicta americana 496
  atristrigatus 494
  fragilis 496
  funeralis 494
  grisea 498
  hasta 500
  impleta 500
  impressa 500
  innotata 498
  insita 496
  insularis 500
  lepusculina 496
  lupini 502

  mansueta 498
  marmorata 496
  morula 498
  oblinita 502
  othello 500
  perdita 500
  radcliffei 498
  strigulata 498
  thoracica 498
  vulpina 496
Actebia fennica 632
Adaina ambrosiae 140
Adela eldorada 20
  flammeusella 20
  punctiferella 20
  septentrionella 20
  singulella 20
  thorpella 22
  trigrapha 20
Adelphagrotis indeterminata 654
  stellaris 652
Admetovis oxymorus 610
  similaris 610
Aemilia, Rosy 412
Aethaloida packardaria 352
Aethalura intertexta 336
Aethes, Smeathmann's 50
Aethes smeathmanniana 50
Aetole bella 36
  extraneella 36
  tripunctella 36
  unipunctella 36
Afilia oslari 380
Agapeta zoegana 50
Agathodes monstralis 186
Aglossa acallalis 148
  cacamica 148
  caprealis 148
  electalis 148
  pinguinalis 148
Agnorisma bugrai 652
Agonopterix, Canadian 112
  Poison Hemlock 112
Agonopterix alstroemeriana 112
  canadensis 112
  nervosa 112
Agrochola purpurea 584
  antica 644
  ipsilon 644
  malefida 644
  obliqua 644
  orthogonia 642
  vancouverensis 644
  venerabilis 642
  vetusta 642
  volubilis 644

Alastria, Bicolored 566
Alastria chico 566
Alberada franclemonti 166
Albuna pyramidalis 96
Aleptina inca 538
Allerastria albiciliatus 434
Alpheias, Transposed 146
Alpheias transferens 146
Alsophila pometaria 344
Alucita montana 136
Alypia langtoni 536
  mariposa 536
  octomaculata 534
  ridingsii 536
Alypiodes bimaculata 534
  flavilinguis 534
  geronimo 534
Ambesa, Red-patched 156
Ambesa laetella 156
Amblyptilia pica 138
Amorbia cuneanum 62
Amphipoea americana 576
  lunata 576
Amphipyra pyramidoides 506
  tragopoginis 506
Amydria, Banded 24
  Brown-blotched 24
  One-blotched 24
Amydria curvistrigella 24
  effrenatella 24
  obliquella 24
Amyelois transitella 152
Amyna stricta 476
Anagrapha falcifera 474
Anania, Brown 170
  Crowned 168
  Golden 170
Anania funebris 170
  hortulata 168
  labeculalis 170
  mysippusalis 170
  tertialis 168
Anaplectoides prasina 650
  pressus 650
Anarta crotchii 604
  farnhami 604
  mutata 602
  trifolii 602
Anathix puta 584
Anatrachyntis badia 116
Anatralata versicolor 196
Anavitrinella atristrigaria 332
  pampinaria 332
Ancylis, Similar, complex 68
Ancylis columbiana 68
  mediofasciana 68
  simuloides 68

Ancylosis, Calico 162
Ancylosis morrisonella 162
  undulatella 162
Androloma maccullochii 536
Andropolia aedon 588
  diversilineata 588
  theodori 590
Anemosella, Green 144
Anemosella viridalis 144
Angle, Adonis 314
  Birch 314
  Broad-lined 320
  Buff-banded 312
  Common 314
  Curve-lined 318
  Deceptive 312
  Dispatched 310
  Dot-lined 308
  Downward 318
  Dusky 312
  Faded 318
  Fawn 310
  Gray-banded 314
  Lorquin's 312
  Maple 312
  Pale-lined 322
  Pale-marked 314
  Plain 308
  Refined 310
  Rusty-edged 322
  Smoky 320
  Split-lined 310
  Strange 318
  Yellow-banded 312
  Yellow-winged 310
Angle Shades, American 566
  Brown 566
Anicla, Poetic 630
Anicla biformata 630
  espoetia 630
  exuberans 630
  infecta 630
Annaphila, Astral 514
  Illustrated 514
  Tenth 516
  White 516
Annaphila astrologa 514
Annaphila decia 516
  depicta 514
  diva 516
Anopina, Triangular 52
Anopina triangulana 52
Anstenoptilia
  marmarodactyla 136
Antepirrhoe semiatrata 280
Anticlea vasiliata 290
Antigastra catalaunalis 186
Anvil-wing Moth 432

Apamea, Ashy 568, 570
  Bordered 568
  Burgess' 570
  Faint-spotted 572
  Fox 568
  Frosted 570
  Snowy-veined 572
  Thoughtful 572
  Traced 568
  Western 570
  White-spurred 568
  Wood-colored 570
Apamea alia 568
  amputatrix 570
  antennata 570
  burgessi 570
  cinefacta 570
  cogitata 572
  cuculliformis 568
  devastator 572
  digitula 568
  inficita 572
  lignicolora 570
  niveivenosa 572
  occidens 570
  scoparia 572
  sordens 568
  spaldingi 568
  unanimis 568
Apantesis arge 396
  blakei 400
  carlotta 402
  f-pallida 398
  figurata 398
  incorrupta 400
  nevadensis 400
  ornata 400
  parthenice 396
  phyllira 398
  proxima 400
  ursina 400
  virgo 396
  virguncula 398
  williamsii 398
Aphid Moth, Contrasting
  134
Aphomia sociella 146
Apilocrocis, Pima 186
Apilocrocis pimalis 186
Aplectoides condita 650
Aplocera plagiata 266
Apomyelois, Two-striped
  152
Apomyelois bistriatella 152
Apotomis removana 66
Apple Moth, Greenish 60
  Light Brown 60
Arachnis aulaea 406
  picta 406
Arches, Ashen 622
  Bronze 620

Brown 620
Brown-patched Golden
  616
Clever 606
Coastal 618
Collared 622
Collared, complex 622
Configured 608
Crossed Golden 616
Cuneate 618
Emerald 620
Flowing 608
Four-lined 622
Friendly 618
Fused 622
Girdled 620
Granite 604
Green 650
Hitched 606
Intricate 608
Lined Golden 616
Little 452
Nevada 606
Northern 452
Olive 620
Otter 606
Pendant 618
Perplexed 452
Purple 604
Red-spot 620
Royal 604
Schiniesque 616
Serrated 618
Shadowy 452
Simple 602
Singular Golden 616
Smoky 450
Stormy 604
Tacoma 608
Weathered 622
Windy 604
Wooden 604
Archiearis infans 300
Archips argyrospila 58
  cerasivorana 58
  rosana 58
Archirhoe neomexicana 284
Arcobara multilineata 258
Arctia caja 402
  parthenos 402
  plantaginis 402
  virginalis 402
Areniscythris brachypteris
  134
Argentostiria koebelei 522
Argyrotaenia coloradana 54
  dorsalana 56
  franciscana 54
  niscana 56
  provana 54
Aristotelia, Elegant 122

Gray-banded 120
Pink-washed 122
Silver-banded 120
Six-lined 122
Sloped 120
Aristotelia argentifera 120
  calens 120
  devexella 120
  elegantella 122
  hexacopa 122
  roseosuffusella 122
Armyworm, Beet 558
  Fall 558
  Western Yellow-striped
    558
  Wheat Head 610
  Yellow-striped 560
Aroga, Skunk-backed 130
  Staring 128
  Striped 128
  Tan-backed 128
Aroga morenella 128
  paraplutella 128
  paulella 128
  unifasciella 130
Arta, Gallant 146
  Posturing 144
Arta epicoenalis 146
  statalis 144
Asaphocrita aphidiella 134
Ascalapha odorata 436
Aseptis binotata 584
  characta 586
  perfumosa 584
Astiptodonta wymola 380
Atteva aurea 38
Autographa ampla 472
  californica 470
  corusca 470
  mappa 470
  metallica 470
  pseudogamma 470
Autoplusia egenoides 470
Autumnal Moth 274
Azenia edentata 524
  implora 524
  obtusa 524
Azeta, Moonlight 462
Azeta schausi 462

B
Bactra, Rush 64
Bactra furfurana 64
  verutana 66
Bagisara buxea 476
Bagworm, Creosote Bush 22
  Plaster 26
Bantam, Amber 196
  Belted 196
  Brown 192
  Cloaked 196

Desert 196
Pink 194
Sunflower 194
Sunshine 194
Tawny 194
Venerable 196
Zigzag 192
Barbara colfaxiana 70
Basilodes chrysopis 518
Batia lunaris 110
Battaristis, Elegant 118
  concinnusella 118
Beauty, Antique 352
  Enigmatic 352
  Mountain 472
  Pale 344
  Pine 350
  Willow 350
Bedellia somnulentella 34
Bee Moth 146
Beehive Honey Moth 154
Beggar, Orange 266
Behrensia conchiformis 526
Bellura obliqua 578
Bertholdia, Grote's 416
Bertholdia trigona 416
Besma, Oak 366
Besma quercivoraria 366
Bicolored Moth 394
Bigwing, Powdered 264
Birch-miner, Purplish 14
Bird Nest Moth 28
Bird-dropping Moth,
  Arizona 484
  Broken-spotted 486
  Brown 484
  Brown-backed 490
  Brown-ringed 482
  Chalky 490
  Exposed 484
  Eyed 486
  Four-patched 486
  Four-spotted 486
  Gray-ringed 482
  Half-circle 482
  Intense 484
  Large 488
  Narrow-winged 488
  Olive-shaded 482
  One-spotted 482
  Polished 488
  Prairie 480
  Pretty 480
  Rusty-patched 488
  Scalloped 484
  Silvery 486
  Speckled 484
  Sunny 482
  Virgin 480
  White-banded 488
  Zebra 488

Bird's-Wing, Black 556
Biston betularia 336
  sinuaria 338
Black, California 282
  Spear-marked, complex
  276
  White-striped 282
Blackneck, Levant 434
Blastobasis glandulella 134
Bleptina caradrinalis 424
Bondia comonana 142
Borer, Beardtongue 102
  Blue Cactus 164
  Buckwheat 100
  Cattail 578
  Chrysanthemum Flower
  52
  Coronopus 102
  Erythrina 188
  Glorious Squash Vine 98
  Gray Squash Vine 98
  Knotweed Root 98
  Lesser Cornstalk 160
  Maple Twig 80
  Peachtree 100
  Poplar Branchlet 84
  Prune Limb 142
  Ragwort Stem 576
  Raspberry Crown 96
  Reddish Cholla 166
  Snakeweed 76
  Sugarbeet Crown 162
  Sweetclover Root 116
  Sycamore 100
  Thin-lined Pricklypear
  166
  Toothed Pricklypear 164
  Umbellifer 576
  Western Pine Shoot 70
  Wild Geranium 102
Borer-Mimic, Serrated 628
Brachylomia populi 586
Brindle, American 580
  Small Clouded 568
Brocade, Discordant 590
  Exposed 574
  Great 648
  Mossy 590
  Oblique 574
  Raspberry 574
  Toadflax 526
  Violet 574
  Wandering 590
Broken-twig Moth, Similar
  610
Brother, The 494
Brown, Olive 516
  Swift 478
  White-shouldered 516
  White-spotted 186
Bruceia hubbardi 394

Brymblia quadrimaculella
  110
Bryolymnia, Green 560
  Half-banded 560
Bryolymnia semifascia 560
  viridata 560
Bucculatrix albertiella 30
Bud Moth, Eye-spotted 70
  Spruce 80
  Sunflower 80
  Verbena 64
Budworm, Juniper 58
  Spruce 56
  Tobacco 542
  Western Black-headed 48
  Western Spruce 56
Budworm Moth, Green 68
Bulia deducta 448
  similaris 448
Burdock Seedhead Moth
  120
Button, Marbled 48

C
Cabbageworm,
  Purple-backed 200
Cabera erythemaria 340
Cacoecimorpha pronubana
  58
Cacozelia, Yellow-based 146
Cacozelia basiochrealis 146
Cactobrosis fernaldialis 164
Caenurgia togataria 456
Caenurgina caerulea 454
  crassiuscula 454
  erechtea 454
Cahela Moth 166
Cahela ponderosella 166
Calliprora, Hook-winged
  120
Calliprora sexstrigella 120
Callistege diagonalis 454
  intercalaris 454
Callizzia amorata 256
Calophasia lunula 526
Caloptilia, Poison Oak 32
  Poplar 32
  Reticulated 32
  Sugar Bush 32
Caloptilia azaleella 32
  diversilobiella 32
  ovatiella 32
  reticulata 32
  stigmatella 32
Cameraria sp. 32
Campaea perlata 344
Cankerworm, Fall 344
  Winter 338
Caphys, Rosy 142
Caphys arizonensis 142
Capsule Moth, Variable 610

Caradrina meralis 564
  montana 564
  morpheus 564
Carales arizonensis 414
Carcina quercana 114
Cargida pyrrha 380
Caripeta aequaliaria 364
  angustiorata 364
  divisata 364
Carmenta engelhardti 102
  giliae 102
  mimuli 102
Carpenterworm, Aspen 94
  Black-veined 94
  Desert 92
  Henry's 94
  Margaret's 92
  Pine 92
  Poplar 94
  Robin's 94
  Smudged 92
  Theodore 92
  White-striped 94
Carpet, Alpine Brown 288
  Beaded 298
  Bent-line 286
  Black-banded 280
  Brown-banded 280
  Brown Bark 268
  Brown-veined 280
  Calico 286
  Dark Marbled 278
  Double-banded 290
  Falcate Brown 286
  Forgotten 266
  Formosa 278
  George's 282
  Guenée's 282
  Juniper 282
  Labrador 284
  Marbled 278
  Mottled Gray 264
  New Mexico 284
  Northern 284
  Ochre Brown 288
  Orange-barred 278
  Orange-winged 288
  Pearsall's 266
  Rocky 282
  Sharp-angled 284
  Somber 282
  Sperry's Toothed 288
  Ten-spotted
    Rhododendron 278
  Toothed Brown 286
  Twelve-lined 266
  Variable 290
  Western 334
  Western White-Ribboned
    288
  White-banded 290

White-banded Toothed
  288
White-ribboned 288
Carphoides incopriarius 334
Casebearer, Household 26
  Large Clover 132
Casemaker, Oak Ribbed 30
Catabena lineolata 526
Catabenoides terminellus
  526
Catastia, Brown-banded 156
Catastia actualis 156
Caterpillar Moth,
  Red-humped 382
  Ugly-nest 58
  Western Red-Humped
  382
  Zebra 606
Catocala aholibah 438
  amatrix 442
  briseis 440
  californica 440
  chelidonia 442
  desdemona 442
  grotiana 440
  ilia 438
  irene 440
  junctura 442
  neogama 436
  parta 438
  piatrix 436
  relicta 438
  semirelicta 440
  ultronia 444
  unijuga 438
  verrilliana 442
  violenta 444
Catoptria, Oregon 214
  Two-banded 214
Catoptria latiradiellus 214
  oregonicus 214
Cauchas simpliciella 22
Cecrita lunata 380
Celypha cespitana 66
Celypha Moth 66
Cenopis reticulatana 64
Cerastis enigmatica 646
Ceratodalia gueneata 282
Ceratomia undulosa 244
Chabulina onychinalis 188
Chalcoela, Sooty-winged
  200
Chalcoela iphitalis 200
Chalcopasta fulgens 522
  howardi 520
  territans 520
Chamaeclea pernana 492
Chestnut Y, Wavy 470
Chionodes, Banded 126
  Black-smudged 128
  Coastal 126

Large-toothed 126
Small-toothed 126
Two-banded 128
White-patched 126
Yellow-headed 128
Chionodes abella 126
 dentella 126
 fructuaria 126
 lophosella 126
 lugubrella 128
 mediofuscella 128
 phalacra 128
 pinguicula 126
Chloridea virescens 542
Chlorochlamys appellaria 306
 chloroleucaria 306
 phyllinaria 306
 triangularis 306
Chloronycta tybo 494
Chlorosea banksaria 300
 margaretaria 300
Chocolate-tip, Bruce's 374
Choreutis, Diana's 44
Choreutis diana 44
 pariana 44
 sexfasciella 44
Choristoneura conflictana 56
 fumiferana 56
 houstonana 58
 lambertiana 58
 occidentalis 56
 pinus 58
 retiniana 56
 rosaceana 56
Choristostigma, Elegant 182
 Lead-marked 182
 Pink-bordered 182
 Zephyr 182
Choristostigma elegantalis 182
 plumbosignalis 182
 roseopennalis 182
 zephyralis 182
Chrysodeixis includens 468
Chrysoecia atrolinea 518
 gladiola 518
 scira 518
Chytolita morbidalis 424
Cinnabar Moth 410
Cirrhophanus dyari 518
Cisseps fulvicollis 420
Cissusa, Indiscrete 444
 Vigorous 444
Cissusa indiscreta 444
 valens 444
Cisthene angelus 390
 barnesii 390
 deserta 388
 dorsimacula 388
 faustinula 388
 juanita 390

liberomacula 388
 martini 390
 perrosea 390
 schwarziorum 390
Cladara limitaria 264
Clearwing, Canyon 102
 Currant 98
 Fireweed 96
 Hummingbird 252
 Nebraska 96
 Rocky Mountain 254
 Snowberry 254
 Western Poplar 96
 Western Willow 98
Clemensia umbrata 394
Clepsis, Clemens' 60
Clepsis clemensiana 60
 consimilana 60
 peritana 60
 persicana 60
 virescana 60
Clostera albosigma 374
 apicalis 374
 brucei 374
 inornata 376
 ornata 374
Clothes Moth, Case-bearing 28
 Webbing 28
 Western 28
 Yellow-headed 28
Clover Moth, Spotted 544
Cloverworm, California 426
 Decorated 426
 Green 426
Cobubatha, Dark-banded 478
 Divided 478
 Sharp-banded 478
Cobubatha dividua 478
 lixiva 478
 orthozona 478
Cochylid, Broad-patch 52
 Cone 50
Codling Moth 90
Coelodasys unicornis 382
Coenochroa, Western 166
Coenochroa californiella 166
Coleophora, Metallic 132
 Peppered 132
 Streaked 132
 White-edged 130
Coleophora accordella 130
 cratipennella 132
 glaucella 132
 mayrella 132
 parthenica 132
 trifolii 132
Coleotechnites, Conifer, complex 122
Coleotechnites spp. 122

Comadia albistriga 94
 henrici 94
Concealer Moth, Eucalyptus 112
 Four-spotted 110
 White-edged 112
 Yellow-spotted 110
Conchylodes, Zebra 182
Conchylodes octonalis 182
 ovulalis 182
Condica albolabes 538
 discistriga 538
 temecula 536
Cone Moth, Douglas-Fir 70
Coneborer, Fir 72
 Lodgepole 70
 Pinyon 70
 Ponderosa 70
Coneworm, Douglas-fir 158
 Evergreen 158
 Ponderosa Pine 160
 Spruce 158
Copanarta, Gold-winged 528
Copanarta aurea 528
Coptodisca sp. 18
 arbutiella 18
 cercocarpella 18
 powellella 18
 quercicolella 18
 saliciella 18
Cosipara, Tricolored 208
Cosipara tricoloralis 208
Cosmet, Shy 116
Cosmia calami 586
 praeacuta 586
Costaconvexa centrostrigaria 286
Cotton Stem Moth 120
Crambid, Pondside 206
 Rufous-banded 196
Crambidia casta 396
 cephalica 396
Crambodes talidiformis 538
Cream, Yellow-dusted 340
Creosote Moth 316
Crescent, Lesser Tawny 110
 Sandy 512
Crimson, Clouded 546
Crocidosema plebejana 82
Cryphia, Camouflaged 556
 Eyed 556
"*Cryphia*" *cuerva* 556
 olivacea 556
Cryptocala acadiensis 648
Ctenucha, Brown 420
 Red-shouldered 420
 Thin-lined 420
 Veined 420
 Virginia 420
 White-margined 420

Ctenucha brunnea 420
 cressonana 420
 multifaria 420
 rubroscapus 420
 venosa 420
 virginica 420
Cucullia antipoda 504
 asteroides 502
 charon 504
 dorsalis 504
 eccissica 504
 eulepis 504
 florea 502
 intermedia 504
 laetifica 504
 lilacina 504
 montanae 502
 pulla 506
 serraticornis 506
 speyeri 504
 strigata 506
Cutworm, Army 636
 Bristly 618
 Bronzed 602
 Citrus 600
 Dingy 642
 Glassy 572
 Green 630
 Olive Green 610
 Pale Western 642
 Red-backed 640
 Shield-backed 584
 Speckled 606
 Triplex 538
 Variegated 630
 Well-marked 656
 Western Bean 632
 Western W-marked 648
 Yellow-headed 570
Cyclophora dataria 256
 nanaria 258
 pendulinaria 256
Cycnia, Delicate 416
 Oregon 416
Cycnia oregonensis 416
 tenera 416
Cydia, Aspen 88
Cydia latiferreana 90
 piperana 90
 pomonella 90
 populana 88

D
Dagger, American 496
 Black-striped 494
 Cherry 500
 Cottonwood 496
 Eastern Ochre 498
 Fragile 496
 Funerary 494
 Gentle 498

Dagger *continued*
Gray 498
Impressed 500
Large Gray 496
Lupine 502
Marbled Oak 496
Marsh 500
Miller 496
Othello 500
Pacific 500
Pale Green 494
Radcliffe's 498
Smeared 502
Strigulose 498
Tawny 498
Unmarked 498
Yellow-haired 500
Dalcerides ingenita 106
Dalmation, Variable 368
Dame's Rocket Moth 40
Dancing Moth, Hawaiian 22
Dargida diffusa 610
  procinctus 610
  tetera 610
Dart, Adaptable 636
  Antique 644
  Aurous 638
  Black Cedar 656
  Blurry 636
  Brown-edged 654
  Catocaline 648
  Chestnut 656
  Clandestine 648
  Collared 652
  Creaky 632
  Dappled 650
  Divergent 638
  Double 648
  Enigmatic 646
  Exuberant 630
  Finnish 632
  Flame-shouldered 644
  Formal 654
  Four-toothed 640
  Great Brown 648
  Hairy 640
  Hungry 646
  Indefinite 654
  Insular 654
  Ipsilon 644
  Knotwood 652
  Labrador 646
  Laughing 634
  Lesser Black-letter 652
  Luteous 656
  Master's 642
  Median-banded 638
  Miller's 634
  Murdock's 640
  Obelisk 640
  Oblique 644

Old Man 642
Olive 636
Pale-edged 656
Parental 634
Perpendicular 632
Quivering 650
Rascal 644
Reaper 636
Red-breasted 654
Red-washed 652
Rosy 652
Serano 634
Skirted 634
Smith's 650
Soothsayer 648
Stellar 652
Subterranean 642
Tessellate 638, 652
Tricolored 650
Umber 646
Vancouver 644
Venerable 642, 644
Vibrant 654
Violaceous Brown 656
Voluble 644
White-winged 638
Yellow 634
Yellow-based 640
Yellow-eyed 630
Zigzag 638
Daschira grisefacta 386
  vagans 386
Dasyfidonia avuncularia 308
Dasypyga, Gray-based 162
  Gray-shouldered 162
Dasypyga alternosquamella 162
  salmocolor 162
Datana, Large 378
  Scalloped 378
  Spotted 378
Datana californica 378
  perfusa 378
  perspicua 378
Daviscardia coloradella 30
Decantha stonda 110
Decaturia, Fox-faced 146
Decaturia pectinalis 146
Decodes basiplagana 48
  fragariana 48
Deilephila elpenor 254
Dejongia californicus 138
Delta, Belted 334
  Enigmatic 334
  Green 334
  Smoky 334
  Streaked 334
Desert Marigold Moth 550
Desert Willow Destroyer 144

Desmia maculalis 190
Destutia excelsa 366
  flumenata 364
Diamondback, Blushing 122
Diamondback Moth 42
Diaphania hyalinata 188
Diarsia esurialis 646
  rosaria 646
Diastictis fracturalis 186
  sperryorum 186
  ventralis 186
Diathrausta harlequinalis 190
Dichagyris capota 634
  cataclivis 632
  grotei 632
  variabilis 634
Dichomeris, Sunset 118
Dichomeris simpliciella 118
Dichorda illustraria 304
  rectaria 304
Dichordophora phoenix 306
Dichrorampha simulana 88
  vancouverana 88
Dicymolomia,
  Dusky-patched 200
  Peppery 200
Dicymolomia metalliferalis 200
  opuntialis 200
Digrammia atrofasciata 320
  californiaria 314
  cinereola 320
  colorata 316
  continuata 318
  curvata 318
  decorata 320
  denticulata 316
  excurvata 318
  imparilata 318
  irrorata 322
  muscariata 320
  neptaria 320
  nubiculata 316
  pallorata 318
  pervolata 316
  pictipennata 316
  subminiata 320
  triviata 318
  ubiquitata 316
  yavapai 320
Dioryctria abietivorella 158
  auranticella 160
  cambiicola 160
  pseudotsugella 158
  reniculelloides 158
  zimmermani 160
Diphthera festiva 464
Diptychophora harlequinalis 214

Disclisioprocta stellata 282
Ditula angustiorana 62
Donacaula, Brown-edged 204
  Dark-striped 204
  Delightful 204
  Wandering 204
  White-edged 204
Donacaula n. sp. 204
  albicostellus 204
  dispersellus 204
  melinellus 204
Double Lobed Moth 572
Drasteria adumbrata 452
  divergens 452
  edwardsii 450
  fumosa 450
  howlandii 452
  hudsonica 452
  inepta 450
  mirifica 448
  ochracea 450
  pallescens 450
  perplexa 452
  petricola 452
  sabulosa 450
  scrupulosa 448
  stretchii 452
Drepanulatrix bifilata 342
  carnearia 342
  falcataria 342
  foeminaria 342
  hulstii 342
  monicaria 344
  quadraria 342
  secundaria 344
  unicalcararia 342
Dryadaula terpsichorella 22
Dun-bar, American 586
  Variable 586
Duponchelia fovealis 190
Dyotopasta yumaella 30
Dypterygia patina 556
Dyseriocrania auricyanea 14
Dysodia, Netted 142
Dysodia granulata 142
Dysschema howardi 410
Dysstroma brunneata 280
  citrata 278
  formosa 278
  hersiliata 278
  mancipata 280
  sobria 278
  truncata 278

E
Eana argentana 48
Ear Moth, American 576
  Crescent 576
Earworm, Corn 542
Ecliptopera silaceata 280

Ectropis crepuscularia 332
Ectypia bivittata 418
  clio 418
Egira cognata 600
  crucialis 598
  curialis 600
  dolosa 600
  februalis 600
  hiemalis 598
  perlubens 600
  rubrica 600
  simplex 598
  variabilis 598
Eichlinia gloriosa 98
  snowii 98
Eight-spot 476
Elaphria alapallida 560
Elasmopalpus lignosella 160
Elder Moth 588
Elophila icciusalis 206
  obliteralis 206
Emarginea, Beloved 512
  Dark-collared 512
Emarginea dulcinea 512
  percara 512
Embola powelli 36
Emerald, Ardent 302
  Arizona 302
  Bank's 300
  Columbian 302
  Common 306
  Day 306
  Festive 302
  Honest 304
  Illustrated 304
  Lovely 302
  Margaret's 300
  Oblique-striped 304
  Phoenix 306
  Rosy 302
  Single-lined 300
  Slanted 302
  Southern 304
  Thick-lined Ivory 306
  Thin-lined Ivory 306
  Wavy-lined 304
  Western Ivory 306
  White-winged 304
  Yellow-fringed 304
Emmelina monodactyla 140
Enargia, Pale 588
Enargia decolor 588
  infumata 588
Enchoria lacteata 286
Endothenia hebesana 64
Endrosis sarcitrella 110
Engrailed, Small 332
Ennomos magnaria 368
Ephestia kuehniella 156
Ephestiodes erythrella 154
  gilvescentella 154

Ephippiphora lunatana 88
Epiblema, Ragweed 80
Epiblema strenuana 80
Epilechia catalinella 118
Epinotia, Alder 84
  Currant 82
  Delta 86
  Madrone 86
  Manzanita 84
  Ocean Spray 84
  Pacific 84
  Serpentine 86
  Sigmoid 88
  Summer Holly 86
  Variable 86
Epinotia albangulana 84
  arctostaphylana 86
  castaneana 82
  emarginana 86
  johnsonana 84
  kasloana 86
  lomonana 86
  nigralbana 86
  nisella 84
  radicana 82
  signiferana 88
  subplicana 84
  subviridis 84
Epiphyas postvittana 60
Epirrhoe alternata 288
  plebeculata 288
  sperryi 288
Epirrita autumnata 274
Episimus argutana 66
Eralea, White-lined 116
Eralea albalineella 116
Erannis tiliaria 338
  vancouverensis 338
Eremobina claudens 574
Erinnyis ello 250
  obscura 250
Eriocrania semipurpurella
  14
Eriplatymetra coloradaria
  362
  grotearia 364
Esperia sulphurella 110
Estigmene acrea 404
  albida 404
Eteobalea, Silver-spotted
  116
Eteobalea iridella 116
Ethmia, Gray 114
  Marbled 114
  Mountain-mahogany
  114
  Mourning 114
  Shaded 114
  Stone 114
Ethmia arctostaphylella 114
  delliella 488

discostrigella 114
  marmorea 114
  monticola 114
  semilugens 114
  semitenebrella 114
Etiella, Gold-banded 162
Etiella zinckenella 162
Euacidalia sericearia 258
Eubaphe unicolor 266
Eubarnesia ritaria 328
Eublemma, Everlasting
  432
Eublemma minima 432
  recta 432
Eucalantica polita 38
Eucaterva variaria 368
Euceratia castella 38
  securella 38
Euchaetes antica 416
  egle 416
  zella 416
Euchlaena, Johnson's 348
  Mottled 350
  Scrub 350
  Soft 350
Euchlaena johnsonaria 348
  madusaria 350
  mollisaria 350
  tigrinaria 350
Euchromius californicalis
  212
  ocellea 212
Euclea incisa 106
  obliqua 106
Euclidia ardita 456
  cuspidea 456
Eucopina bobana 70
  ponderosa 70
  rescissoriana 70
  siskiyouana 72
  sonomana 70
Eucosma, Eccentric 72
  Pale-striped 72
  Saddle-backed 72
Eucosma apacheana 72
  offectalis 72
  pallidarcis 72
Eudactylota, Rusty 124
Eudactylota iobapta 124
Eudesmia, Arid 388
Eudesmia arida 388
Eudonia, Four-eyed 210
  Gold-lined 210
  Hook-lined 210
  Long-lined 210
  Straight-lined 210
Eudonia commortalis 210
  echo 210
  rectilinea 210
  spenceri 210
  torniplagalis 210

Eudrepanulatrix rectifascia
  340
Eufernaldia cadarellus 212
Euhagena nebraskae 96
Eulia, Ferruginous 52
Eulia ministrana 52
Eulithis, Currant 280
  Variable 280
Eulithis destinata 280
  propulsata 280
  xylina 280
Eulithosia, Beautiful 520
  Striped 520
Eulithosia composita 520
  discistriga 520
  plesioglauca 520
Eulogia, Broad-banded 152
Eulogia ochrifrontella 152
Eumacaria madopata 308
Eumorpha achemon 250
  typhon 250
  vitis 250
Euparthenos nubilis 462
"Euphyia" implicata 298
  intermediata 284
Eupithecia, Common 268
  Tawny 274
Eupithecia absinthiata 272
  acutipennis 272
  annulata 270
  behrensata 270
  cretaceata 270
  edna 268
  gilvipennata 272
  graefii 272
  gypsata 274
  maestosa 268
  miserulata 268
  misturata 268
  mutata 270
  nevadata 272
  olivacea 270
  ravocostaliata 274
  rotundopuncta 270
  subapicata 272
  unicolor 268
  zelmira 270
Euplexia benesimilis 566
Euproserpinus phaeton 252
Eupsilia tristigmata 582
Eurois astricta 648
  occulta 648
Eusarca, Confused 358
  Falcate 358
  Rectangular 358
Eusarca confusaria 358
  falcata 358
  geniculata 358
Euscirrhopterus cosyra 534
  gloveri 534
Eutricopis nexilis 540

Euxoa albipennis 638
  auxiliaris 636
  basalis 640
  bochus 636
  comosa 640
  divergens 638
  hollemani 638
  medialis 638
  messoria 636
  murdocki 640
  obeliscoides 640
  ochrogaster 640
  olivia 636
  pluralis 638
  quadridentata 640
  septentrionalis 636
  tessellata 638
Euzophera semifuneralis 152
Evergestis, Brown-patched
  202
  Frosted 202
  Gray 202
  Singed 200
  Sleeping 202
  Slender 202
  Streaked 202
Evergestis consimilis 200
  funalis 202
  lunulalis 202
  obliqualis 202
  pallidata 200
  simulatilis 202
  subterminalis 202
  vinctalis 202

F

Faculta inaequalis 130
Fairy Moth, California 22
  Dark 22
  Desert 20
  Flaming 20
  Gilia 20
  Ocean Spray 20
  Orange-headed 20
  Three-striped 20
Fan-Foot, Wavy-lined 422
Fawn Moth, Bordered 348
Feltia herilis 642
  jaculifera 642
  subterranea 642
Feralia comstocki 508
  deceptiva 508
  februalis 508
  jocosa 508
Filatima, Bent-line 128
Filatima albilorella 128
Filbertworm Moth 90
Fishia, Dark Gray 590
Fishia discors 590
  illocata 590
  yosemitae 590

Fissicrambus intermedius
  214
  profanellus 214
  quadrinotellus 214
Flag Moth, Northern Giant
  410
Flannel Moth,
  Brown-patched 104
  Pink-spotted 104
Flightless Moth, Oso Flaco
  134
Flour Moth, Mediterranean
  156
Flower Moth, Acute-lined
  550
  Agoseris 554
  Approachable 552
  Arcigera 548
  Beautiful 546
  Black-etched 518
  Boneset 546
  Brickellbush 546
  Caramel 518
  Carefree 518
  Chicory 554
  Chrysellus 552
  Eyed 552
  Familiar 544
  Gray-lined 554
  Green 544
  Gumweed 548
  Hoary 518
  Hulst's 554
  Jaguar 544
  Little 540
  Lynx 548
  Mead's 546
  Mexican 546
  Palm 468
  Pink-washed 554
  Rabbitbrush 552
  Shining 544
  Silver-spotted 552
  Snakeweed 552
  Spanish Needles 550
  Swan Plant 188
  Third 550
  Toothed 550
  Wandering 548
  White-banded 552
  White-spotted 518
Forester, Blue-spotted 534
  Eight-spotted 534
  Geronimo 534
  Langton's 534
  MacCulloch's 536
  Mariposa 536
  Ridings' 536
  Two-spotted 534
Forsebia cinis 448
Four-spotted Moth 524

Frechinia helianthiales 194
  laetalis 196
Frederickia cyda 322
  hypaethrata 322
  nigricomma 322
  s-signata 322
Friseria, White-dusted 124
Friseria caieta 124
  cockerelli 124
Frothy Moth 516
Fruitworm, Madrone 140
  Reticulated 64
  Sparganothis 62
Fruitworm Moth,
  Gooseberry 164
  Speckled Green 598
Fungus Moth, Bicolored 432
  Miserable 430
  Poplar 26
Furcula, Ashy 378
  Gray 378
  Modest 378
  Zigzag 378
Furcula cinerea 378
  cinereoides 378
  modesta 378
  scolopendrina 378

G

Gabriola dyari 352
Galasa nigrinodis 144
  nigripunctalis 144
Galenara lixaria 334
Galgula partita 560
Gall Moth, Coffeeberry
  Midrib 116
  Coyote Brush Stem 130
  Rabbitbrush Stem 130
Galleria mellonella 146
Garella, Small 464
Garella nilotica 464
Gazoryctra mathewi 14
  novigannus 14
Gelechia, White-lined 126
Gelechia desiliens 126
Gem, Citrus 544
  Little Dark 548
  Oregon 542
  Silver-banded 554
Gem Moth 286
Genista Broom Moth 172
Geometer, Barberry 276
  Black-dashed 346
  Brown-bordered 308
  Chickweed 258
  Dog-face 344
  Eight-lined 340
  Falcate 364
  Hollow-spotted 256
  Scallop-lined 364
  Silvery-gray 338

Sweetfern 256
  Thick-lined 362
Gerra sevorsa 532
Gerrodes minatea 532
Gesneria, Smoky 208
Gesneria centuriella 208
Ghost Moth, Four-spotted
  16
  Mathew's 14
  Orange-lined 14
  Smooth 14
Gillmeria pallidactyla 136
Gilt-edged Moth 30
Girdle, Dark-banded 372
  Golden 372
  Gray 372
  Hoary 360
  Large 360
  Packard's 372
  Sagebrush 362
  Sharp-toothed 360
  Smooth-lined 360
  Variable 372
  White-striped 370
Girdle Moth, Red 364
Givira lotta 92
  marga 92
  mucidus 92
  theodori 92
Glassy-wing, Edwards' 414
  Freckled 414
  Mottled 414
  Striped 414
Glaucina epiphysaria 328
  erroraria 326
  eupetheciaria 328
  gonia 328
  interruptaria 328
  macdunnoughi 326
  nephos 328
  platia 328
Glena grisearia 324
  nigricaria 324
  quinquelinearia 324
Globia oblonga 576
Gluphisia, Banded 376
  Common 376
Gluphisia septentrionis 376
  severa 376
Glyph, Pale 478
  Pink-banded 478
Glyphidocera, Four-spotted
  108
Glyphidocera septentrionella
  108
Glyphipterix, Two-banded
  42
Glyphipterix bifasciata 42
Gnophaela discreta 410
  latipennis 410
  vermiculata 410

Gnorimoschema bacchariisella 130
octomaculella 130
Gold Spot, Shaded 470
Golden Moth, Beautiful 520
Shining 522
Splendid 520
Gorse Tip Moth 112
Gothic, American 602
Grain Moth, European 26
Granite, Brown-lined 320
Californian 314
Dark-bordered 320
Decorated 320
Humble 320
Painted 316
Pearl 316
Shaded 318
Tawny-veined 316
Three-patched 318
Toothed 316
Ubiquitous 316
Vermilion 320
Graphic, Ashy 448
Deduced 448
Divergent 452
Handsome 450
Indomitable 446
Inept 450
Merry 446
Novel 446
Ochre 450
Paler 450
Perpendicular 446
Royal Poinciana 448
Sandy 450
Scrupulous 448
Similar 448
Streaked 452
Stretched 452
Wonderful 448
Graphiphora augur 648
Grapholita, One-barred 88
Grass Moth, Julia 542
Paradoxical 542
Pink-banded 544
Spotted 428
Grass-Tubeworm,
Bold-dashed 24
Divided 24
Gray 24
Rusty 26
Short-dashed 24
Variable 26
Grass-veneer, Belted 212
California 212
Dimorphic 212
Intermediate 214
Profane 214
White-veined 212
Woolly 214

Gray, Bent-line 330
Brindled 334
Broad-winged 336
Brown-winged 328
Cinnamon 332
Common 332
Cross-lined 328
Exalted 326
Five-lined 324
Four-barred 336
Fragile 330
Gulf Coast 332
Linear 328
Orange-washed 324
Porcelain 332
Russet 326
Rusty-patched 326
Scalloped 434
Sharp-toothed 324
Six-spotted 434
Straight-line 324
Tarnished 330
Twin-dotted 324
Gray-Streak, Blushing 610
Green, Pink-margined 302
Greya, California 18
Obscure 16
Polite 16
Reticulated 18
Greya obscura 16
politella 16
reticulatus 18
solenobiella 18
Grotella, Spotted 522
Thick-lined 522
Tricolored 524
White 522
Grotella binda 522
blanca 522
sampita 522
tricolor 524
Groundling, Black 538
Black-dashed 536
Gray-striped 538
Six-spotted 124
Gyros muirii 196

H
Hadena variolata 610
Haematopis grataria 258
Hahncappsia, Persistent 170
Hahncappsia pergilvalis 170
Half-Spot, Brown 566
Half-yellow Moth 480
Brown-headed 480
White-lined 480
Halysidota davisii 412
Haploa, LeConte's 410
Haploa lecontei 410

Harrisina americana 102
metallica 104
Harrow Moth 530
Hawkmoth, Elephant 254
Spurge 254
Hayworm, Brown 150
Clover 150
Hecatera dysodea 610
Hedya, Off-white 66
Hedya nubiferana 68
ochroleucana 66
Helcystogramma, Sandy 118
Helcystogramma badia 118
Helicoverpa zea 542
Heliocheilus julia 542
paradoxus 542
toralis 544
Heliolonche carolus 554
modicella 554
pictipennis 554
Heliothis oregonica 542
phloxiphaga 542
Heliothodes, Small 540
Heliothodes diminutiva 540
Hellinsia grandis 140
homodactylus 138
Hellula aqualis 198
rogatalis 198
Helotropha reniformis 576
Hemaris diffinis 254
thetis 254
thysbe 252
Hemeroplanis, Black-marked 432
Pale-lined 434
Rusty 432
Hemeroplanis historialis 432
incusalis 432
punitalis 432
rectalis 434
Hemibryomima, Masked 562
Hemibryomima chryselectra 562
Hemieuxoa rudens 632
Heminocloa mirabilis 492
Hemiplatytes epia 212
Hemithea aestivaria 306
Henricus, Brown-shouldered 50
Contrasting 50
Henricus edwardsiana 50
fuscodorsana 50
umbrabasana 50
Herald Moth 428
Herpetogramma, Northern 188
Herpetogramma aquilonalis 188
bipunctalis 188

Hesperumia fumosaria 336
latipennis 336
sulphuraria 336
Heteranassa mima 458
Heterocampa averna 380
Hexorthodes accurata 628
nipana 628
Hieroglyphic Moth 464
Highflier, Oak Winter 292
July 292
Hileithia magualis 190
Hofmannophila
pseudospretella 110
Holochroa dissociarius 352
Homoeosoma electella 166
Homoglaea, Black-barred 580
Twin-spotted 580
Homoglaea carbonaria 580
dives 580
Homorthodes communis 624
fractura 624
furfurata 624
hanhami 624
Honeysuckle Moth, Brown 38
Speckled 38
Honora, Honeyed 162
Tricolored 164
Honora mellinella 162
montinatatella 164
Hooded Owlet,
Brown-patched 504
Dark 506
Dimorphic 506
Goldenrod 502
Gray 502
Intermediate 504
Mountain 502
Rabbitbrush 504
Small 504
Speyer's 504
Spotted 506
Streaked 506
Tansyaster 504
Horisme intestinata 268
Hornet Moth, Pacific 96
House Moth, Brown 110
European 110
White-shouldered 110
Huckleberry Moth, White 38
Hydraecia medialis 576
Hydriomena, Beautiful 292
Coastal 290
Eden 292
Five-banded 292
Furious 290
Manzanita 294
Purged 290
White-banded 292

Hydriomena albifasciata 292
  edenata 292
  expurgata 290
  furcata 292
  irata 290
  manzanita 294
  marinata 290
  nubilofasciata 292
  quinquefasciata 292
  speciosata 292
Hydroeciodes serrata 628
Hyles euphorbiae 254
  gallii 254
  lineata 254
Hymenia perspectalis 190
Hypena abalienalis 426
  bijugalis 424
  californica 426
  decorata 426
  humuli 426
  palparia 426
  scabra 426
Hyphantria cunea 404
Hypocala andremona 430
Hypocala Moth 430
Hypocrisias, Least 412
Hypocrisias minima 412
Hypoprepia cadaverosa 392
  fucosa 392
  inculta 394
  miniata 392
Hypopta palmata 92
Hypotrix lunata 628
Hyppa, Summer 586
Hyppa contrasta 586
Hypsocompe permaculata 406
Hypsopygia, Yellow-fringed 150
Hypsopygia costalis 150
  olinalis 150
  phoezalis 150
Hystrichophora vestaliana 68

I

Ianassa pallida 384
Idaea asceta 260
  basinta 260
  bonifata 260
  demissaria 260
  dimidiata 262
  eremiata 260
  gemmata 260
  occidentaria 260
Idia, American 422
  Common 422
  Glossy Black 422
  Western 422
Idia aemula 422
  americalis 422
  lubricalis 422
  occidentalis 422

Infant, The 300
Inga, Chalky 108
Inga cretacea 108
Inguromorpha itzalana 94
Ipimorpha pleonectusa 588
Iridopsis clivinaria 330
  dataria 328
  emasculatum 330
  fragilaria 330
  larvaria 330
  obliquaria 330
Isa schaefferana 106
Isogona, Framed 434
Isogona segura 434
Isophrictis, Black-dashed 120
Isophrictis magnella 120
Ivory-edged Moth 432
Ixala desperaria 344

J

Jativa castanealis 196
Javelin Moth 66
Jester, Brown 294
  Checkered 298
  Chestnut 294
  Dark-patched 294
  Festive 294
  Formal 296
  Marbled 296
  Orange 296
  Painted 296
  Smooth 298
  Topaz 296
  White-patched 294
Joker, The 508

K

Kidney Moth, Three-spotted 198
Knapweed Moth, Sulphur 50

L

Lacanobia nevadae 606
  subjuncta 606
Lace-border, Large 262
Lacinipolia acutipennis 618
  buscki 622
  circumcincta 620
  comis 620
  cuneata 618
  davena 620
  dimocki 618
  illaudabilis 622
  lepidula 620
  martini 622
  olivacea 620
  patalis 622
  pensilis 618
  quadrilineata 622

renigera 618
  sareta 618
  stricta 620
  strigicollis 622
Laetilia coccidivora 164
  dilatifasciella 164
  zamacrella 164
Lambdina fiscellaria 366
Lantana Stick Moth 526
Lantanophaga pusillidactylus 138
Lateroligia ophiogramma 572
Lawn Moth, Western 214
Leafblotch Miner, Manzanita 34
  White-striped 32
  Willow 32
Leafcutter, Waterlily 206
Leaffolder, Serviceberry 68
  White-headed Grape 190
Leafminer, Aspen Serpentine 34
  Azalea 32
  Citrus 34
  Morning-glory 34
  Orange-headed 16
  Silver 18
Leafroller, Early Aspen 82
  Fruit-tree 58
  Green Aspen 66
  Live-Oak 82
  Oblique-banded 56
  Omnivorous 64
  Pandemis 54
  Poplar 66
  Sesame 186
Leafroller Moth, Western Avocado 62
Leaftier, Boxwood 144
  Celery 184
  False Celery 184
  Rusty 144
  Sumac 66
Leaftier Moth, Maple 46
Leptarctia californiae 408
Leptostales rubromarginaria 262
Lesmone griseipennis 456
Leucania commoides 612
  dia 614
  farcta 612
  imperfecta 612
  insueta 614
  multilinea 612
  oaxacana 614
  oregona 614
  stolata 614
Leucogoniella californica 118
Leucoma salicis 386

Lichen Moth, Angel 390
  Barnes' 390
  Black-and-yellow 392
  Cadaver 392
  Crescent-marked 388
  Crimson 394
  Gray 388
  Juanita's 390
  Little Shaded 394
  Martin's 390
  Painted 392
  Pearly-winged 396
  Peppered 394
  Pink-lined 394
  Rosy 390
  Royal 392
  Scarlet-winged 392
  Schwarz's 390
  Shining 392
  Splendid 392
  Three-spotted 388
  Vermilion 394
  White-patched 388
  Yellow-headed 396
Limnaecia phragmitella 116
Lineodes, Eggplant 186
  Interrupted 186
Lineodes integra 186
  interrupta 186
Lineostriastiria hachita 516
  sexseriata 516
Lintneria istar 246
Lipocosma, Fancy 200
Lipocosma albinibasalis 200
Lithariapteryx abroniaeella 34
  jubarella 36
Litholomia napaea 580
Lithomoia germana 580
Lithophane georgii 582
  innominata 580
  pertorrida 582
  petulca 582
Litocala Moth 446
Litocala sexsignata 446
Litoprosopus coachella 468
Lobocleta ossularia 258
  peralbata 258
  plemyraria 258
Lobophora nivigerata 264
Lomographa semiclarata 346
  vestaliata 346
Looper, Alfalfa 470
  Bilobed 472
  Black-veined Conifer 366
  Blackberry 306
  Brown-lined 336
  Brown Pine 364
  Buckskin 322
  Cabbage 468

Caramel 470
Celery 474
Cerulean 454
Clover 454
Dyar's 352
Epigaea 472
Forage 454
Gray 470
Gray Spruce 364
Green Broomweed 322
Hemlock 366
Hoary Conifer 366
Large 472
Lesser Bean 470
Linden 338
Mesquite 322
Mountain-mahogany 330
Oblique 330
Omnivorous 370
Pacific Mountain 474
Pale-lined 456
Phantom Hemlock 366
Putnam's 474
Rannoch 310
Rocky Mountain 474
Saddled 472
Salt-and-Pepper 474
Signate 322
Soybean 468
Spruce False 472
Tamarack 268
Vancouver 338
Western False Hemlock 366
Lophocampa argentata 412
  ingens 412
  maculata 414
  mixta 412
  pura 414
  roseata 412
Lorita scarificata 52
Lotisma trigonana 140
Loxostege, Charming 172
  Gray-banded 174
  Tan-edged 172
  Wolfberry 174
Loxostege albiceralis 172
  allectalis 174
  cerealis 174
  lepidalis 172
  munroealis 174
  typhonalis 174
Lucerne Moth 192
Lychnosea helveolaria 364
Lycia ursaria 338
Lycomorpha fulgens 392
  pholus 392
  regulus 392
  splendens 392
Lygephila victoria 434
Lygropia, Eight-barred 182

M

Macaria adonis 314
  aemulataria 314
  amboflava 310
  austrinata 312
  bitactata 310
  brunneata 310
  colata 310
  deceptrix 312
  guenearia 312
  lorquinaria 312
  marcescaria 308
  metanemaria 310
  occiduaria 310
  pallipennata 314
  plumosata 312
  quadrilinearia 312
  signaria 314
  truncataria 308
  ulsterata 314
Macrorrhinia aureofasciella 160
Magician Moth 190
Magpie, Small 168
Magusa divaricata 558
  orbifera 558
Mallow Moth, Pale Yellow 476
Mamestra configurata 608
Manulea bicolor 394
Marathyssa, Dark 462
Marathyssa inficita 462
Marmara arbutiella 34
Maroonwing 608
Marsh Moth, Henry's 500
Martania grandis 268
Matigramma, Brownish 460
  Reddish 458
Matigramma emmilta 460
  rubrosuffusa 458
Meal Moth, Common 148
  Indian 154
Mecyna, Pale-spotted 192
Mecyna mustelinalis 192
Megalographa biloba 472
Megalopyge lapena 104
Meganola, Confused 464
  Dusky 464
Meganola fuscula 464
  minuscula 464
Melanchra adjuncta 606
  picta 606
Melanolophia imitata 334
Meleneta antennata 492
Melipotis acontioides 448
  indomita 446
  jucunda 446
  novanda 446
Melipotis perpendicularis 446

Melitara dentata 164
  subumbrella 166
Melonworm Moth 188
Mericisca gracea 332
Meroptera pravella 158
Mesapamea secalis 574
Mesogona olivata 582
  subcuprea 582
Mesoleuca gratulata 288
  ruficillata 288
Mesophleps adustipennis 118
Mesothea incerta 306
Metalectra edilis 432
  miserulata 430
Metaname, Pale 348
Metanema inatomaria 348
Metaponpneumata rogenhoferi 524
Metarranthis, Ruddy 346
Metarranthis duaria 346
Metzneria lappella 120
Michaelophorus indentatus 138
Micrathetis costiplaga 538
  triplex 538
Microhelia angelica 540
Microtheoris ophionalis 194
  vibicalis 194
Micrurapteryx salicifoliella 32
Midget, Gray 466
  Pale-winged 560
  White-spotted 540
Mime, The 458
Mimorista, Brown-shaded 192
Mimorista subcostalis 192
Mimoschinia rufofascialis 196
Miner, Madrone Skin 34
  Russian Thistle Stem 132
Mint Moth, Orange 178
  Southern Purple 180
Miodera stigmata 628
Miracavira, Brilliant 508
Miracavira brillians 508
Miranda Moth 564
Mirificarma eburnella 130
Mojavia achemonalis 194
Mompha, One-lined 134
  Red-streaked 134
  White-based 134
Mompha albocapitella 134
  eloisella 134
  unifasciella 134
Monitor, The 492
Monopis crocicapitella 28
Moon Moth, Black 430
Moonseed Moth 428
Mountain Bluebell Moth 410

Mountain Moth, Harlequin 214
  Snowy 364
Mouse Moth 506
Mutant, The 602
Mycterophora, Blushing 430
  Long-palped 430
Mycterophora longipalpata 430
  rubricans 430
Myelopsis, Gray 152
  Minute 152
Myelopsis minutulella 152
  subtetricella 152
Mythimna oxygala 612
  unipuncta 612

N

Nadata gibbosa 380
  oregonensis 380
Nannobotys commortalis 194
Narraga fimetaria 322
Narrow-wing, Orbed 558
  Variable 558
Narrowwing, Brown 326
  Golden 326
  Hoary 326
  Interrupted, complex 328
  Scalloped 328
  Streaked 326
Narthecophora pulverea 522
Nedra stewarti 566
Needleworm, Larch 82
  Red-Striped 82
Nemapogon defectella 26
  granella 26
  molybdanellus 26
Nematocampa brehmeata 372
  resistaria 372
Nemoria arizonaria 302
  darwiniata 302
  festaria 302
  glaucomarginaria 304
  leptalea 302
  obliqua 302
  pistaciaria 302
  pulcherrima 300
  unitaria 300
  zelotes 302
Neoalbertia constans 104
Neoalcis californiaria 336
Neogalea sunia 526
Neoheliodines vernius 36
Neohelvibotys, Arizona 172
Neohelvibotys arizonensis 172
Neoilliberis fusca 104
Neoterpes edwardsata 360
  ephelidaria 358
  trianguliferata 358

Nephelodes minians 602
Nephrogramma separata 198
Nepytia freemani 366
  phantasmaria 366
  swetti 366
  umbrosaria 366
Nettle Moth, Spectacled 468
Neumoegenia poetica 522
Niditinea fuscella 28
Nocloa, Chihuahuan 510
  Golden 510
  Lemon 510
  Masked 510
  Pale 510
  Rivulet 510
Nocloa alcandra 510
  aliaga 510
  cordova 510
  nanata 510
  pallens 510
  rivulosa 510
Noctua comes 646
  pronuba 646
Noctueliopsis aridalis 192
  brunnealis 192
  bububattalis 194
Nola, Ceanothus 464
Nola minna 464
Nomenia duodecemlineata 266
  obsoleta 266
Nomophila nearctica 192
Norape tener 104
Notela jaliscana 384
Numia, Bicolored 344
Numia bicoloraria 344
Nutmeg, The 602
Nychioptera noctuidalis 434
Nycteola n. sp. 466
  cinereana 466
  columbiana 466
  frigidana 466

O
Oak Moth, California 384
Oak-Slug Moth, Green 106
  White-lined 106
Oakminer, Golden 14
Ochropleura implecta 644
Ocotillo Moth 328
Odontoptila obrimo 262
Oecophora bractella 112
Oedemasia concinna 382
  salicis 382
  semirufescens 384
Oegoconia novimundi 108
Ofatulena, Luminous 88
  Twelve-lined 88
Ofatulena duodecemstriata 88
  luminosa 88

Oidaematophorus grisescens 140
Oinophila v-flava 30
"Oligia" divesta 574
  rampartensis 574
  violacea 574
Olybria, Hollow-spotted 156
Olybria aliculella 156
Operophtera bruceata 274
  brumata 274
  danbyi 276
  occidentalis 276
Opogona omoscopa 28
Orange, Black-banded 308
Orangeworm Moth, Navel 152
Orgyia antiqua 386
  pseudotsugata 386
  vetusta 386
Ornativalva erubescens 122
Orthonama obstipata 286
Orthosia arthrolita 596
  behrensiana 596
  erythrolita 594
  ferrigera 596
  hibisci 598
  mys 594
  pacifica 596
  praeses 594
  pulchella 594
  revicta 596
  transparens 594
"Oruza" albocostata 432
Oslaria, Green 508
Oslaria viridifera 508
Owlet, Astronaut 434
  Bent-winged 424
  Broken-spotted 434
  Brown-speckled 456
  Columbia 466
  Frigid 466
  Gold-lined 522
  Gray-winged 456
  Morbid 424
  Powdered 522
  Rusty 466
  Three-lined 430
Oxycilla tripla 428
Oxynemis advena 512
  fusimacula 514
Ozarba propera 478

P
Pachysphinx modesta 248
  occidentalis 248
Paectes, Barrens 462
Paectes abrostolella 462
Paleacrita longiciliata 338
Palpita, Four-spotted 188
  Gracile 188

Palpita atrisquamalis 188
  quadristigmalis 188
Palthis, Dark-spotted 424
Palthis angulalis 424
Pandemis cerasana 54
  pyrusana 54
Panopoda, Black-patched 462
Panopoda rigida 462
Panthea, Giant 492
  Southern 494
  Western 492
Panthea gigantea 492
  virginarius 492
Paonias excaecata 248
  myops 248
Papaipema insulidens 576
Parabagrotis exsertistigma 654
  formalis 654
  insularis 654
  sulinaris 654
Parachma, Ochre 144
Parachma ochracealis 144
Paradiarsia littoralis 646
Paramiana perissa 512
  smaragdina 510
Paranthrene robiniae 96
Parapheromia cassinoi 334
Parasa, Smaller 108
Parasa chloris 108
Paraseptis adnixa 586
Pasiphila rectangulata 274
Pelochrista, Avalon 74
  Black-circle 76
  Brown-chevron 78
  Curly 74
  Fernald's 76
  Galena 74
  Marbled 76
  Morning 78
  Morrison's 74
  Pied 72
  Reversed 76
  Silver-lined 78
  Skewbald 74
  Spangled 78
  Thin-lined 74
  Two-barred 76
  Variegated 78
  Wonderful 78
Pelochrista agassizii 72
  agricolana 74
  argenteana 78
  avalona 74
  bolanderana 74
  canana 76
  comatulana 74
  corosana 78
  curlewensis 76
  eburata 76

fernaldana 76
galenapunctana 74
matutina 78
mirosignata 78
morrisoni 74
ragonoti 78
reversana 76
ridingsana 76
scintillana 78
Pennisetia marginatum 96
Penstemonia clarkei 102
Pepper Moth, European 190
Peppered Moth 336
Peridroma saucia 630
Perigonica angulata 592
Perigonica pectinata 592
  tertia 592
Perispasta caeculalis 168
"Perizoma" costiguttata 298
  curvilinea 298
  custodiata 284
  epictata 298
Pero, Behr's 356
  Honest 354
  McDunnough's 356
  Meske's 354
  Mizon 354
  Morrison's 354
  Straight-lined 354
  Western 356
  Yellow-brown 354
Pero behrensaria 356
  flavisaria 354
  honestaria 354
  macdunnoughi 356
  meskaria 354
  mizon 354
  morrisonaria 354
  occidentalis 356
  radiosaria 354
Petrophila, Confusing 206
  Jalisco 206
  Kearfott's 206
  Schaeffer's 206
Petrophila confusalis 206
  jaliscalis 206
  kearfottalis 206
  schaefferalis 206
Phaeoura aetha 352
  belua 352
  cristifera 350
  mexicanaria 350
  perfidaria 352
Pheosia californica 376
  rimosa 376
Pheosidea elegans 376
Phereoeca praecox 26
  uterella 26
Pherne, Bold-lined 362
  Brown-lined 362

Shaded 362
Thin-lined 362
Pherne parallelia 362
placeraria 362
sperryi 362
subpunctata 362
Phigalia plumogeraria 338
Philedia punctomacularia
346
Phlogophora periculosa 566
Phobolosia anfracta 430
Phoenicophanta bicolor 478
hampsonii 418
Phoenix, Northwestern 280
Small 280
Phragmatobia fuliginosa
406
Phryganidia californica 384
Phycitodes, White-edged
166
Phycitodes mucidella 166
Phyllocnistis citrella 34
populiella 34
Phyllonorycter manzanita 34
Phymatopus hectoides 14
Pima, White-edged 156
Pima fosterella 156
Pine Moth, Western 160
Zimmerman 160
Pinion, False 580
Large Gray 582
Nameless 580
Torrid 582
Wanton 582
Pitch Moth, Douglas-fir 100
Sequoia 100
Plagiomimicus dimidiata
516
spumosum 516
tepperi 516
Plagodis, Straight-lined 348
Plagodis phlogosaria 348
pulveraria 348
Plataea calcaria 360
californiaria 360
diva 360
personaria 360
trilinearia 362
ursaria 360
Platphalonidia felix 52
Platyedra subcinerea 120
Platynota, Toasted 64
Platynota labiosana 64
stultana 64
Platyptilia carduidactylus
136
Plemyria georgii 282
Pleromella opter 508
Pleromelloida bonuscula 528
cinerea 528
conserta 528

Pleurota albastrigulella 112
Plodia interpunctella 154
Plum Borer, American 152
Plume Moth, Artichoke
136
Coyote Brush Borer 140
Geranium 138
Gumweed 138
Hourglass 138
Lantana 138
Morning-glory 140
Plain 138
Ragweed 140
Sage 136
Snapdragon 138
Wormwood 140
Yarrow 136
Plusia, West Coast 474
Plusia nichollae 474
putnami 474
Plusiodonta compressipalpis
428
Plutella porrectella 40
xylostella 42
Poetry Moth 522
Polia nimbosa 604
nugatis 606
piniae 604
purpurissata 604
Police Car Moth 410
Polix coloradella 110
Ponometia acutus 484
altera 484
bicolorata 480
binocula 480
candefacta 482
clausula 480
cuta 482
elegantula 484
fasciatella 484
hutsoni 484
libedis 482
nannodes 482
phecolisca 482
semiflava 480
tortricina 482
venustula 480
virginalis 480
Prairie Moth, Pink 540
Primrose Moth 546
Reginia 550
Prionomelia spododea 334
Prionoxystus robiniae 94
Probole, Friendly 348
Probole amicaria 348
Prochoerodes amplicineraria
370
forficaria 370
truxaliata 370
Prolimacodes trigona 106
Prolita, Variable 124

Prolita sexpunctella 124
variabilis 124
Prominent, Apical 374
Black-crescent 380
Black-rimmed 376
Brown-rimmed 376
Elegant 376
Gray-streaked 384
Hollow-spotted 384
Mimosa 380
Morning-glory 382
Olive-patched 380
Orange-tailed 380
Oregon 380
Ornate 374
Pale 384
Red-washed 384
Russet 376
Sigmoid 374
Smiling 380
Steel-gray 382
Unicorn 382
White-dotted 380
White-lined 382
Promylea lunigerella 160
Properigea, Gray 562
Mottled 562
Tricolored 562
White-spotted 562
Properigea albimacula 562
continens 562
niveirena 562
suffusa 562
Proserpinus clarkiae 252
flavofasciata 252
juanita 252
lucidus 252
terlooii 252
Proteoteras aesculana 80
Prothrinax luteomedia 512
Protitame subalbaria 340
virginalis 340
Protoboarmia porcelaria
332
Protodeltote albidula 478
Protogygia milleri 634
Protolampra rufipectus 654
Protoschinia nuchalis 544
Proxenus miranda 564
Psamatodes abydata 308
Psammopolia wyatti 618
Psara, Spiderling 190
Psara dryalis 190
Psectrotarsia suavis 540
Pseudanarta caeca 616
crocea 616
flava 616
singula 616
Pseudanthoecia tumida 616
Pseudeustrotia, Pink-barred
556

Pseudeustrotia carneola 556
Pseudexentera habrosana 82
oregonana 82
Pseudhapigia brunnea 384
Pseudobryomima, Deceptive
564
Mossy 564
Pseudobryomima fallax 564
muscosa 564
Pseudochelaria, Saddled
124
White-lined 124
Pseudochelaria manzanitae
124
scabrella 124
Pseudohemihyalea ambigua
414
edwardsii 414
labecula 414
splendens 414
Pseudopanthea palata 494
Pseudorgyia russula 430
Pseudorthodes irrorata 626
Pseudoschinia elautalis 196
Pseudosciaphila duplex 66
Psychedelic Young Moth 90
Pterotaea cariosa 328
lamiaria 328
Ptichodis ovalis 456
Ptychoglene coccinea 394
phrada 394
Pug, Brown 268
Brown-patched 272
Chalky 270
Dun-striped 272
Edna's 268
Graef's 272
Gray-lined 270
Gray-patched 274
Green 274
Larch 270
Mournful 268
Olive 270
Red-banded 270
Sharp-winged 272
Spruce Cone 270
Wormwood 272
Yellow-spotted 272
Zelmira 270
Purslane Moth 534
Pygarctia, Mousey 418
Rose-headed 418
Sprague's 418
Pygarctia murina 418
roseicapitis 418
spraguei 418
Pyralid, Hoary 148
Titian Peale's 168
White-trimmed Brown,
complex 198
Pyralis farinalis 148

Pyrausta, California 178
  Chestnut 176
  Coffee-loving 180
  Coyote-mint 180
  Desert 178
  Elegant 176
  Frosted 176
  Fulvous-edged 174
  Hoary 178
  Inornate 178
  Lethal 176
  Mottled 178
  One-banded 180
  Pink-banded 180
  Raspberry 176
  Rusty 180
  Shasta 178
  Sociable 180
  Volupial 176
Pyrausta californicalis 178
  dapalis 178
  fodinalis 180
  grotei 176
  inornatalis 178
  laticlavia 180
  lethalis 176
  napaealis 176
  nexalis 174
  nicalis 176
  orphisalis 178
  perrubralis 178
  pseudonythesalis 178
  scurralis 180
  semirubralis 180
  signatalis 176
  socialis 180
  subsequalis 178
  tyralis 180
  unifascialis 180
  volupialis 176
Pyrrharctia isabella 406
Pyrrhia exprimens 540

Q
Quaker, Accurate 628
  Alder 624
  Angled 592
  Arched 592
  Beautiful 594
  Bicolored 626
  Black Moon 628
  Dark-spotted 626
  Dark-winged 574
  Fancy 594
  Ferruginous 596
  Fractured 624
  Gray-veined 628
  Grote's Black-tipped 632
  Hanham's 624
  Humble 596
  Light-spotted 626

  Lined 572
  Normal 592
  Northern Scurfy 624
  Nutmeg 624
  Orange-rimmed 592
  Orbiculate 626
  Pacific 596
  Pink-legged 594
  Protector 594
  Rare Sand 564
  Ruddy 624
  Rufous 626
  Sandy 596
  Sharp-spotted 628
  Spalding's 568
  Subdued 596
  Third 592
  Transparent 594
  V-lined 625
  Varied 626
  White-specked 628
Quasisalebria admixta 156

R
Rachiplusia ou 470
Raisin Moth, Dusky 154
Ranunculus, Small 610
Raphia frater 494
Recycler Moth, Cactus 30
  White-horned 28
  Yellow-V 30
Red-lined Moth 554
Regal Moth, Psychedelic 90
Renia, Peppered 424
Renia hutsoni 424
Retinia, Fir Twig 68
  Knotty 68
Retinia picicolana 68
  sabiniana 68
Revealer Moth,
  White-patched 532
  White-striped 532
Rheumaptera hastata 276
  meadii 276
  subhastata 276
  undulata 276
Rhigognostis interrupta 42
Richia chortalis 634
  parentalis 634
  serano 634
Rifseria, Ribboned 126
Rifseria fuscotaeniaella 126
Rivula propinqualis 428
Rivulet, Grand 268
  Gray 298
  Spotted 298
  Two-banded 298
  Woodgrain 284
Rosewing 608
Ruacodes, Webbed 512
Ruacodes tela 512

Rudenia, Black-tipped 52
Rudenia leguminana 52
Rustic, Civil 564
  Common 574
  Kidney-spotted 576
  Mottled 564
  Pallid 576
  Square-spot 650

S
Sable, White-spotted 170
Sabulodes aegrotata 370
  edwardsata 372
  griseata 372
  niveostriata 370
  packardata 372
  spoliata 372
  venata 372
Sagenosoma elsa 244
Sallow, Ardent 514
  Ashy 528
  August 530
  Bicolored 584
  Blind 588
  California 530
  Cloaked 528
  Comstock's 508
  Copper-winged 582
  Deceptive 508
  Delta-lined 532
  Dimorphic 526
  Even-lined 588
  Expressive 538
  February 508
  Fervent 514
  Fine-lined 526
  Full Moon 526
  Fused 514
  Garland 530
  Happy 530
  Inca 538
  Long-spotted 528
  Lupine 584
  Observant 588
  Peridot 510
  Purple-lined 540
  Puta 584
  Rogenhofer's 524
  Saddled 530
  Scribbled 528
  Semicollared 532
  Shrouded 528
  Sleeping 590
  Smoked 588
  Smooth-lined 582
  Splendid 512
  Thread-lined 508
  Three-Spotted 582
  Turquoise-banded 512
  Two-lined 530
  White-margined 512

Salt Marsh Moth 404
Satin Moth, White 386
Satole ligniperdalis 144
Satyr, Grote's 590
Saucrobotys, Dogbane 168
  Dusky 168
Saucrobotys fumoferalis 168
  futilalis 168
Scallop Shell 276
Scape Moth, Yellow-collared
  420
Scavenger Moth, Florida
  Pink 116
Schinia, Alluring 548
  Painted 550
  White-banded 548
Schinia accessa 552
  acutilinea 550
  albafascia 552
  arcigera 548
  argentifascia 552
  chrysellus 552
  ciliata 552
  citrinellus 544
  cumatilis 554
  errans 548
  florida 546
  gaurae 546
  grandimedia 546
  hulstia 554
  jaguarina 544
  luxa 544
  lynx 548
  meadi 546
  mexicana 546
  miniana 550
  mortua 548
  niveicosta 550
  oculata 552
  oleagina 546
  pulchripennis 546
  regina 550
  reniformis 554
  scarletina 554
  simplex 544
  siren 548
  suetus 544
  tertia 550
  unimacula 552
  vacciniae 548
  villosa 548
  volupia 550
  walsinghami 550
Schizura ipomaeae 382
Scholar, Interrupted 42
Schreckensteinia festaliella 42
Sciota, Double-banded 158
Sciota bifasciella 158
Scolecocampa atriluna 430
  liburna 430
Scoliopteryx libatrix 428

Scoopwing, Gray 256
Scoparia, Double-striped 208
  Many-spotted 208
  Pale 208
Scoparia basalis 208
  biplagialis 208
  palloralis 208
Scopula ancellata 262
  inductata 264
  junctaria 262
  limboundata 262
  luteolata 264
  plantagenaria 262
  quinquelinearia 264
Scybalistodes, Crescent-
  spotted 198
Scybalistodes periculosalis
  198
Scythris, Banded 132
Scythris trivinctella 132
Sedge Borer, Oblong 576
Seed Moth, Straight-lined
  432
Seedcropper, Gilded 518
Seedworm Moth, Ponderosa
  Pine 90
Selenia alciphearia 346
Sericosema juturnaria 348
Sesia pacificum 96
Setaceous Hebrew Character
  652
Setagrotis pallidicollis 652
Shade, Silver 48
Sheathminer, Pine Needle 38
Shoulder-Knot, Ashy 586
  Brown 586
  Dusky 584
  Poplar 586
  Rusty 584
Sicya crocearia 356
  macularia 356
  morsicaria 358
Sideridis maryx 608
  rosea 608
Silver Y, Delicate 470
Sitochroa, Dimorphic 172
Sitochroa chortalis 172
Six-plume Moth, Montana
  136
Skeletonizer, Apple Leaf 44
  Banyan Leaf 44
  Black-cloaked 104
  Blackberry 42
  Canyon Grapeleaf 104
  Grapeleaf 102
  Oak 114
  Western Grapeleaf 104
Skiff Moth, Western 106
Skunk Moth 110
Sky-pointing Moth 186
Slant-Line, White 368

Slug Moth, Early Button 106
  Jelly 106
  Plain 106
Smerinthus astarte 248
  cerisyi 246
  jamaicensis 246
  ophthalmica 248
  saliceti 248
Snout, Dimorphic 424
  Fractured Western 186
  Hop Vine 426
  Monterey Scale-feeding
    164
  Mottled 426
  Western Scale-feeding 164
  White-lined 426
Snout Moth, Cross-banded
  156
  Red-streaked 154
  Rusty-banded 160
  Whip-marked 194
  White-notched 160
Snowia montanaria 364
Somberwing, Burnt 456
  Toothed 456
Sonia, Broken-spotted 80
Sonia vovana 80
Sorhagenia nimbosus 116
Sosipatra, Black-dotted 154
Sosipatra rileyella 154
Spaelotis bicava 648
  clandestina 648
Spanworm, Bruce 274
  Gray Bruce 276
  Horned 372
  Maple 368
  Spurred 372
  Stout 338
  Walnut 338
  Western Bruce 276
Spargaloma sexpunctata 434
Spargania luctuata 290
  magnoliata 290
Sparganothis, Ancient 62
  One-lined 62
Sparganothis senecionana 62
  sulfureana 62
  unifasciana 62
Sphinx, Achemon 250
  Asellus 244
  Blinded 248
  Clark's Day 252
  Doll's 246
  Elegant 244
  Ello 250
  Elsa 244
  Falcon 254
  Gallium 254
  Great Ash 244
  Istar 246
  Juanita 252

Modest 248
Obscure 250
One-eyed 246
Pacific Green 252
Phaeton Primrose 252
Sequoia 246
Small-eyed 248
Terloo 252
Twin-spotted 246
Typhon 250
Vashti 244
Vine 250
Waved 244
Western Eyed, complex
  248
Western Poplar 248
White-lined 254
Wild Cherry 244
Yellow-banded Day 252
Sphinx asellus 244
  chersis 244
  dollii 246
  drupiferarum 244
  perelegans 244
  sequoiae 246
  vashti 244
Spilonota ocellana 70
Spilosoma pteridis 404
  vagans 404
  vestalis 402
  virginica 404
Spiramater lutra 606
Spodolepis danbyi 344
Spodoptera exigua 558
  frugiperda 558
  ornithogalli 560
  praefica 558
Spoladea recurvalis 190
Spragueia, Iron-banded 490
  Jaguar 490
  Magnificent 490
  Somber 490
Spragueia funeralis 490
  jaguaralis 490
  magnifica 490
  obatra 490
Spring Moth, Bluish 346
  White 346
Staghorn Cholla Moth 534
Stamnoctenis pearsalli 298
Stamnodes affiliata 294
  albiapicata 294
  blackmorei 294
  coenonympha 294
  deceptiva 296
  formosata 294
  marmorata 296
  modocata 296
  seiferti 296
  tessellata 298
  topazata 296

Stegea, Western 198
Stegea salutalis 198
Stem Moth, Sunflower 80
Stenoporpia anastomosaria
  324
  excelsaria 326
  glaucomarginaria 326
  macdunnoughi 326
  pulmonaria 324
Stenoptilodes antirrhina 138
Sthenopis purpurascens 16
Stibaera thyatiroides 538
Stictocsome suffusa 406
Stigmella sp. 16
Stinger Moth, Mesquite 104
Stiria intermixta 520
  rugifrons 520
Stored Grain Moth 148
Stowaway, Toothed 518
Straw Moth, Darker-spotted
  542
Strawberry Crown Moth 98
Stretchia muricina 592
Striacosta albicosta 632
Suleima baracana 80
  helianthana 80
Sulphur Moth 336
  Smoky-eyed 336
Sun Moth, Jewel-studded 34
  Orange-banded 36
Sunflower Moth, American
  166
  Western 520
  Yellow 520
Sunira bicolorago 584
  decipiens 584
Sunspur, Banded 36
  Beautiful 36
  Brilliant 36
  Elegant 36
  Splendid 36
  Spotted 36
Swordgrass, American 578
  Bruce's 578
  Dot-and-Dash 578
  Gray 578
Symmerista, Wavy 384
Symmerista zacualpana 384
Symmoca, Signate 108
Symmoca signatella 108
Sympistis augustus 530
  behrensi 530
  fifia 530
  gracillinea 528
  greyi 530
  occata 530
  perscripta 528
  poliochroa 530
  ragani 532
  semicollaris 532
  umbrifascia 530

Synanthedon albicornis 98
  bibionipennis 98
  chrysidipennis 98
  exitiosa 100
  novaroensis 100
  polygoni 100
  resplendens 100
  sequoiae 100
  tipuliformis 98
Synchlora aerata 304
  bistriaria 304
  faseolaria 304
  frondaria 304
Syngrapha angulidens 474
  celsa 474
  epigaea 472
  ignea 472
  orophila 472
  rectangula 474
  viridisigma 472

T
Tabby, Elected 148
  Jacketed 148
  Large 148
  Wide-banded 148
Tachystola hemisema 112
Tacoma, Pale-patched 162
Tacoma feriella 162
Taeniogramma octolineata 340
Tan, Curve-lined 428
  Dusty-winged 428
  Soft-lined 430
Tansy Root Moth 88
Tarache albifusa 486
  aprica 484
  areli 486
  areloides 486
  arida 486
  augustipennis 488
  expolita 488
  geminocula 486
  huachuca 488
  idella 488
  lanceolata 488
  lucasi 490
  major 488
  quadriplaga 486
  tetragona 486
  toddi 486
Tathorhynchus exsiccata 434
Taygete, Hooked 108
Taygete decemmaculella 108
Tebenna, Everlasting 44
  Jeweled 44
Tebenna gemmalis 44
  gnaphaliella 44
Tegeticula maculata 18
  yuccasella 18
Tehama bonifatella 214

Telethusia, Oval 158
Telethusia ovalis 158
Telphusa, California 122
Telphusa sedulitella 122
Terastia meticulosalis 188
Tetanolita, Orange-spotted 424
Tetanolita palligera 424
Tetracis cachexiata 368
  cervinaria 368
  formosa 370
  hirsutaria 370
  jubararia 368
  pallulata 368
Thallophaga, Northern 346
  Taylor's 346
Thallophaga hyperborea 346
  taylorata 346
Thaumatographa regalis 90
  youngiella 90
Thaumatopsis pexellus 214
Thera juniperata 282
  otisi 282
Theroa zethus 382
Thimble, Cotton 492
  Green 492
  Raspberry 492
Tholera americana 602
Thorn, Besma 364
  Brown-tipped 346
  California 370
  Canary 358
  Elegant 368
  Faded 358
  October 368
  Peppered 370
  Shaded 368
  Speckled 370
  Straight-lined 366
  Wavy-lined 370
  White-dotted 370
  Yellow 360
Thurberiphaga diffusa 492
Thyraylia nana 52
Thyridopteryx ephemeraeformis 22
  meadii 22
Thyris, Spotted 142
Thyris maculata 142
Tiger Moth, Arge 396
  Aulaean 406
  Barred 404
  Blake's 400
  Brown 404
  California 408
  Carlotta's 402
  Clio 418
  Figured 398
  Freckled 418
  Gray-spotted 406
  Great 402

Immaculate 400
  Isabella 406
  Lettered 398
  Little Bear 400
  Little Virgin 398
  Many-spotted 406
  Mexican 400
  Nevada 400
  Ornate 400
  Painted 406
  Parthenice 396
  Phyllira 398
  Ranchman's 402
  Ruby 406
  St. Lawrence 402
  Vestal 402
  Virgin 396
  Virginian 404
  Wandering 404
  Williams' 398
  Wood 402
Tinea niveocapitella 28
  occidentella 28
  pellionella 28
Tineola bisselliella 28
Tipworm Moth, Cotton 82
Tissue Moth 276
  California 278
Toripalpus trabalis 148
Tornos benjamini 326
  erectarius 326
Tortricid, Ashy 48
  Gray-marked 48
  Saddled 88
  Snowy 68
Tortricidia testacea 106
Tortrix, Barred Fruit-tree 54
  Carnation 58
  Chamise 56
  Cloudy-banded 54
  Garden 60
  Garden Rose 46
  Happy 52
  Large Aspen 56
  Orange 54
  Pasqueflower 54
  Privet 60
  Red-barred 62
  Rhomboid 46
  Rose 58
  Sugar Pine 58
  Variable Conifer 56
  White Triangle 60
Toxonprucha, Cruel 458
  Spotted 458
  Sudden 458
  Winged 458
Toxonprucha crudelis 458
  pardalis 458
  repentis 458

volucris 458
Tracheops bolteri 334
Treble-bar Moth 266
Trichocosmia inornata 602
Trichodezia albovittata 282
  californiata 282
Tricholita chipeta 628
Trichoplusia ni 468
Trichopolia alfkenii 626
  curtica 624
  melanopis 626
  oviduca 624
  rufula 626
Trichordestra liquida 608
  prodeniformis 608
  tacoma 608
Tridepia nova 604
Triocnemis saporis 512
Triphosa californiata 278
  haesitata 276
Tripudia, Belted 476
  Bicolored 476
  Luxurious 478
  Playful 478
Tripudia balteata 476
  dimidata 476
  luda 476
  luxuriosa 478
Trosia obsolescens 104
Tubic, Gold-base 112
  Sulphur 110
Tussock Moth, Ancient 416
  Arizona 414
  Davis' 412
  Douglas-fir 386
  Dusky 416
  Great 412
  Grizzled 386
  Milkweed 416
  Modest 414
  Rusty 386
  Silver-spotted 412
  Speckled 412
  Spotted 414
  Variable 386
  Western 386
Twin-Spot, Red 286
  Western Red 284
Twirler, California 118
  Clover 130
Tyria jacobaeae 410
Tyta luctuosa 524

U
Udea, Eight-marked 184
  Gray-eyed, complex 184
  Shaded 184
  Washington 184
Udea abstrusa 184
  itysalis 184
  octosignalis 184

profundalis 184
rubigalis 184
turmalis 184
washingtonalis 184
Ufeus satyricus 590
satyricus sagittarius 590
satyricus satyricus 590
Ulolonche disticha 626
orbiculata 626
Umber, American Barred 348
Unciella flagrantis 514
Unciella primula 514
Underwing, Aholibah 438
Bride 436
Briseis 440
Copper 506
Delicate 442
Desdemona 442
Grote's 440
Ilia 438
Irene's 440
Joined 442
Large Yellow 646
Lesser Yellow 646
Locust 462
Mother 438
Once-married 438
Passionate 444
Pearly 630
Penitent 436
Semirelict 440
Sweetheart 442
Ultronia 444
Verrill's 442
Western 440
White 438
Uresiphita reversalis 172
Urola, Snowy 212
Urola nivalis 212
Ursia noctuiformis 382

V

Venusia cambrica 266
pearsalli 266
Verbena Moth 538
Vinemina opacaria 334
Virbia, Fragile 408
Gray 408
Orange 408
Rusty 408
Showy 408
Virbia aurantiaca 408
costata 408
ferruginosa 408
fragilis 408
ostenta 408
Virgin Moth 340
Freckled 340
Viridiseptis marina 590

Vitula, Gray 154
Vitula insula 154
serratilineella 154

W

Wainscot, Dia 614
Heterodox 614
Imperfect 612
Lesser 612
Many-lined 612
Meadow 612
Oregon 614
Speckled 614
Stola 614
Two-lined 612
Walshia miscecolorella 116
Wasp Moth, Saffron-spotted 418
Wave, Angled 262
Convergent 342
Dark-ribboned 262
Delicate 340
Desperate 344
Distant 352
Drab Brown 258
Dwarf Tawny 258
Falcate 342
Fortunate 260
Gem 260
Gentle 264
Jeweled 260
Many-lined 258
Modest 264
Narrow-winged 258
Orange-banded 342
Orbed 262
Parallel 342
Pebbled 344
Quadrate 342
Red-and-white 260
Red-bordered 260
Red-lined 342
Red-winged 308
Royal 262
Secluded 352
Secondary 344
Signet 260
Simple 262
Single-dotted 262
Soft-lined 264
Speckled 258
Spurred 342
Straight-lined 258
Straw 260
Welsh 266
Wax Moth, Greater 146
Lesser 146
Webworm, Ailanthus 38
Alfalfa 174
Brassica 198

Cabbage 198
Fall 404
Garden 170
Harlequin 190
Hawaiian Beet 190
Lesser Aspen 158
Nearctic Beet 174
Palo Verde 130
Southern Beet 188
Spotted Beet 190
Western Garden 170
Webworm Moth, Mesquite 124
Soybean 118
Wedge Moth, Broken-banded 454
Straight-banded 454
Wedgling, The 560
Whaleback Moth 538
White Cross Moth 118
White-Speck, The 612
Whiteband, Snowy 516
Sunny 516
Wild Forget-me-not Moth 410
Willow Moth, Small Mottled 558
Winter Moth 274
Witch, Black 436
Wood-nymph, Wilson's 532
Woodling, Brown 600
Crucial 598
Familiar 600
Grieving 600
Mottled Oak 600
Peregrine 600
Simple 598
Variable 598
Winter 598

X

Xanthorhoe abrasaria 284
defensaria 284
ferrugata 286
labradorensis 284
lacustrata 286
Xanthostege, Pink-bordered 172
Xanthostege plana 172
Xanthothrix neumoegeni 518
ranunculi 518
Xerociris wilsonii 532
Xestia c-nigrum 652
conchis 650
infimatis 652
oblata 652
smithii 650
xanthographa 650
Xylena brucei 578
cineritia 578

curvimacula 578
nupera 578
Xylesthia, Pale 30
Xylesthia albicans 30
pruniramiella 30
Xylomoia indirecta 574
Xylophanes falco 254

Y

Yellow, Brown-edged 356
Checkered 524
Cinnamon 358
Gray-edged 520
Modest 524
Obtuse 524
Sharp-lined 356
Yellow-veined Moth 194
Yellowneck, Four-spotted 108
Ypsolopha, Canary 40
Fawn 40
Ochre 40
Pennant 40
Scythed 40
Streaked 40
Ypsolopha barberella 40
canariella 40
cervella 40
cockerella 40
dentiferella 40
falciferella 40
ochrella 40
Yucca Moth 18
Chaparral 18

Z

Zale, Colorful 460
Desert Peaks 460
Dusky 460
Lunate 460
Southwestern 460
Zale colorado 460
insuda 460
lunata 460
minerea 460
termina 460
Zaleops, Shadowed 458
Zaleops umbrina 458
Zanclognatha jacchusalis 422
Zeiraphera canadensis 80
improbana 82
Zelicodes linearis 428
Zelleria haimbachi 38
Zenophleps alpinata 288
lignicolorata 286
obscurata 288
Zophodia grossulariella 164
Zosteropoda hirtipes 628
Zotheca tranquilla 588

LIST OF TAXONOMIC GROUPS

Metallic Leaf-miners	14	Typical Sphinx Moths	240
Ghost Moths	14	Eyed Sphinx Moths	246
Pygmy Leafmining Moths	16	Diurnal and Striped Sphinx Moths	250
Yucca and Shield-bearing Moths	16	Scoopwings	256
Fairy (Longhorn) Moths	20	Waves	256
Bagworm Moths	22	Carpets, Pugs, Hydriomenas,	
Dancing, Fungus, Tubeworm, and		and Jesters	264
Clothes Moths	22	Infants	300
Ribbed Cocoon-Maker and Leaf		Emeralds	300
Blotch Miner Moths	30	Angles and Granites	308
Sun Moths	34	Grays and Allies	324
Ermine and Ypsolopha Moths	38	Assorted Geometers	340
Sedge Moths	42	Peros	354
Bristle-legged Moths	42	Broad-winged Geometers	356
Metalmark Moths	44	Prominents	374
Tortrix Leafrollers	46	Tussock Moths	386
Cochylid Moths	50	Lichen Moths	388
Archips Leafrollers	54	Tiger Moths	396
Sparganothid Leafrollers	62	Litter Moths	422
Olethreutine Moths	64	Assorted Owlets	428
Psychedelic Leafrollers	90	Witches	436
Carpenterworm Moths	92	Underwings	436
Clearwing Moths	96	Graphics, Somberwings, and Allies	444
Leaf Skeletonizer Moths	102	Zales and Allies	456
Flannel Moths	104	Marathyssas and Paectes	462
Slug Moths	106	Hieroglyphic Moth	464
Concealer and Scavenger Moths	108	Nolas and Midgets	464
Flat Moths and Allies	112	Palm Borers	468
Cosmet Moths	116	Loopers	468
Twirler Moths	118	Tripudias, Cobubathas, and Glyphs	476
Casebearer Moths	130	Bird-Dropping Moths and Thimbles	480
Scythrid Flower Moths and Allies	132	Pantheas and Brothers	492
Many-plumed Moths	136	Daggers	494
Plume Moths	136	Hooded Owlets	502
Fruitworm Moths and Allies	140	Amphipyrine Sallows	506
Window-winged Moths	142	Annaphilas, Grotellas, and	
Assorted Pyralids	142	Golden Flower Moths	514
Phycitine Moths	150	Oncocnemidine Sallows	526
Pyraustine Moths	168	Wood-nymphs and Foresters	532
Spilomeline Moths	182	Groundlings	536
Assorted Crambids	192	Flower Moths	540
Donacaulas	204	Assorted Noctuids	556
Aquatic Crambids	206	Rustics	562
Moss-eating Crambids	208	Half-spots	566
Grass-veneers	212	Angle Shades	566
Thyatirids and Hooktips	222	Apameas, Brocades, and Allies	568
Tent Caterpillar and Lappet		Swordgrasses	578
Moths, and Tolypes	224	Pinions and Xylenine Sallows	580
Apatelodid Moths	230	Spring Quakers and Woodlings	592
Royal Silkmoths	230	Large Arches	602
Pinemoths, Buckmoths, and		Wainscots	612
Eyed-Silkmoths	232	Small Arches and Summer Quakers	616
Typical Silkmoths	238	Darts	630

Acleris
p. 46

Angle Shades
p. 566

Apamea
p. 568

Caloptilia
p. 32

Clearwing Borer
p. 96

Cochylidae
p. 50

Dagger
p. 494

Dart
p. 630

Flower Moth
p. 540

Geometer
pp. 256–372

Grass-veneer
p. 212

Looper
p. 468

Marathyssa
p. 462

Olethreutinae
p. 64

Owlet
pp. 422–434

Phycitinae
p. 150

Pinion
p. 580

Plume Moth
p. 136

Prominent
p. 374

Pug
p. 268

Pyraustinae
p. 168

Quaker
p. 616

Sallow
p. 506

Scoopwing
p. 256